# Electrochemical Reduction of Carbon Dioxide

## Fundamentals and Technologies

# ELECTROCHEMICAL ENERGY STORAGE AND CONVERSION

**Series Editor: Jiujun Zhang**
National Research Council Institute for Fuel Cell Innovation
Vancouver, British Columbia, Canada

## Published Titles

**Electrochemical Supercapacitors for Energy Storage and Delivery: Fundamentals and Applications**
Aiping Yu, Victor Chabot, and Jiujun Zhang

**Proton Exchange Membrane Fuel Cells**
Zhigang Qi

**Graphene: Energy Storage and Conversion Applications**
Zhaoping Liu and Xufeng Zhou

**Electrochemical Polymer Electrolyte Membranes**
Jianhua Fang, Jinli Qiao, David P. Wilkinson, and Jiujun Zhang

**Lithium-Ion Batteries: Fundamentals and Applications**
Yuping Wu

**Lead-Acid Battery Technologies: Fundamentals, Materials, and Applications**
Joey Jung, Lei Zhang, and Jiujun Zhang

**Solar Energy Conversion and Storage: Photochemical Modes**
Suresh C. Ameta and Rakshit Ameta

**Electrochemical Energy: Advanced Materials and Technologies**
Pei Kang Shen, Chao-Yang Wang, San Ping Jiang, Xueliang Sun, and Jiujun Zhang

**Electrolytes for Electrochemical Supercapacitors**
Cheng Zhong, Yida Deng, Wenbin Hu, Daoming Sun, Xiaopeng Han, Jinli Qiao, and Jiujun Zhang

**Electrochemical Reduction of Carbon Dioxide: Fundamentals and Technologies**
Jinli Qiao, Yuyu Liu, and Jiujun Zhang

# Electrochemical Reduction of Carbon Dioxide

## Fundamentals and Technologies

Edited by
Jinli Qiao · Yuyu Liu · Jiujun Zhang

**CRC Press**
Taylor & Francis Group
Boca Raton London New York

CRC Press is an imprint of the
Taylor & Francis Group, an **informa** business

CRC Press
Taylor & Francis Group
6000 Broken Sound Parkway NW, Suite 300
Boca Raton, FL 33487-2742

First issued in paperback 2019

© 2016 by Taylor & Francis Group, LLC
CRC Press is an imprint of Taylor & Francis Group, an Informa business

No claim to original U.S. Government works

ISBN-13: 978-1-4822-5824-0 (hbk)
ISBN-13: 978-0-367-87083-6 (pbk)

**Visit the Taylor & Francis Web site at**
**http://www.taylorandfrancis.com**

**and the CRC Press Web site at**
**http://www.crcpress.com**

# Contents

Contents

# Series Preface

## ELECTROCHEMICAL ENERGY STORAGE AND CONVERSION

The goal of the *Electrochemical Energy Storage and Conversion* book series is to provide comprehensive coverage of the field, with titles focusing on fundamentals, technologies, applications, and the latest developments, including secondary (or rechargeable) batteries, fuel cells, supercapacitors, $CO_2$ electroreduction to produce low-carbon fuels, electrolysis for hydrogen generation/storage, and photoelectrochemistry for water splitting to produce hydrogen, among others. Each book in this series is self-contained, written by scientists and engineers with strong academic and industrial expertise who are at the top of their fields and on the cutting edge of technology. With a broad view of various electrochemical energy conversion and storage devices, this unique book series provides essential reads for university students, scientists, and engineers and allows them to easily locate the latest information on electrochemical technology, fundamentals, and applications.

**Jiujun Zhang**
*National Research Council of Canada*
*Richmond, British Columbia*

# Preface

As identified, carbon dioxide ($CO_2$) is the most common greenhouse gas in our planet. It is a necessary material for the growth of all the Earth's plants and many industrial processes. Unfortunately, with the intensified industrial activities by mankind, more $CO_2$ is released to the environment, causing Earth–carbon imbalance, leading to possible global warming issue. Therefore, reducing $CO_2$ production and converting $CO_2$ into useful materials at the same time seem to be necessary and also critical to environment protection, which can be signaled by the increased investment from various governments worldwide to address this $CO_2$ issue.

Among different technologies addressing the $CO_2$ issue, there are two major approaches, the first one is to capture and geologically sequestrate $CO_2$, and the other is to convert $CO_2$ into useful low-carbon fuels. Particularly, $CO_2$ conversion and utilization seem rather attractive and promising in today's high energy demanding world. However, at the current state of technology, there are still some barriers hindering the practical applications of $CO_2$ capture, conversion, and utilization, including (1) high costs of $CO_2$ capture, separation, purification, and transportation to user sites; (2) high energy requirements for $CO_2$ chemical/electrochemical conversion; (3) limited market size and investment incentives; (4) lack of industrial commitments for enhancing $CO_2$-based chemicals; and (5) insufficient socioeconomical driving forces. Although there are such challenges, the effort in $CO_2$ capture, conversion, and utilization is still recognized as one of the feasible and promising cutting-edge researches in energy and environment areas.

In recent years, $CO_2$ conversion using electrocatalysis approaches has attracted great attention due to its several advantages, such as (1) the process can be controlled by electrode potentials and the reaction temperature; (2) the supporting electrolytes can be fully recycled so that the overall chemical consumption can be minimized to just water or waste water; (3) the source of electricity used to drive the process can be obtained without generating any new $CO_2$, including solar, wind, hydroelectric, geothermal, tidal, and thermoelectric processes; and (4) the electrochemical reaction systems are compact, modular, on-demand, and easy for scale-up applications. However, some challenges exist such as slow kinetics of $CO_2$ electroreduction even with some electrocatalysts and high electrode reduction potential being applied, and low energy efficiency of the process due to the parasitic or decomposition reaction of the solvent at high reduction potential, as well as high energy consumption.

To assist the training of high-quality people (HQPs) such as undergraduates/graduate students who will be the primary researchers in energy and environment science and technology, a book with focus on the fundamentals and applications of $CO_2$ electroreduction to produce useful small fuels may be necessary. We believe that well-trained HQP will be the major resources in overcoming the challenges in electrochemical energy storage and conversion, including those in $CO_2$ electroreduction. This book is primarily intended to present an overview of the electrochemical reduction of $CO_2$ from fundamentals to applications, as well as the advancements made in recent years. In particular, the fundamental sciences involved in the operation and the technological developments in material science and engineering for materials fabrication are summarized. An in-depth analysis on the electrocatalysts for $CO_2$ electroreduction in terms of reaction mechanisms, catalytic activity/stability, as well as product selectivity, is also presented. This book will serve as a resource benefiting researchers, students, industrial professionals, and manufacturers by providing a systematic overview of the materials, electrocatalyst/electrode design, and related issues for $CO_2$ electroreduction.

We express our deep appreciation to all the chapter authors/coauthors who contributed high-quality chapters to this book. We also thank the CRC staff, Allison Shatkin and Laurie Oknowsky, for their professional assistance and strong support during this project. Finally, we welcome any constructive comments for further improving the quality of this book.

**Jinli Qiao, PhD**
*Donghua University, Shanghai, China*

**Yuyu Liu, PhD**
*Tohoku University, Japan*

**Jiujun Zhang, PhD**
*National Research Council of Canada, Canada*

# Editors

**Jinli Qiao** is a professor, PhD supervisor, and disciplines leader of the College of Environmental Science and Engineering, Donghua University, China. She earned her PhD in electrochemistry from Yamaguchi University, Japan in 2004. After which she joined the National Institute of Advanced Industrial Science and Technology (AIST), Japan, as a research scientist working on both acidic/alkaline polymer electrolyte membranes and non-noble metal catalysts for proton exchange membrane (PEM) fuel cells. From 2004 to 2008, as a project leader/ principal investigator, she carried out seven projects, including two New Energy and Industrial Technology Development Organization (NEDO) projects and three Ministry of Economy, Trade and Industry (METI) of Japan projects on catalyst and membrane materials for fuel cells. From 2008 to the present, she has carried out a total of 12 projects funded by the Chinese Government, including the National Natural Science Foundation of China.

As the first author and corresponding author, Dr. Qiao has published more than 100 peer-reviewed publications and 40 keynote/invited oral presentations, 5 coauthored books/book chapters, and holds 30 patents/patent publications. Professor Qiao is a referee for the project assessment of the National Natural Science Foundation of China, the Scientific Research Foundation of State Education Ministry of China, and the Foundation of the Shanghai Science and Technology Committee. She also serves as the organizer/co-chair and committee member of several international conferences, and is a guest editor for the *Electrochimica Acta/International Journal of Hydrogen Energy/Applied Energy* and also the reviewer editor of *Frontiers in Fuel Cells*. She is now the vice-president of the International Academy of Electrochemical Energy Science (IAOEES) and a board committee member of the Electro-Membrane Association of China. Dr. Qiao has more than 20 years of scientific research experience, particularly in the area of electrochemical material development and energy storage and conversion, including PEM fuel cells, metal–air batteries, supercapacitors, and $CO_2$ electroreduction. She is also an active member of the American Chemical Society, The Electrochemical Society, and the China Association of Hydrogen Energy.

**Yuyu Liu** is an associate professor at the Graduate School of Environmental Studies, Tohoku University, Japan, and is also a professor at the Taiyuan University of Technology, China. Dr. Liu earned his PhD in environmental engineering from Yamaguchi University, Japan in 2003. He then worked at the Kyushu Environmental Evaluation Association, Osaka Institute of Technology, Tokyo University of Agriculture and Technology, and Yokohama National University as a postdoctoral and research fellow. As a principal investigator, he carried out more than 10 projects, including two NEDO projects, one for the Japan Science and Technology Agency (JST), and one for the Ministry of Education, Culture, Sports, Science and Technology (MEXT) of Japan. Dr. Liu has more than 10 years of experience in environmental science and technology, particularly in the areas of air quality monitoring, water and soil research, and their associated instrument development. He has published over 40 research papers in peer-reviewed journals and over 20 conference proceedings. He is now the vice president of the International Academy of Electrochemical Energy Science (IAOEES), and an active member of the Japan Society on Water Environment and the Japan Society of Material Cycles and Waste Management.

**Jiujun Zhang** is a principal research officer and core competency leader at the Energy, Mining & Environment, National Research Council of Canada (NRC-EME). Dr. Zhang earned his BS and MSc in electrochemistry from Peking University in 1982 and 1985, respectively, and his PhD in electrochemistry from Wuhan University in 1988. After completing his PhD, he took a position as an associate professor at the Huazhong Normal University for two years. Starting in 1990, he carried out three terms of postdoctoral research at the California Institute of Technology, York University, and the University of British Columbia. Dr. Zhang has over 30 years of R&D experience in theoretical and applied electrochemistry, including over 40 years of fuel cell R&D (among these 6 years at Ballard Power Systems and 7 years at NRC-IFCI), and 3 years of electrochemical sensor experience. Dr. Zhang holds over 14 adjunct professorships, including one at the University of Waterloo, one at the University of British Columbia, and one at Peking University. Dr. Zhang serves as the editor/editorial board member for several international journals as well as the editor for the book series *Electrochemical Energy Storage and Conversion*, CRC Press. He is a fellow of the International Society of Electrochemistry (ISE), the chairman/president of the International Academy of Electrochemical Energy Science (IAOEES), an active member of the International Society of Electrochemistry (ECS), the American Chemical Society (ACS), and the Canadian Institute of Chemistry (CIC).

# Contributors

**Suddhasatwa Basu**
Department of Chemical Engineering
Indian Institute of Technology Delhi
New Delhi, India

**Rongzhi Chen**
College of Recourse and Environment
University of Chinese Academy
    Sciences
Beijing, China

and

Nanoparticle Functional Design
    Group
Nanomaterials Research Institute
National Institute of Advanced
    Industrial Science and
    Technology
Ibaraki, Japan

**Yu Chen**
School of Materials Science
    and Engineering
Shaanxi Normal University
Xi'an, China

**Mengyang Fan**
College of Environmental Science
    and Engineering
Donghua University
Shanghai, China

**M. Ali Haider**
Department of Chemical Engineering
Indian Institute of Technology Delhi
New Delhi, India

**Makoto Hatakeyama**
Nakamura Laboratory
RIKEN Innovation Center
Saitama, Japan

**Feng Hong**
College of Chemistry, Chemical
    Engineering and Biotechnology
Donghua University
Shanghai, China

**Zhibao Huo**
School of Environmental Science
    and Engineering
Shanghai Jiao Tong University
Shanghai, China

**Fangming Jin**
School of Environmental Science
    and Engineering
Shanghai Jiao Tong University
Shanghai, China

**Neetu Kumari**
Department of Chemical Engineering
Indian Institute of Technology Delhi
New Delhi, India

**Yunjie Liu**
School of Environmental Science
    and Engineering
Shanghai Jiao Tong University
Shanghai, China

**Yuyu Liu**
Institute for Sustainable Energy Storage
    and Conversion
Shanghai University (SU-ISESC)
Shanghai, China

and

Graduate School of Environmental
    Studies
Tohoku University
Sendai, Japan

**Shinichiro Nakamura**
Nakamura Laboratory
RIKEN Innovation Center
Saitama, Japan

**Koji Ogata**
Nakamura Laboratory
RIKEN Innovation Center
Saitama, Japan

**Jinli Qiao**
College of Environmental Science
    and Engineering
Donghua University
Shanghai, China

**Dezhang Ren**
School of Environmental Science
    and Engineering
Shanghai Jiao Tong University
Shanghai, China

**Zhiyuan Song**
School of Environmental Science
    and Engineering
Shanghai Jiao Tong University
Shanghai, China

**Dongmei Sun**
School of Chemistry and Materials
    Science
Nanjing Normal University
Nanjing, China

**Jingyu Tang**
College of Chemistry, Chemical
    Engineering and Biotechnology
Donghua University
Shanghai, China

**Xiaomin Wang**
College of Materical Science
    and Engineering
Taiyuan University of Technology
Taiyuan, China

**Yuanqing Wang**
Nakamura Laboratory
RIKEN Innovation Center
Saitama, Japan

**Xiaozhou Zou**
College of Chemistry, Chemical
    Engineering and Biotechnology
Donghua University
Shanghai, China

# 1 Introduction to $CO_2$ Electroreduction

*Yuyu Liu and Jinli Qiao*

## CONTENTS

## 1.1 CHEMICAL AND PHYSICAL PROPERTIES OF $CO_2$

Carbon dioxide ($CO_2$) (mole weight 44.0) is a linear molecular (O=C=O), in which two oxygen (O) atoms are each covalently double bonded to a single carbon (C) atom. With an average C=O bond energy $E_{C=O}$ [=187 ($2 \times 93.5$) kcal/mole] being much

1

higher than $E_{O=O}$ [=116 (2 × 58)] and $E_{C=C}$ [=145 (2 × 72.5)], $CO_2$ is rather chemically stable unless chemically,[1–10] electrocatalytically, and photocatalytically[11–19] treated under vigorous conditions. It usually exists as a colorless and odorless gas in air with a mean atmospheric concentration of 0.039% (v/v), and is well known as one of the most notorious greenhouse gases. It forms solid dry ice at −78.5°C and it partially dissolves and is even dissociated in water (1.45 g/L at 25°C, 100 kPa).

$$CO_2 + H_2O \rightleftharpoons H_2CO_3 \qquad (1.1)$$

$$H_2CO_3 \xrightarrow{k_{aa1}\,(=2.5\times 10^{-4})} H^+ + HCO_3^- (\text{bicarbonate ion}) \qquad (1.2)$$

$$HCO_3^- \xrightarrow{k_{a2}\,(=4.6910^{-11})} H^+ + CO_3^{2-} (\text{carbonate ion}) \qquad (1.3)$$

where

$$k_{a1} = \frac{[H^+] \times [HCO_3^-]}{[H_2CO_3]}, \quad k_{a2} = \frac{[H^+] \times [CO_3^{2-}]}{[HCO_3^-]} \qquad (1.4)$$

More information of $CO_2$ can be found when necessary.[6]

## 1.2   $CO_2$ PRODUCTION AND CONSUMPTION

Global $CO_2$ is released from both nature and anthropogenic sources. Nature sources include soils,[20] inland waters,[21] the ocean,[22] and even volcanic activity,[23] whereas anthropogenic sources include cement industry[24] (responsible for 8% of the world's $CO_2$ emissions[25]), metrics[26] (approximately 5% from steel[27]), land transport,[28] aviation,[29–32] shipping,[33] and compost reactors.[34] The combustion of fossil fuels (primarily coal, fuel oil, and natural gas) is the main reason of long-term climate change.[35] With the carbon balance being broken, the atmospheric concentration of $CO_2$ is gradually rising,[30,36–38] also causing carbon accumulation in land and oceans[39] and irreversible climate change.[40,41] It is predicted that the daily atmospheric concentration of $CO_2$ will soon surpass 400 ppm at the sentinel spot of Mauna Loa, Hawaii, a value not reached at this key surveillance point for a few million years,[42] and rise to above 750 ppm by 2100.[43]

The $CO_2$ emission varies according to each country and region (Table 1.1).[44] The United States is one of the largest $CO_2$-emitting countries of the world,[45,46] although the emission decreased by 4% to 5.19 × 10⁹ t in 2012, following a 2% decrease in 2011. In 2006, China exceeded the United States in $CO_2$ emission, and became the world's largest emitting country due to its economic growth and high energy consumption.[44,47] In Asian countries, China, India, and Japan emitted 9.86, 1.97, and 1.32(×10⁹ t) $CO_2$ in 2012, respectively, increased from 9.55, 1.84, and 1.24 in 2011, accounting for 38% of the total emission (34.5) of the world in 2012. In EU countries, Germany is the largest $CO_2$-emitting country; the $CO_2$ emission was 0.81(×10⁹ t) in 2012, accounting for 22% of EU countries' total emissions.

**TABLE 1.1**

**The 1990–2012 $CO_2$ Emissions per Region/Country ($\times 10^9$ t)**

| Countries | 1990 | 1995 | 2000 | 2005 | 2010 | 2011 | 2012 |
|---|---|---|---|---|---|---|---|
| United States | 4.99 | 5.26 | 5.87 | 5.94 | 5.50 | 5.39 | 5.19 |
| EU27 | 4.32 | 4.08 | 4.06 | 4.19 | 3.91 | 3.79 | 3.74 |
| • France | 0.39 | 0.39 | 0.41 | 0.41 | 0.39 | 0.37 | 0.37 |
| • Germany | 1.02 | 0.92 | 0.87 | 0.85 | 0.82 | 0.80 | 0.81 |
| • Italy | 0.43 | 0.44 | 0.46 | 0.48 | 0.42 | 0.41 | 0.39 |
| • Spain | 0.23 | 0.25 | 0.31 | 0.36 | 0.28 | 0.29 | 0.29 |
| • United Kingdom | 0.59 | 0.56 | 0.55 | 0.55 | 0.51 | 0.47 | 0.49 |
| • Netherlands | 0.16 | 0.17 | 0.17 | 0.18 | 0.18 | 0.17 | 0.16 |
| Japan | 1.16 | 1.25 | 1.28 | 1.32 | 1.24 | 1.24 | 1.32 |
| Australia | 0.27 | 0.30 | 0.36 | 0.41 | 0.43 | 0.44 | 0.43 |
| Canada | 0.45 | 0.48 | 0.55 | 0.57 | 0.55 | 0.56 | 0.56 |
| Russian Federation | 2.44 | 1.75 | 1.66 | 1.72 | 1.71 | 1.78 | 1.77 |
| China | 2.51 | 3.52 | 3.56 | 5.85 | 8.74 | 9.55 | 9.86 |
| India | 0.66 | 0.87 | 1.06 | 1.29 | 1.78 | 1.84 | 1.97 |
| Total | 22.7 | 23.6 | 25.4 | 29.3 | 33.0 | 34.0 | 34.5 |

*Source:* Selected from Olivier, J. G.; Janssens-Maenhout, G.; Peters, J. A., *Trends in Global $CO_2$ Emissions: 2012 Report.* PBL Netherlands Environmental Assessment Agency: **2012**.

Carbonate precipitation in soils is a sink for atmospheric $CO_2$.[48] Carbon is transferred into the soil through dissolution in rainwater ($Ca^{2+} + 2HCO_3^- = CaCO_3 + H_2O + CO_2$). Photosynthesis also utilizes the sun's energy with chlorophyll in plants as a catalyst to recycle $CO_2$ and water into new plant life.[49] The oceans also take up a considerable amount of $CO_2$.[39,50,51] It was reported that between 1959 and 2008, 43% of each year's $CO_2$ emissions remained in the atmosphere on average and the rest was absorbed by carbon sinks on land and in the oceans.[50]

## 1.3  $CO_2$ CAPTURE, CONVERSION, AND UTILIZATION

At present, certain barriers still hinder the practical application of $CO_2$ capture, conversion, and utilization. These barriers include: (1) the high costs of $CO_2$ capture, separation, purification, and transportation to user sites; (2) the high energy requirements for $CO_2$ chemical/electrochemical conversion; (3) limitations in market size and investment incentives; (4) lack of industrial commitment to enhance $CO_2$-based chemicals; and (5) insufficient socioeconomic driving forces.[52,53] Despite such challenges, $CO_2$ capture, conversion, and utilization are still recognized as feasible and promising cutting-edge areas of exploration in energy and environmental research.

### 1.3.1 $CO_2$ CAPTURE

$CO_2$ may be chemically and physically captured from either fuel gas (precombustion capture) or flue gas (postcombustion and oxy-combustion capture).[54–59] Precombustion $CO_2$ capture is normally carried out to remove $CO_2$ from $H_2$–$CO_2$ mixture gas, and the resultant $H_2$ is supplied to power generation devices which may be fuel cell and power plants. Postcombustion capture is a mature technology. Solvent adsorbents frequently used contain monoethanolamine (MEA), 2-amino-2-methyl-1-propanol, diethanolamine, and methyldiethanolamine, which work based on the following reactions:

$$CO_2 + 2R_1R_2NH \rightleftharpoons R_1R_2NCO_2^- + R_1R_2NH_2^+ \qquad (1.5)$$

$$CO_2 + R_1R_2R_3N + H_2O \rightleftharpoons R_1R_2R_3NH^+ + HCO_3^- \qquad (1.6)$$

Selexol (dimethylether polyethylene glycol), Rectisol (chilled methanol), Fluor (propylene carbonate), and Purisol (*N*-methyl-2-pyrollidone) have often been used for physical adsorption of $CO_2$.

Carbonate looping processes are also a postcombustion technology using limestone called dry sorbents.[60] As shown in Figure 1.1, calcium oxide (CaO) is carbonized in a reversible and exothermic reaction "$CaO_{(s)} + CO_{2(g)} \rightleftharpoons CaCO_{3(s)}$." at 600–700°C and the formed calcium carbonate ($CaCO_3$) calcination in an endothermic reaction at 900°C. The regeneration of CaO is enhanced under reduced pressure.

Oxy-combustion capture is an alternative to postcombustion capture.[61] Pure oxygen is provided instead of air for combustion, producing flue gas streams with high $CO_2$ concentrations. This is beneficial to $CO_2$ condensation (cryogenics separation).

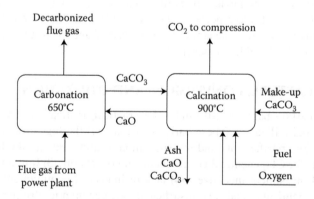

**FIGURE 1.1** Carbonate looping process. (Markewitz, P. et al., Worldwide innovations in the development of carbon capture technologies and the utilization of $CO_2$. *Energ Environ Sci* **2012**, *5*, (6), 7281–7305. Reproduced by permission of The Royal Society of Chemistry.)

However, oxy-combustion capture is not considered as a practical means to separate CO$_2$ from flue gases because of the high energy costs involved.

### 1.3.2 CO$_2$ CONVERSION

CO$_2$ can be converted to other chemicals by chemical, photochemical, electrochemical, and photoelectrochemical methods, which break the C=O bond.

CO$_2$ recovery started early in the late nineteenth century. It was reported that in 1860, Kolbe succeeded in preparing salicylic acid by heating a mixture of phenol and sodium (potassium) under 1 atm CO$_2$ (Kolbe–Schmitt reaction)[62]:

$$C_6H_5ONa(K) + CO_2 \xrightarrow[\text{0.1–1.0 MPa CO}_2]{\text{100–150°C}} C_6H_4(OH)COONa(K) \qquad (1.7)$$

Another CO$_2$ conversion and utilization that should be mentioned is the synthesis of urea with ammonia and CO$_2$, including two steps:

$$2NH_3 + CO_2 \rightleftharpoons H_2NCOONH_4 \quad \text{(fast, exothermic)} \qquad (1.8)$$

$$H_2NCOONH_4 \rightleftharpoons (NH_2)_2CO + H_2O \quad \text{(slower, endothermic)} \qquad (1.9)$$

The overall process (CO$_2$ + 2NH$_3$ → H$_2$NCONH$_2$ + H$_2$O) is exothermic.

Centi and Perathoner concluded that chemical recycling of CO$_2$ to fuels, as a complementary technology to carbon sequestration and storage (CSS), can minimize as much as possible the consumption of hydrogen (or hydrogen sources), produce fuels that can be easily stored and transported, and use renewable energy sources.[7]

### 1.3.3 CO$_2$ UTILIZATION

The use of CO$_2$ for the production of chemicals ranges today around 170 Mt/y (Table 1.2).[6]

## 1.4 ELECTROCHEMICAL REDUCTION OF CO$_2$

Compared with chemical CO$_2$ conversion, both electrocatalytical and photocatalytical CO$_2$ reduction have been developed greatly in the recent decades.[63]

Electrochemical reduction of CO$_2$ can proceed through two-, four-, six-, and eight-electron reduction pathways in gaseous, aqueous, and nonaqueous phases at both low and high temperatures. The major reduction products are carbon monoxide (CO), formic acid (HCOOH) or formate (HCOO$^-$) in basic solution, oxalic acid (H$_2$C$_2$O$_4$) or oxalate (C$_2$O$_4^{2-}$ in basic solution), formaldehyde (CH$_2$O), methanol (CH$_3$OH), methane (CH$_4$), ethylene (CH$_2$CH$_2$), ethanol (CH$_3$CH$_2$OH), as well as others. The thermodynamic electrochemical half-reactions of CO$_2$ reduction and their associated standard electrode potentials are listed in Table 1.3.[64]

**TABLE 1.2**

**Present and Short-Term Uses of $CO_2$**

| Compound | Actual Production | $CO_2$ Used | 2016 Forecast | $CO_2$ Needed |
|---|---|---|---|---|
| Urea | 155 | 114 | 180 | 132 |
| Methanol | 50 | 8 | 60 | 10 |
| Dimethyl ether | 11.4 | 3 | >20 | >5 |
| *tert*-Butyl methyl ether | 30 | 1.5 | 40 | 3 |
| Formaldehyde | 21 | 3.5 | 25 | 5 |
| **Other Fuels** | | | | |
| Higher alcohols | | | | |
| Hydrocarbons | | | | |
| Methane | | | | |
| Carbonates | 0.2 | 0.005 | >2 | 0.5 |
| Polycarbonates | 4 | 0.01 | 5 | 1 |
| Carbamates | 5.3 | 0 | >6 | 1 |
| Polyurethane | >8 | 0 | 10 | 0.5 |
| Acrylates | 2.5 | 0 | 3 | 1.5 |
| Polyacrylates | | | | |
| Formic acid | 0.6 | 0 | 1 | 0.9 |
| Inorganic carbonates | 200 | ca. 50 | 250 | 70 |
| Soda Solvay, pigments | 113.9, 50 | | | |
| Total | | 172 | | 207 |
| Technological | | 28 | | 80 |
| Algae for the production of biodiesel | 0.005 | 0.01 | 1 | 2 |
| | | 200 | | 299 |

*Source:* Adapted from Aresta, M.; Dibenedetto, A., *Dalton Trans* **2007**, (28), 2975–2992.

$CO_2$ conversion using electrochemical catalysis approaches have attracted great attention due to their several advantages[65]:

- The process is controllable by electrode potentials and reaction temperature.
- The supporting electrolytes can be fully recycled so that the overall chemical consumption can be minimized to simply water or waste water.
- The electricity used to drive the process can be obtained without generating any new $CO_2$—sources include solar, wind, hydroelectric, geothermal, tidal, and thermoelectric processes.
- The electrochemical reaction systems are compact, modular, on-demand, and easy for scale-up applications.

**TABLE 1.3**

**Selected Standard Potentials of CO$_2$ in Aqueous Solutions (V vs. SHE) at 1.0 atm and 25°C Calculated According to the Standard Gibbs Energies of the Reactants in Reactions**

| Half-Electrochemical Thermodynamic Reactions | Electrode Potentials (V vs. SHE) at Standard Conditions |
|---|---|
| $CO_2(g) + 4H^+ + 4e^- = C(s) + 2H_2O(l)$ | 0.210 |
| $CO_2(g) + 2H_2O(l) + 4e^- = C(s) + 4OH^-$ | −0.627 |
| $CO_2(g) + 2H^+ + 2e^- = HCOOH(l)$ | −0.250 |
| $CO_2(g) + 2H_2O(l) + 2e^- = HCOO^-(aq.) + OH^-$ | −1.078 |
| $CO_2(g) + 2H^+ + 2e^- = CO(g) + H_2O(l)$ | −0.106 |
| $CO_2(g) + 2H_2O(l) + 2e^- = CO(g) + 2OH^-$ | −0.934 |
| $CO_2(g) + 4H^+ + 4e^- = CH_2O(l) + H_2O(l)$ | −0.070 |
| $CO_2(g) + 3H_2O(l) + 4e^- = CH_2O(l) + 4OH^-$ | −0.898 |
| $CO_2(g) + 6H^+ + 6e^- = CH_3OH(l) + H_2O(l)$ | 0.016 |
| $CO_2(g) + 5H_2O(l) + 6e^- = CH_3OH(l) + 6OH^-$ | −0.812 |
| $CO_2(g) + 8H^+ + 8e^- = CH_4(g) + 2H_2O(l)$ | 0.169 |
| $CO_2(g) + 6H_2O(l) + 8e^- = CH_4(g) + 8OH^-$ | −0.659 |
| $2CO_2(g) + 2H^+ + 2e^- = H_2C_2O_4(aq.)$ | −0.500 |
| $2CO_2(g) + 2e^- = C_2O_4^{2-}(aq.)$ | −0.590 |
| $2CO_2(g) + 12H^+ + 12e^- = CH_2CH_2(g) + 4H_2O(l)$ | 0.064 |
| $2CO_2(g) + 8H_2O(l) + 12e^- = CH_2CH_2(g) + 12OH^-$ | −0.764 |
| $2CO_2(g) + 12H^+ + 12e^- = CH_3CH_2OH(l) + 3H_2O(l)$ | 0.084 |
| $2CO_2(g) + 9H_2O(l) + 12e^- = CH_3CH_2OH(l) + 12OH^-$ | 0.744 |

*Source:* Reprinted with permission from Bard, A. J.; Parsons, R.; Jordan, J., *Standard Potentials in Aqueous Solutions*. CRC Press: **1985**; Vol. 6. Copyright © 1985 CRC Press.

However, challenges remain:

- The slow kinetics of CO$_2$ electroreduction, even when electrocatalysts and high electrode reduction potential are applied.
- The low energy efficiency of the process, due to the parasitic or decomposition reaction of the solvent at high reduction potential.
- The high energy consumption. Researchers have recognized that the biggest challenge in CO$_2$ electroreduction is low performance of the electrocatalysts (i.e., low catalytic activity and insufficient stability).

## 1.5  FARADAIC/CURRENT EFFICIENCY OF $CO_2$ ELECTROREDUCTION

Both current and faradaic efficiencies ($\eta$ and $f$) are two indexes that are being popularly used to measure the utilization efficiency of electricity within an electrochemical reaction system.

$\eta$ can be either the instantaneous current efficiency ($i_{r,t}/i_{total,t}$) or the overall current efficiency ($Q_{r,T}/Q_{total,T}$). $i_{r,t}$ and $i_{total,t}$ are the currents of the $r$th process and all processes that occur simultaneously at an electrode at a time, $t$, respectively. $Q_{r,T}(=\int_0^T i_{r,t}dt)$ and $Q_{total,T}(=\int_0^T i_{total,t}dt)$ are the currents of the $r$th process and all ones at an electrode overall a period of time, $T$, respectively. $\eta$ will be 100% when only one process is occurring. For a desirable electrochemical reduction of $CO_2$, $\eta$ for the formation of target products should be as high as possible with hydrogen evolution being completely inhibited.

$f$ of a specific product of $CO_2$ electroreduction describes the efficiency with which charges (electrons) are transferred in an electrochemical reaction. It was calculated from the number of electrons consumed in the electroreduction process by using the formula

$$f = \frac{mnF}{\int_0^t i_{r,t}dt} \tag{1.10}$$

where $m$ is the number of moles of product harvested, $n$ is the number of electrons required for the formation of the product, $F$ is the Faraday constant ($9.6485 \times 10^4$ C/mol), and $i_{r,t}$ is the circuit current. If we focus on the product selectivity, $f$ will be given.

Experimental results show that $f$ may be affected by

- Catalysts themselves, for example, element[66,67] and chemical composition[68]
- Morphology of catalysts, for example, thickness of the catalyst layer,[69,70] catalyst particle size[69])
- Reaction conditions, for example, temperature,[71,72] water content of solvents,[72] pH value in $CO_2$-saturated solution,[73] reaction time,[73] etc.

The difference in Faraday efficiencies will be found when using different metal electrodes and metal complex catalysts-loaded ones. Various reasons have been given. For example, the thickness independence of $\eta$ had been concluded to be the counteractive effects caused by increasing local proton concentration and decreasing electrical field when catalyst layer thickness was increased.[69] In Figure 1.2, the effect of water contents on faradaic yield is shown during the electrochemical reduction of $CO_2$ to carbon CO, $(COO)^{2-}$, $HCOO^-$ (formate), and $H_2$.[72]

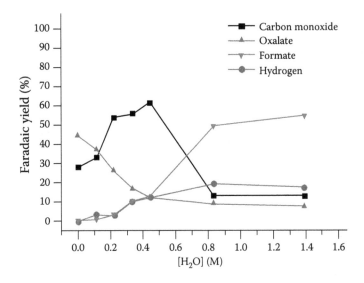

**FIGURE 1.2** Faradaic yield of CO$_2$ reduction as a function of water content. Potentiostatic electrolysis was carried at −2.45 V (Fc/Fc$^+$) at −20°C in 0.1 M (Bu)$_4$NPF$_6$/CH$_3$CN. (Oh, Y. et al., Electrochemical reduction of CO$_2$ in organic solvents catalyzed by MoO$_2$. *Chem Commun* **2014**, *50*, (29), 3878–3881. Reproduced by permission of The Royal Society of Chemistry.)

## 1.6 PRODUCT SELECTIVITY OF CO$_2$ ELECTROREDUCTION

### 1.6.1 AFFECTING FACTORS

As discussed above, the catalyst selectivity to produce desired products in catalyzed CO$_2$ electroreduction is very important for practical applications. Normally, this selectivity is closely related to the reduction mechanism, with different reaction pathways or combinations of different pathways leading to different products. In the initial reduction step of a typical CO$_2$ electroreduction mechanism, CO$_2$ can obtain electrons either directly from the cathode surface (a bare electrode surface or a surface coated catalyst) or indirectly from a medium, such as a soluble catalyst, to produce an intermediate, such as CO$_2^{\bullet-}$, which then absorbs on the cathode for product formation. In general, which kinds of pathways and how many pathways are required for the reduction process will be strongly affected by experimental conditions, such as the catalysts and/or electrodes, electrode potential, electrolyte solution, buffer strength, pH, CO$_2$ concentration, and pressure, as well as temperature.[72,74] The possible reaction pathways are summarized in Table 1.4.[67,75–77]

#### 1.6.1.1 Effects of Single Metal Electrode Type

Regarding product selectivity, single metal electrodes seem to be the most popular type of electrocatalysts for CO$_2$ reduction.[78,79] Two groups can be roughly designated[80]: (1) CO formation metals (Cu, Au, Ag, Zn, Pd, Ga, Ni, and Pt) and (2) HCOO$^-$ formation metals (Pb, Hg, In, Sn, Cd, and Tl). There are other types of catalysts that

**TABLE 1.4**

**Proposed Reaction Pathways in $CO_2$ Electroreduction**

| | | | |
|---|---|---|---|
| $CO_2 + e^-$ | $\cdot CO_{2(ads)}^- + H^+ + e^-$ | $CO + 4H^+ + 4e^-$ | $\cdot CH_{2(ads)} + 2H^+ + 2e^- \rightarrow CH_4$ |
| $\rightarrow \cdot CO_{2(ads)}^-$ | $\rightarrow CO + OH^-$ | $\rightarrow \cdot CH_{2(ads)} + H_2O$ | $2 \cdot CH_{2(ads)} \rightarrow C_2H_4$ |
| or | *Product: CO* | | $2 \cdot CH_{2(ads)} + 2H^+ + 2e^- \rightarrow C_2H_6$ |
| | | | *Products: $CH_4$, $C_2H_4$, $C_2H_6$* |
| $NR_4^+ + e^- \rightarrow \cdot NR_4$ | $\cdot CO_{2(ads)}^- + H_2O$ | $HCO_2 \cdot_{(ads)} + e^- \rightarrow HCO_2^-$ | $HCO_2^- + CH_3OH \rightarrow HCOOCH_3$ |
| $CO_2 + \cdot NR_4$ | $\rightarrow \cdot HCO_{2(ads)} + OH^-$ | *Product: $HCO_2^-$* | $+ OH^-$ |
| $\rightarrow \cdot CO_2^- + NR_4^+$ | | | *Product: $HCOOCH_3$* |
| | $2 \cdot CO_{2(ads)}^- \rightarrow C_2O_4^{2-}$ | $C_2O_4^{2-} + 2H^+ + 2e^- \rightarrow$ | $HC(=O)COO^- + 2e^- + 2H^+$ |
| | *Product: $C_2O_4^{2-}$* | $HC(=O)COO^- + OH^-$ | $\rightarrow H_2C(OH)COO^-$ |
| | | | *Product: $H_2C(OH)COO^-$* |

*Source:* Adapted from Ikeda, S.; Takagi, T.; Ito, K., *Bull Chem Soc Jpn* **1987,** *60*, (7), 2517–2522; Saeki, T. et al., *J Electroanal Chem*, **1995,** *390*, (1–2), 77–82; Li, J. W.; Prentice, G., *J Electrochem Soc* **1997,** *144*, (12), 4284–4288; Magdesieva, T. V. et al., *Russ Chem B+* **2002,** *51*, (5), 805–812.

also have both high selectivity and high current efficiency. In addition, some electrocatalysts have been specifically designed for and show unique catalytic activities toward $CO_2$ reduction to produce desired products with high current efficiencies.

### 1.6.1.2 Effects of Metal Complex, Metal Center, and Ligand Types

It is well known that the catalytic performance of metal complex catalysts for $CO_2$ reduction strongly depends on the chemical properties of the metal center and ligand. Therefore, it is expected that the distribution of the electrolysis products, the current efficiencies, and the reaction mechanism of $CO_2$ electroreduction will also be strongly affected by the types of central metals and ligands in macrocyclic complexes. For example, $CO_2$ electroreduction on a glassy carbon electrode (GCE) modified with polymeric M-tetrakis aminophthalocyanines (M = Co, Ni, Fe) indicated that different metal centers produced different products.[81] When M was Co, HCOOH was the only product; when M was Fe, a mixture of $CH_2O$ and $H_2$ was produced; and when M was Ni, a mixture of HCOOH and $CH_2O$ was observed.[82] Furuya and Matsui[83] investigated the electrocatalytic reduction of $CO_2$ on gas diffusion electrodes (GDEs) modified by 17 kinds of metal phthalocyanine (MPc) catalysts (where M = Co, Ni, Fe, Pd, Sn, Pb, In, Zn, Al, Cu, Ga, Ti, V, Mn, Mg, P) in 0.5 M $KHCO_3$. They found that the distribution and current efficiencies of the electrolysis products were strongly dependent on the nature of the central metal coordinated to the phthalocyanines. With transition metals of Co- and Ni-phthalocyanines, the main electrolysis product was CO, with a current efficiency of ~100%. On the other hand, HCOOH was the main product on phthalocyanines with Sn, Pb, or In metal centers. The highest current efficiency, ~70%, was observed on SnPc around –1.6 V. In the case of Cu-, Ga-, and Ti-phthalocyanines, $CH_4$ was the main product, with the highest current efficiencies being 30%–40%. Analogously, in the simultaneous reduction of $CO_2$ and $NO_3^-$ with various MPc catalysts (where M = Ti, V, Cr, Mo, Fe, Ru, Co,

Ni, Pd, Cu, Zn, Cd, Ga, In, Ge, Sn, Pb), the current efficiencies of CO formation at CoPc and NiPc catalysts in the reduction of $CO_2$ alone were further demonstrated to be far higher than at pure metal catalysts; hence, these M/Pc catalysts were expected to have a fairly high capacity for urea formation.[84] The effects of various aza-macrocyclic ligands on the production of fuels from macrocyclic complexes were also investigated, including simple tetraazamacrocycles, porphyrin, phthalocyanine, and biphenanthrolinic hexaazacyclophanes.[85,86] The mechanism for producing a particular product is normally related to the structural features of the azacyclam framework and its interaction with the central metal and $CO_2$ or CO molecules, whereas the replacement of a $-CH_2$ group in the ligand backbone by an amide residue, for example, does not disturb the catalytic process.

### 1.6.1.3 Effect of Cations and Anions in the Electrolyte

As described previously, alkaline metals and alkaline earth metals cannot be used as electrodes for $CO_2$ catalytic reduction catalysts. However, their salts have commonly been used as supporting electrolytes in electrochemical cells for $CO_2$ electroreduction, and they have different effects on product selectivity. For example, in the case of FeTPP catalyst, the addition of Lewis acid cations such as $Mg^{2+}$, $Ca^{2+}$, $Ba^{2+}$, $Li^+$, or $Na^+$ decreased HCOOH formation as the acidity increased; the order of reactivity of these Lewis acid synergists was $Mg^{2+} = Ca^{2+} > Ba^{2+} > Li^+ > Na^+$.[87] Thorson et al.[88] confirmed that the presence of large cations such as cesium (Cs) and rubidium (Ru) in the electrolyte could enhance the electrochemical conversion of $CO_2$ to CO. This was explained by the interplay between the level of cation hydration and the extent of cation adsorption on the metal electrodes. The effects of anions in the electrolyte on the products of $CO_2$ electroreduction were also investigated using a copper mesh electrode in aqueous solutions containing 3 M of KCl, KBr, and KI as the respective electrolytes.[89] The results showed that the bond between adsorbed halide anion (e.g., $Br^-$, $Cl^-$, or $I^-$) and carbon helped the electron transfer from the adsorbed halide anion to the vacant orbital of $CO_2$, promoting $CO_2$ conversion.[89] The stronger the adsorption of the halide anion to the electrode, the more strongly $CO_2$ was restrained, resulting in a higher $CO_2$ reduction current. Furthermore, the specifically adsorbed halide anions suppressed the adsorption of protons, leading to a higher hydrogen overvoltage. This reaction mechanism was also confirmed by Schizodimou and Kyriacou,[90] who showed that the rate of electrochemical reduction of $CO_2$ increased in the order $Cl^- < Br^- < I^-$.

### 1.6.1.4 Effects of Supporting Electrolytes

In fact, even for the same metal electrode with the same purity, different supporting electrolytes have a great effect on the final products. For example, in the electroreduction of $CO_2$ at a copper electrode (99.999% purity) in $CH_3OH$, Saeki et al.[75] observed two different main products—CO (current efficiency, 48.1%–86.8%) and $CH_3COOH$ (54.5% and 46.7%)—when the supporting electrolyte was either $(Bu)_4N^+$ (TBA) salts (i.e., TBA $\cdot$ $BF_4$ and TBA $\cdot$ $ClO_4$) or lithium salts ($LiBF_4$ and $LiClO_4$). They proposed that the intermediate, $CO_2^{*-}$, was stabilized by forming a $TBA^+ - CO_2^{*-}$ ion pair or by being adsorbed on the electrode surface. By contrast, $NH_4ClO_4$ as a supporting electrolyte in the same electrolysis only led to hydrogen evolution (84.6%). Other electrolytes have also been explored, such as $(CH_3)_4N \cdot ClO_4(TEA)$.[91]

### 1.6.1.5  Effects of Solvent/$CO_2$ Concentration

It should be mentioned that $CO_2$ utilization in aqueous solution can be limited by its low solubility in water at standard temperature and pressure.[92] This is because there are relatively small amounts of $CO_2$ available for the reaction to proceed at the electrode surface. To speed up the reaction process for industrial purposes, pressurized $CO_2$ is normally required,[93,94] which often causes a certain degree of change in product selectivity. Normally, in aqueous solutions, metallic catalysts (or electrodes), such as *sp* group metals (e.g., In or Pb), tend to give higher CO production at pH levels higher than 4, whereas *d* group metals (such as Pd and Cu) can promote HCOOH production.[74] It was concluded that the main products obtained in aqueous media under ambient conditions were strongly dependent on the type of cathode: Cu electrodes mainly yielded mixtures of hydrocarbons (mostly methane and ethylene) and alcohols; Au, Ag, and Zn mainly produced CO, whereas other metals, such as In, Sn, Hg, and Pb, were selective for the production of $HCOOH/HCOO^-$.[95] Compared with water solutions, the application of nonaqueous solvents is relatively popular due to their high solubility for $CO_2$. For example, *N,N*-dimethylformamide (DMF), PC, and $CH_3OH$ may contain up to 20, 8, and 5 times more $CO_2$, respectively, than corresponding amounts of aqueous solutions. Among them, solvents having low proton availability, such as DMF, favor the formation of oxalate and CO, whereas aqueous solution favors $HCOO^-$.[96] Strategies that have been demonstrated experimentally are further described below according to the selective generation of desired products.

### 1.6.2  Selective Production of Carbon Monoxide

CO is one of the important products generated in $CO_2$ electroreduction. The electrochemical conversion of $CO_2$ to CO is a two-electron process. Some metal cathodes, such as Ag, Au, and Zn, are highly selective for the electrocatalytic reduction of $CO_2$ to CO in $KHCO_3$ aqueous solutions.[66] [Ni(cyclam)]$^{2+}$ is a well-known catalyst for $CO_2$ electroreduction to CO on a mercury cathode at −0.9 V in aqueous solutions.[97] Two other electrocatalysts with a similar structure to [Ni(cyclam)]$^{2+}$ were also found to be selective for CO production.[98] In addition, metal polyphosphine complexes, such as Pd(triphosphine)L$^{2+}$ (L = $CH_3CN$, P(OMe)$_3$, PEt$_3$, P(CH$_2$OH)$_3$, and PPh$_3$), exhibited high catalytic activity for the reduction of $CO_2$ to CO in acidic $CH_3CN$ solutions.[99]

As an enzyme-based catalyst, a Ni- and Fe-containing metalloenzyme isolated from *Moorella thermoacetica* showed highly selective activity toward the conversion of $CO_2$ to CO, with a current efficiency as high as ~100% at −0.57 V vs. normal hydrogen electrode (NHE) in a 0.1 M phosphate buffer solution (pH 6.3).[100]

### 1.6.3  Selective Production of Formic Acid (and Formate)

The electrochemical conversion of $CO_2$ to formic acid "HCOOH" (and $HCOO^-$) is also a two-electron process. Electrochemical reduction of $CO_2$ in aqueous solution to HCOOH and $HCOO^-$ was reported as early as 1870.[101] Noda et al.[66] reported that

Zn, Cd, Hg (Group 12), In (Group 13), Sn, and Pb (Group 14) metal cathodes in 0.1 M KHCO$_3$ aqueous solution exhibited high production selectivity for HCOO$^-$ formation. Chen et al.[102] reported that HCOOH was the main product of CO$_2$ electroreduction by both [Ir$_2$(dimen)$_4^{2+}$](PF$_6^-$)$_2$ and [Ir$_2$(dimen)$_4^{2+}$](B(C$_6$H$_5$)$_4^-$)$_2$ (dimen = 1,8-diisocya nomenthane) electrocatalysts. Kang et al.[103] reported selective electrocatalytic reduction of CO$_2$ to HCOO$^-$ by a water-soluble iridium pincer catalyst. HCOO$^-$ was the only reduced carbon product, formed in 93% faradaic yield with no formation of CO. As previously mentioned, to optimize the large-scale (even industrial-scale) electrocatalytic reduction of CO$_2$ to HCOO$^-$, Li and Oloman[104] investigated a series of condition variables that might affect the performance of a reactor. The current efficiency of HCOO$^-$ formation was reportedly as high as 91%. Recently, the engineering and economic feasibility of the large-scale electrochemical reduction of CO$_2$ to HCOOH and HCOO$^-$ has been discussed by Agarwal et al.[65]

Electrocatalytic reduction of CO$_2$ by an enzyme catalyst, formate dehydrogenase enzyme (FDH1), which was isolated from *Syntrophobacter fumaroxidans*, was found (as either a homogeneous or a heterogeneous catalyst) to produce HCOO$^-$ exclusively.[105] Two acetogenic bacteria, *M. thermoacetica* (Mt, formerly *Clostridium thermoaceticum*, ATCC 35,608) and *Clostridium formicoaceticum* (Cf, DSM 92), were explored as catalysts for CO$_2$ electroreduction in 1.0 atm CO$_2$-saturated 0.1 M phosphate buffer solution (pH 7.0) at −0.58 V vs. NHE; the results showed that these catalysts could efficiently convert CO$_2$ to HCOO$^-$ with current efficiencies of 80% for Mt and 100% for Cf.[106]

### 1.6.4 Selective Production of Formaldehyde

There are very few cases of highly selective production of HCHO by the electrochemical reduction. Sende et al.[68] earlier reported that [M(4-v-tpy)$_2$]$^{2+}$ and [M(6-v-tpy)$_2$]$^{2+}$ (M = Cr, Ni, Co, Fe, Ru, or Os), after being electropolymerized onto GCEs to form films, exhibited electrocatalytic activity toward CO$_2$ reduction, with HCHO as virtually the only product. The current efficiency for films of Cr(4-v-tpy)$_2$]$^{2+}$ was as high as 87%. Nakata et al.[107] also achieved a high-yield electrochemical production of HCHO from CO$_2$ and seawater using a boron-doped diamond electrode under ambient conditions. The high faradaic efficiency for the production of HCHO was 74%, using either CH$_3$OH, aqueous NaCl, or seawater as the electrolyte, and attributed to the $sp^3$-bonded carbon of the boron-doped diamond.

### 1.6.5 Selective Production of Methanol

The mechanism of reaction is always a topic of interest. Ganesh[108] suggested a most probable reaction sequence that occurred in the electrochemical reduction of CO$_2$ to CH$_3$OH, which is as follows:

$$CO_2 + 2e^- \text{ (at cathode surface)} \rightarrow CO_2^{2-}$$
$$CO_2^{2-} + H_2O \rightarrow HCOO^- + OH^-$$
$$CO_2^{2-} + 2H^+ + H_2O \rightarrow HCOOH \text{ (formic acid)} + H_2O$$

$HCOOH + 2e^-$ (at cathode surface) $\rightarrow HCOOH^{2-}$
$HCOOH^{2-} + 2H^+ \rightarrow CH_2(OH)_2$ (an unstable compound)
$CH_2(OH)_2$ (an unstable compound) $\rightarrow HCHO + H_2O$
$HCHO + 2e^-$ (at cathode surface) $+ 2H^+ \rightarrow CH_3OH$

Although the mechanism requires further confirmation, the net reaction is certainly a six-electron process: $CO_2 + 6H^+ + 6e^- \rightarrow CH_3OH + H_2O$, and the other compounds such as HCOOH, HCHO, and $CH_4$ will also form. Some electrocatalysts have been found to be fairly selective for $CH_3OH$ production in $CO_2$ electroreduction, as listed in Table 1.5.

### 1.6.6 SELECTIVE PRODUCTION OF OXALIC ACID (OXALATE)

The electroreduction of $CO_2$ in DMF solution showed a current efficiency of 73% for oxalic acid $[(COOH)_2]$, with a little $HCOO^-$ and CO production.[109] The macrocyclic nickel complex Ni–Etn(Me/COOEt)–Etn was found to be one of the most active and persistent homogeneous catalysts for $CO_2$ electroreduction selectively to $(COOH)_2$.[110] Dinuclear copper (I) complexes electrocatalyzed $CO_2$ conversion selectively to $(COOH)_2$ in $CH_3CN$ with a soluble lithium salt, resulting in quantitative precipitation of lithium oxalate.[111] In addition, anion radicals of aromatic esters such as phenyl benzoate and methyl benzoate, and of nitriles such as benzonitrile, in DMF at an inert electrode (e.g., mercury) were able to reduce $CO_2$ exclusively to $(COOH)_2$.[112,113]

### 1.6.7 SELECTIVE PRODUCTION OF LACTIC ACID

Lactic acid $(CH_3CH(OH)COOH)$ is an organic $C_3$ compound that plays an important role in numerous industries, including food, medicine, and cosmetics. Ogura et al.[114] found selective reduction of $CO_2$ to $CH_3CH(OH)COOH$, catalyzed by Fe(II)-4,5-dihydroxybenzene-1,3-disulfonate immobilized on a polyaniline/Prussian blue (PAn/PB)-modified Pt electrode in 0.5 M KCl solution. However, if Co-4,5-dihydroxybenzene-1,3-disulfonate was used, acetaldehyde $(CH_3CHO)$ was also produced.

To summarize the selective generation of desired products in electrocatalytic $CO_2$ reduction, Table 1.5 lists typical examples for several important low-carbon fuels, together with their generation conditions.

## 1.7 CHALLENGES, PERSPECTIVES, AND RESEARCH DIRECTIONS IN CO₂ ELECTROREDUCTION

The challenges in $CO_2$ electroreduction come from various aspects. Generally, lowering the overpotential for the electrochemical conversion of $CO_2$ to useful products is one of the major challenges.[160] For this goal, highly effective catalysts are indispensable. However, it seems that so far we still cannot simultaneously realize high faradaic/current efficiency, product selectivity and yield, and catalyst durability,[136] which are quite essential for industrial-scale processes. Mao et al.[161] also concluded the major issues and challenges associated with electrochemical $CO_2$ reduction to be

**TABLE 1.5**

**A Summary of Selected Electrocatalytic Reductions of $CO_2$ to Selectively Produce Several Important Low-Carbon Fuels**

| Product(s) | Electrode/Electrocatalysts | Potential (V), Reference Electrode | Electrolyte | Temperature/ Pressure, etc. | Current Efficiency | Reference |
|---|---|---|---|---|---|---|
| (1) CO | Cu (99.98%) electrode | −2.0 to 3.5 V (Ag rod quasi-reference electrode) | 0.5 M CsOH/10 atm $CO_2$-saturated $CH_3OH$, (<0.1%) $H_2O$ | −25 ± 0.5°C, H-type cell (in stainless-steel vessel) | 84% | [115] |
| | Ag (99.98%) electrode | −1.6 V vs. Ag/AgCl (saturated with KCl) | 0.1 M $KHCO_3$/water | 25°C | 64.7% | [66] |
| | Au (99.95%) electrode | | | | 81.5% | |
| | Zn metallic electrode | | | | 39.6% | |
| | Cd metallic electrode | | | | 14.4% | |
| | (99.95%) Pd electrode | −1.8 V (Ag/AgCl/ KCl) | 0.1 M $KHCO_3$/50 atm $CO_2$, saturated water | 30°C | 57.9% | [116] |
| | Nanoporous Ag electrocatalyst | −0.50 V (reversible hydrogen electrode [RHE]) | 0.5 M $KHCO_3$/ $CO_2$-saturated water | Room temperature, ~9.0 mA/cm² | 92% | [117] |
| | $Bi^{3+}$ catalyst | −2.0 V (saturated calomel electrode [SCE]) | 0.3 mM [BMIM]$PF_6$ | 82 ± 12 | 31 ± 2 | [118] |
| | Ditto | | 0.3 mM [BMIM]$BF_6$ | 82 ± 11 | 26 ± 4 | |
| | Ditto | | 0.3 mM [BMIM]Cl | 79 ± 12 | 17 ± 2 | |
| | Ditto | | 0.3 mM [BMIM]Br | 74 ± 4 | 20 ± 1% | |
| | Ditto | | 0.3 mM[BMIM]OTf | 87 ± 8 | 25 ± 2% | |
| | GCE | | 0.3 mM [BMIM]OTf /$CH_3CN$ (OTf = ($CF_3SO_3^-$) | None | <0.4 mA/cm² | |

*(Continued)*

**TABLE 1.5 (Continued)**

**A Summary of Selected Electrocatalytic Reductions of $CO_2$ to Selectively Produce Several Important Low-Carbon Fuels**

| Product(s) | Electrode/Electrocatalysts | Potential (V), Reference Electrode | Electrolyte | Temperature/Pressure, etc. | Current Efficiency | Reference |
|---|---|---|---|---|---|---|
| | CsCl | | ~mmol disilane (e.g., $(Ph_2MeSi)_2$, $(PhMe_2Si)_2$, $(Me_3Si)_3SiH$), or silylborane, or (pinacolato)$BSiMe_2Ph$/ $(CH_3)_2$ S=O | 20°C | | [119] |
| | $MoO_2$ microparticle | −2.45 V (Fc/Fc+) | 0.1 M $(Bu)_4N \cdot PF_6$/ $CH_3CN$, 0.4 M $H_2O$ | −20°C | 70% | [72] |
| | Bulk $MoS_2$ | −0.764 V (RHE) | 1-Ethyl-3-methylimidazolium tetrafluoroborate solution | | 100% | [120] |
| | $Re^{(0)}$ (bpy)$(CO)_3$ Cl, glassy carbon electrode (bpy = 2,2'-bipyridine) | −1.25 V (NHE) | 0.1 M $Et_4N \cdot$ Cl/ $CO_2$-saturated $(CH_3)_2NCHO/H_2O$ (4:1–19:1) solutions | 25°C | 91%–98% | [121] |
| | fac-Re(5,5'bisphenylethynyl) 2,2'bipyridyl)$(CO)_3$Cl | −1.75 V (NHE) | 0.1 M $Bu_4N \cdot PF_6$/ $CO_2$-satured $CH_3CN$ | One-compartment cell | 45% | [122] |
| | (1 mM) fac-Re(L)$(CO)_3$Cl, Pt disk electrode (L = pyrrole-substituted 2,2'-bipyridine) | −1.85 V (Ag/ [0.01 M]Ag⁺) | 0.1 M $Bu_4N \cdot ClO_4$/ $CO_2$-saturated $CH_3CN$ | Metrohm cell made airtight with vacuum grease (M Apiezon) | 92% | [123,124] |

*(Continued)*

**TABLE 1.5 (Continued)**
**A Summary of Selected Electrocatalytic Reductions of CO$_2$ to Selectively Produce Several Important Low-Carbon Fuels**

| Product(s) | Electrode/Electrocatalysts | Potential (V), Reference Electrode | Electrolyte | Temperature/Pressure, etc. | Current Efficiency | Reference |
|---|---|---|---|---|---|---|
| | fac–Re(L)(CO)$_3$Cl, carbon felt (WE) | | | 2 h | ~98.5% | [124] |
| | Re film electrodeposited onto a polycrystalline Au support | −1.35 V (SCE) | 0.1 M LiClO$_4$/1 atm CO$_2$-saturated CH$_3$OH | Stirred conditions / Quiescent conditions | 87% / 57% | [125] |
| | poly-Re(CO)$_3$(vbpy)Cl, Pt gauze electrode (vbpy = 4-vinyl-4′-methyl-2,2′-bipyridine) | −1.55 V (sodium saturated calomel electrode [SSCE]) | 0.1 M Bu$_4$N · PF$_6$/CO$_2$-saturated CH$_3$CN | 80 min, gas-tight cell | 92.3% | [126] |
| | Polymeric films formed by coelectropolymerization of cis-[Ru(bpy)$_2$(vpy)$_2$]$^{2+}$ with fac-[Re(CO)$_3$(vbpy)Cl] or fac-[Re(CO)$_3$(vbpy)CH$_3$CN]$^+$ | −1.55 V (SSCE) | 0.1 M Bu$_4$N · OH/CO$_2$-saturated CH$_3$CN | | 90%–98% | [127] |
| | fac-Re(2,2′-bipyridine)(CO)$_3$Cl (homogeneous catalyst), working electrode: glassy carbon | −1.66 V (Fc$^{+/0}$) | 1 atm CO$_2$-saturated 1-ethyl-3-methylimidazolium tetracyanoborate | 25 ± 3°C | 88 ± 10% | [128] |
| | | −2.11 V (Fc$^{+/0}$) | 0.1 M Bu$_4$N · PF$_6$/1 atm CO$_2$-saturated CH$_3$CN | | | |
| | [Mn(bpy)(CO)$_3$]$^+$ | −1.70 V (Ag/[10 mM]Ag$^+$) | 0.1 M Bu$_4$N · ClO$_4$/CO$_2$-satured CH$_3$CN | Room temperature | 85% | [129] |
| | [Mn(dmbpy)(CO)$_3$]$^+$ | | | | 100% | |

*(Continued)*

**TABLE 1.5 (Continued)**

**A Summary of Selected Electrocatalytic Reductions of $CO_2$ to Selectively Produce Several Important Low-Carbon Fuels**

| Product(s) | Electrode/Electrocatalysts | Potential (V), Reference Electrode | Electrolyte | Temperature/ Pressure, etc. | Current Efficiency | Reference |
|---|---|---|---|---|---|---|
| | $[Mn(bpy\text{-}t\text{-}Bu)(CO)_3]^+$ | −2.2 V (SCE) | 0.1 M $Bu_4N \cdot BF_6$/ $CO_2$-saturated $CH_3CN$ | | 100% | [130] |
| | $Mn(mesbpy)(CO)_3Br$, $[Mn(mesbpy)(CO)_3(CH_3CN)]$ (OTf) mesbpy = 6,6'dimesityl-2,2'-bipyridine OTf = $CF_3SO_3^-$ | | 0.1 M $Bu_4N \cdot PF_6$/ $CH_3CN$ | | | [131] |
| | (1) Iron 5,10,15,20-tetrakis (2',6'-dihydroxyphenyl)-porphyrin | −1.333 V (SCE) | (0.23 M) $CO_2$, $(CH_3)_2NCHO$, 0.1 M $Bu_4N \cdot PF_6$/2 M $H_2O$ | A single-compartment cell | 41%–56% | [132] |
| | (2) Iron 5,10,15,20-tetrakis (2',6'-dimethoxyphenyl)-porphyrin Working electrode: glassy carbon | −1.69 V (SCE) | | | 89%–104% | |
| | Carbon monoxide dehydrogenase (CODH) from *M. thermoacetica* Working electrode: glassy carbon disk | −0.57 V (NHE) | 1 atm $CO_2$-saturated 0.1 M phosphate buffer solution (pH 6.3) | 50°C | ~100% | [100] |

(Continued)

**TABLE 1.5 (*Continued*)**

**A Summary of Selected Electrocatalytic Reductions of CO$_2$ to Selectively Produce Several Important Low-Carbon Fuels**

| Product(s) | Electrode/Electrocatalysts | Potential (V), Reference Electrode | Electrolyte | Temperature/ Pressure, etc. | Current Efficiency | Reference |
|---|---|---|---|---|---|---|
| | Ni(cyclam)$^{2+}$, Ni$_2$(biscyclam)$^{4+}$ (cyclam = 1,4,8,11 -tetraazacyclotetradecane) | −1.25 V (SCE) | Water | Room temperature | >93% | [133] |
| | [Ru(bpy)(CO)$_2$Cl$_2$], [Ru(bpy)(CO)$_2$ClCOOCH$_3$], Cl(CO)$_2$(bpy)Ru–Ru(bpy)(CO)$_2$Cl, CH$_3$OOC(CO)$_2$(bpy)Ru–Ru(bpy)(CO)$_2$COOCH$_3$ | −1.5 V (Ag/[0.01 M] Ag$^+$) | 0.1 M (Bu)$_4$N · ClO$_4$/ CO$_2$-saturated CH$_3$CN | | 95%–97% | [134] |
| (2) HCOOH/HCOO⁻ | In electrode Pb electrode Zn electrode Sn electrode | −2.0/2.4 V (Ag/ AgCl) | 0.1 M (CH$_3$CH$_2$)$_3$N · H$_3$PO$_4$/ water | 100°C, normal pressure | 87.6%/83.2% 72.9%/78.9% 46.6%/53.4% 67.5%/37.1% | [67] |
| | Zn electrode Cd electrode Hg electrode | −1.6 V (Ag/AgCl saturated with KCl) | 0.1 M KHCO$_3$/water | 25°C | 20% 39% 94% | [66] |
| | In electrode Sn electrode Pb electrode | ~−1.9 V, ~−1.9 V, −2.17 V, Ag/AgCl (3 M NaCl) | CO$_2$-saturated 1-ethyl-3- methylimidazolium trifluoroacetate with 33% water | | ~100% ~100% ~100% | [135] |
| | Sn (99.9985%) electrode | −1.8 V (SCE) | 2 M KCl/CO$_2$-saturated water | | 60% | [136] |

*(Continued)*

**TABLE 1.5 (Continued)**
**A Summary of Selected Electrocatalytic Reductions of $CO_2$ to Selectively Produce Several Important Low-Carbon Fuels**

| Product(s) | Electrode/Electrocatalysts | Potential (V), Reference Electrode | Electrolyte | Temperature/ Pressure, etc. | Current Efficiency | Reference |
|---|---|---|---|---|---|---|
| | Sn (99.998%) electrode | −1.7 V (SCE) | 0.1 M $Na_2SO_4$/ $CO_2$-saturated water | A conventional three-electrode electrochemical cell | ~95% | [137] |
| | | −2.0 V (SCE) | 0.5 M $KHCO_3$/ $CO_2$-saturated water | | ~63% | |
| | Sn powder-decorated gas diffusion layer electrode | −1.6 V (SCE) | 0.5 M $NaHCO_3$/ambient $CO_2$ pressure, water | 27.3 mA/cm² | 70% | [138] |
| | Nanostructured Sn/graphene | −0.76 V (SCE) | $NaHCO_3$/water | 18 h, >10 mA/cm² | >93% | [139] |
| | Pb granule electrodes | −1.8 V (SCE) | 0.2 M $K_2CO_3$/(50 bar) $CO_2$-saturated water | 0.5 h, 80°C, 0.41 mA/cm² | 94% | [140] |
| | Tinned-copper sheet | | 0.45 M $KHCO_3$ | 0.22 kA/m², ambient conditions | 86% | [141] |
| | Fe (99.5%) electrode | −1.53 to 1.61 V (Ag/ AgCl) saturated KCl | 0.1 M $KClO_4$/30 atm $CO_2$-saturated water | 25°C; 120 mA/cm²; three-compartment glass cell in a stainless autoclave | 59.5%– 59.6% | [93] |
| | $Cu_2O$ dispersed polyaniline (PANI) electrodes | −0.3 V (SCE [sat. KCl]) | 0.1 M $(Bu)_4NClO_4$/ $CO_2$-saturated $CH_3OH$ | 25 ± 5°C | 30.4% | [142] |
| | $MoO_2$ microparticle | −2.45 V (Fc/Fc+) | 0.1 M $(Bu)_4N \cdot PF_6$/ $CH_3CN$, 1.4 M $H_2O$ | −20°C | 55% | [72] |

*(Continued)*

**TABLE 1.5 (Continued)**

**A Summary of Selected Electrocatalytic Reductions of CO$_2$ to Selectively Produce Several Important Low-Carbon Fuels**

| Product(s) | Electrode/Electrocatalysts | Potential (V), Reference Electrode | Electrolyte | Temperature/ Pressure, etc. | Current Efficiency | Reference |
|---|---|---|---|---|---|---|
| | $(2.00 \times 10^4$ M) Ni(cyclam)$^{2+}$/ Ni$_2$(biscyclam)$^{4+}$ | | 0.1 M NaClO$_4$/1 atm CO$_2$-saturated, (CH$_3$)$_2$NCHO | 20°C | 75% | [133] |
| | [Ru(bpy)$_2$(CO)$_2$]$^{2+}$ Working electrode: Hg pool | −1.3 V (SCE) | In the presence of (1) CH$_3$NH$_2$ · HCl (2) (CH$_3$)$_2$NH · HCl (3) C$_6$H$_5$OH CO$_2$-saturated H$_2$O/ (CH$_3$)$_2$NCHO (9:1, v/v) solution | 1.6 mA/cm$^2$ 3.3 mA/cm$^2$ 2.3 mA/cm$^2$ | 64.1% 84.3% 81.0% | [143] |
| | [(bpy)$_2$Ru(dmbbbpy)](PF$_6$)$_2$ [(bpy)$_2$Ru(dmbbbpy)]- Ru(bpy)$_2$(PF$_6$)$_4$ (dmbbbpy = 2,2′-bis(1- methylbenzimidazol-2-yl)- 4,4′-bipyridine) | −1.65 V −1.55 V (Ag/AgCl) | (2.5%) H$_2$O/CO$_2$- saturated CH$_3$CN | | 89% 90% | [144] |
| | 1. Carbon nanotube 2. Nitrogen-doped carbon nanotube 3. Nitrogen-doped graphenated carbon nanotube/glassy carbon (GC) electrode, polyethylenimine (cocatalyst) | −1.8 V (SCE) | 0.1 M KHCO$_3$/ CO$_2$-saturated water | | 59% 85% 87% | [145] |

(Continued)

**TABLE 1.5 (Continued)**

**A Summary of Selected Electrocatalytic Reductions of $CO_2$ to Selectively Produce Several Important Low-Carbon Fuels**

| Product(s) | Electrode/Electrocatalysts | Potential (V), Reference Electrode | Electrolyte | Temperature/ Pressure, etc. | Current Efficiency | Reference |
|---|---|---|---|---|---|---|
| (3) $CH_3COOH$ | Tungsten-containing formate dehydrogenase enzyme (FDH1) from *S. fumaroxidans* Working electrode: rotating disk pyrolytic graphite edge electrode | −0.41 to 0.81 V (Ag/AgCl) | 0.02 M $Na_2CO_3$/water (pH 6.5) | 37°C | 97.3%–102.1% | [105] |
| (4) $CH_3OH$ $$CO_{2(g)} + 6H^+ + 6e^- = CH_3OH_{(l)} + H_2O_{(l)},$$ (0.016) $$CO_{2(g)} + 5H_2O_{(l)} + 6e^- = CH_3OH_{(l)} + 6OH^-,$$ (−0.812) | $Cu_2O$ dispersed polyaniline (PANI) electrodes | −0.3 V (SCE [sat. KCl]) | 0.1 M $(Bu)_4NClO_4$/ $CO_2$-saturated $CH_3OH$ | 25 ± 5°C | 63.0% | [142] |
| | Cu | −1.1 V (Ag/AgCl) | 0.1 M LiCl/(34/68 atm) $CO_2$-saturated $C_2H_5O$-$H_2O$ with 20 m/o $H_2O$ | 80°C, 8 h, 9.0 mA/cm$^2$ | 36%/40% | [76] |
| | Preoxidized Cu foil electrode | −0.4 V (SCE) | 0.5 M $KHCO_3$/ $CO_2$-saturated water (pH = 7.6) | pH 7.5–7.6 at 25°C 33 mA/cm$^2$ | 100% | [146] |
| | Preoxidized Cu foil electrode: electrodeposited cuprous oxide film | −1.1 V (SCE) | 0.5 M $KHCO_3$/ $CO_2$-saturated water | A typical three-electrode cell (30 mL) | 38% | [147] |

*(Continued)*

**TABLE 1.5** *(Continued)*
**A Summary of Selected Electrocatalytic Reductions of CO$_2$ to Selectively Produce Several Important Low-Carbon Fuels**

| Product(s) | Electrode/Electrocatalysts | Potential (V), Reference Electrode | Electrolyte | Temperature/ Pressure, etc. | Current Efficiency | Reference |
|---|---|---|---|---|---|---|
| | Hydrogenated Cu-modified Pd electrode | −1.6 V (SCE) | 0.1 M KHCO$_3$/ CO$_2$-saturated water | | 15% | [148] |
| | Hydrogenated Pd electrode, (0.01 M) pyridine | | 0.5 M NaClO$_4$/ CO$_2$-saturated water | pH 5 | 30% | [149] |
| | Ru | −0.54 V (SCE) | 0.2 M Na$_2$SO$_4$/ CO$_2$-saturated water (pH 3.5–5.5) | 60°C, 0.387 mA | 42% | [150] |
| | Ru | −0.8 V (SCE) | 0.5 M NaHCO$_3$/ CO$_2$-saturated water | 2 h<br>4 h<br>6 h<br>8 h | 30.5%<br>23.6%<br>15.3%<br>17.2% | [151] |
| | Ru/Cu | | | 2 h<br>4 h<br>6 h<br>8 h | 18.2%<br>28.3%<br>32.4%<br>41.3% | |
| | Ru/Cd | | | 2 h<br>4 h<br>6 h<br>8 h | 20.4%<br>21.3%<br>28.4%<br>38.2% | |
| | Hydrochloric acid-pretreated Mo electrode (KOH/HF-pretreated) Mo electrode | −0.8 V (SCE) | 0.2 M Na$_2$SO$_4$/ CO$_2$-saturated water | 22°C/pH 4.2 | 55%<br>84% | [152] |

*(Continued)*

**TABLE 1.5 (*Continued*)**

**A Summary of Selected Electrocatalytic Reductions of $CO_2$ to Selectively Produce Several Important Low-Carbon Fuels**

| Product(s) | Electrode/Electrocatalysts | Potential (V), Reference Electrode | Electrolyte | Temperature/ Pressure, etc. | Current Efficiency | Reference |
|---|---|---|---|---|---|---|
| | $RuO_2/TiO_2$ (i.e., $Ru_3TiO_8$) | −0.035 V (Ag/AgCl/3 M KCl) | 0.2 M $Na_2SO_4$ | 20°C | 30% | [153] |
| | $RuO_2/TiO_2$ nanoparticles composite-modified Pt electrodes | −0.8 V (SCE) | 0.5 M $NaHCO_3$/ $CO_2$-saturated water | | 40.2% | [154] |
| | $RuO_2/TiO_2$ nanotubes composite-modified Pt electrodes | | | | 60.5% | |
| | $K_2Fe^{II}[Fe^{II}(CN)_6]$ film coated on Pt plate electrode | −0.337 to 0.394 V (SCE) | 0.015 M $Na_3[Fe(CN)_5(H_2O)]$/ $CO_2$-saturated water (pH 3.5)/20 mM $CH_3OH$ | | >80% | [155] |
| (5) $H_2C_2O_4/C_2O_4^{2-}$ $2CO_2(g) + 2H^+ + 2e^- =$ $H_2C_2O_4$ (aq.), (−0.500) $2CO_2(g) + 2e^- = C_2O_4^{2-}$ (aq.), (−0.590) | $Ni{-}ET_n(Me/COOET)Et_n$ | | 0.25 M $Bu_4N \cdot ClO_4$/ $CO_2$-saturated $CH_3CN$ | | 98% | [110] |
| | $[(bpy)_2Ru(dmbbbpy)](PF_6)_2$; $[(bpy)_2Ru(dmbbbpy)]{-}$ $Ru(bpy)_2](PF_6)_4$ | −1.65 V, −1.55 V vs. Ag/AgCl | $CO_2$-saturated $CH_3CN$ | | 64% 70% | [144] |
| | Pb electrode | −2.6 V Ag/AgCl | 0.1 M tetraethylammonium (TEAP)/propylene carbonate | 100°C under normal pressure | 73.3% | [67] |

(*Continued*)

**TABLE 1.5 (Continued)**

**A Summary of Selected Electrocatalytic Reductions of $CO_2$ to Selectively Produce Several Important Low-Carbon Fuels**

| Product(s) | Electrode/Electrocatalysts | Potential (V), Reference Electrode | Electrolyte | Temperature/ Pressure, etc. | Current Efficiency | Reference |
|---|---|---|---|---|---|---|
| | $MoO_2$ microparticle | −2.45 V (Fc/Fc+) | 0.1 M $(Bu)_4N \cdot PF_6$/ $CH_3CN$, without $H_2O$ | −20°C | 45% | [72] |
| (6) $CH_4$<br>$CO_2(g) + 8H^+ + 8e^- =$<br>$CH_4(g) + 2H_2O(l)$,<br>(0.169)<br>$CO_2(g) + 6H_2O(l) + 8e^-$<br>$= CH_4(g) + 8OH^-$,<br>(−0.659) | Ppy Re microalloy<br>polypyrrole/<br>Ppy Cu–Re microalloy<br>polypyrrole<br>Au electrodes | −1.35 V | 0.1 M $LiClO_4$/ $CH_3OH$ | 1 atm $CO_2$ | 34%<br>31% | [125] |
| | 99.999% Cu sheet cathode | | 0.5 M $KHCO_3$/ $CO_2$-saturated water | 0°C | 65% | [156] |
| | Polycrystal Cu<br>Cu(100)<br>Cu(110)<br>Cu(111)<br>Single-crystal electrodes | −1.44 V<br>−1.42 V<br>−1.55 V<br>−1.56 V (NHE) | 0.1 M $KClO_4$/water | 18°C/pH 6.8 | 33.3%<br>25.0%<br>49.5%<br>38.9% | [157] |
| (7) $C_2H_4$ | Cu (99.999%) electrode | | 0.5 M $KHCO_3$/ $CO_2$-saturated water | 40°C | 20% | [156] |
| | Cu (99.999%) electrode | −1.40 V (NHE) | 0.1 M $KClO_4$/ $CO_2$-saturated water | 19°C/pH 5.9, 5 mA cm⁻² | 48.1% | [158] |
| | Cu (99.98%) electrode | −3.5 V (Ag/AgCl) | 0.080 M $CsOH/CH_3OH$ | −30°C | 32.3% | [159] |

*(Continued)*

**TABLE 1.5 (Continued)**

**A Summary of Selected Electrocatalytic Reductions of $CO_2$ to Selectively Produce Several Important Low-Carbon Fuels**

| Product(s) | Electrode/Electrocatalysts | Potential (V), Reference Electrode | Electrolyte | Temperature/ Pressure, etc. | Current Efficiency | Reference |
|---|---|---|---|---|---|---|
| | 1. Polycrystal Cu | −1.44 V | 0.1 M $KClO_4$/water | 18°C/pH 6.8 | 25.5% | [157] |
| | 2. Cu(100) | −1.42 V | | | 31.7% | |
| | 3. Cu(110) | −1.55 V | | | 15.1% | |
| | 4. Cu(111) | −1.56 V (NHE) | | | 4.7% | |
| | Single-crystal electrodes | | | | | |
| (8) HCHO | $[Co(4–v–tpy)_2]^{2+}$ | −1.100 V | 0.1 M $NaClO_4$/ $CO_2$-saturated water | | 39% | [68] |
| | $[Fe(4–v–tpy)_2]^{2+}$ | −1.057 V | | | 28% | |
| | $[Cr(4–v–tpy)_2]^{2+}$ electropolymerized on glass carbon electrodes | −1.100 V (Ag/AgCl) | | | 87% | |

an inability to control product distribution, high operation cost, low catalyst activity, insufficient catalyst durability, and a lack of mechanistic understanding. Thus, in the following aspects, we should make more efforts:

- Develop inexpensive and efficient catalysts

  To achieve the desirable product selectivity, more metal complexes with complex structures (of ligands) are being increasingly developed, such as that for CO (Figure 1.3)[162] and $CH_3OH$ (Figure 1.4).[163]

  The drawback is the high cost of catalysts due to the complexity of manufacturing techniques. It has also been considered to reduce the cost of catalysts by using inexpensive metals instead of Pt,[80,164,165] Ru,[144,151,163] Pd,[79,116,166] etc. Besides Fe (complex catalysts),[93,162] Cu,[157–159] Pb,[67,135,140] and Sn[135–137] (single and oxide metal electrodes), Mn is attracting more and more attention as the metal center of complex catalysts.[129,130,167–169] Medina-Ramos et al.[118] developed a bismuth (Bi) – carbon monoxide-evolving catalyst and realized an efficient reduction of $CO_2$ to CO with high current density (Figure 1.5). This novel catalyst was electrodeposited on inexpensive glassy carbon under either aqueous or nonaqueous conditions.

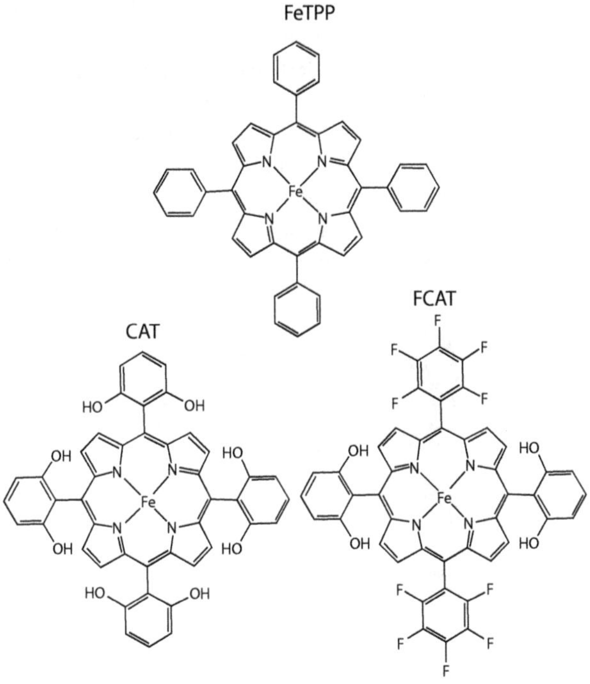

**FIGURE 1.3** Iron-based catalysts for $CO_2$-to-CO reduction. (Reprinted with permission from Costentin, C. et al., Ultraefficient homogeneous catalyst for the $CO_2$-to-CO electrochemical conversion. *Proc Natl Acad Sci USA*, *111*, (42), 14990–14994. Copyright 2014 American Chemical Society.)

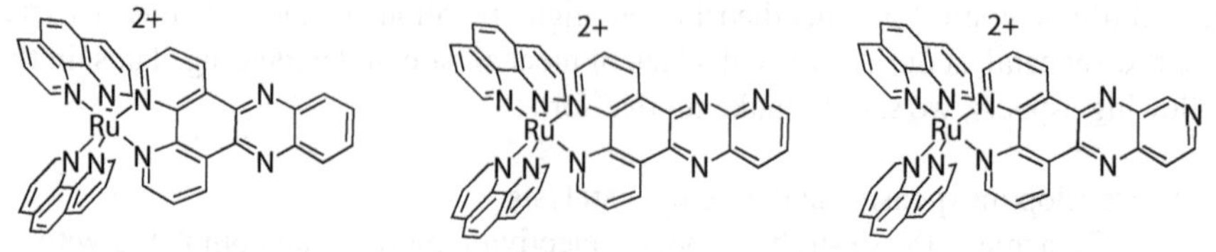

[(phen)$_2$Ru(dppz)]$^{2+}$
**Rudppz**

[(phen)$_2$Ru(ptpbα)]$^{2+}$
**Ruα**

[(phen)$_2$Ru(ptpbα)]$^{2+}$
**Ruβ**

**FIGURE 1.4** Structures of [(phen)$_2$Ru(ptpbα)]$^{2+}$ and [(phen)$_2$Ru(ptpbβ)]$^{2+}$ complexes and [(phen)$_2$Ru(dppz)]$^{2+}$ complex. (Reprinted with permission from Boston, D. J. et al., Electrocatalytic and photocatalytic conversion of CO$_2$ to methanol using ruthenium complexes with internal pyridyl cocatalysts. *Inorg Chem*, 53, (13), 6544–6553. Copyright 2014 American Chemical Society.)

- Nanotechnology applications

  It seems that more attention is being focused on nanotechnology for improved electrochemical performance of catalysts and modified electrodes such as the selectivity toward desired products. Some metal elements (e.g., Ti,[170] Ni,[171] Pt,[172] Cu,[142,173–177] Ag,[178,179] and Au,[176,179] in the state of single metals,[172,173,175,177–179] oxides,[142,170,174] or alloys[176,179]) were directly used as catalysts in the form of nanoparticles, or some others (e.g., Pt,[180] Fe,[180,181] Co,[182] and Ru[154]) were loaded on carbon nanotube,[180,181] nanoporous activated carbon fiber,[182] and TiO$_2$ nanotubes or nanoparticles.[154] Improved performance such as more positive reduction potentials, higher catalytic current density, faradaic efficiency, and catalyst durability have also been observed, all being undoubtedly attributed to the increase in (specific) surface area and active sites. The nanostructure of metal electrode surface can be created by two steps: high-temperature annealing to produce metal

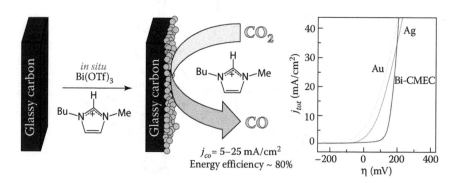

**FIGURE 1.5** Efficient reduction of CO$_2$ to CO with high current density using *in situ* or *ex situ* prepared Bi-based materials. (Reprinted with permission from Medina-Ramos, J.; DiMeglio, J. L.; Rosenthal, J., Efficient reduction of CO$_2$ to CO with high current density using *in situ* or *ex situ* prepared Bi-based materials. *J Am Chem Soc*, 136, (23), 8361–8367. Copyright 2014 American Chemical Society.)

oxides, and reducing electrochemically in aqueous solutions or with $H_2$. It was found that controlled preparation conditions (e.g., temperature) are critical.[173,183]

- Develop novel electrocatalytic technologies

Those so-called novel techniques often seem to be unexpected. For example, Lescot et al.[119] reported an efficient (2 h) fluoride-catalyzed conversion of $CO_2$ to CO at room temperature, in which the oxygen abstraction step was performed with only the presence of catalytic cesium fluoride and a stoichiometric amount of a disilane in dimethyl sulfoxide (DMSO) (Figure 1.6). The nonaqueous solvent was used. Organic molecules such as tetraalkylammonium salts, aromatic nitriles and esters, and especially ionic liquids[118,135,170,184] are popularly used as mediators and catalysts.[185] However, the drawback of using organic molecules is not cheap. The metal-free catalysts such as conducting polymers, pyridinium derivatives, aromatic anion radicals, and heteroatom-doped carbon materials[63,161] are also considered to be the next-generation, renewable materials that promise to be cost-effective, relative to their metal-containing counterparts, particularly relative to noble-metal-based catalysts.[63,185]

- Application of theoretical modeling techniques

Efficient catalysts for $CO_2$ electrochemical reduction should be developed not only by experiments but also by theoretical calculations. The use of first-principles computational techniques,[177,186] such as density functional theory (DFT), to understand electrochemical reactions is witnessing an ever-increasing popularity within the physical electrochemistry community.[187] DFT calculations are thought to be an effective tool,[188–194] which can, for example, (1) determine the potential-dependent reaction free energies and activation barriers for several reaction paths of $CO_2$ electrochemical

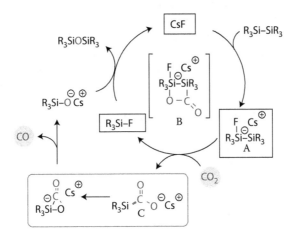

**FIGURE 1.6** Proposed mechanism for $CO_2$ reduction with CsF and disilane. (Reprinted with permission from Lescot, C. et al., Efficient fluoride-catalyzed conversion of $CO_2$ to CO at room temperature. *J Am Chem Soc, 136*, (16), 6142–6147. Copyright 2014 American Chemical Society.)

reduction[193,194]; (2) examine adsorption stabilities of H*, OH*, O*, and H$_2$O* species by calculating the free energy of formation of the adsorbate as a function of electrode potential[191]; and (3) determine new intermediates, searching for new catalysts, and identifying reaction pathways.[188] Challenges and perspectives as to DFT calculations have been outlined in several review papers.[187,188,195] The accuracy of calculation results is to be improved.

## 1.8 APPLICATIONS OF CO$_2$ ELECTROREDUCTION

CO$_2$ electroreduction is applied for the production of low-carbon fuels,[14,63] which may be solar fuel[196] or intermediate materials for the petrochemical industry.[197] However, large-scale processes are quite different with lab-scale ones.

It may be found that most studies toward large-scale CO$_2$ electrochemical reduction have used Sn as catalysts to produce HCOO$^-$.[65,104,141,195,198] Agarwal et al. discussed the engineering and economic feasibility of large-scale electrochemical reduction of CO$_2$ to HCOO$^-$ salts and HCOOH at a gas/solid/ liquid interface, using a flow-through reactor.[65] They concluded that faradaic efficiency, current density, and voltage determine the energy and electrode costs for the large-scale process, and that further work will be needed in the areas of reducing consumable costs, increasing catalyst lifetime, and improved reactor designs to enhance the overall attractiveness of the process.[65]

## 1.9 FURTHER READING AND HOMEWORK

### 1.9.1 MEASUREMENT AND COMPUTATIONAL METHODS

Cyclic voltammetry (CV) is a reversal technique and is the potential-scan equivalent of double potential step chronoamperometry. It has become a very popular technique for initial electrochemical studies of new systems and has proven very useful in obtaining information about fairly complicated electrode reactions.[199] An example is given in Figure 1.7,[131] in which CV spectra of CO$_2$ electrochemical reductions which are catalyzed by Mn(mesbpy)(CO)$_3$Br and Mn(bpy)(CO)$_3$Br indicate that the potential when the electron is transferred varies with ligand. Another is that Sanchez-Sanchez et al.[200] used CV to study the electrocatalytic reduction of CO$_2$ on Pt single-crystal electrodes modified with adsorbed adatoms. To our knowledge, *Cyclic Voltammetry: Simulation and Analysis of Reaction Mechanisms*[201] is one of the early published CV monographs that can still be found, and now the related content is always more or less described in electrochemical textbooks.[202–204] CO$_2$ electroreduction is also sometimes studied at catalysts using linear sweep voltammetry (LSV),[174] rotating ring-disk electrode (RRDE),[179,205,206] and electrochemical impedance spectroscopy (EIS).[179]

Gas chromatograph (GC) is used for quantitative analyses of products such as H$_2$, CO, CO$_2$, hydrocarbons (C$_1$–C$_6$), alcohols and aldehydes (C$_1$–C$_4$), and the common carboxylic acids. Gaseous products[207,208] and reaction intermediates[190] are detected using online mass spectroscopy (MS). Nuclear magnetic resonance (NMR) spectroscopy is sometimes used for detecting trace levels (micromole level) of aqueous

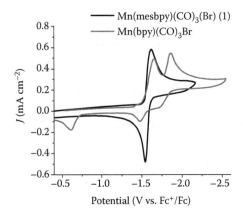

**FIGURE 1.7** Comparison of the cyclic voltammograms of Mn(mesbpy)(CO)$_3$(Br)(1) and Mn(bpy)(CO)$_3$Br under identical conditions (1 mM complex). Each experiment is performed in MeCN with 0.1 M (Bu)$_4$N · PF$_6$ as the supporting electrolyte, under an atmosphere of N$_2$, at a scan rate of 100 mV/s, with a glassy carbon working electrode (3 mm diameter), a platinum wire counter electrode, a Ag/AgCl wire pseudo-reference, and Fc added as an internal reference. (Reprinted with permission from Sampson, M. D. et al., Manganese catalysts with bulky bipyridine ligands for the electrocatalytic reduction of carbon dioxide: Eliminating dimerization and altering catalysis. *J Am Chem Soc*, 136, (14), 5460–5471. Copyright 2014 American Chemical Society.)

electrolysis organic products.[209,210] HCOOH is often measured with NMR[135,211] and ion chromatograph (IC).[90,136] Metal ions in electrolytes are quantified with an inductively coupled plasma–mass spectrometry (ICP–MS). Chemical analysis of electrode surface was performed by x-ray photoelectron spectroscopy (XPS).[136,212] Scanning electron microscopy (SEM),[212] XPS,[136] and x-ray diffraction (XRD)[212] are employed in *ex situ* analyses. More information can be easily found in textbooks.[204]

Theoretical modeling, in particular DFT simulations, provides a powerful and effective tool to discover chemical reaction mechanisms and design new catalysts for the chemical conversion of CO$_2$, overcoming the repetitious and time/labor-consuming trial-and-error experimental processes.[188,213] DFT calculations have been widely used to characterize the reduced species from the reduction of CO$_2$.[194] As described before, many related research has recently been published.[188–194] There have already been a lot of monographs on the theory and application of DFT, such as *Density Functional Theory: A Practical Introduction* by David Sholl,[214] *The Fundamentals of Density Functional Theory*,[215] and *A Chemist's Guide to Density Functional Theory* by Koch et al.,[216] all being suitable for researchers at different levels.

### 1.9.2 PUBLICATIONS

Until now, there is barely a monograph on electrochemical reduction of CO$_2$, although it is already not a new topic. Most of the studies associated with the electrochemical reduction of CO$_2$, especially the early ones, were published in

**TABLE 1.6**
*Magazine Abbreviation/PUBLISHER/Association*

*J Electroanal Chem*/ELSEVIER SCIENCE SA/International Society of Electrochemistry
*Electrochim Acta*//ELSEVIER SCIENCE SA/International Society of Electrochemistry
*Electrochem Commun*//ELSEVIER SCIENCE SA/International Society of Electrochemistry
*J Solid State Electr*/SPRINGER/International Society of Electrochemistry
*J Appl Electrochem*/SPRINGER/International Society of Electrochemistry
*Chem Rev*/AMERICAN CHEMICAL SOCIETY/American Chemical Society
*J Am Chem Soc*/AMERICAN CHEMICAL SOCIETY/American Chemical Society
*J Phys Chem A*/AMERICAN CHEMICAL SOCIETY/American Chemical Society
*J Phys Chem B*/AMERICAN CHEMICAL SOCIETY/American Chemical Society
*J Phys Chem C*/AMERICAN CHEMICAL SOCIETY/American Chemical Society
*Inorg Chem*/AMERICAN CHEMICAL SOCIETY/American Chemical Society
*Organometallics*/AMERICAN CHEMICAL SOCIETY/American Chemical Society
*Proc Natl Acad Sci USA*/NATIONAL ACADEMY OF SCIENCES/National Academy of Sciences of
   United States of America
*Energ Environ Sci*/ROYAL SOCIETY OF CHEMISTRY/The Royal Society of Chemistry
*Chem Commun* and the former journal *J Chem Soc Chem Commun*/ROYAL SOCIETY OF
   CHEMISTRY/The Royal Society of Chemistry
*Phys Chem Chem Phys* and the former journal *J Chem Soc Faraday Trans*/ROYAL SOCIETY OF
   CHEMISTRY/The Royal Society of Chemistry
*Angew Chem Int Ed*/WILEY-VCH VERLAG GMBH/German Chemical Society
*Bull Chem Soc Jpn*/CHEMICAL SOCIETY OF JAPAN/Chemical Society of Japan
*Chem Lett*/CHEMICAL SOCIETY OF JAPAN/Chemical Society of Japan
*Coord Chem Rev*/ELSEVIER SCIENCE SA
*Appl Catal A-Gen*/ELSEVIER SCIENCE SA
*Catal Today*/ELSEVIER SCIENCE SA
*J Power Sources*/ELSEVIER SCIENCE SA

veteran magazines such as *Journal of the American Chemical Society*,[100,217,218] *Inorganic Chemistry*,[219–221] *Journal of the Electrochemical Society*,[146,164] *Journal of Electroanalytical Chemistry*,[222–224] *Electrochimica Acta*,[80,225,226] and *Chemical Letters*.[156,227,228] Some interesting research and high-quality review papers are also published in *Nature*,[229] *Science*,[132] *Chemical Reviews*,[230] *Chemical Society Reviews*,[63,231] *Energy & Environmental Science*,[232] and *Proceedings of the National Academy of Sciences of the United States of America*.[105,162] The typical journals that are noteworthy are recommended in Table 1.6.

Up to now, there still is not a monograph that focuses on electrochemical reduction of $CO_2$. Researchers can only find little incomplete description in several books on $CO_2$ conversion and utilization.[233–235] This book aims to address this issue.

## ACKNOWLEDGMENTS

The authors gratefully acknowledge financial support from Hundred Talent Program of Shanxi province (China) and International Academic Cooperation and Exchange Program of Shanghai Science and Technology Committee (14520721900).

Furthermore, this work was partly conducted by the Division of Multidisciplinary Research on the Circulation of Waste Resources endowed by the Sendai Environmental Development Co., Ltd. Japan. All the financial supports are gratefully acknowledged.

## REFERENCES

1. Meyer, T. J., Chemical approaches to artificial photosynthesis. *Acc Chem Res* **1989**, *22*, (5), 163–170.
2. Leitner, W., The coordination chemistry of carbon dioxide and its relevance for catalysis: A critical survey. *Coord Chem Rev* **1996**, *153*, 257–284.
3. Cheng, M.; Lobkovsky, E. B.; Coates, G. W., Catalytic reactions involving C-1 feedstocks: New high-activity Zn(II)-based catalysts for the alternating copolymerization of carbon dioxide and epoxides. *J Am Chem Soc* **1998**, *120*, (42), 11018–11019.
4. Arakawa, H.; Aresta, M.; Armor, J. N.; Barteau, M. A.; Beckman, E. J.; Bell, A. T.; Bercaw, J. E. et al., Catalysis research of relevance to carbon management: Progress, challenges, and opportunities. *Chem Rev* **2001**, *101*, (4), 953–996.
5. Alstrum-Acevedo, J. H.; Brennaman, M. K.; Meyer, T. J., Chemical approaches to artificial photosynthesis. 2. *Inorg Chem* **2005**, *44*, (20), 6802–6827.
6. Aresta, M.; Dibenedetto, A., Utilisation of CO$_2$ as a chemical feedstock: Opportunities and challenges. *Dalton Trans* **2007**, *36*, (28), 2975–2992.
7. Centi, G.; Perathoner, S., Opportunities and prospects in the chemical recycling of carbon dioxide to fuels. *Catal Today* **2009**, *148*, (3–4), 191–205.
8. Omae, I., Recent developments in carbon dioxide utilization for the production of organic chemicals. *Coord Chem Rev* **2012**, *256*, (13–14), 1384–1405.
9. Concepcion, J. J.; House, R. L.; Papanikolas, J. M.; Meyer, T. J., Chemical approaches to artificial photosynthesis. *Proc Natl Acad Sci USA* **2012**, *109*, (39), 15560–15564.
10. Dagle, R. A.; Hu, J.; Jones, S. B.; Wilcox, W.; Frye, J. G.; White, J. F.; Jiang, J.; Wang, Y., Carbon dioxide conversion to valuable chemical products over composite catalytic systems. *J Energy Chem* **2013**, *22*, (3), 368–374.
11. Sakakura, T.; Choi, J. C.; Yasuda, H., Transformation of carbon dioxide. *Chem Rev* **2007**, *107*, (6), 2365–2387.
12. Oloman, C.; Li, H., Electrochemical processing of carbon dioxide. *Chem Sus Chem* **2008**, *1*, (5), 385–391.
13. Benson, E. E.; Kubiak, C. P.; Sathrum, A. J.; Smieja, J. M., Electrocatalytic and homogeneous approaches to conversion of CO$_2$ to liquid fuels. *Chem Soc Rev* **2009**, *38*, (1), 89–99.
14. Lee, J.; Kwon, Y.; Machunda, R. L.; Lee, H. J., Electrocatalytic recycling of CO$_2$ and small organic molecules. *Chem-Asian J* **2009**, *4*, (10), 1516–1523.
15. Windle, C. D.; Perutz, R. N., Advances in molecular photocatalytic and electrocatalytic CO$_2$ reduction. *Coord Chem Rev* **2012**, *256*, (21–22), 2562–2570.
16. Finn, C.; Schnittger, S.; Yellowlees, L. J.; Love, J. B., Molecular approaches to the electrochemical reduction of carbon dioxide. *Chem Commun* **2012**, *48*, (10), 1392–1399.
17. Inglis, J. L.; MacLean, B. J.; Pryce, M. T.; Vos, J. G., Electrocatalytic pathways towards sustainable fuel production from water and CO$_2$. *Coord Chem Rev* **2012**, *256*, (21–22), 2571–2600.
18. Schneider, J.; Jia, H. F.; Muckerman, J. T.; Fujita, E., Thermodynamics and kinetics of CO$_2$, CO, and H + binding to the metal centre of CO$_2$ reduction catalysts. *Chem Soc Rev* **2012**, *41*, (6), 2036–2051.
19. Mori, K.; Yamashita, H.; Anpo, M., Photocatalytic reduction of CO$_2$ with H$_2$O on various titanium oxide photocatalysts. *RSC Adv* **2012**, *2*, (8), 3165–3172.

20. Raich, J. W.; Potter, C. S., Global patterns of carbon-dioxide emissions from soils. *Global Biogeochem Cycles* **1995**, *9*, (1), 23–36.
21. Raymond, P. A.; Hartmann, J.; Lauerwald, R.; Sobek, S.; McDonald, C.; Hoover, M.; Butman, D. et al., Global carbon dioxide emissions from inland waters. *Nature* **2013**, *503*, (7476), 355–359.
22. Siegenthaler, U.; Sarmiento, J. L., Atmospheric carbon-dioxide and the ocean. *Nature* **1993**, *365*, (6442), 119–125.
23. Hall-Spencer, J. M.; Rodolfo-Metalpa, R.; Martin, S.; Ransome, E.; Fine, M.; Turner, S. M.; Rowley, S. J.; Tedesco, D.; Buia, M. C., Volcanic carbon dioxide vents show ecosystem effects of ocean acidification. *Nature* **2008**, *454*, (7200), 96–99.
24. Worrell, E.; Price, L.; Martin, N.; Hendriks, C.; Meida, L. O., Carbon dioxide emissions from the global cement industry. *Annu Rev Energy Environ* **2001**, *26*, 303–329.
25. Wilson, A., Cement and concrete: Environmental considerations. *Environ. Building News* **1993**, *2*, (2), 1–11.
26. Fuglestvedt, J. S.; Shine, K. P.; Berntsen, T.; Cook, J.; Lee, D. S.; Stenke, A.; Skeie, R. B.; Velders, G. J. M.; Waitz, I. A., Transport impacts on atmosphere and climate: Metrics. *Atmos Environ* **2010**, *44*, (37), 4648–4677.
27. von Scheele, J., Short-term opportunities for decreasing $CO_2$ emissions from the steel industry. *Int J Green Energy* **2006**, *3*, (2), 139–148.
28. Uherek, E.; Halenka, T.; Borken-Kleefeld, J.; Balkanski, Y.; Berntsen, T.; Borrego, C.; Gauss, M. et al., Transport impacts on atmosphere and climate: Land transport. *Atmos Environ* **2010**, *44*, (37), 4772–4816.
29. Lee, D. S.; Pitari, G.; Grewe, V.; Gierens, K.; Penner, J. E.; Petzold, A.; Prather, M. J. et al., Transport impacts on atmosphere and climate: Aviation. *Atmos Environ* **2010**, *44*, (37), 4678–4734.
30. Lee, D. S.; Fahey, D. W.; Forster, P. M.; Newton, P. J.; Wit, R. C. N.; Lim, L. L.; Owen, B.; Sausen, R., Aviation and global climate change in the twenty-first century. *Atmos Environ* **2009**, *43*, (22–23), 3520–3537.
31. Howitt, O. J. A.; Carruthers, M. A.; Smith, I. J.; Rodger, C. J., Carbon dioxide emissions from international air freight. *Atmos Environ* **2011**, *45*, (39), 7036–7045.
32. Masiol, M.; Harrison, R. M., Aircraft engine exhaust emissions and other airport-related contributions to ambient air pollution: A review. *Atmos Environ* **2014**, *95*, 409–455.
33. Eyring, V.; Isaksen, I. S. A.; Berntsen, T.; Collins, W. J.; Corbett, J. J.; Endresen, O.; Grainger, R. G.; Moldanova, J.; Schlager, H.; Stevenson, D. S., Transport impacts on atmosphere and climate: Shipping. *Atmos Environ* **2010**, *44*, (37), 4735–4771.
34. Beck-Friis, B.; Smars, S.; Jonsson, H.; Kirchmann, H., Gaseous emissions of carbon dioxide, ammonia and nitrous oxide from organic household waste in a compost reactor under different temperature regimes. *J Agr Eng Res* **2001**, *78*, (4), 423–430.
35. Schmalensee, R.; Stoker, T. M.; Judson, R. A., World carbon dioxide emissions: 1950–2050. *Rev Econ Stat* **1998**, *80*, (1), 15–27.
36. Keeling, C. D.; Whorf, T. P.; Wahlen, M.; Vanderplicht, J., Interannual extremes in the rate of rise of atmospheric carbon-dioxide since 1980. *Nature* **1995**, *375*, (6533), 666–670.
37. Petit, J. R.; Jouzel, J.; Raynaud, D.; Barkov, N. I.; Barnola, J. M.; Basile, I.; Bender, M. et al., Climate and atmospheric history of the past 420,000 years from the Vostok ice core, Antarctica. *Nature* **1999**, *399*, (6735), 429–436.
38. Shakun, J. D.; Clark, P. U.; He, F.; Marcott, S. A.; Mix, A. C.; Liu, Z. Y.; Otto-Bliesner, B.; Schmittner, A.; Bard, E., Global warming preceded by increasing carbon dioxide concentrations during the last deglaciation. *Nature* **2012**, *484*, (7392), 49–54.
39. Ballantyne, A. P.; Alden, C. B.; Miller, J. B.; Tans, P. P.; White, J. W. C., Increase in observed net carbon dioxide uptake by land and oceans during the past 50 years. *Nature* **2012**, *488*, (7409), 70–72.

40. Serreze, M. C.; Walsh, J. E.; Chapin, F. S.; Osterkamp, T.; Dyurgerov, M.; Romanovsky, V.; Oechel, W. C.; Morison, J.; Zhang, T.; Barry, R. G., Observational evidence of recent change in the northern high-latitude environment. *Clim Change* **2000**, *46*, (1–2), 159–207.

41. Solomon, S.; Plattner, G. K.; Knutti, R.; Friedlingstein, P., Irreversible climate change due to carbon dioxide emissions. *Proc Natl Acad Sci USA* **2009**, *106*, (6), 1704–1709.

42. Monastersky, R., Global carbon dioxide levels near worrisome milestone. *Nature* **2013**, *497*, (7447), 13–14.

43. Wang, M.; Lawal, A.; Stephenson, P.; Sidders, J.; Ramshaw, C., Post-combustion $CO_2$ capture with chemical absorption: A state-of-the-art review. *Chem Eng Res Des* **2011**, *89*, (9), 1609–1624.

44. Olivier, J. G.; Janssens-Maenhout, G.; Peters, J. A., *Trends in Global $CO_2$ Emissions: 2012 Report*. Hague, the Netherlands: PBL Netherlands Environmental Assessment Agency, **2012**.

45. Soytas, U.; Sari, R.; Ewing, B. T., Energy consumption, income, and carbon emissions in the United States. *Ecol Econ* **2007**, *62*, (3–4), 482–489.

46. Gurney, K. R.; Mendoza, D. L.; Zhou, Y. Y.; Fischer, M. L.; Miller, C. C.; Geethakumar, S.; Du Can, S. D., High resolution fossil fuel combustion $CO_2$ emission fluxes for the United States. *Environ Sci Technol* **2009**, *43*, (14), 5535–5541.

47. Zhang, X. P.; Cheng, X. M., Energy consumption, carbon emissions, and economic growth in China. *Ecol Econ* **2009**, *68*, (10), 2706–2712.

48. Renforth, P.; Manning, D. A. C.; Lopez-Capel, E., Carbonate precipitation in artificial soils as a sink for atmospheric carbon dioxide. *Appl Geochem* **2009**, *24*, (9), 1757–1764.

49. Olah, G. A.; Prakash, G. K. S.; Goeppert, A., Anthropogenic chemical carbon cycle for a sustainable future. *J Am Chem Soc* **2011**, *133*, (33), 12881–12898.

50. Le Quere, C.; Raupach, M. R.; Canadell, J. G.; Marland, G.; Bopp, L.; Ciais, P.; Conway, T. J. et al., Trends in the sources and sinks of carbon dioxide. *Nat Geosci* **2009**, *2*, (12), 831–836.

51. Kumar, M. D.; Naqvi, S. W. A.; George, M. D.; Jayakumar, D. A., A sink for atmospheric carbon dioxide in the northeast Indian Ocean. *J Geophys Res Oceans* **1996**, *101*, (C8), 18121–18125.

52. Song, C. S., Global challenges and strategies for control, conversion and utilization of $CO_2$ for sustainable development involving energy, catalysis, adsorption and chemical processing. *Catal Today* **2006**, *115*, (1–4), 2–32.

53. Spinner, N. S.; Vega, J. A.; Mustain, W. E., Recent progress in the electrochemical conversion and utilization of $CO_2$. *Catal Sci Technol* **2012**, *2*, (1), 19–28.

54. Kanniche, M.; Gros-Bonnivard, R.; Jaud, P.; Valle-Marcos, J.; Amann, J. M.; Bouallou, C., Pre-combustion, post-combustion and oxy-combustion in thermal power plant for $CO_2$ capture. *Appl Therm Eng* **2010**, *30*, (1), 53–62.

55. MacDowell, N.; Florin, N.; Buchard, A.; Hallett, J.; Galindo, A.; Jackson, G.; Adjiman, C. S.; Williams, C. K.; Shah, N.; Fennell, P., An overview of $CO_2$ capture technologies. *Energ Environ Sci* **2010**, *3*, (11), 1645–1669.

56. Rubin, E. S.; Mantripragada, H.; Marks, A.; Versteeg, P.; Kitchin, J., The outlook for improved carbon capture technology. *Prog Energy Combust Sci* **2012**, *38*, (5), 630–671.

57. Markewitz, P.; Kuckshinrichs, W.; Leitner, W.; Linssen, J.; Zapp, P.; Bongartz, R.; Schreiber, A.; Muller, T. E., Worldwide innovations in the development of carbon capture technologies and the utilization of $CO_2$. *Energ Environ Sci* **2012**, *5*, (6), 7281–7305.

58. Li, B. Y.; Duan, Y. H.; Luebke, D.; Morreale, B., Advances in $CO_2$ capture technology: A patent review. *Appl Energy* **2013**, *102*, 1439–1447.

59. Kenarsari, S. D.; Yang, D. L.; Jiang, G. D.; Zhang, S. J.; Wang, J. J.; Russell, A. G.; Wei, Q.; Fan, M. H., Review of recent advances in carbon dioxide separation and capture. *RSC Adv* **2013**, *3*, (45), 22739–22773.

60. Strohle, J.; Junk, M.; Kremer, J.; Galloy, A.; Epple, B., Carbonate looping experiments in a 1 MWth pilot plant and model validation. *Fuel* **2014,** *127,* 13–22.
61. Wall, T.; Liu, Y. H.; Spero, C.; Elliott, L.; Khare, S.; Rathnam, R.; Zeenathal, F. et al., An overview on oxyfuel coal combustion—State of the art research and technology development. *Chem Eng Res Des* **2009,** *87,* (8A), 1003–1016.
62. Lindsey, A. S.; Jeskey, H., The Kolbe-Schmitt reaction. *Chem Rev* **1957,** *57,* (4), 583–620.
63. Qiao, J.; Liu, Y.; Hong, F.; Zhang, J., A review of catalysts for the electroreduction of carbon dioxide to produce low-carbon fuels. *Chem Soc Rev* **2014,** *43,* (2), 631–675.
64. Bard, A. J.; Parsons, R.; Jordan, J., *Standard Potentials in Aqueous Solutions.* New York: Marcel Dekker, Inc., **1985;** Vol. 6.
65. Agarwal, A. S.; Zhai, Y. M.; Hill, D.; Sridhar, N., The electrochemical reduction of carbon dioxide to formate/formic acid: Engineering and economic feasibility. *ChemSusChem* **2011,** *4,* (9), 1301–1310.
66. Noda, H.; Ikeda, S.; Oda, Y.; Imai, K.; Maeda, M.; Ito, K., Electrochemical reduction of carbon-dioxide at various metal-electrodes in aqueous potassium hydrogen carbonate solution. *Bull Chem Soc Jpn* **1990,** *63,* (9), 2459–2462.
67. Ikeda, S.; Takagi, T.; Ito, K., Selective formation of formic-acid, oxalic-acid, and carbon-monoxide by electrochemical reduction of carbon-dioxide. *Bull Chem Soc Jpn* **1987,** *60,* (7), 2517–2522.
68. Sende, J. A. R.; Arana, C. R.; Hernandez, L.; Potts, K. T.; Keshevarzk, M.; Abruna, H. D., Electrocatalysis of $CO_2$ reduction in aqueous-media at electrodes modified with electropolymerized films of vinylterpyridine complexes of transition-metals. *Inorg Chem* **1995,** *34,* (12), 3339–3348.
69. Wu, J. J.; Sharma, P. P.; Harris, B. H.; Zhou, X.-D., Electrochemical reduction of carbon dioxide: IV Dependence of the faradaic efficiency and current density on the microstructure and thickness of tin electrode. *J Power Sources* **2014,** *258,* 189–194.
70. Wu, J. J.; Risalvato, F. G.; Ma, S. G.; Zhou, X. D., Electrochemical reduction of carbon dioxide III. The role of oxide layer thickness on the performance of Sn electrode in a full electrochemical cell. *J Mater Chem A* **2014,** *2,* (6), 1647–1651.
71. Ougitani, Y.; Aizawa, T.; Sonoyama, N.; Sakata, T., Temperature dependence of the probability of chain growth for hydrocarbon formation by electrochemical reduction of $CO_2$. *Bull Chem Soc Jpn* **2001,** *74,* (11), 2119–2122.
72. Oh, Y.; Vrubel, H.; Guidoux, S.; Hu, X. L., Electrochemical reduction of $CO_2$ in organic solvents catalyzed by $MoO_2$. *Chem Commun* **2014,** *50,* (29), 3878–3881.
73. Lv, W. X.; Zhang, R.; Gao, P. R.; Lei, L. X., Studies on the faradaic efficiency for electrochemical reduction of carbon dioxide to formate on tin electrode. *J Power Sources* **2014,** *253,* 276–281.
74. Chaplin, R. P. S.; Wragg, A. A., Effects of process conditions and electrode material on reaction pathways for carbon dioxide electroreduction with particular reference to formate formation. *J Appl Electrochem* **2003,** *33,* (12), 1107–1123.
75. Saeki, T.; Hashimoto, K.; Kimura, N.; Omata, K.; Fujishima, A., Electrochemical reduction of $CO_2$ with high-current density in a $CO_2$ plus methanol medium. 2. Co formation promoted by tetrabutylammonium cation. *J Electroanal Chem* **1995,** *390,* (1–2), 77–82.
76. Li, J. W.; Prentice, G., Electrochemical synthesis of methanol from $CO_2$ in high-pressure electrolyte. *J Electrochem Soc* **1997,** *144,* (12), 4284–4288.
77. Magdesieva, T. V.; Zhukov, I. V.; Kravchuk, D. N.; Semenikhin, O. A.; Tomilova, L. G.; Butin, K. P., Electrocatalytic $CO_2$ reduction in methanol catalyzed by mono-, di-, and electropolymerized phthalocyanine complexes. *Russ Chem B+* **2002,** *51,* (5), 805–812.
78. Hori, Y.; Kikuchi, K.; Suzuki, S., Production of Co and $CH_4$ in electrochemical reduction of $CO_2$ at metal-electrodes in aqueous hydrogencarbonate solution. *Chem Lett* **1985,** *14,* (11), 1695–1698.

79. Azuma, M.; Hashimoto, K.; Hiramoto, M.; Watanabe, M.; Sakata, T., Electrochemical reduction of carbon-dioxide on various metal-electrodes in low-temperature aqueous $KHCO_3$ media. *J Electrochem Soc* **1990**, *137*, (6), 1772–1778.

80. Hori, Y.; Wakebe, H.; Tsukamoto, T.; Koga, O., Electrocatalytic process of Co selectivity in electrochemical reduction of $CO_2$ at metal-electrodes in aqueous-media. *Electrochim Acta* **1994**, *39*, (11–12), 1833–1839.

81. Yoshida, T.; Kamato, K.; Tsukamoto, M.; Iida, T.; Schlettwein, D.; Wohrle, D.; Kaneko, M., Selective electrocatalysis for $CO_2$ reduction in the aqueous-phase using cobalt phthalocyanine/poly-4-vinylpyridine modified electrodes. *J Electroanal Chem* **1995**, *385*, (2), 209–225.

82. Isaacs, M.; Armijo, F.; Ramirez, G.; Trollund, E.; Biaggio, S. R.; Costamagna, J.; Aguirre, M. J., Electrochemical reduction of $CO_2$ mediated by poly-M-aminophthalocyanines (M = Co, Ni, Fe): Poly-Co-tetraaminophthalocyanine, a selective catalyst. *J Mol Catal A—Chem* **2005**, *229*, (1–2), 249–257.

83. Furuya, N.; Matsui, K., Electroreduction of carbon-dioxide on gas-diffusion electrodes modified by metal phthalocyanines. *J Electroanal Chem* **1989**, *271*, (1–2), 181–191.

84. Shibata, M.; Furuya, N., Simultaneous reduction of carbon dioxide and nitrate ions at gas-diffusion electrodes with various metallophthalocyanine catalysts. *Electrochim Acta* **2003**, *48*, (25–26), 3953–3958.

85. Costamagna, J.; Canales, J.; Vargas, J.; Ferraudi, G., Electrochemical reduction of carbon-dioxide by Hexa-Azamacrocyclic complexes. *Pure Appl Chem* **1995**, *67*, (7), 1045–1052.

86. Costamagna, J.; Ferraudi, G.; Canales, J.; Vargas, J., Carbon dioxide activation by aza-macrocyclic complexes. *Coord Chem Rev* **1996**, *148*, 221–248.

87. Bhugun, I.; Lexa, D.; Saveant, J. M., Catalysis of the electrochemical reduction of carbon dioxide by iron(0) porphyrins. Synergistic effect of Lewis acid cations. *J Phys Chem* **1996**, *100*, (51), 19981–19985.

88. Thorson, M. R.; Siil, K. I.; Kenis, P. J. A., Effect of cations on the electrochemical conversion of $CO_2$ to CO. *J Electrochem Soc* **2013**, *160*, (1), F69–F74.

89. Ogura, K.; Ferrell, J. R.; Cugini, A. V.; Smotkin, E. S.; Salazar-Villalpando, M. D., $CO_2$ attraction by specifically adsorbed anions and subsequent accelerated electrochemical reduction. *Electrochim Acta* **2010**, *56*, (1), 381–386.

90. Schizodimou, A.; Kyriacou, G., Acceleration of the reduction of carbon dioxide in the presence of multivalent cations. *Electrochim Acta* **2012**, *78*, 171–176.

91. Eneau-Innocent, B.; Pasquier, D.; Ropital, F.; Leger, J. M.; Kokoh, K. B., Electroreduction of carbon dioxide at a lead electrode in propylene carbonate: A spectroscopic study. *Appl Catal B-Environ* **2010**, *98*, (1–2), 65–71.

92. Carroll, J. J.; Slupsky, J. D.; Mather, A. E., The solubility of carbon-dioxide in water at low-pressure. *J Phys Chem Ref Data* **1991**, *20*, (6), 1201–1209.

93. Hara, K.; Kudo, A.; Sakata, T., Electrochemical reduction of high-pressure carbon-dioxide on Fe electrodes at large current-density. *J Electroanal Chem* **1995**, *386*, (1–2), 257–260.

94. Hara, K.; Kudo, A.; Sakata, T., Electrochemical $CO_2$ reduction on a glassy carbon electrode under high pressure. *J Electroanal Chem* **1997**, *421*, (1–2), 1–4.

95. Hansen, H. A.; Varley, J. B.; Peterson, A. A.; Norskov, J. K., Understanding trends in the electrocatalytic activity of metals and enzymes for $CO_2$ reduction to CO. *J Phys Chem Lett* **2013**, *4*, (3), 388–392.

96. Amatore, C.; Saveant, J. M., Mechanism and kinetic characteristics of the electrochemical reduction of carbon-dioxide in media of low proton availability. *J Am Chem Soc* **1981**, *103*, (17), 5021–5023.

97. Beley, M.; Collin, J. P.; Ruppert, R.; Sauvage, J. P., Nickel(Ii) cyclam—An extremely selective electrocatalyst for reduction of $CO_2$ in water. *J Chem Soc Chem Commun* **1984**, *13*, (19), 1315–1316.

98. Schneider, J.; Jia, H. F.; Kobiro, K.; Cabelli, D. E.; Muckerman, J. T.; Fujita, E., Nickel(II) macrocycles: Highly efficient electrocatalysts for the selective reduction of CO₂ to CO. *Energ Environ Sci* **2012**, *5*, (11), 9502–9510.

99. Dubois, D. L.; Miedaner, A.; Haltiwanger, R. C., Electrochemical reduction of CO₂ catalyzed by [Pd(Triphosphine)(Solvent)](Bf4)2 complexes—Synthetic and mechanistic studies. *J Am Chem Soc* **1991**, *113*, (23), 8753–8764.

100. Shin, W.; Lee, S.; Shin, J.; Lee, S.; Kim, Y., Highly selective electrocatalytic conversion of CO₂ to CO at −0.57 V (NHE) by carbon monoxide dehydrogenase from *Moorella thermoacetica*. *J Am Chem Soc* **2003**, *125*, (48), 14688–14689.

101. Royer, M., Reduction de l'acide carbonique en acid formique. *Compt Rend* **1870**, *70*, 731–732.

102. Cheng, S. C.; Blaine, C. A.; Hill, M. G.; Mann, K. R., Electrochemical and IR spectroelectrochemical studies of the electrocatalytic reduction of carbon dioxide by [Ir-2(dimen)(4)](2+) (dimen equals 1,8-diisocyanomenthane). *Inorg Chem* **1996**, *35*, (26), 7704–7708.

103. Kang, P.; Meyer, T. J.; Brookhart, M., Selective electrocatalytic reduction of carbon dioxide to formate by a water-soluble iridium pincer catalyst. *Chem Sci* **2013**, *4*, (9), 3497–3502.

104. Li, H.; Oloman, C., Development of a continuous reactor for the electro-reduction of carbon dioxide to formate—Part 2: Scale-up. *J Appl Electrochem* **2007**, *37*, (10), 1107–1117.

105. Reda, T.; Plugge, C. M.; Abram, N. J.; Hirst, J., Reversible interconversion of carbon dioxide and formate by an electroactive enzyme. *Proc Natl Acad Sci USA* **2008**, *105*, (31), 10654–10658.

106. Song, J.; Kim, Y.; Lim, M.; Lee, H.; Lee, J. I.; Shin, W., Microbes as electrochemical CO₂ conversion catalysts. *Chemsuschem* **2011**, *4*, (5), 587–590.

107. Nakata, K.; Ozaki, T.; Terashima, C.; Fujishima, A.; Einaga, Y., High-yield electrochemical production of formaldehyde from CO₂ and seawater. *Angew Chem Int Ed* **2014**, *53*, (3), 871–874.

108. Ganesh, I., Conversion of carbon dioxide into methanol—a potential liquid fuel: Fundamental challenges and opportunities (a review). *Renew Sust Energ Rev* **2014**, *31*, 221–257.

109. Tezuka, M.; Yajima, T.; Tsuchiya, A.; Matsumoto, Y.; Uchida, Y.; Hidai, M., Electroreduction of carbon-dioxide catalyzed by iron sulfur clusters [Fe4s4(Sr)4]2-. *J Am Chem Soc* **1982**, *104*, (24), 6834–6836.

110. Rudolph, M.; Dautz, S.; Jager, E. G., Macrocyclic [N-4(2-)] coordinated nickel complexes as catalysts for the formation of oxalate by electrochemical reduction of carbon dioxide. *J Am Chem Soc* **2000**, *122*, (44), 10821–10830.

111. Angamuthu, R.; Byers, P.; Lutz, M.; Spek, A. L.; Bouwman, E., Electrocatalytic CO₂ conversion to oxalate by a copper complex. *Science* **2010**, *327*, (5963), 313–315.

112. Gennaro, A.; Isse, A. A.; Severin, M. G.; Vianello, E.; Bhugun, I.; Saveant, J. M., Mechanism of the electrochemical reduction of carbon dioxide at inert electrodes in media of low proton availability. *J Chem Soc Faraday Trans* **1996**, *92*, (20), 3963–3968.

113. Gennaro, A.; Isse, A. A.; Saveant, J. M.; Severin, M. G.; Vianello, E., Homogeneous electron transfer catalysis of the electrochemical reduction of carbon dioxide. Do aromatic anion radicals react in an outer-sphere manner? *J Am Chem Soc* **1996**, *118*, (30), 7190–7196.

114. Ogura, K.; Sugihara, H.; Yano, J.; Higasa, M., Electrochemical reduction of carbon-dioxide on dual-film electrodes modified with and without cobalt(II) and iron(II) complexes. *J Electrochem Soc* **1994**, *141*, (2), 419–424.

115. Kaneco, S.; Iiba, K.; Katsumata, H.; Suzuki, T.; Ohta, K., Electrochemical reduction of high pressure carbon dioxide at a Cu electrode in cold methanol with CsOH supporting salt. *Chem Eng J* **2007**, *128*, (1), 47–50.

116. Nakagawa, S.; Kudo, A.; Azuma, M.; Sakata, T., Effect of pressure on the electro-chemical reduction of $CO_2$ on group-Viii metal-electrodes. *J Electroanal Chem* **1991**, *308*, (1–2), 339–343.

117. Lu, Q.; Rosen, J.; Zhou, Y.; Hutchings, G. S.; Kimmel, Y. C.; Chen, J. G. G.; Jiao, F., A selective and efficient electrocatalyst for carbon dioxide reduction. *Nat Commun* **2014**, *5*, 6.

118. Medina-Ramos, J.; DiMeglio, J. L.; Rosenthal, J., Efficient reduction of $CO_2$ to CO with high current density using *in situ* or *ex situ* prepared Bi-based materials. *J Am Chem Soc* **2014**, *136*, (23), 8361–8367.

119. Lescot, C.; Nielsen, D. U.; Makarov, I. S.; Lindhardt, A. T.; Daasbjerg, K.; Skrydstrup, T., Efficient fluoride-catalyzed conversion of $CO_2$ to CO at room temperature. *J Am Chem Soc* **2014**, *136*, (16), 6142–6147.

120. Asadi, M.; Kumar, B.; Behranginia, A.; Rosen, B. A.; Baskin, A.; Repnin, N.; Pisasale, D. et al., Robust carbon dioxide reduction on molybdenum disulphide edges. *Nat Commun* **2014**, *5*, 8.

121. Hawecker, J.; Lehn, J. M.; Ziessel, R., Electrocatalytic reduction of carbon-dioxide mediated by Re(Bipy)(Co)3cl(Bipy=2,2′-Bipyridine). *J Chem Soc Chem Commun* **1984**, *13*, (6), 328–330.

122. Portenkirchner, E.; Oppelt, K.; Ulbricht, C.; Egbe, D. A. M.; Neugebauer, H.; Knor, G.; Sariciftci, N. S., Electrocatalytic and photocatalytic reduction of carbon dioxide to car-bon monoxide using the alkynyl-substituted rhenium(I) complex (5,5′-bisphenylethynyl-2,2′-bipyridyl)Re(CO)(3)Cl. *J Organomet Chem* **2012**, *716*, 19–25.

123. Cosnier, S.; Deronzier, A.; Moutet, J. C., Electrochemical coating of a platinum-electrode by a Poly(Pyrrole) film containing the Fac-Re(2,2′-Bipyridine)(Co)3cl sys-tem—Application to electrocatalytic reduction of $CO_2$. *J Electroanal Chem* **1986**, *207*, (1–2), 315–321.

124. Cosnier, S.; Deronzier, A.; Moutet, J. C., Electrocatalytic reduction of $CO_2$ on elec-trodes modified by Fac-Re(2,2′-Bipyridine)(Co)3cl complexes bonded to polypyrrole films. *J Mol Catal* **1988**, *45*, (3), 381–391.

125. Schrebler, R.; Cury, P.; Herrera, F.; Gomez, H.; Cordova, R., Study of the electro-chemical reduction of $CO_2$ on electrodeposited rhenium electrodes in methanol media. *J Electroanal Chem* **2001**, *516*, (1–2), 23–30.

126. Otoole, T. R.; Margerum, L. D.; Westmoreland, T. D.; Vining, W. J.; Murray, R. W.; Meyer, T. J., Electrocatalytic reduction of $CO_2$ at a chemically modified electrode. *J Chem Soc Chem Commun* **1985**, *14*, (20), 1416–1417.

127. Otoole, T. R.; Sullivan, B. P.; Bruce, M. R. M.; Margerum, L. D.; Murray, R. W.; Meyer, T. J., Electrocatalytic reduction of $CO_2$ by a complex of rhenium in thin polymeric films. *J Electroanal Chem* **1989**, *259*, (1–2), 217–239.

128. Grills, D. C.; Matsubara, Y.; Kuwahara, Y.; Golisz, S. R.; Kurtz, D. A.; Mello, B. A., Electrocatalytic $CO_2$ reduction with a homogeneous catalyst in ionic liquid: High cata-lytic activity at low overpotential. *J Phys Chem Lett* **2014**, *5*, (11), 2033–2038.

129. Bourrez, M.; Molton, F.; Chardon-Noblat, S.; Deronzier, A., [Mn(bipyridyl)(CO)(3)Br]: An abundant metal carbonyl complex as efficient electrocatalyst for $CO_2$ reduction. *Angew Chem Int Ed* **2011**, *50*, (42), 9903–9906.

130. Smieja, J. M.; Sampson, M. D.; Grice, K. A.; Benson, E. E.; Froehlich, J. D.; Kubiak, C. P., Manganese as a substitute for rhenium in $CO_2$ reduction catalysts: The impor-tance of acids. *Inorg Chem* **2013**, *52*, (5), 2484–2491.

131. Sampson, M. D.; Nguyen, A. D.; Grice, K. A.; Moore, C. E.; Rheingold, A. L.; Kubiak, C. P., Manganese catalysts with bulky bipyridine ligands for the electrocatalytic reduc-tion of carbon dioxide: Eliminating dimerization and altering catalysis. *J Am Chem Soc* **2014**, *136*, (14), 5460–5471.

132. Costentin, C.; Drouet, S.; Robert, M.; Saveant, J. M., A local proton source enhances $CO_2$ electroreduction to CO by a molecular Fe catalyst. *Science* **2012**, *338*, (6103), 90–94.

133. Collin, J. P.; Jouaiti, A.; Sauvage, J. P., Electrocatalytic properties of Ni(Cyclam)2+ and Ni2(Biscyclam)4+ with respect to $CO_2$ and $H_2O$ reduction. *Inorg Chem* **1988**, *27*, (11), 1986–1990.

134. Chardon-Noblat, S.; Deronzier, A.; Ziessel, R.; Zsoldos, D., Selective synthesis and electrochemical behavior of trans(Cl)- and cis(Cl)-[Ru(bpy)(CO)(2)Cl-2] complexes (bpy = 2,2′-bipyridine). Comparative studies of their electrocatalytic activity toward the reduction of carbon dioxide. *Inorg Chem* **1997**, *36*, (23), 5384–5389.

135. Watkins, J. D.; Bocarsly, A. B., Direct reduction of carbon dioxide to formate in high-gas-capacity ionic liquids at post-transition-metal electrodes. *Chemsuschem* **2014**, *7*, (1), 284–290.

136. Anawati; Frankel, G. S.; Agarwal, A.; Sridhar, N., Degradation and deactivation of Sn catalyst used for $CO_2$ reduction as function of overpotential. *Electrochim Acta* **2014**, *133*, 188–196.

137. Wu, J. J.; Risalvato, F. G.; Ke, F. S.; Pellechia, P. J.; Zhou, X. D., Electrochemical reduction of carbon dioxide I. Effects of the electrolyte on the selectivity and activity with Sn electrode. *J Electrochem Soc* **2012**, *159*, (7), F353–F359.

138. Prakash, G. K. S.; Viva, F. A.; Olah, G. A., Electrochemical reduction of $CO_2$ over Sn-Nafion (R) coated electrode for a fuel-cell-like device. *J Power Sources* **2013**, *223*, 68–73.

139. Zhang, S.; Kang, P.; Meyer, T. J., Nanostructured tin catalysts for selective electrochemical reduction of carbon dioxide to formate. *J Am Chem Soc* **2014**, *136*, (5), 1734–1737.

140. Koleli, F.; Balun, D., Reduction of $CO_2$ under high pressure and high temperature on Pb-granule electrodes in a fixed-bed reactor in aqueous medium. *Appl Catal A-Gen* **2004**, *274*, (1–2), 237–242.

141. Li, H.; Oloman, C., The electro-reduction of carbon dioxide in a continuous reactor. *J Appl Electrochem* **2005**, *35*, (10), 955–965.

142. Grace, A. N.; Choi, S. Y.; Vinoba, M.; Bhagiyalakshmi, M.; Chu, D. H.; Yoon, Y.; Nam, S. C.; Jeong, S. K., Electrochemical reduction of carbon dioxide at low overpotential on a polyaniline/$Cu_2O$ nanocomposite based electrode. *Appl Energy* **2014**, *120*, 85–94.

143. Ishida, H.; Tanaka, H.; Tanaka, K.; Tanaka, T., Selective formation of $HCOO^-$ in the electrochemical $CO_2$ reduction catalyzed by [Ru(Bpy)2(Co)2]2+ (Bpy = 2,2′-Bipyridine). *J Chem Soc Chem Commun* **1987**, *16*, (2), 131–132.

144. Meser Ali, M.; Sato, H.; Mizukawa, T.; Tsuge, K.; Haga, M.-A.; Tanaka, K., Selective formation of $HCO_2^-$ and $C2O42-$ in electrochemical reduction of $CO_2$ catalyzed by mono- and di-nuclear ruthenium complexes. *Chem Commun* **1998**, *0*, (2), 249–250.

145. Zhang, S.; Kang, P.; Ubnoske, S.; Brennaman, M. K.; Song, N.; House, R. L.; Glass, J. T.; Meyer, T. J., Polyethylenimine-enhanced electrocatalytic reduction of $CO_2$ to formate at nitrogen-doped carbon nanomaterials. *J Am Chem Soc* **2014**, *136*, (22), 7845–7848.

146. Frese, K. W., Electrochemical reduction of $CO_2$ at intentionally oxidized copper electrodes. *J Electrochem Soc* **1991**, *138*, (11), 3338–3344.

147. Le, M.; Ren, M.; Zhang, Z.; Sprunger, P. T.; Kurtz, R. L.; Flake, J. C., Electrochemical reduction of $CO_2$ to $CH_3OH$ at copper oxide surfaces. *J Electrochem Soc* **2011**, *158*, (5), E45–E49.

148. Ohkawa, K.; Noguchi, Y.; Nakayama, S.; Hashimoto, K.; Fujishima, A., Electrochemical reduction of carbon-dioxide on hydrogen-storing materials. 3. The effect of the absorption of hydrogen on the palladium electrodes modified with copper. *J Electroanal Chem* **1994**, *367*, (1–2), 165–173.

149. Seshadri, G.; Lin, C.; Bocarsly, A. B., A new homogeneous electrocatalyst for the reduction of carbon-dioxide to methanol at low overpotential. *J Electroanal Chem* **1994**, *372*, (1–2), 145–150.

150. Frese, K. W.; Leach, S., Electrochemical reduction of carbon-dioxide to methane, methanol, and Co on Ru electrodes. *J Electrochem Soc* **1985**, *132*, (1), 259–260.

151. Popic, J. P.; Avramovlvic, M. L.; Vukovic, N. B., Reduction of carbon dioxide on ruthenium oxide and modified ruthenium oxide electrodes in 0.5 M NaHCO$_3$. *J Electroanal Chem* **1997**, *421*, (1–2), 105–110.

152. Summers, D. P.; Leach, S.; Frese, K. W., The electrochemical reduction of aqueous carbon-dioxide to methanol at molybdenum electrodes with low overpotentials. *J Electroanal Chem* **1986**, *205*, (1–2), 219–232.

153. Bandi, A.; Kuhne, H. M., Electrochemical reduction of carbon-dioxide in water—Analysis of reaction-mechanism on ruthenium-titanium-oxide. *J Electrochem Soc* **1992**, *139*, (6), 1605–1610.

154. Qu, J. P.; Zhang, X. G.; Wang, Y. G.; Xie, C. X., Electrochemical reduction of CO$_2$ on RuO$_2$/TiO$_2$ nanotubes composite modified Pt electrode. *Electrochim Acta* **2005**, *50*, (16–17), 3576–3580.

155. Ogura, K.; Uchida, H., Electrocatalytic reduction of CO$_2$ to methanol. 8. Photoassisted electrolysis and electrochemical photocell with N-TiO$_2$ anode. *J Electroanal Chem* **1987**, *220*, (2), 333–337.

156. Hori, Y.; Kikuchi, K.; Murata, A.; Suzuki, S., Production of methane and ethylene in electrochemical reduction of carbon-dioxide at copper electrode in aqueous hydrogen-carbonate solution. *Chem Lett* **1986**, *15*, (6), 897–898.

157. Hori, Y.; Wakebe, H.; Tsukamoto, T.; Koga, O., Adsorption of Co accompanied with simultaneous charge-transfer on copper single-crystal electrodes related with electrochemical reduction of CO$_2$ to hydrocarbons. *Surf Sci* **1995**, *335*, (1–3), 258–263.

158. Hori, Y.; Murata, A.; Takahashi, R.; Suzuki, S., Enhanced formation of ethylene and alcohols at ambient-temperature and pressure in electrochemical reduction of carbon-dioxide at a copper electrode. *J Chem Soc Chem Commun* **1988**, *17*, (1), 17–19.

159. Kaneco, S.; Iiba, K.; Hiei, N.; Ohta, K.; Mizuno, T.; Suzuki, T., Electrochemical reduction of carbon dioxide to ethylene with high faradaic efficiency at a Cu electrode in CsOH/methanol. *Electrochim Acta* **1999**, *44*, (26), 4701–4706.

160. Rosen, B. A.; Haan, J. L.; Mukherjee, P.; Braunschweig, B.; Zhu, W.; Salehi-Khojin, A.; Dlott, D. D.; Masel, R. I., In situ spectroscopic examination of a low overpotential pathway for carbon dioxide conversion to carbon monoxide. *J Phys Chem C* **2012**, *116*, (29), 15307–15312.

161. Mao, X.; Hatton, T. A., Recent advances in electrocatalytic reduction of carbon dioxide using metal-free catalysts. *Ind Eng Chem Res* **2015**, *54*, (16), 4033–4042.

162. Costentin, C.; Passard, G.; Robert, M.; Saveant, J. M., Ultraefficient homogeneous catalyst for the CO$_2$-to-CO electrochemical conversion. *Proc Natl Acad Sci USA* **2014**, *111*, (42), 14990–14994.

163. Boston, D. J.; Pachon, Y. M. F.; Lezna, R. O.; de Tacconi, N. R.; MacDonnell, F. M., Electrocatalytic and photocatalytic conversion of CO$_2$ to methanol using ruthenium complexes with internal pyridyl cocatalysts. *Inorg Chem* **2014**, *53*, (13), 6544–6553.

164. Tomita, Y.; Teruya, S.; Koga, O.; Hori, Y., Electrochemical reduction of carbon dioxide at a platinum electrode in acetonitrile-water mixtures. *J Electrochem Soc* **2000**, *147*, (11), 4164–4167.

165. Dhar, K.; Cavallotti, C., Investigation of the initial steps of the electrochemical reduction of CO$_2$ on Pt electrodes. *J Phys Chem A* **2014**, *118*, (38), 8676–8688.

166. Hoshi, N.; Noma, M.; Suzuki, T.; Hori, Y., Structural effect on the rate of CO$_2$ reduction on single crystal electrodes of palladium. *J Electroanal Chem* **1997**, *421*, (1–2), 15–18.

167. Zeng, Q.; Tory, J.; Hartl, F., Electrocatalytic reduction of carbon dioxide with a manganese(I) tricarbonyl complex containing a nonaromatic alpha-Diimine ligand. *Organometallics* **2014**, *33*, (18), 5002–5008.
168. Grice, K. A.; Kubiak, C. P., Recent studies of rhenium and manganese bipyridine carbonyl catalysts for the electrochemical reduction of $CO_2$. In *$CO_2$ Chemistry*, Aresta, M.; Eldik, R. V., Eds. **2014**; Vol. 66, pp. 163–188, San Diego: Elsevier Academic Press.
169. Grills, D. C.; Farrington, J. A.; Layne, B. H.; Lymar, S. V.; Mello, B. A.; Preses, J. M.; Wishart, J. F., Mechanism of the formation of a Mn-based $CO_2$ reduction catalyst revealed by pulse radiolysis with time-resolved infrared detection. *J Am Chem Soc* **2014**, *136*, (15), 5563–5566.
170. Chu, D.; Qin, G. X.; Yuan, X. M.; Xu, M.; Zheng, P.; Lu, J., Fixation of $CO_2$ by electrocatalytic reduction and electropolymerization in ionic liquid-$H_2O$ solution. *Chemsuschem* **2008**, *1*, (3), 205–209.
171. Ali, I.; Ullah, N.; Omanovic, S., Electrochemical conversion of $CO_2$ into aqueous-phase organic molecules employing a Ni-nanoparticle-modified glassy carbon electrode. *Int J Electrochem Sci* **2014**, *9*, (12), 7198–7205.
172. Centi, G.; Perathoner, S.; Wine, G.; Gangeri, M., Electrocatalytic conversion of $CO_2$ to long carbon-chain hydrocarbons. *Green Chem* **2007**, *9*, (6), 671–678.
173. Qiao, J. L.; Jiang, P.; Liu, J. S.; Zhang, J. J., Formation of Cu nanostructured electrode surfaces by an annealing-electroreduction procedure to achieve high-efficiency $CO_2$ electroreduction. *Electrochem Commun* **2014**, *38*, (0), 8–11.
174. Fan, M.; Bai, Z.; Zhang, Q.; Ma, C.; Zhou, X.-D.; Qiao, J., Aqueous $CO_2$ reduction on morphology controlled CuxO nanocatalysts at low overpotential. *RSC Adv* **2014**, *4*, (84), 44583–44591.
175. Tang, W.; Peterson, A. A.; Varela, A. S.; Jovanov, Z. P.; Bech, L.; Durand, W. J.; Dahl, S.; Norskov, J. K.; Chorkendorff, I., The importance of surface morphology in controlling the selectivity of polycrystalline copper for $CO_2$ electroreduction. *Phys Chem Chem Phys* **2012**, *14*, (1), 76–81.
176. Kim, D.; Resasco, J.; Yu, Y.; Asiri, A. M.; Yang, P. D., Synergistic geometric and electronic effects for electrochemical reduction of carbon dioxide using gold-copper bimetallic nanoparticles. *Nat Commun* **2014**, *5*, 8.
177. Lim, D. H.; Jo, J. H.; Shin, D. Y.; Wilcox, J.; Ham, H. C.; Nam, S. W., Carbon dioxide conversion into hydrocarbon fuels on defective graphene-supported Cu nanoparticles from first principles. *Nanoscale* **2014**, *6*, (10), 5087–5092.
178. Ma, S. C.; Lan, Y. C.; Perez, G. M. J.; Moniri, S.; Kenis, P. J. A., Silver supported on titania as an active catalyst for electrochemical carbon dioxide reduction. *Chemsuschem* **2014**, *7*, (3), 866–874.
179. Lates, V.; Falch, A.; Jordaan, A.; Peach, R.; Kriek, R. J., An electrochemical study of carbon dioxide electroreduction on gold-based nanoparticle catalysts. *Electrochim Acta* **2014**, *128*, 75–84.
180. Gangeri, M.; Perathoner, S.; Caudo, S.; Centi, G.; Amadou, J.; Begin, D.; Pham-Huu, C. et al., Fe and Pt carbon nanotubes for the electrocatalytic conversion of carbon dioxide to oxygenates. *Catal Today* **2009**, *143*, (1–2), 57–63.
181. Arrigo, R.; Schuster, M. E.; Wrabetz, S.; Girgsdies, F.; Tessonnier, J. P.; Centi, G.; Perathoner, S.; Su, D. S.; Schlogl, R., New insights from microcalorimetry on the FeOx/CNT-based electrocatalysts active in the conversion of $CO_2$ to fuels. *Chemsuschem* **2012**, *5*, (3), 577–586.
182. Magdesieva, T. V.; Yamamoto, T.; Tryk, D. A.; Fujishima, A., Electrochemical reduction of $CO_2$ with transition metal phthalocyanine and porphyrin complexes supported on activated carbon fibers. *J Electrochem Soc* **2002**, *149*, (6), D89–D95.
183. Li, C. W.; Ciston, J.; Kanan, M. W., Electroreduction of carbon monoxide to liquid fuel on oxide-derived nanocrystalline copper. *Nature* **2014**, *508*, (7497), 504–507.

184. Doherty, A. P.; Diaconu, L.; Marley, E.; Spedding, P. L.; Barhdadi, R.; Troupel, M., Application of clean technologies using electrochemistry in ionic liquids. *Asia-Pacific J Chem Eng* **2012**, *7*, (1), 14–23.

185. Oh, Y.; Hu, X. L., Organic molecules as mediators and catalysts for photocatalytic and electrocatalytic CO₂ reduction. *Chem Soc Rev* **2013**, *42*, (6), 2253–2261.

186. Leung, K.; Nielsen, I. M. B.; Sai, N.; Medforth, C.; Shelnutt, J. A., Cobalt-porphyrin catalyzed electrochemical reduction of carbon dioxide in water. 2. Mechanism from first principles. *J Phys Chem A* **2010**, *114*, (37), 10174–10184.

187. Calle-Vallejo, F.; Koper, M. T. M., First-principles computational electrochemistry: Achievements and challenges. *Electrochim Acta* **2012**, *84*, 3–11.

188. Cheng, D. J.; Negreiros, F. R.; Apra, E.; Fortunelli, A., Computational approaches to the chemical conversion of carbon dioxide. *Chemsuschem* **2013**, *6*, (6), 944–965.

189. Agarwal, J.; Fujita, E.; Schaefer, H. F.; Muckerman, J. T., Mechanisms for CO production from CO₂ using reduced rhenium tricarbonyl catalysts. *J Am Chem Soc* **2012**, *134*, (11), 5180–5186.

190. Schouten, K. J. P.; Kwon, Y.; van der Ham, C. J. M.; Qin, Z.; Koper, M. T. M., A new mechanism for the selectivity to C-1 and C-2 species in the electrochemical reduction of carbon dioxide on copper electrodes. *Chem Sci* **2011**, *2*, (10), 1902–1909.

191. Nie, X. W.; Griffin, G. L.; Janik, M. J.; Asthagiri, A., Surface phases of Cu₂O(111) under CO₂ electrochemical reduction conditions. *Catal Commun* **2014**, *52*, 88–91.

192. Bernstein, N. J.; Akhade, S. A.; Janik, M. J., Density functional theory study of carbon dioxide electrochemical reduction on the Fe(100) surface. *Phys Chem Chem Phys* **2014**, *16*, (27), 13708–13717.

193. Nie, X. W.; Luo, W. J.; Janik, M. J.; Asthagiri, A., Reaction mechanisms of CO₂ electrochemical reduction on Cu(111) determined with density functional theory. *J Catal* **2014**, *312*, 108–122.

194. Clark, M. L.; Grice, K. A.; Moore, C. E.; Rheingold, A. L.; Kubiak, C. P., Electrocatalytic CO₂ reduction by M(bpy-R)(CO)(4) (M = Mo, W; R = H, tBu) complexes. Electrochemical, spectroscopic, and computational studies and comparison with group 7 catalysts. *Chem Sci* **2014**, *5*, (5), 1894–1900.

195. Lu, X.; Leung, D. Y. C.; Wang, H. Z.; Leung, M. K. H.; Xuan, J., Electrochemical reduction of carbon dioxide to formic acid. *Chemelectrochem* **2014**, *1*, (5), 836–849.

196. Genovese, C.; Ampelli, C.; Perathoner, S.; Centi, G., Electrocatalytic conversion of CO₂ on carbon nanotube-based electrodes for producing solar fuels. *J Catal* **2013**, *308*, 237–249.

197. Kaneco, S.; Iiba, K.; Ohta, K.; Mizuno, T., Reduction of carbon dioxide to petrochemical intermediates. *Energy Sources* **2000**, *22*, (2), 127–135.

198. Li, H.; Oloman, C., Development of a continuous reactor for the electro-reduction of carbon dioxide to formate—Part 1: Process variables. *J Appl Electrochem* **2006**, *36*, (10), 1105–1115.

199. Bard, A. J.; Faulkner, L. R., *Electrochemical Methods: Fundamentals and Applications*, 2nd edn. Hoboken: Wiley and Sons: **2001**.

200. Sanchez-Sanchez, C. M.; Souza-Garcia, J.; Herrero, E.; Aldaz, A., Electrocatalytic reduction of carbon dioxide on platinum single crystal electrodes modified with adsorbed adatoms. *J Electroanal Chem* **2012**, *668*, 51–59.

201. Gosser, D. K., *Cyclic Voltammetry: Simulation and Analysis of Reaction Mechanisms*. New York: VCH: **1993**.

202. Holze, R., *Experimental Electrochemistry*. Weinheim, Germany: John Wiley & Sons, **2009**.

203. Hamann, C. H.; Hamnett, A.; Vielstich, W., *Electrochemistry*, 2nd Completely Revised and Updated edn. Weinhem, Germany: Wiley-VCH: **2007**.

204. Rouessac, F.; Rouessac, A., *Chemical Analysis: Modern Instrumentation Methods and Techniques*. West Sussex, England: John Wiley & Sons, **2007**.

205. Zhang, J. J.; Pietro, W. J.; Lever, A. B. P., Rotating ring-disk electrode analysis of $CO_2$ reduction electrocatalyzed by a cobalt tetramethylpyridoporphyrazine on the disk and detected as CO on a platinum ring. *J Electroanal Chem* **1996**, *403*, (1–2), 93–100.

206. Aoki, A.; Nogami, G., Rotating-ring-disk electrode study on the fixation mechanism of carbon-dioxide. *J Electrochem Soc* **1995**, *142*, (2), 423–427.

207. Kas, R.; Kortlever, R.; Milbrat, A.; Koper, M. T. M.; Mul, G.; Baltrusaitis, J., Electrochemical $CO_2$ reduction on $Cu_2O$-derived copper nanoparticles: Controlling the catalytic selectivity of hydrocarbons. *Phys Chem Chem Phys* **2014**, *16*, (24), 12194–12201.

208. Wonders, A. H.; Housmans, T. H. M.; Rosca, V.; Koper, M. T. M., On-line mass spectrometry system for measurements at single-crystal electrodes in hanging meniscus configuration. *J Appl Electrochem* **2006**, *36*, (11), 1215–1221.

209. Charlton, A. J.; Donarski, J. A.; Jones, S. A.; May, B. D.; Thompson, K. C., The development of cryoprobe nuclear magnetic resonance spectroscopy for the rapid detection of organic contaminants in potable water. *J Environ Monit* **2006**, *8*, (11), 1106–1110.

210. Detweiler, Z. M.; White, J. L.; Bernasek, S. L.; Bocarsly, A. B., Anodized indium metal electrodes for enhanced carbon dioxide reduction in aqueous electrolyte. *Langmuir* **2014**, *30*, (25), 7593–7600.

211. Li, C. W.; Kanan, M. W., $CO_2$ Reduction at low overpotential on Cu electrodes resulting from the reduction of thick $Cu_2O$ films. *J Am Chem Soc* **2012**, *134*, (17), 7231–7234.

212. Wang, Q. N.; Dong, H.; Yu, H. B.; Yu, H.; Liu, M. H., Enhanced electrochemical reduction of carbon dioxide to formic acid using a two-layer gas diffusion electrode in a microbial electrolysis cell. *RSC Adv* **2015**, *5*, (14), 10346–10351.

213. Haghighi, F. H.; Hadadzadeh, H.; Farrokhpour, H.; Serri, N.; Abdi, K.; Rudbari, H. A., Computational and experimental study on the electrocatalytic reduction of $CO_2$ to CO by a new mononuclear ruthenium(II) complex. *Dalton Trans* **2014**, *43*, (29), 11317–11332.

214. Sholl, D.; Steckel, J. A., *Density Functional Theory: A Practical Introduction*. Hoboken, New Jersey: John Wiley & Sons, **2011**.

215. Eschrig, H., *The Fundamentals of Density Functional Theory*. Hoboken, New Jersey: John Wiley & Sons, **1996**; Vol. 32.

216. Koch, W.; Holthausen, M. C.; Holthausen, M. C., *A Chemist's Guide to Density Functional Theory*. Weinheim: Wiley-VCH: **2001**; Vol. 2.

217. Slater, S.; Wagenknecht, J. H., Electrochemical reduction of $CO_2$ catalyzed by Rh(Diphos)2cl. *J Am Chem Soc* **1984**, *106*, (18), 5367–5368.

218. Silvia, J. S.; Cummins, C. C., Ligand-based reduction of $CO_2$ to CO mediated by an anionic niobium nitride complex. *J Am Chem Soc* **2010**, *132*, (7), 2169–2171.

219. Froehlich, J. D.; Kubiak, C. P., Homogeneous $CO_2$ reduction by Ni(cyclam) at a glassy carbon electrode. *Inorg Chem* **2012**, *51*, (7), 3932–3934.

220. Haines, R. J.; Wittrig, R. E.; Kubiak, C. P., Electrocatalytic reduction of carbon-dioxide by the binuclear copper complex [Cu-2(6-(Diphenylphosphino)-2,2′-Bipyridyl)(2) (Mecn)(2)][Pf6](2). *Inorg Chem* **1994**, *33*, (21), 4723–4728.

221. Smieja, J. M.; Kubiak, C. P., Re(bipy-tBu)(CO)(3)Cl-improved catalytic activity for reduction of carbon dioxide: IR-spectroelectrochemical and mechanistic studies. *Inorg Chem* **2010**, *49*, (20), 9283–9289.

222. Tanaka, H.; Aramata, A., Aminopyridyl cation radical method for bridging between metal complex and glassy carbon: Cobalt(II) tetraphenylporphyrin bonded on glassy carbon for enhancement of $CO_2$ electroreduction. *J Electroanal Chem* **1997**, *437*, (1–2), 29–35.

223. Hoshi, N.; Kato, M.; Hori, Y., Electrochemical reduction of $CO_2$ on single crystal electrodes of silver Ag(111), Ag(100) and Ag(110). *J Electroanal Chem* **1997**, *440*, (1–2), 283–286.

224. Hoshi, N.; Uchida, T.; Mizumura, T.; Hori, Y., Atomic arrangement dependence of reduction rates of carbon-dioxide on iridium single-crystal electrodes. *J Electroanal Chem* **1995**, *381*, (1–2), 261–264.

225. Hoshi, N.; Hori, Y., Electrochemical reduction of carbon dioxide at a series of platinum single crystal electrodes. *Electrochim Acta* **2000**, *45*, (25–26), 4263–4270.

226. Hoshi, N.; Mizumura, T.; Hori, Y., Significant difference of the reduction rates of carbon-dioxide between Pt(111) and Pt(110) single-crystal electrodes. *Electrochim Acta* **1995**, *40*, (7), 883–887.

227. Takahashi, K.; Hiratsuka, K.; Sasaki, H.; Toshima, S., Electrocatalytic behavior of metal porphyrins in the reduction of carbon-dioxide. *Chem Lett* **1979**, *7*, (4), 305–308.

228. Hori, Y.; Murata, A.; Ito, S., Enhanced evolution of Co and suppressed formation of hydrocarbons in electroreduction of $CO_2$ at a copper electrode modified with cadmium. *Chem Lett* **1990**, *19*, (7), 1231–1234.

229. Appel, A. M., Electrochemistry: Catalysis at the boundaries. *Nature* **2014**, *508*, (7497), 460–461.

230. Keith, J. A.; Carter, E. A., Electrochemical reactivities of pyridinium in solution: Consequences for $CO_2$ reduction mechanisms. *Chem Sci* **2013**, *4*, (4), 1490–1496.

231. Costentin, C.; Robert, M.; Saveant, J. M., Catalysis of the electrochemical reduction of carbon dioxide. *Chem Soc Rev* **2013**, *42*, (6), 2423–2436.

232. Rees, N. V.; Compton, R. G., Electrochemical $CO_2$ sequestration in ionic liquids; a perspective. *Energ Environ Sci* **2011**, *4*, (2), 403–408.

233. Centi, G.; Perathoner, S., *Green Carbon Dioxide: Advances in $CO_2$ Utilization*. Hoboken, New Jersey: John Wiley & Sons, **2014**.

234. Bhanage, B. M., *Transformation and Utilization of Carbon Dioxide*. Heidelberg, Germany: Springer, **2014**.

235. Aresta, M., *Carbon Dioxide Recovery and Utilization*. Dordrecht, the Netherlands: Kluwer Academic Publishers, **2003**.

# 2 Thermodynamics of CO$_2$ Electroreduction

*Zhibao Huo, Zhiyuan Song,*
*Dezhang Ren, and Yunjie Liu*

## CONTENTS

## 2.1   THERMODYNAMICS OF ELECTROCHEMISTRY

### 2.1.1   Basic Definitions

Electrochemistry is a branch of chemistry concerning the interrelation of electrical and chemistry effects. In this section, we will introduce some important terms and concepts which are employed in briefly describing electrode reactions.

A galvanic cell is a device in which reactions occur spontaneously at the electrodes when they are connected externally by a conductor. An electrolytic cell is a device in which reactions are affected by the imposition of an external voltage greater than the open-circuit potential of the cell. In an electrochemical reaction, redox reactions occur at the two electrodes, and the whole chemical reaction is composed of two independent half reactions. In an electrochemical reaction, the reduction reactions occur at the cathode and the oxidation reactions occur at the anode. Unlike homogeneous reactions, electrochemical reactions are always heterogeneous, electrochemical reactions always occur at the electrode–electrolyte interface. Where the reactions occur is called the working electrode; the other half of the cell which can be standardized with constant composition is called the reference electrode. The most commonly used reference electrodes are the standard hydrogen electrode, saturated calomel electrode, and silver–silver chloride electrode.

An electrolyte is a material in which the mobile species are ions and the free movement of electrons is blocked. An electrolyte is electroneutral, which obeys the principles of conservation of charge. Electrical conductivity ($G$) is a measure of the electrolyte's ability to conduct an electric current:

$$G = \frac{\kappa S}{l}$$

where

$\kappa$ represents conductivity, which is an intrinsic property of the electrolyte (S/cm)
$S$ represents conductor cross section ($cm^2$)
$l$ represents conductor length (cm); the unit of $G$ is S

Mobility ($u_i$) is the limiting velocity of the ion in an electric field of unit strength. The ion transference number is the contribution to conductivity made by the species divided by the total conductivity:

$$t_i = \frac{|z_i| u_i C_i}{\sum j |z_j| u_j C_j}$$

where

$z$ represents ion charge
$u$ represents mobility ($cm^2$/Vs)
$C$ represents concentration (mol/L)

### 2.1.2 CHEMICAL POTENTIAL AND ELECTROCHEMICAL POTENTIAL

The concept of reversibility is very important in thermodynamics as the thermo-dynamic process can strictly encompass systems only at equilibrium. Reversibility includes chemical reversibility, thermodynamic reversibility, and practical revers-ibility. Strict thermodynamic reversibility is not possible for all processes, but it can be in accordance with the thermodynamic equation within a certain precision. For a thermodynamic equilibrium electrochemical system, the whole reaction is divided into two half reactions: the right electrode reaction and the left electrode reaction. The energy change of the whole reaction is given by the change in Gibbs free energy ($\Delta G$):

$$\Delta G = \left( \sum_i s_i \mu_i \right)_{right} - \left( \sum_i s_i \mu_i \right)_{left}$$

where

$\mu_i$ represents the electrochemical potential of species $i$ (kJ/mol)

$s_i$ represents the stoichiometric coefficient of species $i$; the unit of $\Delta G$ is kJ/mol

We recognize that the Gibbs free energy change could associate with the electric potential difference of the cell reaction:

$$\Delta G = -nFE$$

This equation is very important for it indicates the quantitative relationship between the chemical energy and electrical energy in cell reactions. Other ther-modynamic quantities can be derived from this equation. For example, the entropy change is given by $\Delta G$:

$$\Delta S = -\left( \frac{\partial \Delta G}{\partial T} \right)_P$$

so

$$\Delta S = nF \left( \frac{\partial E}{\partial T} \right)_P$$

and

$$\Delta H = \Delta G + T\Delta S = nF \left[ T \left( \frac{\partial E}{\partial T} \right)_P - E \right]$$

The equilibrium constant of the reaction can also be derived from this equation:

$$RT \ln K = -\Delta G = nFE$$

This equation is a bridge to connect the electrochemical properties and thermo-dynamic properties. We can also predict the electrochemical data from the thermo-chemical data.

### 2.1.3 EQUILIBRIUM ELECTRODE POTENTIAL AND IRREVERSIBLE ELECTRODE POTENTIAL

The Nernst equation is one of the most fundamental theories for electrochemicals. The Nernst equation is a concrete form of thermodynamic equilibrium applicable to electrochemical reaction processes, by which the quantitative value of the reversible electrode potential, the standard electrode potential, and the reactant activity can be given.

Considering that a galvanic cell is a reversible cell, we can use the Nernst equation to calculate the electrodynamic potential of the galvanic cell, for example, $Zn|ZnSO_4$ $(a_{Zn^{2+}})||CuSO_4 (a_{Cu^{2+}})|Cu$, the electrode reaction is

anode    $Zn \rightleftharpoons Zn^{2+} + 2e$

cathode    $Cu + 2e \rightleftharpoons Cu^{2+}$

cell reaction    $Zn + Cu^{2+} \rightleftharpoons Cu + Zn^{2+}$

According to the chemical equilibrium isotherm formulas, the change of Gibbs free energy ($\Delta G$) is

$$-\Delta G = RT \ln K - RT \ln \frac{a_{Cu} a_{Zn^{2+}}}{a_{Zn} a_{Cu^{2+}}}$$

where $K$ represents the thermodynamic equilibrium constant and $a$ represents the activity, hence

$$-\Delta G = nFE$$

$$nFE = RT \ln K - RT \ln \frac{a_{Cu} a_{Zn^{2+}}}{a_{Zn} a_{Cu^{2+}}}$$

$$E = \frac{RT}{nF} \ln K - \frac{RT}{nF} \ln \frac{a_{Cu} a_{Zn^{2+}}}{a_{Zn} a_{Cu^{2+}}}$$

The substances participating in this electrochemical reaction are in standard state, namely, the activity of the substances in the solution is 1, and fugacity is also 1, then

$$E^0 = \frac{RT}{nF} \ln K$$

$E^0$ represents standard electromotive force (EMF), in this circumstance, in non-standard conditions,

$$E = E^0 - \frac{RT}{nF} \ln \frac{a_{Zn^{2+}} a_{Cu}}{a_{Zn} a_{Cu^{2+}}}$$

The above formula can be written as a general formula:

$$E = E^0 - \frac{RT}{nF} \ln \frac{a_{product}^{v'}}{a_{reactant}^{v'}}$$

or

$$E = E^0 + \frac{RT}{nF} \ln \frac{a_{product}^{v'}}{a_{reactant}^{v'}}$$

The above formula is a Nernst equation where
$R$ represents perfect gas constant (J)
$T$ represents temperature (K)
$n$ represents the amount of moles of electrons in an electrochemical reaction (mol)
$F$ represents the Faraday constant (C/mol)
$a$ represents activity (mol/L)
$v$ represents the stoichiometric number of the reactant and product
$E^0$ represents standard EMF (V)

Using a galvanic cell which is composed of a zinc reversible electrode and a standard hydrogen electrode as an example, we can use this to derive the equilibrium by thermodynamic calculation formulas.

$$Zn \mid Zn^{2+}(a^{2+}) \parallel H^+(a^+ = 1) \mid H_2 (p_{H_2} = 101325 \text{ Pa}), Pt$$

anode $\quad Zn \rightleftharpoons Zn^{2+} + 2e$
cathode $\quad 2H^+ + 2e \rightleftharpoons Cu^{2+}$
cell reaction $\quad Zn + 2H^+ \rightleftharpoons Zn^{2+} + H_2$

According to the Nernst equation, the potential of this galvanic cell is

$$E = E^0 - \frac{RT}{2F} \ln \frac{a_{Zn^{2+}} p_{H_2}}{a_{Zn} a_{H^+}^2}$$

and

$$E = \varphi_+ - \varphi_-$$

$$E^0 = \varphi_+^0 - \varphi_-^0$$

The above formula could be written as

$$E = (\varphi_{H_2/H^+}^0 - \varphi_{Zn/Zn^{2+}}^0) - \left( \frac{RT}{2F} \ln \frac{p_{H_2}}{a_{H^+}^2} + \frac{RT}{2F} \ln \frac{a_{Zn^{2+}}}{a_{Zn}} \right)$$

$$= \left( \varphi_{H_2/H^+}^0 + \frac{RT}{2F} \ln \frac{a_{H^+}^2}{p_{H_2}} \right) - \left( \varphi_{Zn/Zn^{2+}}^0 + \frac{RT}{2F} \ln \frac{a_{Zn^{2+}}}{a_{Zn}} \right)$$

And according to the regulations, the potential of the standard hydrogen electrode equals zero, then the above formula can be simplified as

$$\varphi_{Zn/Zn^{2+}} = -E$$

$$= \varphi_{Zn/Zn^{2+}}^0 + \frac{RT}{2F} \ln \frac{a_{Zn^{2+}}}{a_{Zn}}$$

The following formula can be used to describe an electrode reaction in general:

$$O + ne \rightleftharpoons R$$

Then, the above formula can be written as a general formula:

$$\varphi = \varphi^0 + \frac{RT}{nF} \ln \frac{a_O}{a_R}$$

or

$$\varphi = \varphi^0 + \frac{RT}{nF} \ln \frac{a_{Oxidation}}{a_{Reduction}}$$

The above formula is a Nernst electrode potential equation; this formula applies to the reversible electrode. There are four major types of reversible electrode.

The first type of reversible electrode is cationic electrode. This kind of electrode immerges into the same kind soluble metal ion solution, such as

$$Zn|ZnSO_4, Cu| CuSO_4, Ag | AgNO_3$$

The second type of reversible electrode is the anion reversible electrode. This kind of electrode is covered by the same kind of metal ion hard soluble salt, and immerges into the solution containing the same kind of anion which is composed of the hard soluble metal salt, such as

$$Hg \mid HgCl_2(solid), KCl\left(a_{Cl^-}\right)$$

$$Ag \mid AgCl \ (solid), KCl\left(a_{Cl^-}\right)$$

The third type of reversible electrode is composed of Pt or other inert metal, and immerges into the solution containing ions of different valencies of the same metal, such as

$$Pt \mid Fe^{2+}\left(a_{Fe^{2+}}\right), Fe^{3+}\left(a_{Fe^{3+}}\right)$$

$$Pt \mid Sn^{2+}\left(a_{Sn^{2+}}\right), Sn^{4+}\left(a_{Sn^{4+}}\right)$$

$$Pt \mid Fe(CN)_6^{4-} (a_1), Fe(CF)_6^{3-} (a_2)$$

The fourth type of reversible electrode is the gas electrode. Gas is adsorbed on the surface of the inert metal electrode, such as Pt electrode, and on the surface of metal electrode, oxidation–reduction reactions occurs. The corresponding ions, such as $Pt^{2+}$ dissolved in electrolyte.

$$Pt, H_2 \left(p_{H_2}\right) \mid H^+\left(a_{H^+}\right)$$

$$Pt, O_2 \left(p_{O_2}\right) \mid OH^- \left(a_{OH^-}\right)$$

In a real-world situation, the irreversible electrode is in the vast majority. The irreversible electrode can also be sorted into four types.

The first type of irreversible electrode is made of metals that immerge into the solution without containing the same metal ion, such as

$$Zn \mid HCl \ and \ Zn \mid NaCl$$

The second type of irreversible electrode is made of metals, which have higher standard potential, that immerge into the solution that could generate hard soluble salts or oxides of the metal, such as

$$Cu \mid NaOH, Ag \mid NaCl$$

The third type of irreversible electrode is made of metals that immerge into some kind of oxidizing agent solution, such as

$$Fe \mid HNO_3, Fe \mid K_2Cr_2O_7$$

The fourth type of irreversible electrode is the gas irreversible electrode in which metals that have lower hydrogen overpotential immerge into water.

For overpotential, we can use Tafel equation to calculate

$$\eta = a + b \log i$$

The factors that affect the electrode potential include the nature of electrode, the surface state of metals, the mechanical deformation and the stress of metals, pH value of the solution, oxidizing agent in solution, complexing agent in solution, and the solution itself.

### 2.1.4 POTENTIAL–pH

The equilibrium potential reflects the materials' ability of the redox potential, and can be used to judge the feasibility of an electrochemical reaction. The value of equilibrium potential relates to the activity of the reactant; for an electrochemical reaction with $H^+$ and $OH^-$ involved, the electrode potential will change over the pH value. A potential–pH diagram can be used for measuring the feasibility of an electrochemical reaction.

For an electrode reaction with $H^+$ involved without electron participation, the general formula can be written as

$$aA + cH_2O \rightleftharpoons bB + mH^+$$

According to thermochemistry, for a reaction at equilibrium state

$$\sum v_i \ln a_i = \ln K$$

where
$v_i$ represents stoichiometric constant of reactants
$a_i$ represents activity of reactants
$K$ represents equilibrium constant; hence

$$-a \log a_A - c \log a_{H_2O} + b \log a_B + m \log a_{H^+} = \log K$$

$$-\log a_{H^+} = pH$$

so

$$\log \frac{a_B^b}{a_A^a} = \log K + m\text{pH}$$

or

$$\text{pH} = -\frac{1}{m}\log K - \frac{1}{m}\log \frac{a_A^a}{a_B^b}$$

For such a reaction, the equilibrium condition is mainly related to the pH value, but has nothing to do with the electrode potential; likewise for an electrode reaction without H⁺ involved.

$$a\text{A} + m\text{H}^+ + ne \rightleftharpoons b\text{B} + c\text{H}_2\text{O}$$

$$\varphi = \varphi^0 + \frac{2.3RT}{nF} \log \frac{a_A^a}{a_B^b}$$

Apparently, the equilibrium state of this type of reaction has nothing to do with the pH value, but relates to the electrode potential. And

$$a\text{A} + m\text{H}^+ + ne \rightleftharpoons b\text{B} + c\text{H}_2\text{O}$$

$$\varphi = \varphi^0 - \frac{2.3RT}{nF} m\text{pH} + \frac{2.3RT}{nF} \log \frac{a_A^a}{a_B^b}$$

We can see that the equilibrium state of this type of reaction is associated with the pH value of the solution and electrode potential. For more on potential–pH diagram, refer to books about thermodynamics and kinetics of metal electrochemical corrosion.

## 2.1.5 Research Methods

The main research methods of electrochemistry are given as follow:

### 2.1.5.1 Cyclic Voltammetry (CV)

The research method that currently is recorded as a function of potential is called linear sweep voltammetry (LSV) (Figure 2.1a). When the potential is sweeping in both positive and negative directions, this method is called CV (Figure 2.1b).

For a Nernst (reversible) system,

$$E_p - E^0 = -1.11 \frac{RT}{nF}$$

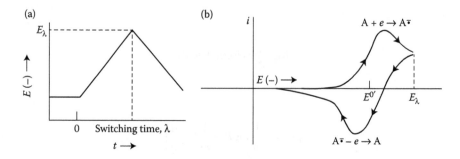

**FIGURE 2.1** (a) Cyclic potential sweep. (b) Resulting cyclic voltammogram. (Bard, A. J. et al.: *Electrochemical Methods Fundamentals and Applications*. P227. Copyright Wiley-VCH Verlag GmbH & Co. KGaA. Reproduced with permission.)

$$i_P = 0.4463nFCD^{1/2}v^{1/2}\left(\frac{nF}{RT}\right)^{1/2}$$

For an irreversible system,

$$E_P - E^0 = \left(-0.783 + \ln\frac{k^0}{D^{1/2}} - \frac{1}{2}\ln\frac{\alpha vnF}{RT}\right)\frac{RT}{\alpha nF}$$

$$i_P = 0.4958nFCD^{1/2}v^{1/2}\left(\frac{\alpha nF}{RT}\right)^{1/2}$$

where

$E_p$ represents peak potential (V)

$i_p$ represents peak current (A)

$C$ represents substrate concentration (mol/L)

$A$ represents area (cm²)

$\alpha$ represents transfer coefficient

$D$ represents diffusion coefficient (cm²/s)

$k$ represents electron transfer rate constant (cm/s)

$v$ represents sweep rate (V/s)

$T$ represents temperature (K)

$F$ represents the Faradic constant (C/mol)

$R$ represents molar gas constant (J/mol K)

### 2.1.5.2 Chronoamperometry (CC)

The research method that currently is recorded as a function of time is called chronoamperometry (Figure 2.2).

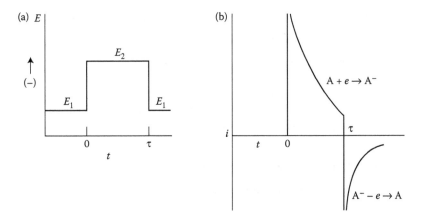

**FIGURE 2.2** Double potential step chronoamperometry. (a) Typical waveform. (b) Current response. (Bard, A. J. et al.: *Electrochemical Methods Fundamentals and Applications*. P159. Copyright Wiley-VCH Verlag GmbH & Co. KGaA. Reproduced with permission.)

As shown in Figure 2.3, when the integral of current versus time is coulomb, chronocoulometry and double potential step chronocoulometry are more useful. The step waveform is similar to Figure 2.2.

$$Q = \frac{2nFACD^{1/2}}{\pi^{1/2}}t^{1/2} + Q_{dl}$$

where

$F$ represents the Faraday constant (C/mol)

$A$ represents area (cm²)

$C$ represents substrate concentration (mol/L)

$D$ represents diffusion coefficient (cm²/s)

$Q_{dl}$ represents charge devoted to double-layer capacitance (C)

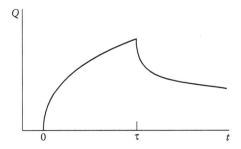

**FIGURE 2.3** Response curve for double potential step chronocoulometry. (Bard, A. J. et al.: *Electrochemical Methods Fundamentals and Applications*. P160. Copyright Wiley-VCH Verlag GmbH & Co. KGaA. Reproduced with permission.)

## 2.2 THERMODYNAMICS OF CO₂ ELECTROREDUCTION (NERNST POTENTIALS)

There are two major ways for electrochemical conversion and utilization of $CO_2$: one is the electrochemical reduction of $CO_2$ to some useful products, such as CO, $CH_4$, HCOOH, and $H_2C_2O_4$, and the other is the electrochemical reduction of $CO_2$ with other organic substrates, to produce new valuable organic materials. We will illustrate these two methods in the following sections.

### 2.2.1 DIRECT ELECTROLYTIC REDUCTION OF CO₂

#### 2.2.1.1 Principles of Electroreduction of CO₂

Possible routes of direct electrolytic reduction of $CO_2$ are as follows [1]:

The above standard electrode has potential relative to the saturated calomel electrode. A fact that variety of products can be obtained under different experimental conditions illustrates that these reactions undergo different reaction mechanisms. For example, Kaneco et al. [2,3] proposed the electrochemical reduction mechanism of $CO_2$ at Cu and Ag electrodes (Figure 2.4).

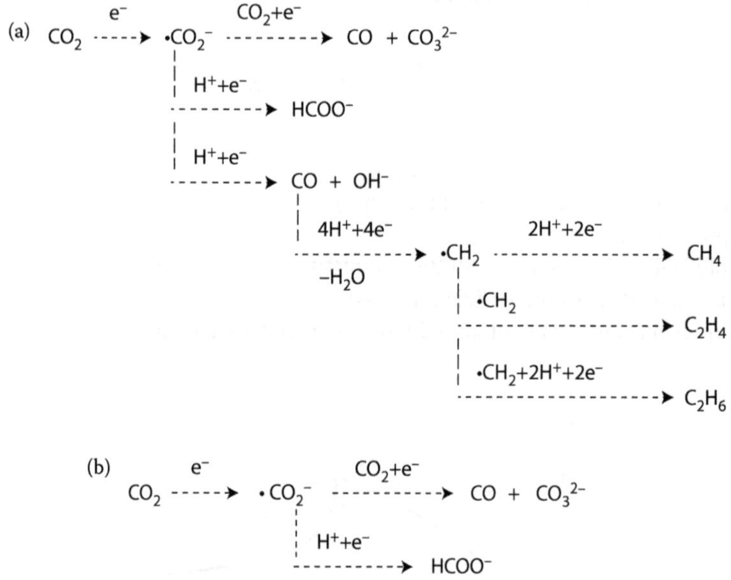

**FIGURE 2.4** Electrochemical reduction mechanism of $CO_2$ at (a) Cu electrode in methanol and at (b) Ag electrode in KOH–methanol. (Reprinted from *Electrochimica Acta*, 44, Kaneco, S. et al., Electrochemical reduction of carbon dioxide to ethylene with high Faradaic efficiency at a Cu electrode in CsOH/methanol, 4701–4706, Copyright 1999; *Electrochimica Acta*, 44, Kaneco, S. et al., Electrochemical reduction of $CO_2$ at an Ag electrode in KOH–methanol at low temperature, 573–578, Copyright 1999, with permission from Elsevier.)

The above mechanisms are proposed for reactions in anhydrous solvents; there are mechanisms proposed for electrochemical reduction for CO$_2$ in aqueous solution, and the mechanism that is widely accepted is as follows [4–6]:

$$CO_2 + e^- \rightarrow CO_2^{\bullet-}$$

$$CO_{2\,ads}^{\bullet-} + H_2O \rightarrow HCO_{2\,ads}^{\bullet} + OH^-$$

$$HCO_{2\,ads}^{\bullet} + e^- \rightarrow HCO_2^-$$

$$CO_2^{\bullet-} + BH + e^- \rightarrow HCO_2^{\bullet} + B^-$$

where BH represents substances that provide proton.

As shown in Figure 2.5, Chaplin et al. [7] summarized all routes of reduction of CO$_2$ by using a graph display.

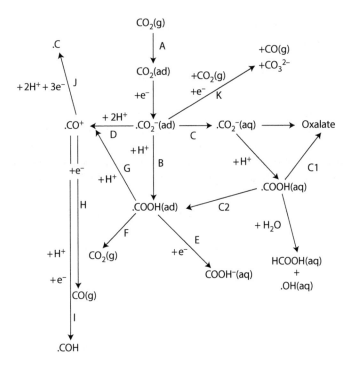

**FIGURE 2.5** CO$_2$ reduction routes commonly proposed for an acid system. (With kind permission from Springer Science + Business Media: *Journal of Applied Electrochemistry*, 33(12), 2003, 1107–1123, Chaplin R. P. S. et al.)

Reaction route A describes the surface adsorption of $CO_2$: $CO_2$ reacts with electrolytes and then produces chemical species in transition state.

$$CO_2(g) \rightarrow CO_2(ag)$$

or

$$CO_2(g) + e^- \rightarrow CO_2^-(g)$$

then

$$CO_2^{\cdot-}(g) + H_2O \rightarrow \cdot COOH(ad) + OH^-$$

The absorption of $CO_2$ on metal electrode surface can reduce the activation energy of the electrochemical reduction (the letter M represents metal catalysts):

$$M + CO_2 + e^- \rightarrow M - CO_2^-$$

$$M - H + CO_2 \rightarrow M - CO_2^- + H^+$$

Reaction route B describes $CO_2$ after being protonated, then generating MCOOH. The process can be expressed as

$$M - CO_2 + e^- + H^+ \rightarrow M - COOH$$

For the electrodes whose surface can easily absorb $CO_2$ and release $CO_2$, group $CO_2^{\cdot-}$ can be generated following reaction route C [8]. But when the pH value is low, reaction routes follow B and D. Group $CO_2^{\cdot-}$ can react with $H^+$ in electrolytes [9,10]. The generated groups can be absorbed again by metal electrodes or can further react to produce formic acid. In a nonaqueous solvent, $\cdot COOH$ follows route C1 to generate oxalate.

Reaction route D describes how O atom is removed from $M - CO_2^-$:

$$M - CO_2^- + e^- + H_2O \rightarrow MCO^+ + 2OH^-$$

or

$$M - CO_2^- + e^- + 2H^+ \rightarrow MCO^+ + 2H_2O$$

In reaction routes E, F, and G, the reduction of adsorption species $\cdot COOH$ (ad), if the product M–COOH is not stable, it would be shredded into M–H and $CO_2$, or lose $OH^-$ through nucleophilic reaction to generate $M-CO^+$:

$$M - COOH + H^+ \rightarrow M - CO^+ + H_2O$$

Formic acid is released into the solution and may undergo the following process:

$$\cdot COOH + e^- \rightarrow COOH^- (ad)$$

Or in the form of hydrate

$$\cdot COOH + e^- + H_2O \rightarrow C(OH)_2 + OH^-$$

and

$$C(OH)_2 \rightarrow HCOOH$$

Reaction routes H, I, and J describe further reduction reactions.

Researchers found that Cu, Rh, and several transition metal electrode reactions followed these three reaction routes [11,12], species M–CO$^+$ react with an electron to generate CO (reaction route H).

According to reaction route I, M–CO$^+$ reacts with a proton and an electron to generate M–CHO; starting from this substrate can help obtain CH$_2$CO, CH$_3$CHO, C$_2$H$_5$OH, C$_2$H$_4$, C$_2$H$_6$, and C$_3$H$_7$OH.

According to reaction route J, M–C=O$^+$ can react with three electrons and two protons; then the O atom is removed to generate •C. After that, this group will react with electrons and protons to generate C$_2$H$_2$ and CH$_4$. Reaction J can easily happen at low temperature. Reaction routes I and J are a pair of competing reactions.

Reaction route K is a disproportionation process; this process usually exists in nonaqueous circumstances. In this process, CO$_2^{\cdot-}$ (ad) absorbs oxygen atom and then generates CO$_3^{2-}$ and CO(g) [13].

### 2.2.1.2 Factors Influencing Electroreduction of CO$_2$

There are many affecting factors influencing the electroreduction of CO$_2$, such as types of electrode, electrolytes, catalysts, electrode potential (current density), temperature, pressure, pH, and reactors. Because the electrode and electrolyte involve more content, we will discuss this later, and now mainly focus on how the other factors affect the electroreduction of CO$_2$.

#### 2.2.1.2.1 Temperature

The factors that affect the electroreduction of CO$_2$, including electron transfer, absorption, and desorption on the surface of the electrode, and the diffusion rate of CO$_2$ are all limited by the solubility of CO$_2$; the solubility of CO$_2$ is higher under a lower temperature. Current efficiency and current density are also affected by temperature. So temperature is very important to the electroreduction of CO$_2$ processes.

Mizuno et al. [14] found that with temperature rising (293–373 K), the current efficiency of the Pb electrode first increased and then decreased. But current efficiency of In and Sn electrodes showed the trend to decrease. Ryu et al. [15] found that the current density improved about 20 times with increasing temperature (275–333 K). Kaneco et al. [16] investigated the reduction of CO$_2$ on a Cu electrode in aqueous $NaHCO_3$ at low temperature. The temperature during the electrolysis of $CO_2$ decreased stepwise from 288 to 271 K. The tendency of current efficiency in this experiment proved that lower temperature was conducive to the electrochemical reduction of $CO_2$.

Most researchers [17,18] agree that lower temperature is available for the reduction process. Low temperature not only increases the solubility of $CO_2$ but also shifts the absorption equilibrium to the direction that is beneficial to the electroreduction of $CO_2$.

### 2.2.1.2.2  pH Value

We know that pH is an important factor in electrochemical thermodynamics. pH is also very important in the process of electrochemical reduction of $CO_2$. Figure 2.6 shows how pH and potential influence the electrochemical reduction of $CO_2$, $H_2O$, and relative materials.

As we can see in Figure 2.6, when the pH value is less than 8, $CO_2$ can steadily exist in solution; when the pH value is less than 7, a hydrogen evolution reaction is more likely to occur. According to Figure 2.6, only in a very limited range (pH is about 7–8) can the electrochemical reduction of $CO_2$ have an advantage.

Most researchers selected 0.5 mol/L $KHCO_3$ for electrolyte (pH 6–8), but there were also some researchers who selected $Na_2SO_4$ (pH = 2) or $K_2CO_3$ (pH = 12) as the electrolyte, and gained a better effect.

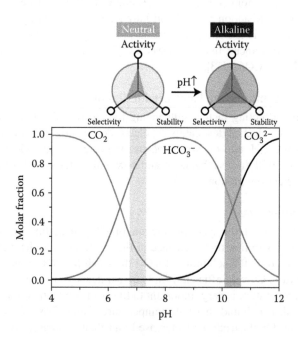

**FIGURE 2.6** Electrolyte pH determining the equilibrium concentrations of carbon dioxide–bicarbonate–carbonate system. (Reprinted with permission from Lee, S. et al. *Journal of Physical Chemistry C*; 119(9), 4884–4890. Copyright 2015 American Chemical Society.)

Li et al. [19] investigated the effects of concentration of KHCO$_3$, and found that current efficiency was the highest when the concentration is 0.5 mol/L; when the concentration continued to rise (reaching 1.96 mol/L), hydrogen evolution reaction was the dominating reaction in solution.

### 2.2.1.2.3 Pressure

The solubility of CO$_2$ in water increases with increasing pressure, which enhances the electrochemical reduction rate. Most researchers tended to believe that increasing pressure could reduce the cathodic superpotential and current density of the hydrogen evolution reaction.

Nakagawa et al. [20] investigated the electrochemical reduction of CO$_2$ under high pressure (up to 60 atm) on group VIII metal electrodes. Even for metals such as Ni and Pt, a hydrogen evolution reaction was more likely to happen under the ordinary conditions of pressure, and could detect reducing products. Hara et al. [21,22] found that the selectivity of products could be improved with an increase in CO$_2$ gas pressure. For example, in KHCO$_3$ aqueous solution, when Fe, Co, Rh, Ni, Pd, and Pt were selected as the metal electrode, the main products were formic acid and CO under high pressure (<6 MPa). When the pressure of CO$_2$ was lower than 0.1 MPa, hydrogen evolution reaction was the main reaction. Todoroki et al. [23] found that when the pressure reached 20 atm, the mass transfer effect was no longer main limiting factor of the reaction rate. When the pressure reached 40 atm, interactions between CO$_2$ and electrolyte changed a lot. When the pressure reached 60 atm, the excessively dissolved CO$_2$ in the solution inhibited the process of reaction.

The above research results indicate that pressure only affect products of electrochemical reduction reaction on several metal electrodes and increasing CO$_2$ pressure in solution has a positive role on the increase of local current density on several electrodes.

### 2.2.1.2.4 Electrode Potential (Current Density)

In an electrochemical reaction, electrode potential is an important parameter. The electrode potential will be changed by changing the reaction condition (electrolyte, temperature, pressure, and so on) and the product distribution and the product current efficiency by controlling the potential. Current density has the same effects.

Koeleli et al. [24] investigated the Pb granule electrode at high pressure and high temperature conditions (electrode potentials between −1.6 and −2.1 V). Current efficiency increased at the beginning and then reduced. The best reduction potential was −1.5 V (vs. standard hydrogen electrode [SHE]). Kaneco et al. [16] investigated the electrochemical reduction of CO$_2$ on a Cu electrode; methane, formic acid, and hydrogen were the main products. As the potential turned negative, the faradaic efficiency of methane increased from 24% to 46%, and the faradaic efficiency of hydrogen went from 72% to 51%. Innocent et al. [25] found that the maximum faradaic efficiency of formic acid was 65% when the potential was −1.6 V on a Pb electrode in KHCO$_3$ electrolyte. Ikeda et al. [26] found that the faradaic efficiency of the main product (CO) was 90% on an electrode at −2.8 V (vs. Ag/AgCl) in 0.1 mol/L tetraethylammonium perchlorate (TEAP)-polycarbonate (PC). Hara et al. [21] found that with the electrochemical reduction of CO$_2$ at a pressure of 30 atm at large current density (1000–5000 A/m²), the main products were CO and formic acid on the Rh and Ag electrode. They also investigated the relationship between the main products and current density on the Rh and Ag electrode and the relationship between the main product and current density on the Pt gas diffusion electrode [27].

### 2.2.2 ELECTROCATALYTIC CARBOXYLATION OF $CO_2$ AND ORGANIC COMPOUNDS

$\cdot CO_2^-$ is a kind of strong alkali. There is a popular belief that $CO_2$ is reduced to $\cdot CO_2^-$ and then as an nucleophile reacts with organics. The qualitative reaction scheme for $CO_2$ conversion is shown as Figure 2.7. According to Figure 2.7, we can see that catalysts and electrolytes acting as cocatalysts can lower the energy of the intermediate and thus improve the energy efficiency of conversion.

#### 2.2.2.1 Electrocarboxylation of Unsaturated Hydrocarbons with $CO_2$

*2.2.2.1.1 Electrocarboxylation of Alkene*

In 1967, when Neikam et al. [28,29] first began to study the electrocarboxylation of styrene with $CO_2$, the yield of monocarboxylic acid and dicarboxylic acid was 25%. In 1973, Gambino et al. [30] investigated the reaction of ethylene with $CO_2$, selected stainless steel AISI 316, with mercury or Au as the cathode, Al as anode at 20 atm in dimethylformamide (DMF), which gave a mixture of oxalate and succinate. In 1984, Gambino et al. [31,32] investigated the electrocarboxylation of styrene and the electrodimerization of $CO_2$ to oxalate. The yield of phenylsuccinic acid was 85%. In 1992, Derien et al. [33] investigated the electroreduction reaction of styrene in DMF containing $Bu_4NBF_4$ under 1 atm $CO_2$ in the presence of Ni (0) complexes as catalyst; the yield of phenylsuccinic acid was 85%. In 2000, Ballivet-Tkatchenko et al. [34] investigated the electroreduction of $[CpFe(CO)_2]_2(Fp)_2$, which was catalystically reduced at −1.8 V (vs. standard calomel electrode [SCE]) liberated $[Cp(CO)_2Fe]^-$ and $\cdot CO_2^-$, making regioselective carboxylation of styrene and isoprene possible (as shown in Figure 2.8). In 2001, Bringmann et al. [35] investigated the electrochemical

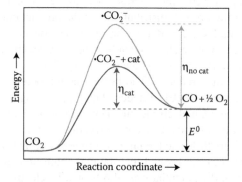

$$\varepsilon_{Faradaic} = \frac{n \cdot F \cdot n}{Q}$$

$$\varepsilon_{energetic} = \frac{E^0}{E^0 + \eta} \times \varepsilon_{Faradaic}$$

where   $\varepsilon_{energetic}$ = energetic efficiency
         $\varepsilon_{Faradaic}$ = Faradaic efficiency
         $n$ = number of electrons
         F = Faraday's constant
         $n$ = number of moles of product
         Q = charge passed
         $E^0$ = standard potential
         $\eta$ = overpotential

**FIGURE 2.7** Qualitative of reaction scheme for $CO_2$ conversion. $\varepsilon$ represents the symbol for the faradaic and energetic efficiencies. (Reprinted with permission from Whipple, D. T. et al. *Journal of Physical Chemistry Letters*; 1, 3451–3458. Copyright 2010 American Chemical Society.)

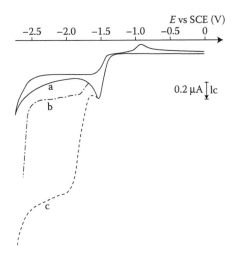

**FIGURE 2.8**   Voltammogram (scan rate 0.1 V/s) in DMF–0.1 M TBABF$_4$ on a gold micro-electrode of (a) Fp$_2$ (10$^{-2}$ M), (b) Fp$_2$–CO$_2$, and (c) with more CO$_2$. (Ballivet-Tkachenko, D., Folest, J. C. and Tanji, J.: Electrocatalytic reduction of CO$_2$ for the selective carboxylation of olefins. *Applied Organometallic Chemistry*. 2000. 14. 847–849. Copyright Wiley-VCH Verlag GmbH & Co. KGaA. Reproduced with permission.)

reductive coupling reactions between alkenes and CO$_2$ with nickel–organic cata-lyst; there were two main products: 2-methyl-butanedioic acid and 3-butenoic acid. In 2001, Senboku et al. [36] selected Pt as the cathode and Mg as the anode, and dimethyl formamide (DMF)-Et$_4$NClO$_4$ as the supporting electrolyte, and inves-tigated the electrochemical carboxylation reaction of styrene and derivatives with CO$_2$. In the form of constant current electrolysis, the yield of phenylsuccinic acid was 66%–91%. As shown in Figure 2.9, in 2004, Chowdhury et al. [37] investigated the electrochemical carboxylation of bicyclo[$n$.1.0]-alkylidene derivatives in a dry DMF solution containing TEAP using a cell equipped with a Pt plate cathode and a Zn plate anode. The reduction potential of bicyclo[$n$.1.0]alkylidene derivatives and CO$_2$ were −2.35 and −2.90 V (vs. Ag/Ag$^+$), respectively (Figure 2.10).

### 2.2.2.1.2   Electrocarboxylation of Alkyne

In 1988, Dunach et al. [38] investigated the regioselective functionalization of the 2-position of terminal alkynes with carbon dioxide with Ni(bipy)$_3$(BF)$_3$ as an active electrochemical reduction. The yield of α-substituted acrylic acids was 65%–90%. Then, in 1989, Dunch et al. [39] investigated the electrochemical reduction reaction of disubstituted alkynes with CO$_2$ to yield mono- and dicarboxylated derivatives. The reaction was performed under mild conditions using Ni(bipy)$_3$(BF$_4$)$_2$ as an activ-ity catalyst. In 1990, Dérien et al. [40] investigated the electrochemical reduction reaction of α, ω-diynes with CO$_2$ using electrogenerated nickel(0) complexes as active catalysts. We can see the cyclic voltammograms in Figure 2.11. These experi-ments were carried out with a fresh vitreous carbon microelectrode (3 mm$^2$) at 20°C at a scan rate of 100 mV/s for a solution of this kind of catalyst (0.3 mmol) in DMF

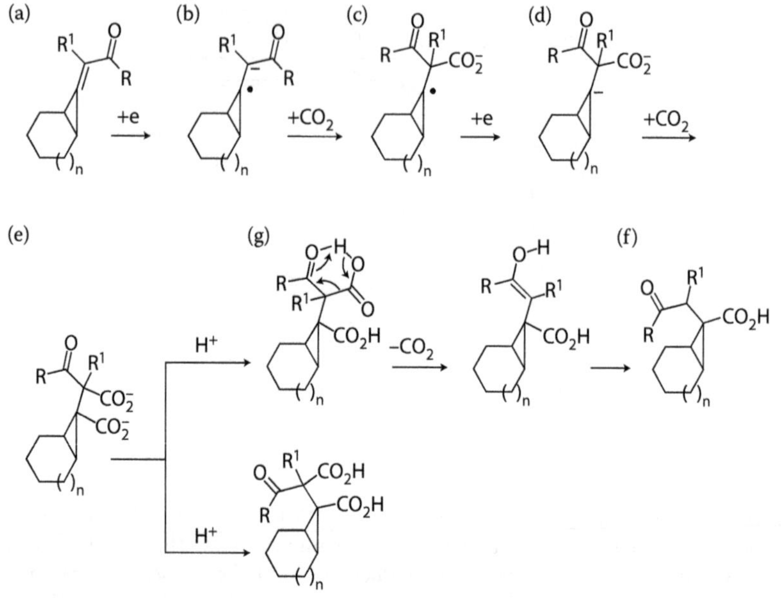

**FIGURE 2.9** Reaction mechanism of electrochemical carboxylation of bicyclo[$n$.1.0]alkylidene. (Reprinted from *Tetrahedron*, 60, Chowdhury, M. A., Senboku, H. and Tokuda, M., Electrochemical carboxylation of bicyclo[$n$.1.0]alkylidene derivatives, 475–481, Copyright 2004, with permission from Elsevier.)

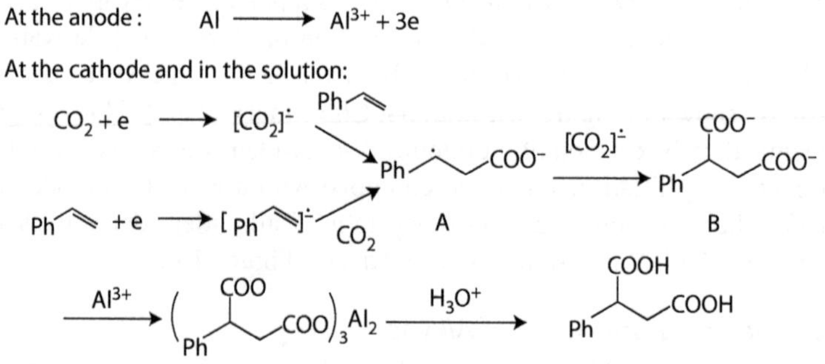

**FIGURE 2.10** Electrochemical reaction mechanism of styrene with $CO_2$. (Reprinted from *Electrochimica Acta*, 53, Yuan et al., 2170–2176, Copyright 2008, with permission from Elsevier.)

(30 mL) containing 0.1 M tetrabutylammonium tetrafluoroborate as the supporting electrode. In this figure, (a) represents solution under argon; (b) represents solution after addition of 4-octyne (0.3 mmol); (c) represents a solution saturated with $CO_2$; and (d) represents solution b saturated with $CO_2$. In 1991, Dérien et al. [41] investigated the electroreductive coupling of alkynes and $CO_2$ using nickel–bipyridine

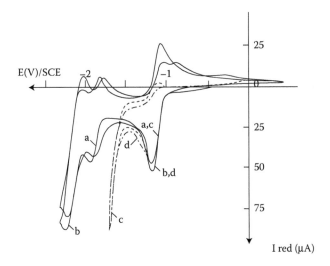

**FIGURE 2.11** Electroreductive coupling of alkynes and CO$_2$ using nickel–bipyridine complexes as catalyst. (Reprinted with permission from Dérien, S., Dunach, E. and Périchon, J. From stoichiometry to catalysis: Electroreductive coupling of alkynes and carbon dioxide with nickel-bipyridine complexes. Magnesium ion as the key for catalysis. *Journal of American Chemical Society*; 113(12), 8447–8454. Copyright 1991 American Chemical Society.)

complexes as a catalyst in the presence of a sacrificial Mg anode. The yield of α,β-unsaturated acids was ideal (Figure 2.12).

In 2008, Yuan et al. [42] investigated the electrochemical decarboxylation of arylacetylenes with CO$_2$ using Ni as the cathode and Al as the anode with n-Bu$_4$NBr–DMF as the supporting electrolyte. The corresponding aryl-maleic anhydrides and 2-arylsuccinic acids were 82%–94%.

## 2.2.2.2 Electrocarboxylation of Aldehydes and Ketones with CO$_2$

In 1984, Ikeda et al. [43] investigated the electrochemical reduction of p-isobutylacetophenone in DMF containing 0.5 M Bu$_4$NI on Hg pool cathode by passing 2.0 F/mol electricity with a constant current density of 48.2 mA/cm$^2$. The product 2-hydroxy-2-(p-isobutylphenyl)-propionic acid was synthesized in 85% yield. In 1985, Ikeda et al. [44] investigated the electrochemical reduction of benzophenones in DMF containing potassium iodide (KI) as the electrolyte on Hg cathode by passing 2.3 F/mol electricity with a constant current density of 2.5 mA/cm$^2$. The highest yield of benzilic acids was 86%. In 1986, Silvestri et al. [45] investigated the electrochemical carboxylation of aldehydes and ketones to the corresponding acids by using diaphragmless cells and sacrificial Al anodes. In 1995, Chan et al. [46] reported an effective synthesis of 2-aryllactic acids via the electrocarboxylation of methyl aryl ketones using an undivided cell with a sacrificial aluminum anode. The yield of 2-aryllactic acids was over 95% based on converted starting material. In 1998, Datta et al. [47] reported the electrochemical carboxylation of 2-acetyl-6-methoxynaphthalene to 2-hydroxy-2-(6-methoxy-2-naphthyl) propionic acid in an undivided flow

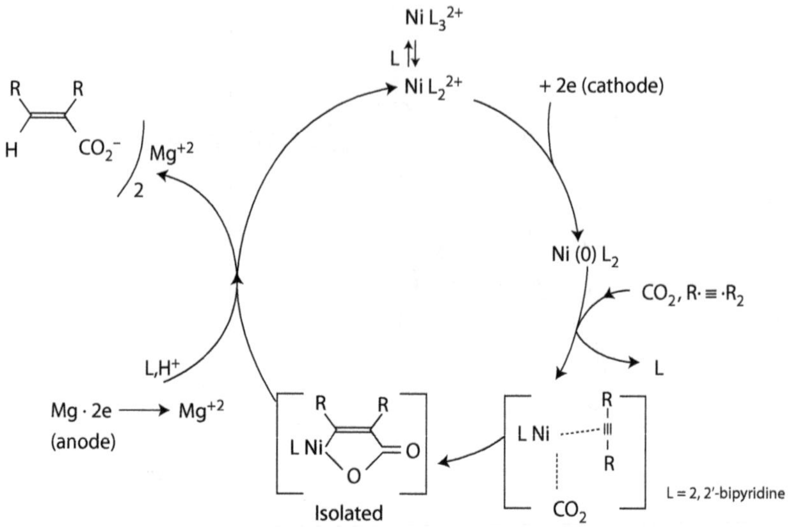

**FIGURE 2.12** Scheme of electroreductive coupling of alkynes and carbon dioxide with nickel–bipyridine complexes and magnesium ions. (Reprinted with permission from Dérien, S., Dunach, E. and Périchon, J. From stoichiometry to catalysis: Electroreductive coupling of alkynes and carbon dioxide with nickel-bipyridine complexes. Magnesium ion as the key for catalysis. *Journal of American Chemical Society*; 113(12), 8447–8454. Copyright 1991 American Chemical Society.)

cell using lead as the cathode and aluminum as the dissolving anode. Using DMF as the solvent and tetraethylammonium chloride hydrate as the electrolyte, the yield of 2-hydroxy-2-(6-methoxy-2-naphthyl) propionic acid was 89% in the electrolysis and 75% as isolated dried product. In 2002, Isse et al. [48] investigated the electrochemical reduction of a series of halogenated benzophenones in an undivided cell using a compact graphite cathode and aluminum sacrificial anode; they also investigated the mechanism of electroreduction in DMF, and an ideal yield was obtained (Figure 2.13).

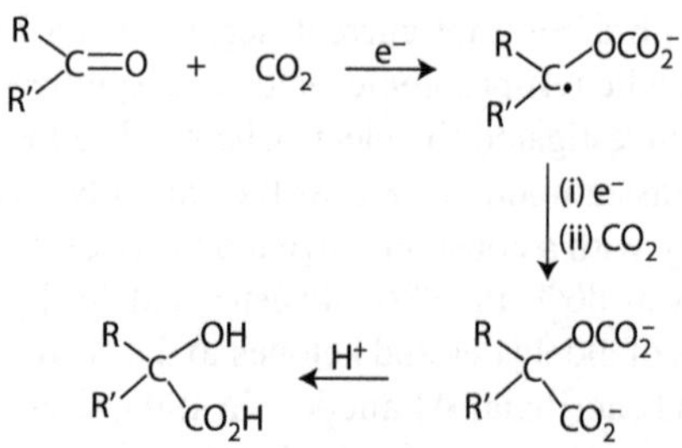

**FIGURE 2.13** Mechanism of electroreduction of benzophenones in DMF. (Reprinted from *Journal of Electroanalytical Chemistry*, 526(1–2), Isse, A. A. et al., Electrochemical reduction and carboxylation of halobenzophenones, 41–52, Copyright 2002, with permission from Elsevier.)

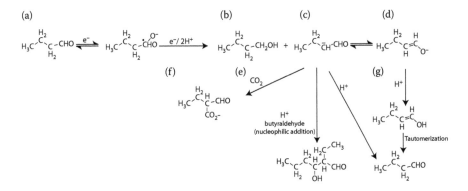

**FIGURE 2.14** Reaction scheme accounting for the presence of enolate and aldol products upon butyraldehyde reduction and suppression in the presence of $CO_2$. (Reprinted from *Electrochimica Acta*, 47(18), Doherty, A. P., Electrochemical reduction of butyraldehyde in the presence of $CO_2$, 2963–2967, Copyright 2002, with permission from Elsevier.)

In 2002, Doherty [49] investigated the electrochemical reduction of butyraldehyde in acetonitrile, and also studied the reaction mechanism. The major product was butanol, enolate anion equilibrium species, and the aldol. (Figure 2.14).

In 2005, Scialdone et al. [50] investigated the electrocarboxylation of halogenated acetophenones and benzophenones to the corresponding hydrocarboxylic acids in an undivided cell equipped with Al sacrificial anode and using methyl-2-pyrrolidinone (NMP) or DMF with 0.1 M Bu₄NBr. They also investigated the mechanism of the reaction that leads to the formation of the corresponding hydroxycarboxylic acid, alcohol, pinacol, and the dehalogenated parent ketone. In 2006, Scialdone et al. [51] investigated the theoretical dependence of the electrocarboxylation of aromatic ketones with $CO_2$ from operative parameters (Figure 2.15).

In 2007, Scialdone et al. [52] investigated the effect of operational parameters such as the water content and the ratio between the carbon dioxide and the ketone concentration on the competition between the formation of the target 2-hydroxy-2-arylpropanoic acid and ring carboxylation in the electrocarboxylation of aromatic ketones.

### 2.2.2.3 Electrocarboxylation of Halohydrocarbons with CO₂

In 1974, Wagenknecht [53] investigated the electrochemical reduction reaction of alkyl halides with $CO_2$ in DMF at Hg and graphite electrodes; the main products formed at graphite electrode were butane, octane, butyl valerate, butyl, *N,N*-dimethoxamate, and di-butyl-2-methylmalonate. The main products at the Hg electrode were Bu₂Hg, dibutyl oxalate, and butyl valerate. In 1985, Sock et al. [54] investigated the electrocarboxylation of a large variety of organic halides in simple and mild condition in diaphragmless cells. In 1988, Fauvarque et al. [55] reported that aryl-2 propionic acids such as fenoprofen and ibuprofen were obtained by the electrocarboxylation of ArCH(CH₃)Cl with $CO_2$. Using tetramethylurea as the solvent and nickel catalysts, the conversion rates were close to 80% and chemical yields up to 85%. On this basis, Fauvarque et al. [56] reported the electrosynthesis of ibuprofen and fenoprofen in a special type of electrolyzer and an undivided cell in the form of constant voltage. In the presence of lithium

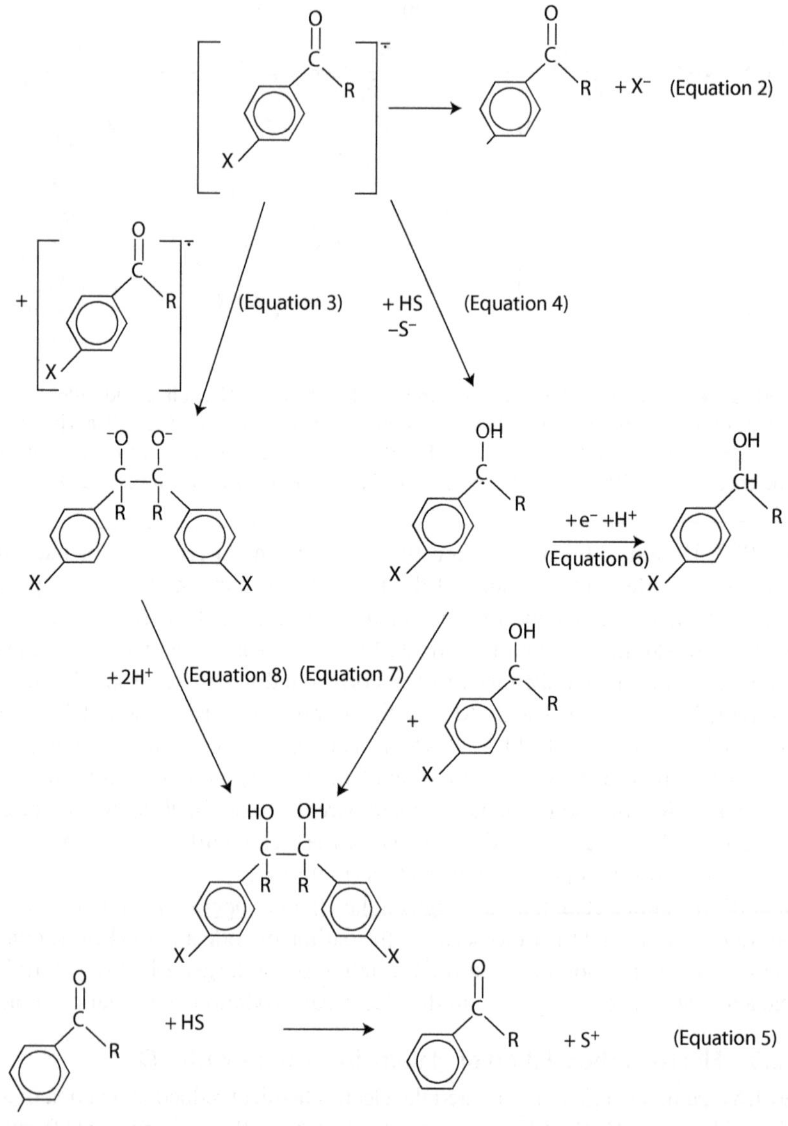

**FIGURE 2.15** Mechanism of electrochemical reduction of halogenated acetophenones and benzophenones. (Reprinted from *Electrochimica Acta*, 50, Scialdone, O. et al., Influence of the nature of the substrate and of operative parameters in the electrocarboxylation of halogenated acetophenones and benzophenones, 3231–3242, Copyright 2005, with permission from Elsevier.)

oxalate or metal powder in suspension with a gas bubbling in the solution, Zn powder added to the solution can be easily oxidized. This was a good alternative to the use of sacrificial anodes. In 1991, Amatore et al. [57] reported the electrocarboxylation of bromobenzene in tetrahydrofuran (THF). They also studied the mechanisms of nickel-catalyzed electrocarboxylation of aromatic halides (Figure 2.16).

(a)    Scheme I (X = Br)

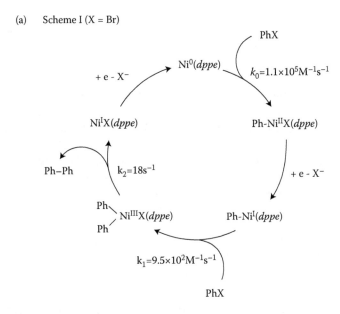

(b)    Scheme II

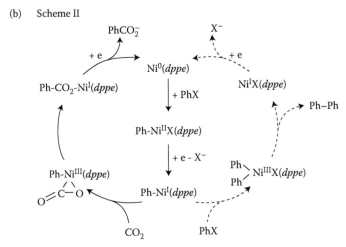

**FIGURE 2.16** Mechanism of nickel-catalyzed electrocarboxylation of aromatic halides. (a) In the absence of CO$_2$ and (b) CO$_2^-$ saturated. (Reprinted with permission from Amatore, C. and Jutand, A. Activation of carbon dioxide by electron transfer and transition metals. Mechanism of nickel-catalyzed electrocarboxylation of aromatic halides. *Journal of the American Chemical Society*, 113, 2819–2826. Copyright 1991 American Chemical Society.)

In 2001, Damodar et al. [58] reported that the electrochemical carboxylation of benzylchlorides with CO$_2$ catalyzed by Pd(II) complexes afforded 2-arylpropionic acids in good yields at a constant current of 10 mA/cm$^2$. In 2002, Isse et al. [59] reported the electrochemical carboxylation of chloroacetonitrile in DMF and MeCN in the form of controlled-potential electrolysis; the yield of the desired products was

73% and 93% in DMF and MeCN, respectively. In 2005, Isse et al. [60] investigated the electrocarboxylation of some arylethyl chlorides at Ag and glassy carbon (GC) electrodes in $CH_3CN$ and DMF. The acid yields were 61%–73% and 70%–81% at 298.15 and 273.15 K, respectively.

#### 2.2.2.4    Electrocarboxylation of Other Organisms with $CO_2$

In 2001, Tascedda et al. [61] reported the electrochemical incorporation of carbon dioxide into expoxides catalyzed by nickel(II) complexes under mild conditions. The best yield of cyclic carbonates was over 90% (Figure 2.17). In 2000, Casadei et al. [62] investigated the electrocarboxylation of alcohols with $CO_2$ to produce organic carbonates; they found that the reaction between primary alcohols and $CO_2$ could generate organic carbonates in good amount, secondary alcohols were converted in moderate amount, and tertiary alcohols and phenols were unreactive. 1,2-Diols gave a mixture of both cyclic and linear di- and monocarbonates. In 2009, Yuan et al. [63] reported the electrochemical synthesis of dimethyl carbonate from methanol and $CO_2$ with Pt electrodes in a dialkylimidazolium ionic liquid–basic compounds–methanol system under ambient conditions (Figure 2.18).

### 2.3    ELECTRODES AND ELECTROLYTES AND SOLUTIONS FOR $CO_2$ REDUCTION (ELECTRODE POTENTIAL WINDOWS)

#### 2.3.1    Electrodes

#### 2.3.1.1    Metal Electrodes

In 1990, Azuma et al. [64] investigated the electrochemical reduction of $CO_2$ on 32 metal electrodes in an aqueous $KHCO_3$ medium. They summarized a periodic table for $CO_2$ reduction (Figure 2.19).

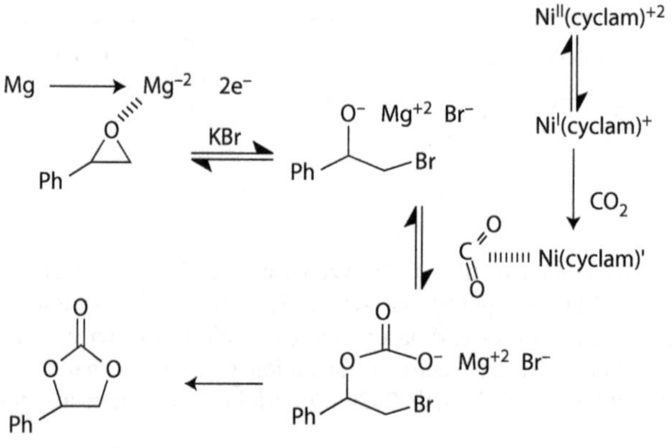

**FIGURE 2.17**    Mechanism of reduced nickel species in the activation of $CO_2$ and the influence of Mg ions as Lewis acids in the activation of the oxirane ring. (Tascedda, P. et al.: Nickel-catalyzed electrochemical carboxylation of expoxides: Mechanistic aspects. *Applied Organometallic Chemistry*. 2001. 15. 141–144. Copyright Wiley-VCH Verlag GmbH & Co. KGaA. Reproduced with permission.)

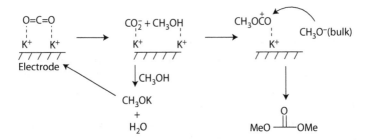

**FIGURE 2.18** Proposed reaction mechanism of electrochemical activation of CO₂ for the synthesis of dimethyl carbonate in an ionic liquid. (Reprinted from *Electrochemical Acta*, 54(10), Yuan, D. D. et al., Electrochemical activation of carbon dioxide for synthesis of dimethyl carbonate in an ionic liquid, 2912–2915, Copyright 2009, with permission from Elsevier.)

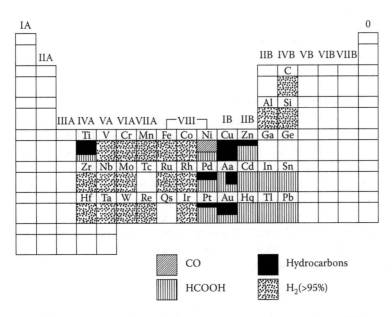

**FIGURE 2.19** Periodic table for CO₂ reduction products at −2.2 V vs. SCE in a low-temperature 0.05 mol/dm³ solution. (Reproduced from Azuma, M. et al. *Journal of the Electrochemical Society*. 1990; 137(6): 1772–1778. With permission from ECS.)

According to this, most of the light transition metals are not so effective for CO₂ reduction except Ti. Light metals in the IIIB or IVB group are not effective. Heavy metals in the IIB, IIIB, and IVB groups, such as Cd, In, Sn, Pb, Tl, Hg, Zn, and Pb, reduced CO₂ into HCOOH effectively. Some of VIIIB and IB metals, such as Ni, Ag, and Au, are effective for CO production. Cu is an important metal that has very high electrocatalytic activity for CO₂ electroreduction into hydrocarbons, and is located between metals effective for CO and HCOOH production.

In their article, they also investigated the current efficiencies for $CO_2$ reduction products on different metal electrodes. These experiments were carried out at $-2.2$ V vs. SCE (193°C) in a $CO_2$-saturated 0.05 M $KHCO_3$ aqueous solution at about 0°C (Table 2.1). Under the same operating conditions, Azuma et al. also investigated the current efficiencies (%) for hydrogen evolution ($\eta[H_2]$) and total $CO_2$ reduction and the distributions (%) into different products (Table 2.2). This chapter indicated the main metal electrodes used for $CO_2$ electroreduction.

Table 2.3 shows the distribution of products of the electroreduction of $CO_2$ on metal electrodes under high pressure (30 atm). This is again the proof that pressure is an important influencing factor in an electrochemical reaction. The change of distribution of products on metal electrodes for $CO_2$ electroreduction is obviously.

### 2.3.1.2   Gas Diffusion Electrodes

In 1987, Mahood et al. [65] investigated lead-impregnated polytetrafluoroethylene-boned carbon gas diffusion electrodes at 115 mA/cm$^2$ in an aqueous acidic electrolyte (pH 2), and the current efficiency of the selective product formic acid was nearly 100%. In 1997, Furuya et al. [66] reported the electrochemical reduction reaction of $CO_2$ at gas diffusion electrodes made of Ru, Pd, and their alloy (Ru:Pd = 1:1), and the current efficiency for the formation of the desired product formic acid was 90% at $-1.1$ V vs. normal standard electrode (NHE).

### 2.3.1.3   Semiconductor Electrodes

In 1984, Frese et al. [67] investigated the electrochemical reduction of $CO_2$ to $CH_3OH$ on n-GaAs (111 As) single-crystal electrodes at 140 µA/cm$^2$ at $-1.2$ to $-1.4$ vs. SCE, and the yield of $CH_3OH$ was 100% for current densities. In 2006, Kaneco et al. [68] investigated the photo-electrochemical reduction of $CO_2$ at metal-modified p-InP photoelectrodes in LiOH/methanol-based electrolytes. The main products at lead, silver, gold, and copper-modified p-InP photocathode were CO and formic acid. The product at palladium-deposited p-InP was only CO. Hydrocarbons were obtained at nickel-deposited p-InP electrode. But semiconductor electrodes need high electrode potential and they exhibit poor corrosion resistance; therefore, these two flaws limit the application of semiconductor electrodes.

### 2.3.1.4   Modified Electrodes

In 1998, Ogura et al. [69] investigated the electrocatalytic reduction of $CO_2$ on a metal complex-immobilized polyaniline (PAn)/Prussian blue-laminated electrode. The main products were lactic, acetic, and formic acids and methanol and ethanol. In 2003, Hori et al. [70] investigated the electrochemical reduction of $CO_2$ at silver-coated ion exchange membrane electrodes with 0.2 M $K_2SO_4$ as the electrolyte with a Pt plate as the counter electrode.

### 2.3.2   ELECTROLYTES AND SOLUTIONS

Sometimes the medium and electrolytes determine the types of products, so the choice of electrolytes and solutions is very important, and the solubility of $CO_2$ in various solvents at 298 K, 1 atm is shown in Table 2.4.

## TABLE 2.1
## Current Efficiencies (%) for CO$_2$ Reduction Products

| Metal | T (°C) | CH$_4$ | CO | C$_2$H$_4$ | C$_2$H$_6$ | HCOOH | H$_2$ | Sum |
|---|---|---|---|---|---|---|---|---|
| Cd[a] | 0 | 0.015 | 3.7 | 0.002 | 0.00056 | 55.9 | 35.7 | 95 |
|  | 20 | 0.0073 | 1.8 | 0.001 | 0.00040 | 35.5 | 63.2 | 100 |
| In[a] | 0 | 0.001 | 3.0 | 0.00035 | 0.0006 | 70.0 | 25.0 | 98 |
|  | 20 | 0.050 | 14.7 | 0.0046 | 0.0067 | 33.3 | 56.5 | 105 |
| Sn | 0 | 0.65 | 1.4 | 0.068 | 0.44 | 28.5 | 67.5 | 99 |
|  | 0 | 0.84 | 0.35 | 0.95 | 0.69 | 5.2 | 94.9 | 102 |
| Pb[a] | 0 | 0.39 | 0.12 | 0.008 | 0.0014 | 16.5 | 82.9 | 100 |
|  | 20 | 0.06 | 0.10 | 0.001 | 0.0003 | 9.9 | 93.3 | 103 |
| Tl | 0 | 0.20 | 0.16 | 0.003 | 0.0010 | 53.4 | 46.2 | 100 |
| Hg | 0 | 0.0004 | 0.20 | t | t | 90.2 | 9.5 | 100 |
|  | 20 | 0.0035 | 0.64 | 0.0002 | 0.00006 | 87.6 | 7.9 | 96 |
| Zn | 0 | 0.23 | 9.8 | t | nm | 19.5 | 68.1 | 98 |
| Pd[a] | 0 | 0.083 | 11.6 | 0.011 | 0.014 | 16.1 | 73.3 | 101 |
|  | 20 | 0.31 | 3.2 | 0.061 | 0.078 | 8.6 | 90.3 | 103 |
| Ti | 0 | t | 13.5 | t | nm | 5.2 | 69.4 | 83 |
| Ni | 0 | 0.71 | 21.0 | 0.069 | 0.18 | 13.7 | 61.7 | 97 |
|  | 20 | 0.13 | 0.60 | 0.010 | 0.021 | 0.10 | 98.8 | 100 |
| Ag | 0 | 1.4 | 40.7 | 0.0052 | 0.013 | 20.5 | 32.6 | 95 |
|  | 20 | 1.1 | 30.0 | 0.0090 | 0.0027 | 16.0 | 50.0 | 98 |
| Au | 0 | t | 16.9 | t | nm | 10.3 | 73.4 | 101 |
| Cu | 0 | 24.7 | 16.5 | 6.5 | 0.015 | 3.0 | 49.3 | 100 |
|  | 20 | 17.8 | 5.4 | 12.7 | 0.039 | 10.2 | 52.0 | 98 |
| C | 0 | 0.11 | t | 0.0064 | 0.0070 | 0.31 | 92.5 | 93 |
| Al | 0 | 0.012 | t | 0.00022 | 0.00040 | 0.78 | 95.7 | 96 |
| Si | 0 | 0.025 | 0.08 | t | t | 1.6 | 102.2 | 104 |
| V | 0 | 0.02 | 1.3 | t | nm | 2.6 | 91.9 | 96 |
| Cr | 0 | 0.74 | 0.49 | 0.050 | 0.18 | 0.15 | 92.2 | 94 |
| Mn | 0 | 1.5 | 0.34 | 0.093 | 0.29 | 0.03 | 90.9 | 93 |
| Fe | 0 | 0.07 | 2.2 | t | nm | 1.1 | 89.8 | 93 |
| Co | 0 | 0.13 | 0.47 | 0.0057 | 0.032 | 0.85 | 92.9 | 94 |
| Zr | 0 | 0.49 | 0.42 | 0.021 | 0.055 | t | 99.9 | 101 |
| Nb | 0 | 0.16 | 0.46 | 0.0088 | 0.042 | 0.03 | 97.3 | 98 |
| Mo[a] | 0 | 0.010 | t | 0.0028 | 0.0015 | 0.21 | 99.9 | 100 |
|  | 20 | 0.031 | 0.02 | 0.00077 | 0.0057 | 0.19 | 98.6 | 99 |
| Ru | 0 | 0.043 | 0.65 | t | t | 0.08 | 99.1 | 100 |
| Rh | 0 | 0.031 | 2.5 | 0.00067 | 0.0036 | 1.35 | 99.3 | 103 |
|  | 20 | 0.053 | 0.66 | 0.0030 | 0.011 | 2.4 | 99.3 | 103 |
| Hf | 0 | 0.0046 | 1.14 | 0.00027 | 0.0010 | 0.35 | 99.2 | 101 |
|  | 20 | 0.0073 | 0.08 | 0.00057 | 0.0005 | 0.21 | 100.9 | 101 |
| Ta[b] | 0 | 0.0015 | 0.09 | 0.0015 | 0.0002 | t | 100.7 | 101 |
|  | 20 | 0.0039 | t | 0.0039 | 0.0001 | t | 102.2 | 102 |
| W[a] | 0 | 0.015 | 0.06 | 0.0043 | 0.0056 | 1.3 | 96.3 | 98 |
|  | 20 | 0.055 | 0.21 | 0.0022 | 0.010 | 2.6 | 96.9 | 100 |
| Re[a] | 0 | 0.044 | t | 0.00022 | 0.0056 | 2.0 | 99.0 | 101 |
|  | 20 | 0.038 | t | 0.00024 | 0.0048 | 1.4 | 95.3 | 97 |
| Ir | 0 | 0.051 | 0.53 | 0.0035 | 0.0072 | 1.0 | 98.8 | 100 |
|  | 20 | 0.086 | t | 0.0057 | 0.015 | 0.58 | 100.3 | 101 |
| Pt | 0 | 0.29 | 1.2 | t | nm | 5.5 | 92.6 | 100 |

*Source:* Azuma, M. et al. *Journal of the Electrochemical Society.* 1990; 137(6): 1772–1778. With permission from ECS.

*Note:* t, trace; nm, not measured.

[a] At −2.0 V vs. SCE.

[b] At −2.8 V vs. SCE.

## TABLE 2.2

**Current Efficiencies (%) for Hydrogen Evolution ($\eta[H_2]$) and Total $CO_2$ Reduction ($\eta[CO_2]$) on Various Metal Electrodes and the Distributions (%) into CO ($F_{CO}$), HCOOH($F_{COOH}$), and Hydrocarbons ($F_{CxHy}$)**

| Electrodes | $\eta(H_2)$ | $\eta(red\text{-}CO_2)$ | $F_{CO}$ | $F_{HCOOH}$ | $F_{CrHy}$ | Note (V) |
|---|---|---|---|---|---|---|
| Cd | 35.7 | 59.6 | 6.2 | 93 8 | t | −2.0 |
| In | 25.0 | 73.0 | 4.1 | 95.3 | t | −2.0 |
| Sn | 67.5 | 31.1 | 4.5 | 91.8 | 3.7 | |
| Pb | 82.9 | 17.0 | 0.7 | 96.9 | 2.4 | −2.0 |
| Tl | 46.2 | 53.8 | t | 99.8 | t | |
| Hg | 9.5 | 90.4 | 0.2 | 99.8 | t | |
| Zn | 68.1 | 29.5 | 33.2 | 66.0 | 0.8 | |
| Pd | 73.3 | 27.8 | 41.7 | 57.9 | 0.4 | −2.0 |
| Ti | 69.4 | 18.7 | 72.2 | 27.8 | t | |
| Ni | 61.7 | 35.7 | 58.9 | 38.4 | 2.7 | |
| Ag | 32.6 | 62.6 | 65.0 | 32.7 | 2.2 | |
| An | 73.4 | 27.2 | 62.1 | 37.9 | t | |
| Cu | 49.3 | 50.7 | 32.5 | 5.9 | 61.5 | |
| C | 92.5 | 0.43 | t | 71.5 | 28.5 | |
| Al | 95.7 | 0.79 | t | 98.4 | 1.6 | |
| Si | 102.2 | 1.7 | 4.77 | 93.8 | 1.5 | |
| V | 91.9 | 3.9 | 33.2 | 66.3 | 0.5 | |
| Cr | 92.2 | 1.6 | 30.4 | 9.3 | 60.2 | |
| Mn | 90.9 | 2.3 | 15.1 | 1.3 | 83.6 | |
| Fe | 89.8 | 3.4 | 65.3 | 32.6 | 2.1 | |
| Co | 92.9 | 1.5 | 31.6 | 57.1 | 11.3 | |
| Zr | 99.9 | 0.99 | 42.6 | t | 57.4 | |
| Nb | 97.3 | 0.70 | 65.6 | 4.3 | 30.1 | |
| Mo | 99.9 | 0.22 | t | 94.7 | 5.3 | −2.0 |
| Ru | 99.1 | 0.77 | 84.1 | 10.4 | 56 | |
| Rh | 99.3 | 3.9 | 84.3 | 34.7 | 0.9 | |
| Hf | 99.2 | 1.5 | 76.2 | 23.4 | 0.4 | |
| Ta | 100.7 | 0.30 | 29.7 | t | 70.3 | −2.8 |
| W | 96.3 | 1.4 | 4.3 | 94.1 | 1.6 | −2.0 |
| Re | 99.0 | 2.0 | t | 97.6 | 2.4 | −2.0 |
| Ir | 98.8 | 1.6 | 33.3 | 62.8 | 3.9 | |
| Pt | 92.6 | 7.9 | 17.2 | 78.7 | 4.1 | |

*Source:* Azuma, M. et al. *Journal of the Electrochemical Society.* 1990; 137(6): 1772–1778. With permission from ECS.

*Note:* t, trace.

**TABLE 2.3**

**Electrochemical Reduction of CO$_2$ under a Pressure of 30 atm on Various Electrodes at 163 mA/cm$^2$**

| Group | Electrode | E (V) | Faradaic Efficiency (%) CH$_4$ | C$_2$H$_6$ | C$_2$H$_4$ | CO | HCOOH | H$_2$ | CO$_2$ red. | Total | PCD(CO$_2$ red.) (mA cm$^{-2}$) |
|---|---|---|---|---|---|---|---|---|---|---|---|
| 4 | Ti | -1.57 | 0.18 | 0.01 | 0.08 | Trace | 4.6 | 80.8 | 4.9 | 85.7 | 8.0 |
|   | Zr | -1.73 | 0.13 | 0.01 | 0.01 | 32.5 | 7.6 | 44.2 | 40.3 | 84.5 | 65.7 |
| 5 | Nb | -1.45 | 0.56 | 0.05 | 0.01 | n | 3.5 | 81.4 | 4.1 | 85.5 | 6.7 |
|   | Ta | -1.51 | 0.55 | 0.05 | Trace | Trace | 7.6 | 74.4 | 8.2 | 82.6 | 13.4 |
| 6 | Cr | -1.49 | 0.53 | 0.05 | 0.07 | 11.8 | 8.2 | 68.6 | 20.7 | 89.3 | 33.7 |
|   | Mo | -1.34 | 0.40 | 0.05 | 0.03 | n | 6.5 | 83.3 | 7.0 | 90.3 | 11.4 |
|   | W | -1.61 | 0.38 | 0.04 | 0.01 | Trace | 31.9 | 53.1 | 32.3 | 85.4 | 52.6 |
| 7 | Mn | -1.69 | 0.68 | 0.10 | 0.06 | 2.8 | 2.8 | 78.8 | 6.5 | 85.3 | 10.6 |
| 8 | Fe | -1.63 | 2.03 | 0.40 | 0.16 | 4.2 | 28.6 | 51.6 | 35.4 | 87.0 | 57.7 |
| 9 | Co | -1.54 | 3.09 | 0.17 | 0.38 | 15.8 | 21.9 | 46.9 | 41.5 | 88.4 | 67.6 |
|   | Rh | -1.41 | 0.26 | 0.03 | 0.01 | 61.0 | 19.5 | 13.1 | 80.8 | 93.9 | 131.7 |
|   | Ir | -1.55 | 0.62 | 0.05 | 0.05 | 17.5 | 22.3 | 48.3 | 40.5 | 88.8 | 66.0 |
| 10 | Ni | -1.59 | 0.72 | 0.08 | 0.11 | 33.5 | 31.3 | 26.0 | 65.7 | 91.7 | 107.1 |
|   | Pd | -1.56 | 0.13 | 0.01 | Trace | 46.1 | 35.6 | 12.8 | 81.8 | 94.6 | 133.3 |
|   | Pd | -1.76 | 0.21 | 0.01 | 0.02 | 35.2 | 44.0 | 13.8 | 79.4 | 93.2 | 397.0 |
|   | Pt | -1.48 | 0.22 | 0.02 | Trace | 6.1 | 50.4 | 33.6 | 56.7 | 90.3 | 92.4 |
| 11 | Cu | -1.64 | 9.95 | 0.06 | 3.74 | 20.1 | 53.7 | 2.5 | 87.6 | 90.1 | 142.8 |
|   | Ag | -1.48 | 0.20 | 0.01 | Tract | 75.6 | 16.8 | 3.9 | 92.6 | 96.5 | 150.9 |
|   | Au | -1.30 | 0.21 | 0.02 | 0.11 | 64.7 | 11.8 | 15.4 | 76.8 | 92.2 | 125.2 |

(Continued)

**TABLE 2.3 (Continued)**

**Electrochemical Reduction of $CO_2$ under a Pressure of 30 atm on Various Electrodes at 163 mA/cm$^2$**

| Group | Electrode | $E$ (V) | Faradaic Efficiency (%) | | | | | | | | PCD($CO_2$ red.) (mA cm$^{-2}$) |
|---|---|---|---|---|---|---|---|---|---|---|---|
| | | | $CH_4$ | $C_2H_6$ | $C_2H_4$ | CO | HCOOH | $H_2$ | $CO_2$ red. | Total | |
| 12 | Zn | −1.70 | 0.31 | 0.03 | Trace | 48.7 | 40.5 | 2.8 | 89.5 | 92.3 | 145.9 |
| 13 | Al | −1.97 | 0.66 | 0.01 | n | n | 1.3 | 86.5 | 2.0 | 88.5 | 3.3 |
| | In | – | 0.28 | Trace | 0.04 | 3.8 | 90.1 | 5.6 | 90.5 | 99.1 | 147.5 |
| 14 | C | −1.68 | 0.45 | 0.03 | 0.04 | 44.0 | 30.2 | 15.6 | 74.7 | 90.3 | 37.4 |
| | C | −2.14 | 0.66 | 0.02 | 0.05 | 3.6 | 6.8 | 75.5 | 11.2 | 86.7 | 18.3 |
| | n-Si | −2.04 | 0.87 | 0.01 | 0.02 | 2.0 | 46.3 | 40.6 | 49.2 | 89.8 | 80.2 |
| | Sn | −1.39 | 0.06 | Trace | Trace | 8.0 | 92.3 | 1.3 | 100.4 | 101.7 | 163.0 |
| | Pb | −1.57 | 0.20 | 0.01 | Trace | Trace | 95.5 | 1.2 | 95.7 | 96.9 | 156.0 |
| 15 | Bi | −1.42 | 0.17 | 0.01 | Trace | 3.3 | 82.7 | 6.3 | 86.2 | 92.5 | 140.5 |

*Source:* Reprinted from *Journal of Electroanalytical Chemistry*, 391(1–2), Hara, K., Kudo, A. and Sakata, T., Electrochemical reduction of carbon dioxide under high pressure on various electrode in an aqueous electrolyte, 141–147, Copyright 1995, with permission from Elsevier.

**TABLE 2.4**
**Solubility of $CO_2$ in Various Solvents**

| Solvent | $CO_2$ Concentration (M) |
|---|---|
| Water | 0.033 |
| Methanol | 0.06 |
| Dimethylsulfoxide (DMSO) | 0.135 |
| Propylene carbonate (PPC) | 0.14 |
| $N,N$-Dimethylformamide (DMF) | 0.175 |
| Tetrahydrofuran (THF) | 0.211 |
| Acetonitrile (AN) | 0.28 |

*Source:* Lide, D.R. et al., *CRC Handbook of Chemistry and Physics*, 76 edn, 5–28. With permission from CRC Press.

### 2.3.2.1 Electroreduction in Aqueous Solution

In 1995, Yoshitake et al. [71] investigated the electrochemical reduction of $CO_2$ in 0.1 M $KHCO_3 + H_2O$ on a Pd membrane surface, using isotope distributions to study the mechanism of the electrochemical reduction of $CO_2$. In the same year, Yoshitake et al. [72] investigated the electrochemical reduction of $CO_2$ in 0.05 M $Li_2CO_3$, 0.1 M $NaHCO_3$, and 0.1 M $KHCO_3$ aqueous solution, and discussed the selectivity of the products of the electrochemical reduction of $CO_2$. In 1997, Jitaru et al. [73] grouped the flat metallic cathodes according to the nature of the cathode (sp or d group metal electrodes). They studied sp group metals in aqueous and nonaqueous electrolytes and d group metals in aqueous and nonaqueous electrolytes. As shown in Figure 2.20, they were divided into 11 groups, (A) Cu, Zn, Sn; (B) In, C, Si, Sn, Pb, Bi, Cu, Zn, Cd, Hg; (C) In, Sn, Pb, Cu, Au, Zn, Cd; (D) In, Sn, Au, Hg; (E) In, Tl, Sn, Pd, Zn, Hg; (F) Mo, W, Ru, Os, Pd, Pt; (G) Ni, Pd, Rh, Ir; (H) Fe, Ru, Ni, Pd, Pt; (I) Zr, Cr, Mn, Fe, Co, Rh, Ir; (J) Ni, Pt; and (K) Ti, Nb, Cr, Mo, Fe, Pd. The products were formic acid (formic), hydrocarbons (h. c.): methane, ethane, ethylene, propane, *n*- and *i*-butane; oxalic acid (oxalic); glyoxylic acid (glyox); methanol (MeOH) along with superior alcohols (ethanol, *i*-propanol). In 2000, Hoshi et al. [74] used the method of voltammetry to investigate the electrochemical reduction of $CO_2$ to adsorbed CO at a Pt single-crystal electrode (Figure 2.21). The order of the activity series was obtained for the stepped surface: Pt(111)Pt(100)Pt(S)-[*n*(111) × (100)] Pt(S)-[*n*(111) × (111)]Pt(110). The voltammograms of the low index planes of Pt in 0.1 M $HClO_4$ is shown in Figure 2.21. The voltammograms of Pt electrode in 0.1 M $HClO_4$ which is saturated with Ar, please refer to the original documents.

In 2001, Tryk et al. [6] investigated the electrochemical reduction of $CO_2$ on high-area transition metal catalysts supported on nanoporous activated carbon fiber in the form of gas diffusion electrodes in aqueous solution. These experiments were carried out under high pressure $CO_2$ (40 atm) in methanol (0.1 M) TBAP at 25°C. As we can see in Figure 2.22, curve a represents the ohmic drop-corrected curve under illumination, curve b represents dark current on the same electrode, curve c

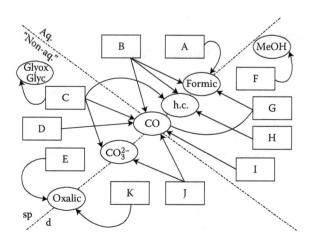

**FIGURE 2.20** Classification of the procedure of $CO_2$ electroreduction based upon the nature of cathode material and the solvent used for the supporting electrolyte (aqueous or nonaqueous solution). (With kind permission from Springer Science + Business Media: *Journal of Applied Electrochemistry*, Electrochemical reduction of carbon dioxide on flat metallic cathodes, 27, 1997, 875–889, Jitaru, M., Lowy, D. A., Toma, M., Toma, B. C. and Oniciu, L.)

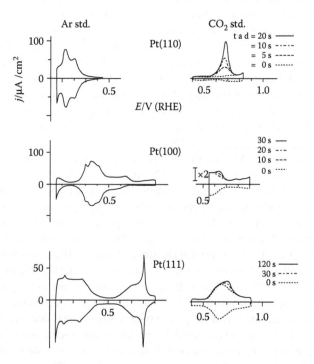

**FIGURE 2.21** Voltammograms of the low index planes of Pt in 0.1 M $HClO_4$. (Reprinted from *Electrochimica Acta*, 45, Hoshi, N. and Hori, Y., Electrochemical reduction of carbon dioxide at a series of platinum single crystal electrodes, 4263–4270, Copyright 2000, with permission from Elsevier.)

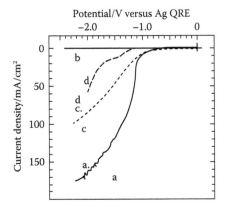

**FIGURE 2.22** Current–potential curves for CO$_2$ reduction on p-InP. (Tryk, D. A. et al.: Recent developments in electrochemical and photoelectrochemical CO$_2$ reduction: Involvement of the (CO$_2$)$_2^{\cdot-}$ dimer radical anion. *Applied Organometallic Chemistry*. 2001. 15. 113–120. Copyright Wiley-VCH Verlag GmbH & Co. KGaA. Reproduced with permission.)

represents noncorrected curve, and curve d represents the curve obtained for a solid copper electrode in the dark.

In 2001, De Jesús-Cardona et al. [12] investigated the electrochemical reduction of CO$_2$ over Cu surface in KHCO$_3$ solution at different temperature, KHCO$_3$ concentrations, and CO$_2$ pressure (Figures 2.23 and 2.24).

In 2001, Lee et al. [75] investigated the electrochemical reduction of CO$_2$ to CH$_4$ and C$_2$H$_4$ in 1 h at constant cathodic potential in CO$_2$-saturated 0.1 M KHCO$_3$ at 5°C, and studied the reduction mechanism (Figure 2.25).

In 2001, Ohmori et al. [76] investigated the electrochemical reduction of CO$_2$ in KCl and KHCO$_3$ aqueous solution using an Au electrode. In 2002, Stevens et al.

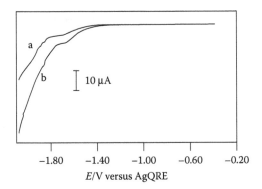

**FIGURE 2.23** Voltammetry of CO$_2$ in a KHCO$_3$ solution saturated with CO$_2$ at 20°C (scan rate 20 mV/s, (a) 0.01 and (b) 0.5 M). (Reprinted from *Journal of Electroanalytical Chemistry*, 513(1), De Jesús-Cardona, H., Del Moral, C. and Cabrera, C. R., Voltammetric study of CO$_2$ reduction at Cu electrode under different KHCO$_3$ concentration, temperature and CO$_2$ pressure, 45–51, Copyright 2001, with permission from Elsevier.)

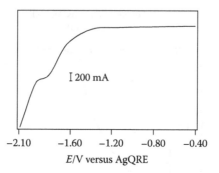

**FIGURE 2.24** Voltammogram at 250 psi of $CO_2$ in a solution of 0.5 M $KHCO_3$ at 20°C (scan rate 100 mV/s). (Reprinted from *Journal of Electroanalytical Chemistry*, 513(1), De Jesús-Cardona, H., Del Moral, C. and Cabrera, C. R., Voltammetric study of $CO_2$ reduction at Cu electrode under different $KHCO_3$ concentration, temperature and $CO_2$ pressure, 45–51, Copyright 2001, with permission from Elsevier.)

[77] investigated the electrochemical reduction of $CO_2$ to CO at an Au film electrode in aqueous $KHCO_3$ in Ar (dotted line) and $CO_2$ at 10 kPa (dashed line), and 50 kPa (solid line) at a scan rate of 100 mV/s (Figure 2.26).

In 2002, Takahashi et al. [78] investigated the electrochemical reduction of $CO_2$ using two single-crystal electrodes at a constant current density of 5 mA/cm² in 0.1 M $KHCO_3$ aqueous solution. These experiments were carried out in $CO_2$-saturated 0.1 M $K_2HPO_4$ + 0.1 M $KH_2PO_4$ (pH 6.8) at 0°C (Figures 2.27 and 2.28). According to their study, the products changed significantly with the crystal orientation. In 2013, Costentin et al. [79] investigated the electrochemical reduction of $CO_2$ at a platinum electrode in the presence of pyridinium ions. These experiments were

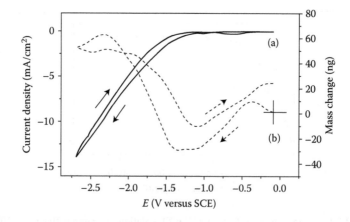

**FIGURE 2.25** $j/E$ (solid line) and $\Delta m/E$ (dotted line) of a CV with scan rate of 10 mV/s on Cu electrode. (Reprinted from *Electrochimica Acta*, 46, Lee, J. Y. and Tak, Y. S., Electrocatalytical activity of Cu electrode in electroreduction of $CO_2$, 3015–3022, Copyright 2001, with permission from Elsevier.)

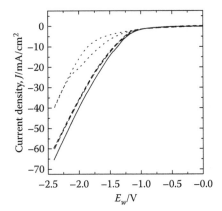

**FIGURE 2.26**   Cyclic voltammograms of a porous Au electrode. (Reprinted from *Journal of Electroanalytical Chemistry*, 526, Stevens, G. B., Reda, T. and Raguse, B., Energy storage by the electrochemical reduction of CO$_2$ to CO at a porous Au film, 125–133, Copyright 2002, with permission from Elsevier.)

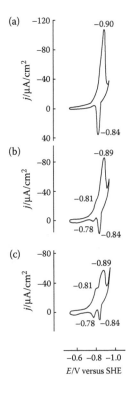

**FIGURE 2.27**   Cyclic voltammograms of a Cu(110) electrodes. (Reprinted from *Journal of Electroanalytical Chemistry*, 533, Takahashi, I. et al., Electrochemical reduction of CO$_2$ at a copper single crystal Cu(S)-[$n$(111) × (111)] and Cu(S)-[$n$(110) × (100)] electrodes, 135–143, Copyright 2002, with permission from Elsevier.)

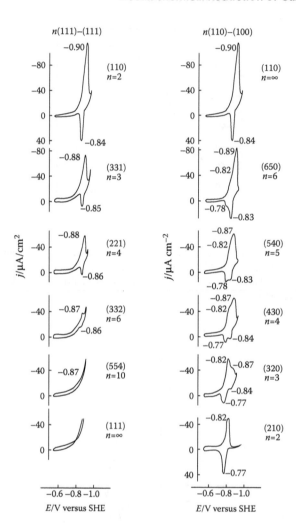

**FIGURE 2.28** Cyclic voltammograms of two series of copper single-crystal electrodes Cu(S)-[$n$(111) × (111)] and Cu(S)-[$n$(110) × (100)]. (Reprinted from *Journal of Electroanalytical Chemistry*, 533, Takahashi, I. et al., Electrochemical reduction of $CO_2$ at a copper single crystal Cu(S)-[$n$(111) × (111)] and Cu(S)-[$n$(110) × (100)] electrodes, 135–143, Copyright 2002, with permission from Elsevier.)

carried out in 3 mM pyridine (a, a′) and 3 mM acetic acid (b, b′) in the presence of 0.1 M $KNO_3$. T = 295 K, at pH = 5.15 (a, a′) and 4 (b, b′). Scan rate: 0.1 V/s (a, a′) and 0.2 V/s (b, b′). The main product was carbonic acid detected without methanol or formate (Figure 2.29). In 2013, Ertem et al. [80] studied the electrochemical reduction of $CO_2$ to formic acid, formaldehyde, and methanol at low overpotentials on a Pt electrode in acidic pyridine aqueous solution (Figure 2.30). In 2013, Yan et al. [81] studied the mechanism of aqueous pyridinium reduced at a Pt electrode, by which $CO_2$ can be reduced to methanol. In 2013, Jia et al. [82] investigated the

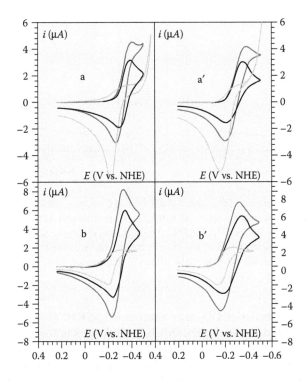

**FIGURE 2.29** CV of a CO$_2$-saturated solution on a platinum electrode. Black: acid alone; light gray: CO$_2$ alone (pH = 4.5 in a, a′ and 4 in b, b′); and dark gray: acid +CO$_2$. a, b: experimental and a′, b′: simulations. (Reprinted with permission from Costentin, C. et al. Electrochemical of acids on platinum. Application to the reduction of carbon dioxide in the presence of pyridinium ion in water. *Journal of the American Chemical Society*; 135, 17671–17674. Copyright 2013 American Chemical Society.)

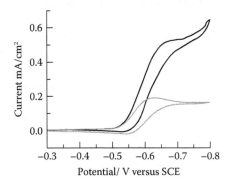

**FIGURE 2.30** Cyclic voltammograms of PyrH$^+$ reduction in the presence of Ar (gray) and CO$_2$ (black) at pH = 5.3, obtained with a Pt disk electrode. (Reprinted with permission from Ertem, M. Z. et al. Functional role of pyridinium during aqueous electrochemical reduction of CO$_2$ on Pt(111). *Journal of Physical Chemistry Letters*; 4, 745–748. Copyright 2013 American Chemical Society.)

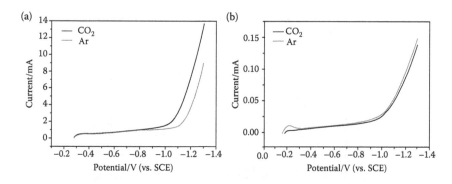

**FIGURE 2.31**  LSV curves of $Cu_{63.9}Au_{36.1}$/NCF (a) and Cu nanoparticles (b) in 0.2 M polybu-tylenes succinate (PBS) solution saturated with $CO_2$ (black line) and Ar (gray line). (Reprinted from *Journal of Power Source*, 252, Jia, F. L., Yu, X. X. and Zhang, L. Z., Enhanced selectivity for the electrochemical reduction of $CO_2$ to alcohols in aqueous solution with nanostructure Cu–Au alloy as catalyst, 85–89, Copyright 2014, with permission from Elsevier.)

electrochemical reduction of $CO_2$ in an aqueous 0.5 M $KHCO_3$ solution at a Cu–Au nanostructured alloys electrode. The faradaic efficiencies of the products depended on the nanostructures and compositions of Cu–Au alloys (Figure 2.31). In 2014, Lan et al. [83] investigated the electrochemical reduction of $CO_2$ with different Cu(core)/Cu(shell) catalyst loading in a standard three-electrode cell in 1 M $KHCO_3$ aqueous solution. The sweep rate was 0.1 V/s (Figure 2.32). In 2015, Zhang et al. [84] investigated the electrochemical reduction of $CO_2$ in $CO_2$-saturated 0.1 mol/L $KHCO_3$ aqueous solution, using $SnO_2$ nanoparticles decorated on the surface of nitrogen-doped multiwalled carbon nanotubes (N-MWCNTs) as catalyst (Figure 2.33).

### 2.3.2.2   Electroreduction in Nonaqueous Solution

In 1997, Ohta et al. [85] investigated the electrochemical reduction of $CO_2$ in methanol-based supporting electrolytes, using benzalkonium chloride, $[RN(CH_3)_2CH_2C_6H_5]^+Cl^-$, where $R = C_8$–$C_{18}$, as the ionophore of the catholyte. Benzalkonium chloride in double-distilled methanol and aqueous $KHCO_3$ solution was the supporting electrolyte. The main products were CO, methanol, and ethane (Figure 2.34). In 2000, Tascedda et al. [86] reported the electrosynthesis of five-membered ring cyclic carbamates involving nickel-catalyzed $CO_2$ incorporation into aziridines in DMF. In 2002, Aydin et al. [87] investigated the electrochemical reduction of $CO_2$ at low overpotential on polyaniline electrodes under ambient conditions and under high pressure in methanol + $LiClO_4$. The faradaic efficiencies of the products were 26.5%, 13.1%, and 57.0% for HCHO, HCCOH, and $CH_3COOH$, respectively (Figure 2.35).

In 2001, Yang et al. [88] reported the electrochemical reduction of $CO_2$ ion liquid [BMIm][$BF_4$] at mild conditions. As we can see in Figure 2.36, curve 1 represents an experiment carried out in pure [BMIm][$BF_4$]; curve 2 represents an experiment carried out in [BMIm][$BF_4$] saturated with $CO_2$ at room temperature; and curve

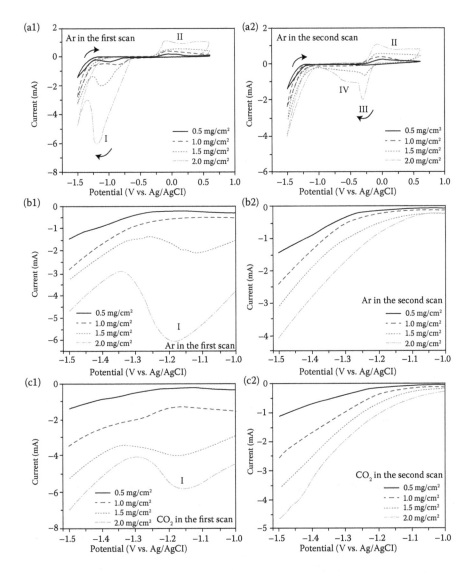

**FIGURE 2.32** Cyclic voltammograms of different Cu(core)/Cu(shell) loadings in 1 M KHCO$_3$ in the presence of (a1–b2) Ar and (c1–c2) CO$_2$. (Reproduced from Lan, Y.C. et al., *Int. J. Electrochem. Sci.*, 9, 8097–8105, 2014. With permission from ESG.)

3 represents an experiment carried out in [BMIm][BF$_4$] saturated with CO$_2$. Scan rate = 0.05 V/s.

In 2004, Zhao et al. [89] reported the electrochemical reduction of supercritical CO$_2$ in ionic liquid BmimPF$_6$. The main products were CO, H$_2$, and formic acid. In 2005, Dubé et al. [90] investigated the electrochemical reduction of CO$_2$ in sulfuric and perchloric acid electrolytes on a polycrystalline Cu surface. The main products were CH$_4$, C$_2$H$_4$, CH$_3$OH, and HCHO. They also proposed a possible mechanism

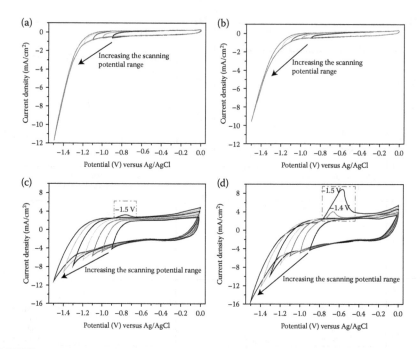

**FIGURE 2.33** CV curves obtained on the N-MWCNTs (a) and (b) and SnO$_2$/N-MWCNTs (c) and (d) modified glassy carbon electrodes (GCEs) in 0.1 mol/L KHCO$_3$ solution saturated with N$_2$ (a) and (c) and CO$_2$ (b) and (d) at a scan rate of 0.05 V/s. (Reprinted from *Materials Letters*, 141, Zhang, R. et al., Electrochemical reduction of CO$_2$ on SnO$_2$/nitrogen-doped multiwalled carbon nanotubes composites in KHCO$_3$ aqueous solution, 63–66, Copyright 2015, with permission from Elsevier.)

(Figure 2.37). In 2005, Pérez et al. [91] investigated the electrochemical reduction of CO$_2$ in DMF + TBAF solution at a gold interface. The main products were CO, formate, and carbonate. The yield of carbamates was 72%–96% (Figure 2.38).

In 2007, Feroci et al. [92] investigated the electrosynthesis of organic carbamates from amines and CO$_2$ in CO$_2$-saturated BMIm–BF$_4$ solutions under mild conditions. The desired products were obtained in good yield. Similarly, in 2007, Wang et al. [93] reported the electrochemical reduction of olefins with CO$_2$ in CO$_2$-saturated ionic liquid DMIMBF$_4$ solution in mild conditions. The yield of the main product monocarboxylic acids were 35%–55% (Figure 2.39).

In 2011, Feroci et al. [94] investigated the electrochemical reaction of amines and CO$_2$ to produce organic carbamates in O$_2$/CO$_2$-saturated ionic liquid (Figure 2.40).

### 2.3.3 MOLECULAR CATALYST IN SOLUTION

#### 2.3.3.1 Macrocycle Metal Complexes

In 1980, Fisher et al. [95] reported the electrochemical reduction of CO$_2$ using macrocycles of Ni and Co at a potential between −1.3 and −1.6 V (vs. SCE). CO$_2$ could be reduced to CO, and the faradaic efficiency was 98%.

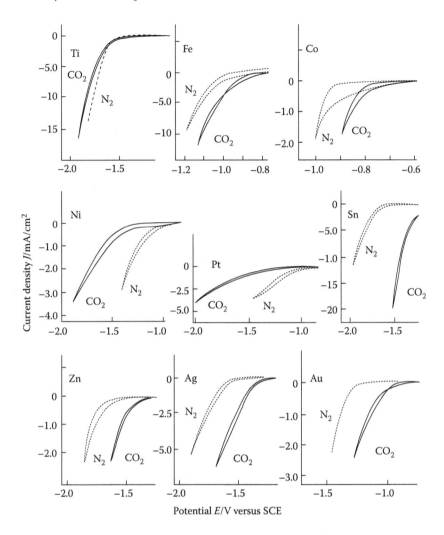

**FIGURE 2.34** Cyclic voltammograms. Solid line: under $CO_2$ atmosphere; broken line: under $N_2$ atmosphere; catholyte: $10^{-2}$ mol/dm³ benzalkonium chloride in methanol; and anolyte: $10^{-1}$ mol/dm³ $KHCO_3$ in water. (With kind permission from Springer Science + Business Media: *Journal of Applied Electrochemistry*, Electrochemical reduction of carbon dioxide in methanol at ambient temperature and pressure, 28, 1998, 717–724, Ohta, K. et al.)

### 2.3.3.2 Bypyridyl Metal Complexes

In 2010, Smieja et al. [96] reported the electrochemical reduction of $CO_2$ in the presence of Re(bipy-tBu)(CO)₃Cl as a catalyst. The main product was CO, and the faradaic efficiency was 99%.

In 2010, Cole et al. [97] reported the electrochemical reduction of $CO_2$ to methanol using pyridinium-based catalysts, which exhibited high faradaic yields for methanol.

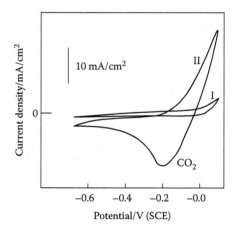

**FIGURE 2.35** Cyclic voltammograms of $CO_2$ reduction on a PAn electrode in MeOH + LiClO$_4$ + H$^+$ + H$_2$O; I—in the absence of $CO_2$; II—$CO_2$ saturated. $v = 50$ mV/s. (Reprinted from *Journal of Electroanalytical Chemistry*, 535, Aydin, R. and Köleli, F., Electrochemical reduction of $CO_2$ on a polyaniline electrode under ambient conditions and at high pressure methanol, 107–112, Copyright 2002, with permission from Elsevier.)

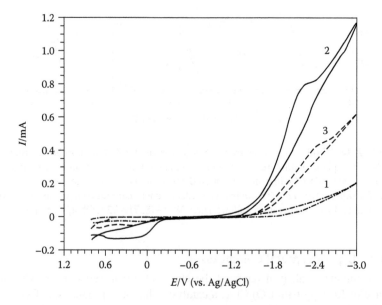

**FIGURE 2.36** Cyclic voltammograms of electroreduction of $CO_2$ in [BMIm][BF$_4$]. (Yang, H. Z. et al. Electrochemical activation of carbon dioxide in ionic liquid: Synthesis of cyclic carbonates at mild reaction conditions. *Chemical Communications*. 2000; 274–275. Reproduced by permission of The Royal Society of Chemistry.)

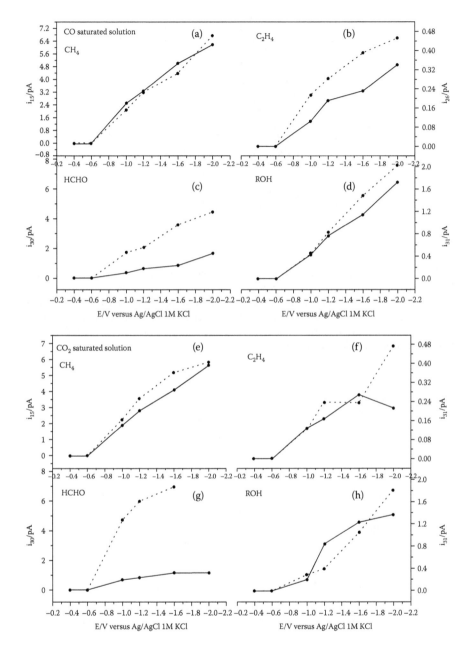

**FIGURE 2.37** Mass current versus potential for CO- and CO$_2^-$ saturated solutions in 0.25 M H$_2$SO$_4$ (solid line) and 0.5 M HClO$_4$ (dotted line) on a copper polycrystalline electrode. (Reprinted from *Journal of Electroanalytical Chemistry*, 582, Dubé, P. and Brisard, G. M., Influence of adsorption processes on the CO$_2$ electroreduction: An electrochemical mass spectrometry study, 230–240, Copyright 2005, with permission from Elsevier.)

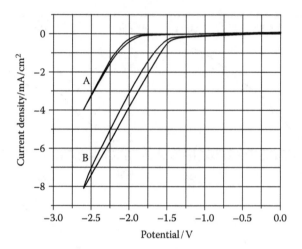

**FIGURE 2.38** CV of the (A) DMF + TBAF solution without $CO_2$ and (B) saturated with $CO_2$. Potentials vs. SCE. Scan rate: 0.05 V/s. Temperature: 25°C. Specific conductivity of the electrolyte solution: 5.63 mS/cm. (Reprinted from *Journal of Electroanalytical Chemistry*, 578, Pérez, E. R. et al., *In situ* FT-IR and *ex situ* EPR analysis for the study of the electroreduction of carbon dioxide in *N,N*-dimethylformamide on a gold interface, 87–94, Copyright 2005, with permission from Elsevier.)

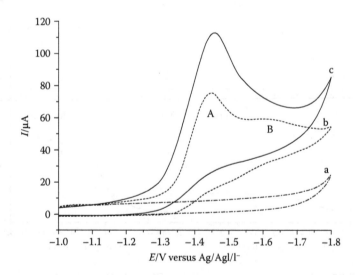

**FIGURE 2.39** Cyclic voltammograms recorded at 0.1 V/s in $BMIMBF_4$ with GC electrode at 25°C: (a) neat $BMIMBF_4$, (b) as (a) +20 mM ethyl cinnamate, and (c) as (b) saturated with $CO_2$. (Reprinted from *Electrochemistry Communications*, 9, Wang, H. et al., Electrocarboxylation of activated olefins in ionic liquid $BMIMBF_4$, 2235–2239, Copyright 2007, with permission from Elsevier.)

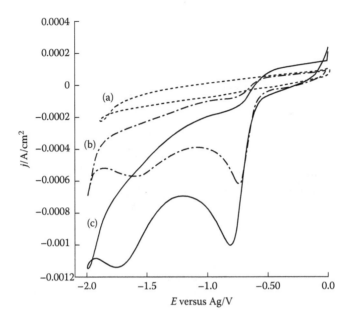

**FIGURE 2.40** Cyclic voltammetric curves (vitreous carbon cathode) in BMIm–BF$_4$ at atmospheric pressure, T = 60°C: (a) after 15 min N$_2$ bubbling (v = 200 mV/s); (b) after 15 min O$_2$/CO$_2$ bubbling (v = 50 mV/s); and (c) after 15 min O$_2$/CO$_2$ bubbling (v = 200 mV/s). (Reprinted from the *Electrochimica Acta*, 56, Feroci, M. et al., Carbon dioxide as carbon source: Activation via electrogenerated O$_2$$^{•-}$ in ionic liquids, 5823–5827, Copyright 2011, with permission from Elsevier.)

### 2.3.3.3 Phosphine Metal Complexes

In 1984, Slater et al. [98] reported the electrochemical reduction of CO$_2$ to formate, using complex Rh(diphos)$_2$Cl (diphos = 1,2-bis(diphenylphosphino)ethane) as an efficient catalyst.

## 2.4 POSSIBILITY OF CO$_2$ REDUCTION BASED ON THERMODYNAMIC POTENTIALS

In general, CO$_2$ is stable and chemically inactive. The requirement for a chemical reaction to occur is $\Delta G < 0$. But for a chemical reaction that is not spontaneous, that can be forced to occur on an electrode, this is what distinguishes the electrochemical reactions from other chemical reactions. We have discussed electrochemical reduction of CO$_2$, as the previous section explained. Now we will move on to discuss the thermodynamic potentials (Figure 2.41).

As shown above [99], when the products were CH$_4$, C$_2$H$_4$, and CH$_3$OH, the electrochemical reduction of CO$_2$ could occur spontaneously. When the product was HCOOH, this reaction could not happen spontaneously. For more data, refer to Lide, D. R. in the *CRC Handbook of Chemistry and Physics*.

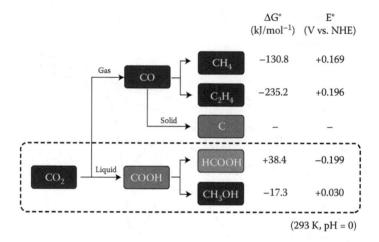

(293 K, pH = 0)

**FIGURE 2.41** Selected major reaction products and intermediates for $CO_2$ reduction reactions with their associated free energies and standard potentials. (Lee, J. Y. et al.: *Chemistry— An Asian Journal*. 2009. 4. 1516–1523. Copyright Wiley-VCH Verlag GmbH & Co. KGaA. Reproduced with permission.)

Compared with other methods, the electrochemical reduction of $CO_2$ has these superiorities:

1. The electroreduction of the $CO_2$ process is completed through reactants gaining or losing electrons at an electrode, and neither oxidizing agent nor reducing agent is needed. Although required in several processes, a small amount of catalyst is necessary. Therefore, this is a process without trash being produced at its essence. With the enhancement of environmental consciousness, the electrochemical reduction of $CO_2$ has become increasingly important.
2. Without adding an oxidizing agent or a reducing agent into the solution, the consumption of substances can be reduced, and the side reactions can be greatly reduced. The separation process can be simplified, and high-purity products can usually be obtained.
3. This process offers the advantages of technological process and is simple, with a few reaction steps.
4. The reactions usually run under ordinary temperature and pressure, and the energy consumption can decrease significantly.
5. As mentioned before, the products can be controlled accurately via controlling the metal electrodes, current, voltage, and electrolytes.
6. Carbon dioxide is the ultimate source of $C_1$-materials around us, and is one kind of cheap raw material.
7. The reaction that may not occur under usual conditions can be carried out through electrochemical reaction.

Of course, there are several limits to the electrochemical reduction of CO$_2$. It is a contradiction that the energy for the reduction process comes from the combustion of a fossil source, from which more CO$_2$ will be produced; hence, the energy for this process can only be obtained from clean energy, such as wind energy and nuclear energy.

The electrolyzer for this process is usually very complicated compared to other reactors. Its maintenance is high cost, which will limit its application in industry. At the same time, the electroreduction process needs to consume large amounts of electricity.

We still have reasons to believe that CO$_2$ is an inexhaustible source, and the electroreduction of CO$_2$ has broad application prospects in energy storage and in producing chemical products with high added value.

Five broad scenarios seem to be possible in the electrochemical reduction of CO$_2$:

1. Finding a good organic solvent to dissolve CO$_2$
2. Finding a better metal electrode to control the constitution of products
3. Using gaseous diffusion electrode to increase the pressure of CO$_2$
4. Using modified electrode to produce more complicated organics
5. Finding an efficient separation technology to isolate products immediately

In this chapter, we discussed the electrochemical conversion of carbon dioxide to produce important chemicals. We first gave a brief introduction of the thermodynamics of electrochemistry, including basic definitions, basic formulas, and main research methods. For the majority of the electrochemical reactions that take place in aqueous solution, the potential–pH diagram has been highlighted.

Next, we introduced the electrochemical reduction of CO$_2$ directly and with other organics. The main products of the electrochemical reduction of CO$_2$ were CO, HCOOH, CH$_4$, CH$_3$OH, and so on, which could have a very high yield. The main products of the electroreduction of CO$_2$ with other organics were the relevant carboxylation products. We also talked about the influence factors.

Electrode and electrolyte were the most complicated and important influencing factors in an electrochemical reaction, which was next. The metal electrodes were mainly applied to the electrochemical reactions, and the Cu electrode was a conspicuous one. We also provided some discussion of other electrodes. Later, we introduced aqueous solutions, nonaqueous solutions, and homogeneous catalysis in solution as an activation species. In this chapter, we classified molecular electrocatalysts in ligands (other researchers [99,100] have classified them based on the type of metal), and found that Cu, Pd, Ni, and Co metal complexes were available to catalyze the electrochemical reduction of CO$_2$. Finally, we discussed the possibility of CO$_2$ reduction of CO$_2$ based on thermodynamic potentials and the advantages and disadvantages of the electrochemical reduction of CO$_2$. The electroreduction of CO$_2$ has wonderful future prospects.

## ABBREVIATIONS

DMF    Dimethyl formamide
GCE    Glassy carbon electrode

IB      The first subgroup
KI      Potassium iodide
NHE     Normal standard electrode
NMP     Methyl-2-pyrrolidinone
PBS     Poly butylenes succinate
PC      Polycarbonate
SCE     Standard calomel electrode
SHE     Standard hydrogen electrode
TBAB    Tetrabutylammonium bromide
TEAP    Tetraethylammonium perchlorate
THF     Tetrahydrofuran

## REFERENCES

1. Collin, J. P. and Sauvage, J. P. Electrochemical reduction of carbon dioxide mediate by molecular catalysts. *Coordination Chemistry Reviews.* 1989; 93(2): 245–268.
2. Kaneco, S., Iiba, K., Hiei, N. H., Ohta, K., Mizuno, T. and Suzuki, T. Electrochemical reduction of carbon dioxide to ethylene with high Faradaic efficiency at a Cu electrode in CsOH/methanol. *Electrochimica Acta.* 1999; 44: 4701–4706.
3. Kaneco, S., Iiba, K., Ohta, K., Mizuno, T. and Saji. A. Electrochemical reduction of $CO_2$ at an Ag electrode in KOH-methanol at low temperature. *Electrochimica Acta.* 1999; 44: 573–578.
4. Vassiliev, Y. B., Bagotsky, V. S., Osetrova, N. V., Khazova, O. A. and Mayorova, N. A. Electroreduction of carbon dioxide: Part I. The mechanism and kinetics of electroreduction of $CO_2$ in aqueous solution on metals with high and moderate hydrogen overvoltages. *Journal of Electroanalytical Chemistry and Interfacial Electrochemistry.* 1985; 189(2): 271–294.
5. Orgura, K., Endo, N. and Nakayama, M. Mechanistic studies of $CO_2$ reduction on a mediated electrode with conducting polymer and inorganic conductor films. *Journal of the Electrochemical Society.* 1998; 145(11): 3801–3809.
6. Tryk, D. A., Yamamoto, T., Kokubun, M., Hirota, K., Hashimoto, K., Okawa, M. and Fujishima, A. Recent developments in electrochemical and photoelectrochemical $CO_2$ reduction: Involvement of the $(CO_2)_2{}^{-}$ dimer radical anion. *Applied Organometallic Chemistry.* 2001; 15: 113–120.
7. Chaplin, R. P. S. The reduction of carbon dioxide to formate in a solid polymer electrolyte reactor, PhD thesis, Exeter: Exeter University, 2001.
8. Aylmer-Kelly, A. W. B., Bewick, A., Cantrill, P. R. and Tuxford, A. M. Studies of electrochemically generated reaction intermediate using modulated specular reflectance spectroscopy. *Faraday Discussions of the Chemical Society.* 1973; 56: 96–107.
9. Ayers, W. M. An overview of electrochemical carbon dioxide reduction. *Special Publication—Royal Society of Chemistry.* 1994; 153: 365–374.
10. Udupa, K. S., Subramanian, G. S. and Udupa, H. V. K. Electrolytic reduction of carbon dioxide to formic acid. *Electrochimica Acta.* 1971; 16(9): 1593–1598.
11. Hori, Y., Murata, A., Takahashi, R. and Suzuki, S. Electroreduction of carbon monoxide to methane and ethylene at a copper electrode in aqueous solution at ambient temperature and pressure. *Journal of the American Chemical Society.* 1987; 109(16): 5022–5023.
12. De Jesús-Cardona, H., Del Moral, C. and Cabrera, C. R. Voltammetric study of $CO_2$ reduction at Cu electrode under different $KHCO_3$ concentration, temperature and $CO_2$ pressure. *Journal of Electroanalytical Chemistry.* 2001; 513(1): 45–51.

13. Christian, A. and Michel, S. J. Mechanism and kinetic characteristics of the electrochemical reduction of carbon dioxide in media of low proton availability. *Journal of the American Chemical Society*. 1981; 103(17): 5021–5023.

14. Takayuki, M., Kiyohisa, O., Akira, S., Toru, A., Masaki, H. and Atsusi, K. Effect of temperature on electrochemical reduction of high-pressure CO$_2$ with In, Sn and Pb electrodes. *Energy Source*. 1995; 17(5): 503–508.

15. Ryu, J., Andersen, T. N. and Eyring, H. Electrode reduction kinetics of carbon dioxide in aqueous solution. *Journal of Physical Chemistry*. 1972; 76(22): 3278–3286.

16. Kanecon, S., Hiei, N. H., Xing, Y., Katsumata, H., Ohnishi, H., Suzuki, T. and Ohta, K. Electrochemical conversion of carbon dioxide to methane in aqueous NaHCO$_3$ solution at less than 273 K. *Electrochimica Acta*. 2002; 48(1): 51–55.

17. Kaneco, S., Iiba, K., Ohta, K., Mizuno, T. and Saji, A. Electrochemical reduction of CO$_2$ on Au in KOH + methanol at low temperature. *Journal of Electroanalytical Chemistry*. 1998; 441(1–2): 215–220.

18. Mizuno, T., Naitoh, A. and Ohta, K. Electrochemical reduction of CO$_2$ in methanol at −30°C. *Journal of Electroanalytical Chemistry*. 1995; 391(1–2): 199–201.

19. Li, H. and Oloman, C. Development of a continuous reactor for electro-reduction of carbon dioxide to formate—Part I: Process variables. *Journal of Applied Electrochemistry*. 2006; 36(10): 1105–1115.

20. Nakagawa, S., Kudo, A., Azuma, M. and Sakata, T. Effect of pressure on the electrochemical reduction of carbon dioxide on group VIII metal electrodes. *Journal of Electroanalytical Chemistry and Interfacial Electrochemistry*. 1991; 308(1–2): 339–343.

21. Hara, K., Kudo, A. and Sakata, T. Electrochemical reduction of carbon dioxide under high pressure on various electrode in an aqueous electrolyte. *Journal of Electroanalytical Chemistry*. 1995; 391(1–2): 141–147.

22. Hara, K., Kudo, A. and Sakata, T. Electrochemical reduction of high pressure carbon dioxide on Fe electrodes at large current density. *Journal of Electroanalytical Chemistry*. 1995; 386(1–2): 257–260.

23. Todoroki, M., Hara, K., Kudo, A. and Sakaata, T. Electrochemical reduction of high pressure CO$_2$ at Pb, Hg and In electrodes in an aqueous KHCO$_3$ solution. *Journal of Electroanalytical Chemistry*. 1995; 394(1–2): 199–203.

24. Koeleli, F. and Balun, D. Reduction of CO$_2$ under high pressure and high temperature on Pb-granule electrode in a fixed-bed reactor in aqueous medium. *Applied Catalysis A: General*. 2004; 274(1–2): 237–242.

25. Innocent, B., Liaigre, D., Pasquier, D., Ropital, F., Leger, J.-M. and Kokoh, K. B. Electro-reduction of carbon dioxide to formate on lead electrode in aqueous medium. *Journal of Applied Electrochemistry*. 2009; 39(2): 227–232.

26. Ikeda, S., Takagi, T. and Ito K. Selective formation of formic acid, oxalic acid and carbon monoxide by electrochemical reduction of carbon dioxide. *Bulletin of the Chemical Society of Japan*. 1987; 60(7): 2517–2522.

27. Hara, K., Kudo, A., Sakata, T. and Watanabe, M. High efficiency electrochemical reduction of carbon dioxide under high pressure on a gas diffusion electrode containing Pt catalysts. *Journal of the Electrochemical Society*. 1995; 142(4): L57–L59.

28. Neikam, W. C. Electrolytic preparation of carboxylic acids. US Patent 3, 344, 045, 1967.

29. Neikam, W. C. Electrolytic preparation of organic carbonates. US Patent 3, 344, 046, 1967.

30. Gambino, S. and Silvestri, G. Electrochemical reduction of carbon dioxide and ethylene. *Tetrahedron Letters*. 1973; 32: 3025–3028.

31. Gambino, S., Gennaro, A., Filardo, G., Silvestri, G. and Vianello, E. Electrochemical carboxylation of styrene. *Journal of the Electrochemical Society*. 1987; 134(9): 2172–2175.

32. Filardo, G., Gambino, S., Silvestri, G., Gennaro, A. and Vianello, E. Electrocarboxylation of styrene through homogenous redox catalysis. *Journal of Electroanalytical Chemistry and Interfacial Electrochemistry.* 1984; 177(1–2): 303–309.

33. Derien, S., Clinet, J. C., Dunach, E. and Perichon, J. Electrochemical incorporation of carbon dioxide into alkenes by nickel complexes. *Tetrahedron.* 1992; 48(25): 5235–5248.

34. Ballivet-Tkachenko, D., Folest, J. C. and Tanji, J. Electrocatalytic reduction of $CO_2$ for the selective carboxylation of olefins. *Applied Organometallic Chemistry.* 2000; 14: 847–849.

35. Bringmann, J. and Dinjus, J. Electrochemical synthesis of carboxylic acid from alkenes using various nickel-organic mediator: $CO_2$ as C1-synthon. *Applied Organometallic Chemistry.* 2001; 15(2): 135–140.

36. Senboku, H., Komatsu, H. and Fujimura, Y. Efficient electrochemical dicarboxylation of phenyl-substituted alkenes: Synthesis of 1-phenylalkane-1, 2-dicarboxylic acids. *Synlett.* 2001; 3: 418–420.

37. Chowdhury, M. A., Senboku, H. and Tokuda, M. Electrochemical carboxylation of bicyclo[n.1.0]alkylidene derivatives. *Tetrahedron.* 2004; 60: 475–481.

38. Dunach, E. and Périchon. Electrochemical carboxylation of terminal alkynes catalyst by nickel complexes: Unusual regioselectivity. *Journal of Organometallic Chemistry.* 1988; 352(1–2): 239–246.

39. Dunach, E., Dérien, S. and Périchon, J. Nickel-catalyzed reductive electrocarboxylation of disubstituted alkynes. *Journal of Organometallic Chemistry.* 1989; 364(3): C33–C36.

40. Dérien, S., Dunach, E. and Périchon, J. Electrogenerated nickel(0) catalyzed carbon dioxide incorporation into α, ω-diynes. *Journal of Organometallic Chemistry.* 1990; 385(3): C43–C46.

41. Dérien, S., Dunach, E. and Périchon, J. From stoichiometry to catalysis: Electroreductive coupling of alkynes and carbon dioxide with nickel-bipyridine complexes. Magnesium ion as the key for catalysis. *Journal of American Chemical Society.* 1991; 113(12): 8447–8454.

42. Yuan, G. Q., Jiang, H. F. and Lin, C. Efficient electrochemical dicarboxylations of arylacetylenes with carbon dioxide using nickel as the cathode. *Tetrahedron.* 2008; 64: 5866–5872.

43. Ikeda, Y. and Manda, E. An electrochemical synthesis of 2-hydroxy-2-(p-isobutylphenyl)-propionic acid. *Chemistry Letters.* 1984; 13(3): 453–454.

44. Ikeda, Y. and Eiichiro, M. Synthesis of benzilic acids through electrochemical reductive carboxylation of benzophenones in the presence of carbon dioxide. *Bulletin of the Chemical Society of Japan.* 1985; 58(6): 1723–1726.

45. Silvestri, G., Gambino, S. and Filardo, G. Electrochemical carboxylation of aldehydes and ketones with sacrificial aluminum anodes. *Tetrahedron Letters.* 1986; 27(29): 3429–3430.

46. Chan, A. S. C., Huang, T. T., Wagenknecht, J. H. and Miller, R. E. A novel of 2-aryllactic acids via electrocarboxylation of methyl aryl ketones. *Journal of Organic Chemistry.* 1995; 60: 742–744.

47. Datta, A. K., Marron, P. A., King, C. J. H. and Wagenknecht, J. H. Process development for electrocarboxylation of 2-acetyl-6-methoxynaphthalene. *Journal of Applied Electrochemistry.* 1998; 28(6): 569–577.

48. Isse, A. A., Galia, A., Belfiore, C., Silvestri, G. and Gennaro, A. Electrochemical reduction and carboxylation of halobenzophenones. *Journal of Electroanalytical Chemistry.* 2002; 526(1–2): 41–52.

49. Doherty, A. P. Electrochemical reduction of butyraldehyde in the presence of $CO_2$. *Electrochimica Acta.* 2002; 47(18): 2963–2967.

50. Scialdone, O., Alessandro, G., Chiara, L. R. and Giuseppe, F. Influence of the nature of the substrate and of operative parameters in the electrocarboxylation of halogenated acetophenones and benzophenones. *Electrochimica Acta*. 2005; 50: 3231–3242.

51. Scialdone, O., Amatore, C., Gilia, A. and Filardo, G. CO$_2$ as a C$_1$-organic building block: Electrocarboxylation of aromatic ketones. A quantitative study of the effect of the concentration of substrate and of carbon dioxide on the selectivity of the process. *Journal of Electroanalytical Chemistry*. 2006; 592: 163–174.

52. Scialdone, O., Galia, A., Isse, A. A., Gennaro, A., Sabatino, M. A., Leone, R. and Filardo, G. Electrocarboxylation of aromatic ketones: Influence of operative parameters on the competition between ketyl and ring carboxylation. *Journal of Electroanalytical Chemistry*. 2007; 609: 8–16.

53. Wagenknecht, J. H. Electroreduction of alkyl halides in the presence of carbon dioxide. *Journal of Electroanalytical Chemistry and Interfacial Electrochemistry*. 1974; 52(3): 489–492.

54. Sock, O., Troupel, M. and Perichon, J. Electrosynthesis of carboxylic acids from organic halides and carbon dioxide. *Tetrahedron Letters*. 1985; 26(2): 1509–1512.

55. Fauvarque, J. F., Jutand, A., Francois, M. Nickel catalysed electrosynthesis of anti-inflammatory agents. Part I—Synthesis of aryl-2 propionic acids, under galvanostatic conditions. *Journal of Applied Electrochemistry*. 1988; 18(1): 109–115.

56. Fauvarque, J. F., De Zelicourt, Y., Amatore, C. and Jutand, A. Nickel-catalysed electrosynthesis of anti-inflammatory agents. III. A new electrolyser for organic solvents; oxidation of metal powder as an alternative to sacrificial anodes. *Journal of Applied Electrochemistry*. 1990; 20: 338–340.

57. Amatore, C. and Jutand, A. Activation of carbon dioxide by electron transfer and transition metals. Mechanism of nickel-catalyzed electrocarboxylation of aromatic halides. *Journal of the American Chemical Society*. 1991; 113: 2819–2826.

58. Damodar, J., Mohan, S. R. K. and Reddy, S. R. J. Synthesis of 2-arylpropionic acids by electrocarboxylation of benzylchlorides catalysed by PdCl$_2$(PPh$_3$)$_2$. *Electrochemistry Communication*. 2001; 3: 762–766.

59. Isse, A. A. and Gennaro, A. Electrochemical synthesis of cyanoacetic acid from chloro-acetonitrile and carbon dioxide. *Journal of the Electrochemical Society*. 2002; 148(8): D113–D117.

60. Isse, A. A., Ferlin, M. Z. and Gennaro, A. Electrocatalytic reduction of arylethyl chlorides at silver cathodes in the presence of carbon dioxide: Synthesis of 2-arylpropanoic acids. *Journal of Electroanalytical Chemistry*. 2005; 581: 38–45.

61. Tascedda, P., Weidmann, M., Dinjus, E. and Dunach, E. Nickel-catalyzed electrochemical carboxylation of epoxides: Mechanistic aspects. *Applied Organometallic Chemistry*. 2001; 15: 141–144.

62. Casadei, M. A., Cesa, S. and Leucio, R. Electrogenerated-base-promoted synthesis of organic carbonates from alcohols and carbon dioxide. *European Journal of Organic Chemistry*. 2000; 13: 2445–2448.

63. Yuan, D. D., Yan, C. H., Lu, B., Wang, H. X., Zhong, C. M. and Cai, Q. H. Electrochemical activation of carbon dioxide for synthesis of dimethyl carbonate in an ionic liquid. *Electrochemical Acta*. 2009; 54(10): 2912–2915.

64. Azuma, M., Hashimoto, K., Hiramoto, M., Watanabe, M. and Sakata, T. Electrochemical reduction of carbon dioxide on various metal electrodes in low-temperature aqueous KHCO$_3$ media. *Journal of the Electrochemical Society*. 1990; 137(6): 1772–1778.

65. Mahmood, M. N., Masheder, D. and Harty, C. J. Use of gas-diffusion electrodes for high-rate electrochemical reduction of carbon dioxide. Reduction at lead, indium- and tin-impregnated electrodes. *Journal of Applied Electrochemistry*. 1987; 17(6): 1159–1170.

66. Furuya, N., Yamazaki, T. and Shibata, M. High performance Ru–Pd catalysts for $CO_2$ reduction at gas-diffusion electrodes. *Journal of Electroanalytical Chemistry.* 1997; 431(1): 39–41.

67. Frese, K. W. Jr. and Canfield, D. Reduction of $CO_2$ on n-GaAs and selective methanol synthesis. *Journal of the Electrochemical Society.* 1984; 131(11): 2518–2522.

68. Kaneco, S., Katsumata, H., Suzuki, T. and Ohta, K. Photoelectrocatalytic reduction of $CO_2$ in LiOH/methanol at metal-modified p-InP electrodes. *Applied Catalysis B: Environmental.* 2006; 64(1–2): 139–145.

69. Ogura, K., Endo, N. and Nakayama, M. Mechanistic studies of $CO_2$ reduction on a mediated electrode with conducting polymer and inorganic conductor films. *Journal of the Electrochemical Society.* 1998; 145(11): 3801–3809.

70. Hori, Y., Ito, H., Okano, K., Nagasu, S. and Sato, S. Silver-coated ion exchange membrane electrode applied to electrochemical reduction of carbon dioxide. *Electrochimica Acta.* 2003; 48: 2651–2657.

71. Yoshitake, H., Kikkawa, T. and Ota, K. I. Isotopic product distributions of $CO_2$ electrochemical reduction on a D flowing-out Pd surface in protonic solution and reactivities of "subsurface" hydrogen. *Journal of Electroanalytical Chemistry.* 1995; 390: 91–97.

72. Yoshitake, H., Kikkawa, T., Muto, G. and Ota, K. I. Poisoning of surface hydrogen process on a Pd electrode during electrochemical reduction of carbon dioxide. *Journal of Electroanalytical Chemistry.* 1995; 396: 491–498.

73. Jitaru, M., Lowy, D. A., Toma, M., Toma, B. C. and Oniciu, L. Electrochemical reduction of carbon dioxide on flat metallic cathodes. *Journal of Applied Electrochemistry.* 1997; 27: 875–889.

74. Hoshi, N. and Hori, Y. Electrochemical reduction of carbon dioxide at a series of platinum single crystal electrodes. *Electrochimica Acta.* 2000; 45: 4263–4270.

75. Lee, J. Y. and Tak, Y. S. Electrocatalytical activity of Cu electrode in electroreduction of $CO_2$. *Electrochimica Acta.* 2001; 46: 3015–3022.

76. Ohmori, T., Nakayama, A., Mametsuka, H. and Suzuki, E. Influence of sputtering parameters on electrochemical $CO_2$ reduction in sputtered Au electrode. *Journal of Electroanalytical Chemistry.* 2001; 514: 51–55.

77. Stevens, G. B., Reda, T. and Raguse, B. Energy storage by the electrochemical reduction of $CO_2$ to CO at a porous Au film. *Journal of Electroanalytical Chemistry.* 2002; 526: 125–133.

78. Takahashi, I., Koga, O., Hoshi, N. and Hori, Y. Electrochemical reduction of $CO_2$ at a copper single crystal Cu(S)-[$n$(111) × (111)] and Cu(S)-[$n$(110) × (100)] electrodes. *Journal of Electroanalytical Chemistry.* 2002; 533: 135–143.

79. Costentin, C., Canales, J. C., Haddou, B. and Savéant, J. M. Electrochemical of acids on platinum. Application to the reduction of carbon dioxide in the presence of pyridinium ion in water. *Journal of the American Chemical Society.* 2013; 135: 17671–17674.

80. Ertem, M. Z., Konezny, S. J., Araujo, C. M. and Batista, V. S. Functional role of pyridinium during aqueous electrochemical reduction of $CO_2$ on Pt(111). *Journal of Physical Chemistry Letters.* 2013; 4: 745–748.

81. Yan, Y., Zeitler, E. L., Gu, J., Hu, Y. and Bocarsly, A. B. Electrochemistry of aqueous pyridinium: Exploration of a key aspect of electrocatalytic reduction of $CO_2$ to methanol. *Journal of the American Chemical Society.* 2013; 135: 14020–14023.

82. Jia, F. L., Yu, X. X. and Zhang, L. Z. Enhanced selectivity for the electrochemical reduction of $CO_2$ to alcohols in aqueous solution with nanostructure Cu–Au alloy as catalyst. *Journal of Power Source.* 2014; 252: 85–89.

83. Lan, Y. C., Ma, S. C., Keins. P. J. A. and Lu, J. X. Hydrogen evolution in the presence of $CO_2$ in an aqueous solution during electrochemical reduction. *International Journal of Electrochemical Science.* 2014; 9: 8097–8105.

84. Zhang, R., Lv, W. X., Li, G. H. and Lei, L. X. Electrochemical reduction of CO$_2$ on SnO$_2$/nitrogen-doped multiwalled carbon nanotubes composites in KHCO$_3$ aqueous solution. *Materials Letters*. 2015; 141: 63–66.

85. Ohta, K., Kawamoto, M., Mizuno, T. and Lowy, D. A. Electrochemical reduction of carbon dioxide in methanol at ambient temperature and pressure. *Journal of Applied Electrochemistry*. 1998; 28: 717–724.

86. Tascedda, P. and Dunach, E. Electrosynthesis of cyclic carbamates from aziridines and carbon dioxide. *Chemical Communications*. 2000; 6: 449–450.

87. Aydin, R. and Köleli, F. Electrochemical reduction of CO$_2$ on a polyaniline electrode under ambient conditions and at high pressure methanol. *Journal of Electroanalytical Chemistry*. 2002; 535: 107–112.

88. Yang, H. Z., Gu, Y. L., Deng, Y. Q. and Shi, F. Electrochemical activation of carbon dioxide in ionic liquid: Synthesis of cyclic carbonates at mild reaction conditions. *Chemical Communications*. 2000; 3: 274–275.

89. Zhao, G. Y., Jiang, T., Han, B. X., Li, Z. H., Zhang, J. M., Liu, Z. M., He, J. and Wu, W. Z. Electrochemical reduction of supercritical carbon dioxide in ionic liquid 1-$n$-butyl-3-methylimidazolium hexafluorophosphate. *Journal of Supercritical Fluids*. 2004; 32: 287–291.

90. Dubé, P. and Brisard, G. M. Influence of adsorption processes on the CO$_2$ electroreduction: An electrochemical mass spectrometry study. *Journal of Electroanalytical Chemistry*. 2005; 582: 230–240.

91. Pérez, E. R., Garcia, J. R., Cardoso, D. R., McGarvey, B. R., Batista, E. A., Rodrigues-Filho, U. P., Vielstich, W. and Franco, D. W. *In situ* FT-IR and *ex situ* EPR analysis for the study of the electroreduction of carbon dioxide in *N,N*-dimethylformamide on a gold interface. *Journal of Electroanalytical Chemistry*. 2005; 578: 87–94.

92. Feroci, M., Orsini, M., Rossi, L., Sotgui, G. and Inesi, A. Electrochemically promoted C–N bond formation from amines and CO$_2$ in ionic liquid BMIm–BF$_4$: Synthesis of carbamates. *Journal of Organic Chemistry*. 2007; 72: 200–203.

93. Wang, H., Zhang, G. R., Liu, Y. Z., Luo, Y. W. and Lu, J. X. Electrocarboxylation of activated olefins in ionic liquid BMIMBF$_4$. *Electrochemistry Communications*. 2007; 9: 2235–2239.

94. Feroci, M., Chiarotto, I., Orsini, M., Sotgiu, G. and Inesi, A. Carbon dioxide as carbon source: Activation via electrogenerated O$_2$$^{·-}$ in ionic liquids. *Electrochimica Acta*. 2011; 56: 5823–5827.

95. Fisher, B. J. and Eisenberg, R. Electrocatalytic reduction of carbon dioxide by using macrocycles of nickel and cobalt. *Journal of the American Chemical Society*. 1980; 102(24): 7361–7363.

96. Smieja, J. M. and Kubiak, C. P. Re(bipy-tBu)(CO)$_3$Cl-improved catalytic activity for reduction of carbon dioxide: IR-spectroelectrochemical and mechanistic studies. *Inorganic Chemistry*. 2010; 49(20): 9283–9289.

97. Cole, E. B., Lakkaraju, P. S., Rampulla, D. M., Morris, A. J., Abelev, E. and Bocarsly, A. B. Using a one-electron shutter for the multielectron reduction of CO$_2$ to methanol: Kinetic, mechanistic, and structural insights. *Journal of the American Chemical Society*. 2010; 132: 11539–11551.

98. Slater, S. and Wagenknecht, J. H. Electrochemical reduction of carbon dioxide catalyzed Rh(diphos)$_2$Cl. *Journal of the American Chemical Society*. 1984; 106(18): 5367–5368.

99. Lim, R. J., Xie, M. S., Sk, M. A., Lee, J. M., Fisher, A., Wang, X. and Lim, K. H. A review on the electrochemical reduction of CO$_2$ in fuel cells, metal electrodes and molecular catalysts. *Catalysis Today*. 2014; 233: 169–180.

100. Lide, D.R. et al. *CRC Handbook of Chemistry and Physics*, 76th edn, pp. 5–28, ISBN 0-8493-0476-8.

# 3 Electrode Kinetics of CO$_2$ Electroreduction

*Dongmei Sun and Yu Chen*

## CONTENTS

## 3.1 INTRODUCTION

Electrochemical reduction of carbon dioxide (ERC) is an attractive way to convert CO$_2$ because of the following advantages over thermochemical methods: (1) water is the proton source; (2) high equilibrium conversion at ambient temperature; and (3) relatively simple and green [1].

The electrochemical reduction of CO$_2$ has been studied since the late nineteenth century [2]. In 1904 [3], Coehn and Jahn used zinc, amalgamated zinc, and amalgamated copper with high hydrogen overpotential as cathodes to electrolytically reduce carbon dioxide in aqueous $NaHCO_3$ and $K_2SO_4$ solutions, in which formic acid was the only product.

Most of the reduction reaction of $CO_2$ occurs in the potential range in which hydrogen evolution reaction (HER) due to hydrolysis of water occurs [4], as the overall standard redox potentials for ERC are in the same range as that for the reduction of protons to $H_2$. For example [5,6],

$$E^0, V (SHE)@298 K (pH = 14)$$

$$\text{Cathode: } CO_2(aq) + H_2O + 2e^- \rightarrow HCOO^- + OH^- \qquad -1.02 \qquad (3.1)$$

$$2H_2O + 2e^- \rightarrow H_2 + 2OH^- \qquad -0.83 \qquad (3.2)$$

Reaction (3.1) is considered a kinetically "slow" process, occurring at cathode potentials from about $-0.8$ to $-1.8$ V vs. a standard hydrogen electrode (SHE) depending on the $CO_2$ pressure, current density, and catholyte pH, with overpotential ranging from about $-0.4$ to $-1.4$ V. The intrinsic kinetics of Reaction (3.1) is said to be independent of pH ($2 < pH < 8$), but a pH above 6 can have a strong effect on the mass transfer limiting current for this reaction, as $CO_2(aq)$ is depleted through the $CO_2(aq)/HCO_3^-/CO_3^{2-}$ equilibria. Reaction (3.2) is thermodynamically favored over Reaction (3.1) over almost the entire pH range at 298 K. However, Reaction (3.1) is kinetically favored over Reaction (3.2) on a group of "high hydrogen overpotential" cathodes, which leads to the high selectivity of formate/formic acid observed by many researchers. The most studied of this favored group of cathode materials are Hg, In, Pb, and Sn, but it is not clear which of these metals is the best because their performance seems to depend on other variables such as $CO_2$ pressure, catholyte composition, potential, and temperature.

As discussed, increase in the reduction rate of carbon dioxide may also lead to the reduction of water, which will result in hydrogen formation, lowering current efficiency (CE) in ERC. To compete with alternative processes successfully, carbon dioxide reduction must take place at low potential with high efficiency and selectivity at high current densities. The key to this process is to find a stable catalyst with high overpotential for hydrogen evolution and affinity for $CO_2$, allowing further $H^+$ transfer steps selectively toward forming methanol. Alternative ways of suppressing HER include operating at high $CO_2$ pressure and to use solid polymer electrolyte membrane reactor to increase the availability of $CO_2$ at the electrocatalyst. Copper, both metallic and in oxides form, is considered a promising electrode material as it offers intermediate hydrogen overvoltage and produces more reduced form of carbon dioxide, such as methanol, ethane, methane, and ethylene. As the potential becomes more negative, the generated carbon monoxide/formate suppresses hydrogen evolution.

Many efforts have been made to fundamental studies of the mechanisms and kinetics of ERC on a variety of electrode surfaces [5,7–9]. The reaction pathways and resulting product distributions can be very complex, because they are not only related to the energies of adsorption of a whole range of possible species, including reactants, intermediates, and products, but also depend on the electrocatalyst property, proton availability, identity of electrolyte and applied cathode potential, carbon dioxide concentration, mass transport, pH, and temperature.

The lowest unoccupied molecular orbital (LUMO) of $CO_2$ is found predominantly on the carbon atom and most of proposed electrocatalytic reduction mechanisms proceed via high-energy intermediate of $\cdot CO_2^-$ as shown in Figure 3.1 [10]. The transition between $CO_2$(ads) and $\cdot CO_2^-$(ads) has been suggested to be a common critical limiting step seen by the elbow in plots of j against overpotential common to most systems regardless of the end product produced [11]. This high-energy intermediate is associated with the changing geometry of the linear $CO_2$ molecule to the bent $\cdot CO_2^-$ radical ion, resulting in the observed high overpotential for $CO_2$ reduction [12]. The detailed mechanistic pathways for each product are not clear at present, and in many cases several different schemes have been proposed. For example, Figure 3.1 shows the most commonly proposed pathways for ERC at Cu electrode in aqueous solution to carbon monoxide, formic acid, methane, ethylene, and ethane [10,11,13–15].

Step A, the adsorption of $CO_2$ onto the electrode surface was suggested [8,16,17] prior to $CO_2$ reduction. Sometimes, an intermediate hydration step (step $A_1'$) is required before the adsorption (step $A_2'$) occurs. In this case, the low solubility of $CO_2$ in water, that is, 0.070 M at STP conditions, will limit the availability of $CO_2$(aq) and thus limit the $CO_2$(ad) concentration on the cathode surface, especially when $CO_2$ reacts competitively to give $HCO_3^- / CO_3^{2-}$ in alkaline solution. The use of gas phase flow should overcome this limitation by allowing the formation of $CO_2$(ad) directly from the gaseous state (step A) [11], and catalysts that have high affinity to $CO_2$ or low activation energy for adsorbing $CO_2$ should help raise the concentration of $CO_2$(ad). In addition, ways of increasing $CO_2$ solubility can be useful in aqueous

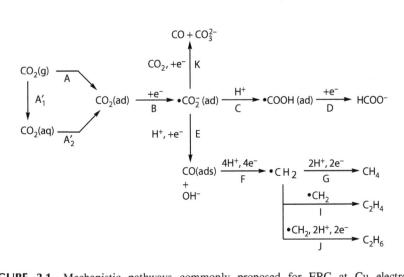

**FIGURE 3.1** Mechanistic pathways commonly proposed for ERC at Cu electrode. (Reproduced from the *Electrochimica Acta*, 10, Kaneco, S. et al. Electrochemical reduction of carbon dioxide to ethylene with high Faradaic efficiency at a Cu electrode in CsOH/methanol, 4701–4706, Copyright 1999, with permission from Elsevier.)

systems, such as operating under high pressure or low temperature and using additives to enhance the solubility of $CO_2$.

Step B is the one-electron reduction of $CO_2(ads)$ to the intermediate radical $\cdot CO_2^-$ species, whose standard redox potential is about $-1.9$ V [18,19] or $-1.85$ V [20] vs. SHE in aqueous media, $-2.21$ V vs. saturated calomel electrode (SCE) in dimethylformamide (DMF) [21].

Adsorbed $\cdot COOH$ species formed by the protonation reaction in carboxylic form, unless stabilized, will tend to revert back to $M-H+CO_2$ or undergo nucleophilic attack to give $M-CO^+$ (i.e., $M-C=O$). This reactive intermediate is thought to be commonly formed for Cu, Ru, and some other d-block metals and is believed to be present in the pathways resulting in highly reduced products.

Step E is the second electronation/protonation to produce a key intermediate species of adsorbed CO for the formation of an adsorbed reactive methylene group, which could be stabilized as $CH_4$, be dimerized to $C_2H_4$ or be transformed to $C_2H_6$.

Pacansky et al. studied self-consistent field (SCF) ab initio molecular orbital energies and atomic population analysis of $\cdot CO_2^-$ at the minimum energy geometry (Figure 3.2). According to their analysis, the unpaired electron density at the highest occupied orbital is localized at C atom at 84% [22]. This result suggests that $\cdot CO_2^-$ is ready to react as a nucleophilic reactant at the carbon atom. Along this line, the weakly adsorbed $\cdot CO_2^-$ goes through a protonation reaction, then there is a second electron transfer to yield formate (steps B–C–D). A parallel disproportionation of the adsorbed $\cdot CO_2^-$ radical anion produces CO (steps B–K) [11,14,23].

It is seen from Table 3.1 that the overall standard redox potentials for ERC are in the same range as that for the reduction of protons to $H_2$ ($E^0 = 0.0$ V [SHE]), which indicates that ERC is not thermodynamically much more difficult than hydrogen evolution. However, ERC is kinetically suppressed because of the high-energy intermediate radical $\cdot CO_2^-$ and the involvement of multielectron transfer, that is, 2 electrons for $HCOO^-$ (steps B–C–D) and 8 electrons for $CH_4$ (steps B–E–F–G).

Research on ERC is still at a stage of fundamental investigations on mechanisms and kinetics using tiny electrodes (e.g., $1 \times 10^{-4}$ m²) with little consideration of the possibility for practical application. Research on the engineering aspects of ERC should be initiated to bridge the gap between the previous laboratory work and industrial reality. Such engineering research includes the design and scale-up of continuous electrochemical reactors, together with the conception, design, and economic projections for complete ERC processes. Akahori et al., who are apparently the first to report continuous operation, used a lead wire bundle cathode in a flow-by reactor with a cation membrane separator. This reactor obtained a formate CE near 100%

**FIGURE 3.2**  Atomic configuration of $\cdot CO_2^-$. (Reproduced with permission from Pacansky, J., Wahlgren, U. and Bagus, P. S. SCF ab-initio ground state energy surfaces for $CO_2$ and $CO_2^-$. *The Journal of Chemical Physics*. 62(7): 2740–2744. Copyright 1975, American Institute of Physics.)

**TABLE 3.1**

**Standard Potential for $CO_2$ Electroreduction Reactions in Aqueous Media at 25°C**

| Reaction | $E^0$/V vs. SHE | $E^{0'}$/V vs. SHE[b] |
|---|---|---|
| $CO_2 + e^- = \cdot CO_2^-$ | −1.90 | – |
| $CO_2 + 2H^+ + 2e^- = CO + H_2O$ | −0.10 | −0.52 |
| $CO_2 + 2H^+ + 2e^- = HCOOH$ | −0.20 | −0.61 |
| $2CO_2 + 2H^+ + 2e^- = H_2C_2O_4$ | −0.475 | −0.889 |
| $CO_2 + 4H^+ + 4e^- = HCHO + H_2O$ | −0.071 | −0.485 |
| $CO_2 + 6H^+ + 6e^- = CH_3OH + H_2O$ | 0.03 | −0.38 |
| $CO_2 + 8H^+ + 8e^- = CH_4 + 2H_2O$ | 0.17 | −0.24 |
| $2CO_2 + 12H^+ + 12e^- = C_2H_4 + 4H_2O$ | 0.07 | −0.34 |
| $2CO_2 + 12H^+ + 12e^- = C_2H_5OH + 3H_2O$ | 0.085 | −0.3287 |
| $3CO_2 + 18H^+ + 18e^- = C_3H_7OH + 5H_2O$ | 0.09 | −0.3237 |
| $2H^+ + 2e^- = H_2$ | 0 | −0.414 |

*Source:* Reproduced from Chaplin, R. P. S. and Wragg, A. A. *Journal of Applied Electrochemistry.* 2003; 33(12): 1107–1123. With permission of Kluwer.

with single-phase flow of a $CO_2$-saturated catholyte solution at 1.4 mL min$^{-1}$ and current about 2 mA (0.02 kA m$^{-2}$). Li and Oloman [24] described the electroreduction of $CO_2$ in a laboratory bench-scale continuous reactor using a flow-by three-dimensional (3D) cathode of 30# mesh tinned copper, with the variables: current (1–8 A), gas phase $CO_2$ concentration (16–100 vol%), and operating time (10–180 min), in operation near ambient conditions (ca. 115 kPa [abs], 300 K). For superficial current densities ranging from 0.22 to 1.78 kA m$^{-2}$, the measured CE, which is one of the performance indicators for electrochemical processes, for HCOO$^-$ = 86%–13%, reactor voltage = 3–6 V, specific energy for HCOO$^-$ = 300–1300 kWh kmol$^{-1}$, space-time yield of HCOO$^-$ = $2 \times 10^{-4}$–$6 \times 10^{-4}$ kmol m$^{-3}$ s$^{-1}$, conversion of $CO_2$ = 20%–80%, and yield of organic products from $CO_2$ = 6%–17%. The initial electroactive species in a primary reaction is subjected to $CO_2$ mass transfer constraint, with CE depending on the current density and partial pressure of $CO_2$ in the gas phase, together with the hydrogen overpotential and mass transfer capacity of the 3D cathode. Figure 3.3 shows the process flow diagram.

## 3.2 ELECTROCHEMICAL KINETICS

### 3.2.1 CERTAIN ASPECTS OF ELECTROCHEMICAL KINETICS

The electrochemical reaction occurs at the interface between the electrode (an electronic conductor) and electrolyte solution (an ionic conductor), composed of a series of steps always includes these three steps: (1) the approach of the reactant species to the electrode surface (first mass transfer); (2) the reaction via heterogeneous electron

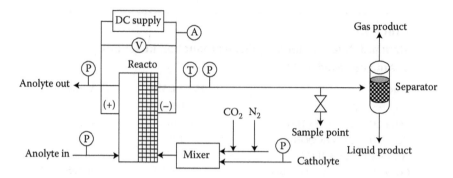

**FIGURE 3.3** Process flow diagram. A = ammeter, p = pressure gauge, T = thermometer, V = voltmeter. (With kind permission from Springer Science+Business Media: *Journal of Applied Electrochemistry*, The electro-reduction of carbon dioxide in a continuous reactor, 35(10), 2005, 955–965, Li, H. and Oloman, C.)

transfer across the interface (actual electrochemical step); and (3) the movement of the product species away from the reaction area into bulk solution (second mass transfer). The actual electrochemical step can be accompanied by different chemical reactions, either in bulk or on the electrode surface. Some of these processes, such as electron transfer at the electrode surface or adsorption, depend upon the electrode potential. Determination of a possible step order and the rate-determining step (r.d.s.) is crucial for the dynamic description of a specific electrode process [25].

For a given electrode reaction:

$$O + ne \underset{k_b}{\overset{k_f}{\rightleftarrows}} R \tag{3.3}$$

where both oxidized form, O, and reduced form, R, of a redox couple are initially present in the bulk solution.

The rate of the forward reduction process, $v_f$, is

$$v_f = k_f C_O(0,t) = \frac{i_c}{nFA} \tag{3.4}$$

whereas the rate of the backward oxidation reaction, $v_b$, is

$$v_b = k_b C_R(0,t) = \frac{i_a}{nFA} \tag{3.5}$$

where $k_f/k_b$ is the forward/backward rate constant, $C_O(0, t)/C_R(0, t)$ is the surface concentration of O/R at time $t$, and $i_c/i_a$ is the cathodic/anodic current. The net reaction rate is

$$v_{net} = v_f - v_b = k_f C_O(0,t) - k_b C_R(0,t) = \frac{i}{nFA} \tag{3.6}$$

in which the overall current is

$$i = i_c - i_a = nFA[k_f C_O(0,t) - k_b C_R(0,t)] \tag{3.7}$$

An accurate kinetic picture of an electrode process is often gained by determining current as a function of potential because there is a proportionality between the current and the net rate of an electrode reaction, and the reaction rate is a strong function of the electrode potential. At equilibrium, the net current is zero, and the electrode potential is defined by the Nernst equation:

$$E = E^{0'} + \frac{RT}{nF} \ln \frac{C_O^*}{C_R^*} \tag{3.8}$$

where $C_O^* / C_R^*$ is the bulk concentration of O/R, and $E^{0'}$ is the former potential. Upon passage of Faradaic current, the electrode potential deviates from its equilibrium value and this is termed electrode polarization. The extent of polarization is determined by the overpotential, $\eta$,

$$\eta = E - E_{eq} \tag{3.9}$$

Based on Butler–Volmer's model of electrode kinetics, for a single one-step, one-electron reversible electrode process without any other chemical step

$$O + e \underset{k_b}{\overset{k_f}{\rightleftharpoons}} R \tag{3.10}$$

The derived current–overpotential equation is

$$i = i_0 \left[ \frac{C_O(0,t)}{C_O^*} e^{-\alpha f \eta} - \frac{C_R(0,t)}{C_R^*} e^{(1-\alpha)f \eta} \right] \tag{3.11}$$

where $f = F/RT$, $\alpha$ is the transfer coefficient, and $i_0$ is the exchange current, which represents the balanced Faradaic current at equilibrium and is equal to either component current, $i_c$ or $i_a$ in magnitude. The value of $i_0$ can be calculated by

$$i_0 = FAk^0 C_O^* e^{-\alpha f (E_{eq} - E^{0'})} \tag{3.12}$$

The curve in Figure 3.4 shows the behavior of Equation 3.11, from which we can see that for the cathodic branch at high $\eta$, the anodic component $i_a$ is insignificant, whereas the anodic contribution $i_c$ is negligible at large positive overpotentials, and the total current is the sum of the components $i_c$ and $i_a$. At extreme $\eta$, the current levels off, reaching a cathodic limiting current, $i_{l,c}$ or an anodic limiting current, $i_{l,a}$,

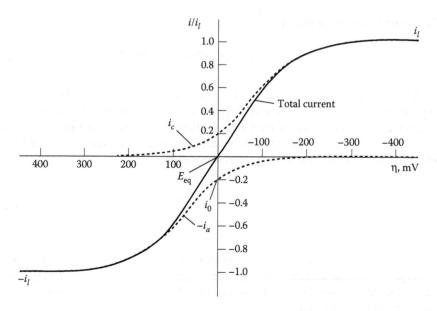

**FIGURE 3.4** Current–overpotential curves for the system with $\alpha = 0.5$, $T = 298$ K, $i_{l,c} = -i_{l,a} = i_l$, and $i_0/i_l = 0.2$. The dashed lines show the component currents $i_c$ and $i_a$. (Bard, A. J. and Faulkner, L. R..: *Electrochemical Methods Fundamentals and Applications*. 2nd ed. p. 100. 2001. Copyright Wiley-VCH Verlag GmbH & Co. KGaA. Reproduced with permission.)

which is limited by mass transfer rather than heterogeneous kinetics and the overpotential is a concentration overpotential. At certain $\eta$, the electrode process could have several slow steps and the current is driven by overpotentials associated with different reaction steps: $\eta_{mt}$ (mass transfer overpotential), $\eta_{ct}$ (charge-transfer overpotential), and $\eta_{rxn}$ (the overpotential associated with a preceding reaction).

The actual relationship between the current and the electrode overpotential usually depends on the nature of the limiting step, which allows hypothesizing about the character of the investigated process.

1. No mass transfer effects
   When mass transfer effects are not included, Equation 3.11 becomes

$$i = i_0[e^{-\alpha f\eta} - e^{(1-\alpha)f\eta}] \quad O + e \underset{k_b}{\overset{k_f}{\rightleftharpoons}} R \tag{3.13}$$

   known as the Butler–Volmer equation historically. The overpotential here is solely for driving the heterogeneous process at the rate reflected by the current. The lower the exchange current, the larger the activation overpotential for any particular net current, indicating more sluggish kinetics.

2. Linear characteristic at small $\eta$
   When the value of $\eta$ is sufficiently small, Equation 3.11 can be expressed as

$$i = -i_0 f\eta \tag{3.14}$$

which indicates that the corresponding $i - \eta$ curve near $E_{eq}$ in a narrow potential range is linear with negative reciprocal slope of $-\eta/i$, which is often called the charge-transfer resistance, $R_{ct}$

$$R_{ct} = \frac{RT}{Fi_0} \tag{3.15}$$

serving as a convenient index of kinetic facility.

3. Tafel behavior at large $\eta$

At large negative/positive overpotentials, one of the bracketed terms in Equation 3.13 becomes negligible. When $\exp(-\alpha f \eta) \gg \exp[(l - \alpha)f\eta]$ for a cathodic branch, Equation 3.13 becomes

$$i = i_0 e^{-\alpha f \eta} \tag{3.16}$$

or expressed as

$$\eta = \frac{RT}{\alpha F} \ln i_0 - \frac{RT}{\alpha F} \ln i \tag{3.17}$$

Compared with the empirical Tafel equation given by Tafel in 1905

$$\eta = a + b \log i \tag{3.18}$$

the constants can now be identified from theory as [25]

$$a = \frac{2.3RT}{\alpha F} \log i_0, \quad b = \frac{-2.3RT}{\alpha F} \tag{3.19}$$

Tafel behavior is an indicator of totally irreversible kinetics. A plot of $\log i$ vs. $\eta$, known as a Tafel plot (Figure 3.6), is useful for the evaluation of kinetic parameters, such as $\alpha$, $i_0$ when applied, though it sharply deviates from linear behavior when $\eta$ approaches zero.

### 3.2.2 OVERPOTENTIALS

#### 3.2.2.1 Basic Concept

The overpotential is the difference between the applied electrode potential, $E_{applied}$, and $E^0$ (products/substrates), at a given current density, reflecting the degree of deviation of the electrode potential from its thermodynamic equilibrium value from the Nernst equation. In other words, an overpotential is the extra amount of nonreversible energy required to overcome the energy barrier which allows the reaction to occur. It shows how far away the electrode is from its equilibrium.

The electrode overpotential mainly consists of three components: concentration overpotential, $\eta_{con}$; activation overpotential, $\eta_{act}$; and ohmic overpotential, $\eta_{ohmic}$, expressed as follows:

$$\eta = \eta_{ohmic} + \eta_{con} + \eta_{act} \tag{3.20}$$

where $\eta$ is the overall overpotential. Other additional overpotentials may exist, such as the chemical reaction overpotential. The cell overpotential cannot be eliminated but can be minimized by proper material modification and good cell design. Furthermore, cell working atmosphere, such as pressure and temperature, can also influence the overpotential.

### 3.2.2.1.1  Ohmic Losses

The ohmic losses are due to resistance created when ions are moving through the electrolyte and electrons moving through the electrode, including the Pt lead wire resistance, electrode resistance, electrolyte resistance, etc. According to Ohm's law:

$$\eta_{ohmic} = IR \tag{3.21}$$

where $R$ is the ohmic resistance. By knowing the conductivity ($\kappa$), conducting length ($l$), and cross-section area ($A_0$), the ohmic resistance of such geometry can be calculated by

$$R = \frac{l}{\kappa A_0} \tag{3.22}$$

Equipotential plane is required to estimate the ohmic resistance of a conductor when the above equation is used. The resistance of a lead wire can be easily evaluated by knowing its conductivity. However, the electrolyte and electrode resistances cannot be directly estimated because the electrochemical reactions at the interface of the electrode and electrolyte cause a nonequipotential plane. An average ohmic resistance is sought. For example, a porous Pt electrode can be viewed as numerous cylindrical pores surrounded by Pt blocks. Each Pt block and electrolyte forms a resistor, as illustrated in Figure 3.5.

The average resistance can therefore be calculated as these numerous resistors are aligned in parallel:

$$R = \frac{l}{\kappa(1-\varepsilon)A_0} \tag{3.23}$$

where $\varepsilon$ is the electrode porosity, and $A_0$ is the electrode physical area. Equation 3.23 is based on assuming the potential field of each single resistor does not interfere with each other. From the above equations, the electrode and electrolyte resistances can be estimated and compared with experimental measurement.

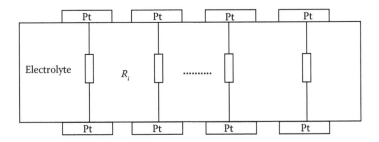

**FIGURE 3.5** Illustration of resistors formed by the electrolyte and electrode. (Reproduced from Tao, G. Investigation of carbon dioxide electrolysis reaction kinetics in a solid oxide electrolyzer, 2003. With permission from ProQuest ILC.)

### 3.2.2.1.2 Concentration Overpotential

A concentration overpotential is caused by the slow diffusion of reactants or products under concentration gradients. It is defined as an external energy required to maintain concentration gradients near the electrode for the sustenance of chemical reactions. This concentration overpotential becomes eminent as mass diffusion effects hinder the entire electrode reaction. For $CO_2$ electroreduction, mass diffusion includes gas bulk diffusion and adsorbates surface diffusion. When the maximum diffusion rate of transporting species to/from the electrode reaction site is slower than the electrochemical reaction rate, diffusion becomes a r.d.s., and a limiting current is reached.

By Fick's diffusion law [27], the limiting current density can be calculated by assuming the reactant concentration at the reaction site is zero:

$$i_L = \frac{nFDC_B}{\delta} \tag{3.24}$$

where $D$ is the diffusivity, $C_B$ is the reactant bulk concentration, and $\delta$ is the thickness of the diffusion layer. Assuming the charge-transfer reaction rate at the electrode is so high that the activation overpotential is negligible compared with the concentration overpotential. The concentration overpotential can therefore be written as

$$\eta_{con} = \frac{RT}{nF} \ln\left(1 - \frac{i}{i_L}\right) \tag{3.25}$$

The concentration overpotential can be reduced by increasing the limiting current, which is a function of concentration and diffusivity as expressed in Equation 3.24. Design of gas bulk diffusion channels and microstructures of an electrode can influence the concentration overpotential.

### 3.2.2.1.3 Activation Overpotential

Activation overpotential describes an external energy required to overcome the maximum activation energy barrier in order to maintain the electrode reaction at an appreciable rate. This irreversible potential occurs when the rate of the electrochemical reaction at the electrode is limited by slow electrode kinetics, or charge transfer. It highly depends on the electrode catalytic characteristics.

Activation overpotentials are affected mostly by the oxidation or reduction rate constants. A decrease in the oxidation or reduction rate constant will increase the activation overpotential. The magnitude of the activation overpotential is more than doubled as the redox rate constant decreases by a magnitude of order two. Therefore, one efficient way to improve the $CO_2$ electroreduction performance is to find a better catalyst with a fast oxidation/reduction rate constant to reduce the activation overpotential. Both the anode and cathode activation overpotentials follow the Tafel behavior, and the charge transfer controls the reaction as discussed below.

### 3.2.2.2 Butler–Volmer Equation

As discussed in Section 3.2.1, for an electrochemical reaction with multiple steps involving one electron without any other chemical step, the activation overpotential can be correlated to current by the Butler–Volmer equation (3.13).

The transfer coefficients are considered as the fraction of the change in polarization, which leads to a change in the reaction rate constant. The transfer coefficients can be given by the Bockris model as [28]

$$\alpha_a = \frac{n - \gamma_c}{\upsilon} - r\beta \tag{3.26}$$

$$\alpha_c = \frac{\gamma_c}{\upsilon} + r\beta \tag{3.27}$$

where $\alpha_a$ and $\alpha_c$ are the anodic and cathodic transfer coefficients, respectively. $\gamma_c$ is the number of charges transferred ahead of the r.d.s., $\upsilon$ is the total number of times of repeating the r.d.s. reaction to complete the reaction, $r$ is the number of charges transferred in the r.d.s., and $\beta$ is the symmetric coefficient (usually 0.5). Apparently, the transfer coefficients have the following relationship:

$$\alpha_a + \alpha_c = \frac{n}{\upsilon} \tag{3.28}$$

The exchange current density is a function of reaction rate constants, the concentration of the anodic and cathodic reactants, etc. It is related to the balanced forward and backward electrode reaction rates at equilibrium. A high exchange current density leads to a fast electrochemical reaction, thus good cell performance is anticipated.

A typical Butler–Volmer equation at large activation overpotential is plotted in Figure 3.6 when the exchange current density is 0.001 mA/cm² and the anodic and cathodic transfer coefficients are unity. The plot shows a high linearity, indicating the Tafel behavior. The plot is symmetric due to the same anodic and cathodic transfer coefficients. Otherwise, it should be asymmetric.

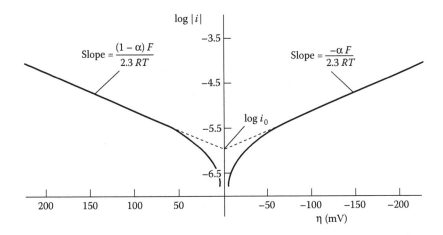

**FIGURE 3.6** Tafel plots for anodic and cathodic branches of the current–overpotential curve for $O + e \underset{k_b}{\overset{k_f}{\rightleftarrows}} R$ with $\alpha = 0.5$, $T = 298$ K, and $j_0 = 10^{-6}$ A/cm$^2$. (Bard, A. J. and Faulkner, L. R.: *Electrochemical Methods Fundamentals and Applications.* 2nd ed. p. 103. 2001. Copyright Wiley-VCH Verlag GmbH & Co. KGaA. Reproduced with permission.)

### 3.2.2.3 Overpotential of CO$_2$ Electroreduction

Direct ERC on most electrode surfaces requires impractically high overpotential which is necessary to drive the process and consequently lowers the conversion efficiency [29].

The standard potential of the reduction of CO$_2$ to different products as shown in Table 3.1, calculated from the thermodynamic data, is much more positive than that the reduction actually takes place at as shown in Table 3.2. Such high overpotential is presumed from involvement of intermediate species, ·CO$_2^-$ anion radical, which requires highly negative potential for the formation. The high overpotential depends on electrode metal, crystal orientation, and reduction product alongside proton availability. CO formation takes place with relatively lower overpotentials than HCOO$^-$ formation. Stabilization of ·CO$_2^-$ adsorption on electrodes may lead to reduction of the overpotential. Among all the electrodes, Au electrode reduces CO$_2$ to CO at remarkably low cathodic potential, $-1.14$ V at 5 mA cm$^{-2}$ (Table 3.2). It is suggested that the formed intermediate species ·CO$_2^-$ on the Au electrode was greatly stabilized by adsorption, leading to decrease of overpotential [30]. Coordination of the CO$_2$ to M can activate ·CO$_2^-$ and promote its reductive disproportionation to form CO + CO$_3^{2-}$.

Clearly, in order to minimize overpotentials, catalysts need to be developed that have formal potentials $E^0(Cat^{n+/0})$ well matched to $E^0(products/CO_2)$, and appreciable rate constants for the electroreduction of CO$_2$ to appropriate products at this potential. In addition, the heterogeneous rate constant for reduction of the electrocatalyst at the electrode must be high for $V_{applied}$ near $E^0(Cat^{n+/0})$.

Transition metal electrodes have been reported to dramatically reduce the overpotential of reduction to at least 500 mV anodic of $E^0$ for CO$_2 \rightarrow$ ·CO$_2^-$. Bruce et al. [32] find $E^0 = +0.1$ V vs. SHE for CO$_2 + 2H^+ + 2e^- \rightarrow$ HCOOH on copper, which is much more positive than that calculated from Gibbs energy of formation.

**TABLE 3.2**

**Potentials and Faradaic Efficiencies of Products of $CO_2$ Reduction at Various Metal Electrodes in 0.1 M $KHCO_3$, $T = 18.5 \pm 0.5°C$**

| Electrode | Potential vs. SHE V | Current Density (mA cm⁻²) | Faradaic Efficiency (%) | | | | | | | |
|---|---|---|---|---|---|---|---|---|---|---|
| | | | $CH_4$ | $C_2H_4$ | EtOH[a] | PrOH[b] | CO | HCOO⁻ | $H_2$ | Total |
| Pb | −1.63 | 5.0 | 0.0 | 0.0 | 0.0 | 0.0 | 0.0 | 97.4 | 5.0 | 102.4 |
| Hg | −1.51 | 0.5 | 0.0 | 0.0 | 0.0 | 0.0 | 0.0 | 99.5 | 0.0 | 99.5 |
| Tl | −1.60 | 5.0 | 0.0 | 0.0 | 0.0 | 0.0 | 0.0 | 95.1 | 6.2 | 101.3 |
| In | −1.55 | 5.0 | 0.0 | 0.0 | 0.0 | 0.0 | 2.1 | 94.9 | 3.3 | 100.3 |
| Sn | −1.48 | 5.0 | 0.0 | 0.0 | 0.0 | 0.0 | 7.1 | 88.4 | 4.6 | 100.1 |
| Cd | −1.63 | 5.0 | 1.3 | 0.0 | 0.0 | 0.0 | 13.9 | 78.4 | 9.4 | 103.0 |
| Au | −1.14 | 5.0 | 0.0 | 0.0 | 0.0 | 0.0 | 87.1 | 0.7 | 10.2 | 98.0 |
| Ag | −1.37 | 5.0 | 0.0 | 0.0 | 0.0 | 0.0 | 81.5 | 0.8 | 12.4 | 94.6 |
| Zn | −1.54 | 5.0 | 0.0 | 0.0 | 0.0 | 0.0 | 79.4 | 6.1 | 9.9 | 95.4 |
| Pd | −1.20 | 5.0 | 2.9 | 0.0 | 0.0 | 0.0 | 28.3 | 2.8 | 26.2 | 60.2 |
| Ga | −1.24 | 5.0 | 0.0 | 0.0 | 0.0 | 0.0 | 23.2 | 0.0 | 79.0 | 102.0 |
| Cu | −1.44 | 5.0 | 33.3 | 25.5 | 5.7 | 3.0 | 1.3 | 9.4 | 20.5 | 103.5[c] |
| Ni | −1.48 | 5.0 | 1.8 | 0.1 | 0.0 | 0.0 | 0.0 | 1.4 | 88.9 | 92.4[d] |
| Fe | −0.91 | 5.0 | 0.0 | 0.0 | 0.0 | 0.0 | 0.0 | 0.0 | 94.8 | 94.8 |
| Pt | −1.07 | 5.0 | 0.0 | 0.0 | 0.0 | 0.0 | 0.0 | 0.1 | 95.7 | 95.8 |
| Ti | −1.60 | 5.0 | 0.0 | 0.0 | 0.0 | 0.0 | not available | 0.0 | 99.7 | 99.7 |

*Source:* Reproduced from *Electrochimica Acta*, 39(11–12), Hori, Y, et al., Electrocatalytic process of CO selectivity in electrochemical reduction of $CO_2$ at metal electrodes in aqueous media. 1833–1839. Copyright 1994, with permission from Elsevier.

[a]  Ethanol.

[b]  n-propanol.

[c]  The total value contains $C_2H_5OH$ (1.4%), $CH_3CHO$ (1.1%), and $C_2H_5CHO$ (2.3%) in addition to the tabulated substances.

[d]  The total value contains $C_2H_6$ (0.2%).

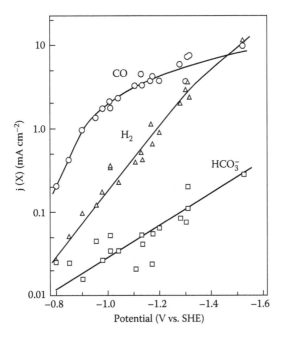

**FIGURE 3.7** Partial current densities of the products $j(X)$ vs. electrode potential in CO₂ reduction at an Au electrode. 0.5 M KHCO₃, 1 atm CO₂, 18°C. (Hori, Y. et al. Electrochemical reduction of carbon dioxides to carbon monoxide at a gold electrode in aqueous potassium hydrogen carbonate. *Journal of the Chemical Society, Chemical Communications.* 1987; (10): 728–729. Reproduced by permission of the Royal Society of Chemistry.)

Fundamental studies will contribute to enhancement of our understanding of the overpotential of the electrodes. A wide range of Tafel slopes for CO₂ reduction from 120 mV decade⁻¹ to 350 mV decade⁻¹ has been reported [5,16,33,34]. These slopes indicate a rate-determining initial electron transfer to CO₂ forming a surface adsorbed ·CO₂⁻ intermediate. The calculated exchange current density for the formation of ·CO₂⁻ radical at 100 kPa and 298 K in CO₂-saturated aqueous solution (~0.036 M) is $4.8 \times 10^{-7}$ kA m⁻² [35]. The exchange current densities for ERC to formate at 293 K in 0.95 M KCl + 0.05 M NaHCO₃ on In, Hg, and Sn are reported, respectively, as $1 \times 10^{-7}$, $1 \times 10^{-10}$, $1 \times 10^{-8}$ kA m⁻² [6].

Gold electrodes yield CO in CO₂ reduction at −0.8 V vs. SHE with a low overpotential in 0.5 M KHCO₃ [30]. The obtained relationship between the potential and logarithm of the partial current of CO formation ($i_c$) (Figure 3.7) showed that the Tafel slope is approximately 130 mV decade⁻¹ in the lower current region, where the transport process of CO₂ does not interfere with the supply of CO₂ to the electrode, the Tafel slope corresponds to the transfer coefficient 0.46. $i_c$ at a constant potential is linear to the CO₂ partial pressure. These data suggest that the CO₂ reduction proceeds in the first order with respect to CO₂, and the r.d.s. of the reaction is the first electron transfer to CO₂. These facts support the reaction scheme discussed in the

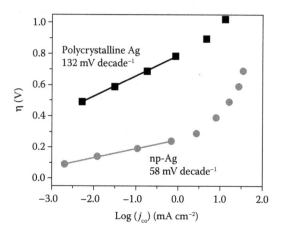

**FIGURE 3.8** Overpotential vs. CO production partial current density on polycrystalline silver and np-Ag. (Reproduced by permission from Macmillan Publishers Ltd. *Nature Communications*, Lu, Q. et al. A selective and efficient electrocatalyst for carbon dioxide reduction. 5: 3242, copyright 2014.)

following. In phosphate buffer solutions of pH 2–6.8 at an Au electrode, $i_c$ is proportional to the pressure of $CO_2$ [36]. The Tafel slope is ca. 120 mV decade$^{-1}$. The potential partial current relation obtained with electrolytes at pH below 4.3 agreed well with that obtained with 0.5 M KHCO$_3$ at pH 7.5 after the double-layer correction due to the difference of the electrolyte concentration [30]. This fact shows that CO formation at Au electrode does not depend on pH of the electrolyte, and that the proton donor is not H$^+$ but H$_2$O molecules.

The intrinsic electrocatalytical activity of nanoporous silver (np-Ag) for CO$_2$ reduction to CO was investigated through Tafel analysis in a CO$_2$-saturated 0.5 M KHCO$_3$, compared with polycrystalline Ag catalyst, as shown in Figure 3.8 [37] As observed, the np-Ag was more preferred for CO$_2$ reduction with a Tafel slope of 58 mV decade$^{-1}$, much smaller than that of 132 mV decade$^{-1}$ for polycrystalline Ag. At a overpotential higher than 0.25 V, the slope increased sharply, indicating another rate-limiting step most probably relating to the issues of the diffusion of reactants and products out of/into the catalyst nanopores. Mechanistically, there are two steps in the two-electron reduction of CO$_2$ to CO on an Ag surface [38]. First, one electron is transferred to a CO$_2$ molecule, forming a ·CO$_2^-$ anion intermediate absorbed on the catalyst surface. Subsequently, the CO$_2^-$ anion takes another electron and two protons, forming a molecule of CO and a molecule of H$_2$O. As reported [39], the first step at a much more negative potential compared with that of the following step is rate determining for the whole process with calculated Tafel slope of 120 mV decade$^{-1}$. This is very similar to that of 132 mV decade$^{-1}$ observed in polycrystalline Ag. For np-Ag, the sharp decrease of Tafel slope at 58 mV decade$^{-1}$ presents a fast step of initial electron transfer before a r.d.s. of later nonelectron transfer [39–41], indicating that the np-Ag with possibly higher index facets supported by the highly curved surface [42,43] are able to stabilize the anion $\cdot CO_2^-$ intermediate much better than a flat one.

Results of studies on the kinetics with a copper foil electrode at 273 and 295 K using two separate electrode pretreatments to remove the oxide layer, that is, 10 wt% HCl and 10 wt% HNO$_3$ [44], is listed in Table 3.3.

The kinetic information/data on ERC is sparse and difficult to summarize because of the variety of approaches and experimental conditions. It has to be noted that the reaction kinetics depend on many interacting factors so results obtained by one researcher under his conditions might not apply under other conditions.

## 3.3  MASS TRANSFER NEAR THE ELECTRODE SURFACE

### 3.3.1  FICK'S LAWS

Diffusion is one of the fundamental processes by which material moves, the macroscopic result of random molecular motion describing the spread of particles, and is an important factor controlling the rate at which many interactions occur. The diffusion coefficient ($D$) is a parameter indicative of the diffusional mobility of species and encountered most famously in Fick's laws (3.29) and (3.30), but also in numerous other equations throughout physics and chemistry. Fick's first law of diffusion states [45]

$$J = -D\frac{\partial C(x)}{\partial x} \tag{3.29}$$

where $J$ is the diffusion flux, $D$ is the diffusion coefficient, $C$ is concentration, and $x$ is the position in the one-dimensional system. This proportionality of flux of matter to the concentration gradient (3.29) shows that the flux of matter is dependent upon the variation in concentration with position. From this relationship, it is suggested that without a concentration gradient there is no net flux and the diffusive flux would

---

**TABLE 3.3**
**Kinetic Information of CO$_2$ Electroreduction to CH$_4$ on a Cu Foil Cathode at Ambient Pressure**

| Temperature (K) | Pretreatment | Tafle Slope V/decade | j$_o$ kA m$^{-2}$ | Highest CE % | CE at −1.46 V (SHE) % |
|---|---|---|---|---|---|
| 273 | 10 wt% HCl | 0.539 | N/A | N/A | 47 |
| 295 | 10 wt% HCl | 0.110 | $1.94 \times 10^{-7}$ | 50 | 20 |
| 295 | 10 wt% HNO$_3$ | 0.170 | N/A | 32 | N/A |

*Source:* Reproduced from *Journal of Electroanalytical Chemistry and Interfacial Electrochemistry*, 245(1–2), Kim, J. J. et al. Reduction of CO$_2$ and CO to methane on Cu foil electrodes, 223–244, Copyright 1988, with permission from Elsevier.

*Note:* Catholyte = 0.5 M KHCO$_3$ at pH = 7.6.

be zero. This first law describes mathematically diffusion within steady-state systems where the concentration is time invariant; however, this is not always the case, which leads to the less specific second law. Fick's second law of diffusion [45]

$$\frac{\partial C(x,t)}{\partial t} = -\frac{\partial J}{\partial x} = D\frac{\partial^2 C}{\partial x^2} \tag{3.30}$$

Fick's second law of diffusion (3.30) predicts the change in concentration (accumulation or depletion) with time as a result of diffusion with $t$ denoting the time. The concentration change is proportional to the diffusion coefficient and the second derivative/curvature of the concentration as illustrated in Figure 3.9. This equation assumes that the diffusion coefficient is independent of composition or the range of compositions is small.

The diffusion coefficient of a species in a given system can be evaluated using the Stokes–Einstein (3.31) and Wilke–Chang (3.32) relations [47,48]:

$$D = \frac{k_B T}{6\pi\eta r} \tag{3.31}$$

$$D = \frac{7.4 \times 10^{-8}(xM)^{1/2}T}{\eta V^{0.6}} \tag{3.32}$$

In the Stokes–Einstein equation (3.31), $k_B$ is the Boltzmann constant, $T$ is the temperature in K, $\eta$ is the dynamic viscosity in Pa·s, and r is the molecular radius of the analyte in m to give $D$ in $m^2 \ s^{-1}$. In the Wilke–Chang relationship (3.32), $x$ is the association parameter of the solvent, M is the molecular mass of the solvent in u, $T$ is the temperature in K, $\eta$ is the dynamic viscosity in cP or mPa·s, and V is the molar volume of analyte at boiling point under standard conditions in $cm^3 \ mol^{-1}$ to give the diffusion coefficient in $cm^2 \ s^{-1}$. Some of the most well utilized electrochemical

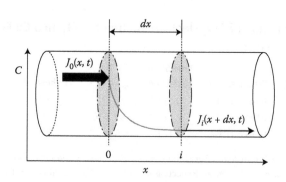

**FIGURE 3.9** Picture of Fick's law showing the change in flux with changing concentration and distance. (Reproduced from Setterfield-Price, B. M. *Electrochemical Reduction of Carbon Dioxide*, 2013. With permission from ProQuest ILC.)

techniques have diffusion as the primary form of mass transport and as such D is a useful parameter to define and can be measured experimentally.

## 3.3.2 SOLUBILITY OF CO$_2$

Traditionally, most of the CO$_2$ reduction research has been carried out in aqueous media. The solubility of CO$_2$ in aqueous electrolyte is quite low, which results in poor mass transport of CO$_2$ to the electrode, and is the limitation to obtain high selectivity at high current density. Under moderate experimental conditions, considering CO$_2$ solubility in water is 30 mM at 1 atm at ambient temperature, the highest current density will be restricted to 20 mA cm$^{-2}$ or so for formation of CO or HCOO$^-$, for example. Such a low transport process must be improved by any means, if the CO$_2$ reduction is utilized for a practical process. Various solutions have been proposed, such as overcoming by operating at elevated pressure, decreased temperature, in alternate organic media, or using 3D electrodes, such as gas diffusion electrodes (GDEs), solid polymer electrolytes, and packed-bed electrodes. As reported, the solubility of CO$_2$ in water increases with the decrease in temperature (Table 3.4), for example, 0.058 mol L$^{-1}$ at 269 K compared with 0.033 mol L$^{-1}$ at 298 K under ambient pressure. At 60 atm and 25°C, the solubility of CO$_2$ in water is over 1 mol L$^{-1}$. Solubility is also increased through alteration of the solvent system. In methanol, the solubility of CO$_2$ is over 1 mol L$^{-1}$ at 25°C and 8 atm, a considerably lower pressure than that required to maintain an equivalent concentration in an aqueous system.

The solubility of CO$_2$ is of great importance for developing the process of ERC because it determines the concentration of CO$_2$ in the liquid phase (CO$_2$(aq)), which affects the rate of CO$_2$ mass transfer to the cathode surface and thus the corresponding product CE.

## 3.3.3 INFLUENCES OF CO$_2$ MASS TRANSFER NEAR THE ELECTRODE SURFACE

When CO$_2$ gas is in contact with an aqueous solution, CO$_2$ is absorbed into the liquid phase as loosely hydrated CO$_2$(aq) (Equation 3.33) that subsequently engages in a relatively slow reaction with water to form carbonic acid, H$_2$CO$_3$ [51–54].

$$CO_2(g) \rightleftarrows CO_2(aq) \tag{3.33}$$

---

**TABLE 3.4**

**Solubility of CO$_2$ in Water (mol/L at 101 kPa CO$_2$ Partial Pressure)**

| T, °C | 15 | 25 | 35 | 45 | 55 | 65 | 75 | 85 | 100 |
|---|---|---|---|---|---|---|---|---|---|
| S | 0.0455 | 0.0336 | 0.0262 | 0.0215 | 0.0175 | 0.0151 | 0.0120 | 0.0090 | 0.0065 |

*Source:* Reproduced from *Progress in Energy and Combustion Science*, 34(2), Bachu, S., CO$_2$ storage in geological media: Role, means, status and barriers to deployment, 254–273, Copyright 2008, with permission from Elsevier.

---

$$CO_2(aq) + H_2O \underset{k_{-0}}{\overset{k_{+0}}{\rightleftharpoons}} H_2CO_3 (K^{\#}) \tag{3.34}$$

For Equation 3.34, the forward reaction ($k_{+0} = 3.0 \times 10^{-2}$ s$^{-1}$) is much slower than the reverse reaction ($k_{-0} = 23.7$ s$^{-1}$) at 298 K. As less than 1% of the dissolved $CO_2$ is present as $H_2CO_3$, Reactions (3.33) and (3.34) are combined to give the pseudo-equilibrium:

$$CO_2(g) + H_2O(l) \rightleftharpoons H_2CO_3^{*} (K_0) \tag{3.35}$$

in which $[HCO_3^{*}] = [CO_2(aq)] + [H_2CO_3]$.

In basic solutions (pH > 7), $CO_2(aq)$ may react directly with OH$^-$:

$$CO_2(aq) + OH^- \underset{k_{-1}}{\overset{k_{+1}}{\rightleftharpoons}} HCO_3^- \quad K_1' \tag{3.36}$$

where the forward reaction ($k_{+1} = 8.5 \times 10^3$ M$^{-1}$ s$^{-1}$) is much faster than the reverse reaction ($k_{-1} = 2.3 \times 10^{-4}$ s$^{-1}$) at 298 K.

The rate of $CO_2$ absorption into the catholyte is determined by the rates of Reactions (3.34) and (3.36) and greatly affected by the physical diffusion rate of $CO_2$ (Equation 3.33, mass transfer step) through the boundary layers at the gas/liquid (G/L) interface, which is strongly influenced by the way $CO_2$ gas is contacted with the liquid.

Combining Equations 3.33, 3.34, and 3.35 gives the mass transfer step as the following equation shows [55,56]. Equations 3.34 and 3.36 are the two thermochemical reaction paths that consume $CO_2(aq)$ [56]:

Mass transfer step:     A.   $CO_2(g) + H_2O(l) \rightleftharpoons CO_2(aq) + H_2O(l) \equiv H_2CO_3^{*}$

Chemical reactions steps:  B.   $CO_2(aq) + H_2O \underset{k_{-0}}{\overset{k_{+0}}{\rightleftharpoons}} H_2CO_3 \rightleftharpoons H^+ + HCO_3^-$

                        C.   $CO_2(aq) + OH^- \underset{k_{-1}}{\overset{k_{+1}}{\rightleftharpoons}} HCO_3^-$

For the mass transfer step A, the rate is given by

$$MT_{rate} = k_{G/L} a p_{CO_2} K_0 \tag{3.37}$$

where:

$k_{G/L}$ = the mass transfer coefficient (m s$^{-1}$)
$a$ = the specific G/L surface area (m$^2$ m$^{-3}$)
$p_{CO_2}$ = the $CO_2$ partial pressure (kPa)
$K_0$ = the equilibrium constant for Equation 3.35, that is, $2.94 \times 10^{-4}$ M$^{-1}$ s$^{-1}$ M kPa$^{-1}$ at 298 K [55].

For the thermochemical reaction steps, for pH < 8, step B dominates and the chemical reaction rate is

$$-\frac{d[CO_2(aq)]}{dt} = k_{+0}h_L\varepsilon[CO_2(aq)] = 3\times10^{-2}h_L\varepsilon[CO_2(aq)] \qquad (3.38)$$

where:

$k_{+0}$ = rate constant of reaction 3.34 ($s^{-1}$)
$h_L$ = liquid hold-up (–)
$\varepsilon$ = voidage of the 3D cathode (–)
$[CO_2(aq)]$ = concentration of dissolved $CO_2$ in the catholyte (M).

At pH > 10, step C is predominant and the rate of $CO_2(aq)$ consumption is

$$-\frac{d[CO_2(aq)]}{dt} = k_{+1}h_L\varepsilon[CO_2(aq)][OH^-] = 8.5\times10^3 h_L\varepsilon[CO_2(aq)][OH^-] \quad (3.39)$$

where:

$k_{+1}$ = rate constant of reaction 3.36 ($s^{-1}$).

In the pH range of 8–10, both steps are important, hence the rate should be

$$-\frac{d[CO_2(aq)]}{dt} = (3.0\times10^{-2}+8.5\times10^3[OH^-])h_L\varepsilon[CO_2(aq)] \qquad (3.40)$$

Table 3.5 shows the calculated mass transfer rate and total thermochemical reaction rates at different pH values where $CO_2$ partial pressure and temperature are 145 kPa and 298 K, respectively.

Comparing the mass transfer and thermochemical reaction rates, it may be seen that when a catholyte solution of carbonate or of low ratio of bicarbonate/carbonate is fed, that is, pH is higher than about 10, the chemical reaction is so

---

**TABLE 3.5**
**Estimated Mass Transfer and Thermo-Chemical Reaction Rates at Different pH**

| pH | 7.5 | 8.5 | 9.5 | 10.5 |
|---|---|---|---|---|
| Mass transfer rate (M $s^{-1}$) | 0.025 | | | |
| Chemical reaction rate (M $s^{-1}$) | 0.0003 | 0.0006 | 0.003 | 0.03 |

*Source:* Reproduced from Li, H. Development of a continuous reactor for the electrochemical reduction of carbon dioxide. 2006. With permission from ProQuest ILC.
*Note:* Assuming a = 7000 $m^2$ $m^{-3}$, $h_L$ = 0.5, $\varepsilon$ = 0.5, $k_{G/L}$ = 1 × $10^{-4}$ m $s^{-1}$.

---

fast that one cannot expect to have both a relatively high concentration of $CO_2(aq)$ and high bulk pH simultaneously. Hence, depending on the $CO_2$ pressure, G/L mass transfer capacity, liquid hold-up, and cathode voltage, the catholyte feed solution should have a pH lower than about 10 and a buffer capacity to maintain the bulk catholyte pH less than 10 throughout the reaction. The $CO_2(aq)$ concentration and pH at the reactive cathode surface depend on these bulk conditions, coupled with the current density and L/S mass transfer coefficient, as detailed in Reference 57. Considered along with the equilibria, these mass transfer and thermochemical reaction rates may favor the current density-dependent reduction of $CO_2$ to formate in the bulk pH range about 7–10 at 298 K. The changes in temperature that occur in a practical case will probably shift this pH window. Kinetic data are not available to calculate the effect of temperature, but applying the rule of thumb, that is, doubling of chemical reaction rates for 10 K increase in T, moves the upper limit of pH down from about 10 to 9 as temperature increases from 298 to 333 K. Apart from the above effects, the electrochemical kinetics of Reactions (3.1) and (3.2) play a major role in determining the formate CE over the whole pH range, and particularly at pH below about 6, where $CO_2$ sequestration as $HCO_3^-/CO_3^{2-}$ is not an issue.

As mentioned above, there exist overpotentials, or polarizations associated with the mass transport processes. For industrial electrochemical processes, the superficial current density and CE should be, respectively, at least 1 kA m$^{-2}$ and 50%. However, the relatively low solubility of $CO_2$ in aqueous solutions (ca. 70 mM at STP), coupled with the $CO_2(aq)/HCO_3^-/CO_3^{2-}$ equilibria, creates a mass transfer constraint on the reduction of $CO_2$. Due to this mass transfer resistance, the resultant current remains very low even at the most active electrocatalyst. The primary current density limits to a maximum value of the order 10 mA cm$^{-2}$ under the typical laboratory reaction conditions (one-phase flow, 2D cathode) with 100 kPa (abs) $CO_2$ pressure at 298 K. Several devices have been suggested to relieve the $CO_2$ mass transfer constraint, including operation at super-atmospheric pressure and/or subambient temperature, using a GDE or using a fixed-bed cathode while providing a "3-phase interface" for the reaction by sparging the cathode chamber with $CO_2$ gas [58–63].

## 3.4 KINETICS AND CATALYSIS OF CO₂ ELECTROREDUCTION

### 3.4.1 STEPS IN CO₂ ELECTROREDUCTION

As stated, most of the $CO_2$ electrocatalytic reduction on the surface of an electrode consists of several steps, including mass transfer of the reactants, surface reactions, interfacial charge transfer, and products involving bulk and surface diffusion, forming a complicated system whose components in those different processes affect each other. Each of them may give rise to a polarization phenomenon as discussed above. Figure 3.10 gives the plausible mass transfer at the electrode–electrolyte boundary layer.

Lack of exact rate constants makes the investigation of $CO_2$ electroreduction mechanisms even more difficult.

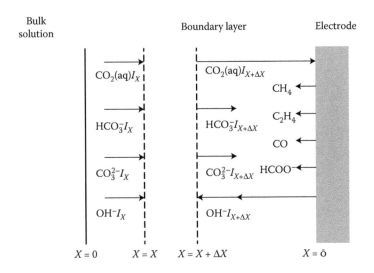

**FIGURE 3.10** Mass transfer across the electrode–electrolyte boundary layer. (With kind permission from Springer Science + Business Media: *Journal of Applied Electrochemistry*, Calculation for the cathode surface concentrations in the electrochemical reduction of CO$_2$ in KHCO$_3$ solutions, 36, 2006, 161–172, Gupta, N., Gattrell, M. and MacDougall, B.)

### 3.4.2 DYNAMIC INFLUENCES ON CO$_2$ ELECTROREDUCTION

CO$_2$ electrochemical reduction is a highly surface-sensitive reaction. The CO$_2$ coordinates to the metal via either a $\eta^1$ or $\eta^2$ bond depending upon the identity of the metal and its surface conditions [12,14], to give adsorbed carbon dioxide. The CO$_2$(ads) then accepts an electron to form the adsorbed radical species, $\cdot$CO$_2^-$(ads) as mentioned above, from which a number of reaction pathways are possible as discussed below.

In addition, the rates of CO$_2$ reduction can also be significantly affected within the system by electrolyte size and charge as well. For example, rates increased Cl$^-$ < Br$^-$ < I$^-$ for CO$_2$ reduction in MeOH/H$_2$O on Cu [64,65], and rates increased upon moving from Na$^+$ to La$^{3+}$ in an aqueous Cu system [64].

### 3.4.3 TURNOVER FREQUENCY OF CO$_2$ HYDROGENATION

The calculated turnover frequency (TOF), as a function of the oxygen adsorption energy, $\Delta E_O$, at ambient pressure and 500 K, is investigated as shown in Figure 3.11 [66]. The optimum in reaction rate is a result of competition between having a too weak interaction with oxygen, resulting in too unstable intermediates and high reaction barriers and a too strong coupling to oxygen, giving rise to surface poisoning by formate, and possibly other species bound through oxygen. As observed, elemental copper is closest to the top, whereas nickel and palladium bind oxygen too strongly and weakly, respectively. Zinc doping in a copper surface step, which can be used to model the active site of Cu/ZnO/Al$_2$O$_3$ industrial catalysts, [67], have close to

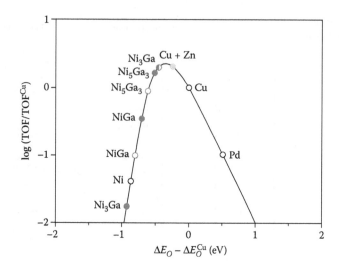

**FIGURE 3.11** Theoretical TOF of $CO_2$ hydrogenation to methanol as a function of $\Delta E_O$, relative to Cu(211). $\Delta E_O$ for Ni–Ga intermetallic compounds is depicted as shown in the left side of the curve. Closed circles indicate nickel-rich sites, open circles gallium-rich sites, and half-open circles mixed sites. Reaction conditions: 500 K, 1 bar, and a $CO_2$:$H_2$ ratio of 1:3. (Reproduced by permission from Macmillan Publishers Ltd. *Nature Chemistry*, Studt, F. et al. Discovery of a Ni-Ga catalyst for carbon dioxide reduction to methanol. 6(4): 320–324, copyright 2014.)

optimal activity. However, the density of such sites is small in a doped system [67], and a more homogeneous catalyst with the same activity per site but more active sites would be advantageous. The model, a good description either in theoretical or experimental view, provides a good starting point for discovering potential catalysts.

Generally, commercial electrochemical processes typically run at a turnover rate of about 1–10 per second.

### 3.4.4 REACTION KINETICS AND MECHANISM OF $CO_2$ ELECTROREDUCTION

#### 3.4.4.1 Metal Electrodes

Although there is no general consensus on the mechanism for hydrogenation at Cu/ZnO catalysts, Cu(I) sites are thought to promote catalytic activity and selectivity toward $CH_3OH$. Although Cu–Zn alloys are considered active sites for $CO_2$ reduction, Cu(I) sites are considered key species for CO adsorption in hydrogenation reactions [68]. Further, Cu(I) sites are believed to stabilize reaction intermediates such as carbonates ($CO_3^{2-}$), formates ($HCOO^-$), and methoxy adsorbates ($H_3CO^-$) due to their higher heats of adsorption [69]. Sheffer et al. showed that alkali metals help stabilize Cu(I) active sites and increase Cu(I) concentrations which significantly improve $CH_3OH$ yield [70]. Oxidation studies at single crystal Cu(I) shows that $H_3CO$ adsorbed at (111) surfaces with Cu(I) atoms at the second atomic layer allows coordinately unsaturated oxygen anions to act as hydrogen abstraction sites for dehydrogenation [71]. In this case, it is possible that the unsaturated oxygen atoms at the (111) surfaces of the cuprous oxide film act as hydrogen donors sites in the reduction reaction.

A theoretical report by Peterson et al. describes a pathway for the electrochemical reduction of $CO_2$ to $CH_4$ at Cu electrodes based on density functional theory (DFT) and computational hydrogen electrode (CHE) models applied to Hori's experimental data [72]. In that work, the authors indicate that the carbon atom of CO adsorbates may be hydrogenated via proton transfer to form HCO at −0.74 V (RHE). Hydrogenation of an adsorbed CO species is proposed to occur directly via proton addition from solution as their availability is significantly greater than hydrogenation from adsorbed hydrogen. Accordingly, once the HCO species is formed, the carbon atom continues proton and electron transfer reactions to form $H_3CO$ adsorbates. Following this pathway, the last proton transfer to the $H_3CO$ species (on Cu(211) surfaces) favors $CH_4$ formation by 0.27 eV. In the case of cuprous oxide electrodes as described here, the reduction reaction may benefit from both improved intermediate stability and the ability of $H^+$ species coordinated with surface-bound oxygen. This surface would allow hydrogen addition to the oxygen atom of the $H_3CO$ adsorbate rather than the carbon atom as shown in Figure 3.12 [73].

The formation of $HCO^-$ intermediate has been reported as the r.d.s. in $CO_2$ reduction to $CH_4$, and $H_2CO^-$ intermediate is detected in both hydrogenation and photoelectrocatalytic process for $CH_3OH$ formation [72,74]. Two possible mechanism pathways (a and b) for the direct electrochemical reduction of $CO_2$ to $CH_3OH$ at Cu/ZnO catalyst are proposed in Figure 3.13 [73]. Although mechanism (a) proceeds through CO pathway, mechanism (b) undergoes formate ($HCOO^-$) intermediate. The last four steps in both (a) and (b) mechanism pathways are the same with HCO adsorbate formation, $H_2CO$ adsorbate formation, $H_3CO$ adsorbate formation, and hydrogenation of $H_3CO$ adsorbate to $CH_3OH$.

Pathway (a) proceeds via CO intermediate, similar to Peterson and Gattrell's mechanisms for $CO_2$ reduction at Cu surfaces [72,75]. The first electron and proton transfer may be associated with the formation of dioxymethylene ($HOCO^-$). When another electron and proton are added to the OH group of the $HOCO^-$ adsorbate, the $C-OH_2$ bond is broken to desorb a $H_2O$ molecule and leave CO adsorbate on the surface. The reaction continues with the formation of HCO species by H addition to the C atom. Once the HCO species are formed, the carbon atom continues proton and electron transfer reactions to form $H_2CO$ and $H_3CO$ adsorbates. Although it is not

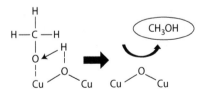

**FIGURE 3.12** Hydrogenation of methoxy adsorbates at $Cu_2O$(111) surfaces. (Le, M. T. H. *Electrochemical Reduction of CO₂ to Methanol.* Louisiana State University, 2011. Reproduced by permission of The Royal Society of Chemistry.)

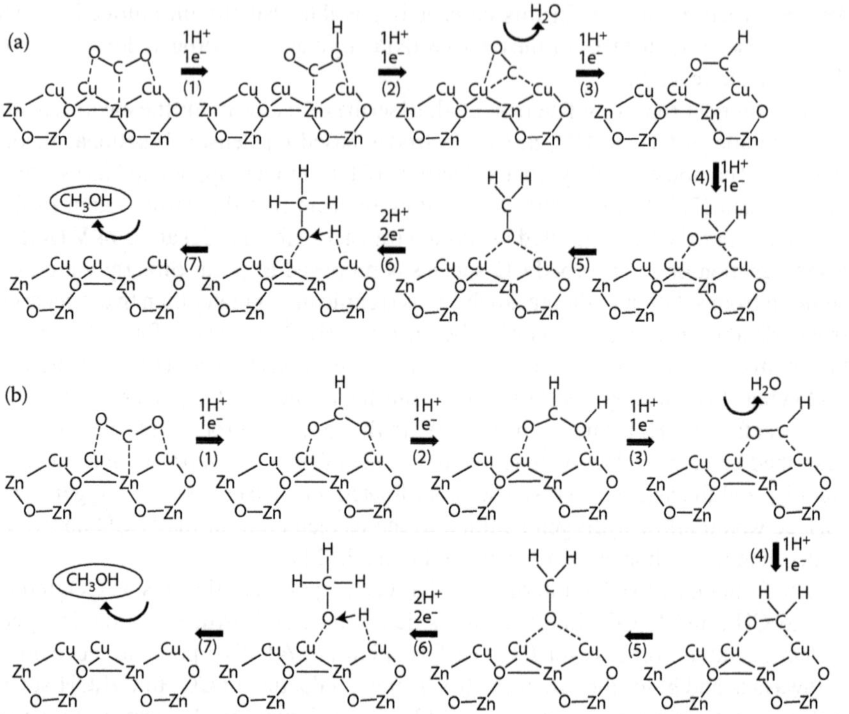

**FIGURE 3.13** Proposed mechanisms through CO pathway (a) and HCOO⁻ intermediate (b) for direct electrochemical reduction of $CO_2$ to $CH_3OH$ using Cu/ZnO catalyst. (Le, M. T. H. *Electrochemical Reduction of $CO_2$ to Methanol*. Louisiana State University, 2011. Reproduced by permission of the Royal Society of Chemistry.)

clear whether these adsorbates would attach to the Cu, Zn, or O bond, ZnO (10–10) surface would promote $CH_3OH$ formation by drawing OH adsorption and allowing the H atom to attach to the O atom of the $H_3CO$ species.

Pathway (b) proceeds via HCOO⁻ intermediate [74]. The H attaches to the C atom instead of the O atom on $CO_2$ adsorbate, hence, forming HCOO⁻ rather than HOCO⁻. The mechanism proceeds with H additions to the O atom of the formate (HCOO) species first to form HCO after an $H_2O$ molecule dissociates from the surface. Similar to mechanism (a), once HCO species are formed, the reaction continues electron and proton transfer until $CH_3OH$ is desorbed from the surface.

In Hg, Tl, Pb, In, Sn, and Cd at aqueous solution, the formation of HCOO⁻ may be initiated by one electron transfer to $CO_2$ at the potential negative of −1.6 V vs. SHE, forming $\cdot CO_2^-$ present mostly freely in the solution close to the electrode as depicted in Figure 3.14 (1). The hydrogen bond by water molecules [76] and the high dielectric constant of water molecule contribute to the stabilization of $\cdot CO_2^-$ in solution. The concentration of $\cdot CO_2^-$ would be ca. $10^{-5}$ mol dm⁻³, if one takes into account the standard potential −1.85 V or −1.90 V and the Nernst relation 59 mV decade⁻¹ at 25°C. Free $\cdot CO_2^-$ will take a proton at the nucleophilic carbon atom from a $H_2O$ molecule

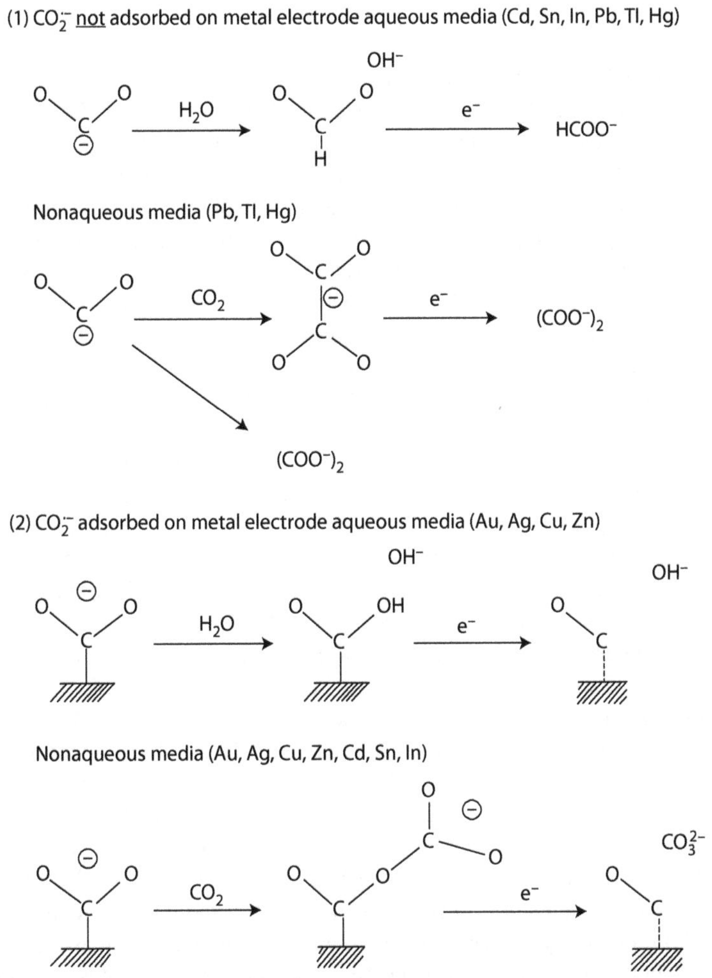

(1) CO$_2^-$ not adsorbed on metal electrode aqueous media (Cd, Sn, In, Pb, Tl, Hg)

Nonaqueous media (Pb, Tl, Hg)

(2) CO$_2^-$ adsorbed on metal electrode aqueous media (Au, Ag, Cu, Zn)

Nonaqueous media (Au, Ag, Cu, Zn, Cd, Sn, In)

**FIGURE 3.14** Reaction scheme of the electrochemical reduction of CO$_2$. (Reproduced from the *Electrochimica Acta*, 39, Hori, Y. et al. Electrocatalytic process of CO selectivity in electrochemical reduction of CO$_2$ at metal electrodes in aqueous media, 1833–1839, Copyright 1994, with permission from Elsevier.)

acting as a Lewis acid, forming HCO$_2$. H$^+$ will not be bonded to the O atom of ·CO$_2^-$ because pKa value of the acid–base couple (·CO$_2^-$/CO$_2$H) is low at 1.4 [77]. HCO$_2$·; is subsequently reduced to HCOO$^-$ at the electrode in aqueous media. The electrode potentials of Pb, Cd, and Tl are −1.6 V or more negative at 5.0 mA cm$^{-2}$ [31]. The potential of Hg is −1.51 V at 0.5 mA cm$^{-2}$, which would be equivalent to −1.7 V at 5.0 mA cm$^{-2}$ with an assumption of the transfer coefficient for HCOO$^-$ formation 0.25 [5,9,65]. In nonaqueous media, H$_2$O molecule is not available and plenty of CO$_2$ is present. A CO$_2$ molecule will play a role of a Lewis acid with the nucleophilic C of ·CO$_2^-$. The coupling of ·CO$_2^-$ and CO$_2$ will lead to the formation of an adduct intermediate $\cdot(CO_2)_2^-$, thus producing oxalate [76], which is more probable than Savéant's coupling mechanism [78]. By contrast, CO is produced by a mechanism different from HCOO$^-$ formation which may proceed with free $\cdot CO_2^-$ intervening. CO is favorably produced from the electrode metals which stabilize $\cdot CO_2^-$ effectively.

On the metal electrode surface of Au, Ag, Zn, Pd, Ga, or Cu, giving CO as the major product, $\cdot CO_2^-$ is stabilized by adsorption both in aqueous and nonaqueous electrolytes. Electrophilic reagents, $H_2O$ in aqueous solution, react with the O atom of adsorbed $\cdot CO_2^-$, forming CO(ad) and OH$^-$ as depicted in Figure 3.14 (2). H$^+$ will not take part in the CO formation as the partial current of CO formation is independent of pH. CO(ad) is readily desorbed from the electrode as a gaseous molecule. The sequence of CO selectivity roughly agrees with that of the electrode potentials shown in Table 3.2. In nonaqueous media, a $CO_2$ molecule reacts as a Lewis acid with adsorbed $\cdot CO_2^-$, and allows a C–O bond of the $\cdot CO_2^-$ to be broken. This process forms CO(ad) and $\cdot CO_2^-$ as postulated by Hammouche and his coworkers [79] for electrochemical reduction of $CO_2$ catalyzed by iron porphyrins. CO(ad) thus formed is easily desorbed, as shown in Figure 3.14 (2). Cd, Sn, and In, of medium CO selectivity, do not strongly adsorb $\cdot CO_2^-$. $\cdot CO_2^-$ will be most freely present in aqueous electrolyte owing to the stabilization by water molecules over hydrogen bond [76] and the high dielectric constant. $\cdot CO_2^-$ stabilized in the electrolyte will be further reduced to HCOO$^-$. However, $\cdot CO_2^-$ is not sufficiently stabilized in nonaqueous electrolyte due to lack of hydrogen bond formation and low dielectric constant of the solvents. Thus, $\cdot CO_2^-$ adsorbed on Cd, Sn, and In may be relatively more stable than $\cdot CO_2^-$ dissolved in the electrolyte. These metals yield CO in nonaqueous media in the same manner as Au, Ag, and Zn in Figure 3.14 (2). The CO selectivity mentioned above will be closely connected to the stability of adsorbed $\cdot CO_2^-$ on the electrode.

In $H_2O$–DMF solutions, the electrochemical reduction of $CO_2$ at Pb and Hg give mainly oxalate or formate, depending on the concentration of $H_2O$, with scarce formation of CO at relatively low yield [80–82]. It is proposed that an electron transfer to $CO_2$ molecule initiates the process, forming $\cdot CO_2^-$, which is not adsorbed on Pb, Hg, and Tl, and is freely present in nonaqueous electrolyte solution as well. And the interactions between the electrode and the reactants, intermediates, and products are negligible as the $CO_2$ reduction at Hg and Pb electrodes proceeds at potentials close to the standard potential of $\cdot CO_2^-/CO_2$ couple. Several competing homogeneous reactions sequentially take place in parallel in the electrolyte solution (Figure 3.15). The product distribution at Pb and Hg electrodes is determined by the current density and the concentration of $CO_2$ and $H_2O$ [80]. Oxalate is formed by coupling of $\cdot CO_2^-$, Figure 3.15 (1), formate is formed in a way similar to that in aqueous electrolyte, Figure 3.15 (5)(5'), and CO is formed by sequential reactions (2) and (3) or (3') in Figure 3.15.

### 3.4.4.2 Molecular Electrocatalysts

In the proposed mechanism of molecular catalyst of [Ni$^{II}$cyclam]$^{2+}$ complex (structure 1, Figure 3.16), the [Ni$^I$cyclam]$^+$ complex is assumed to play a crucial role as shown in Figure 3.17, and in particular, a [Ni$^I$cyclam(CO)]$^+$ intermediate species has been detected in the course of selective electroreduction of $CO_2$ to CO in aqueous solution. Fourteen-membered cyclam framework is a crucial prerequisite for the

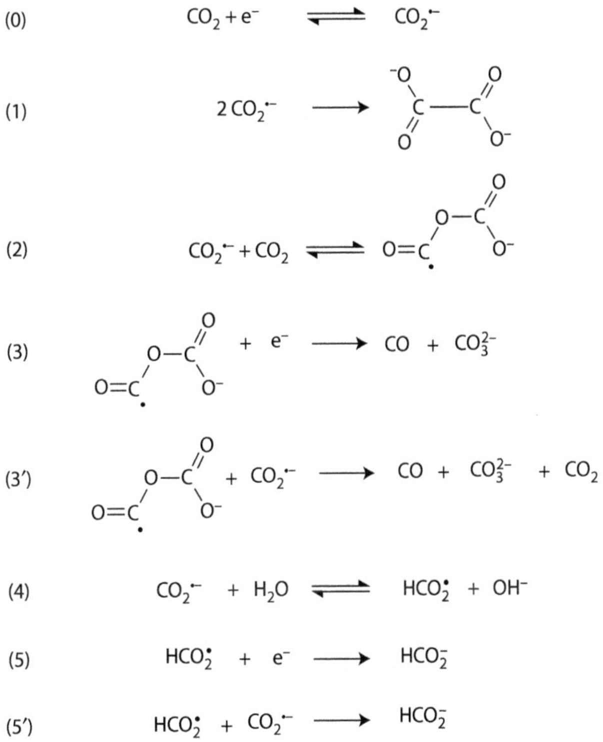

(0) $\quad CO_2 + e^- \rightleftharpoons CO_2^{\cdot -}$

(1) $\quad 2\,CO_2^{\cdot -} \longrightarrow$ [structure: oxalate-type dianion, $^-O$ and $O$ on carbons]

(2) $\quad CO_2^{\cdot -} + CO_2 \rightleftharpoons$ [structure]

(3) [structure] $+ e^- \longrightarrow CO + CO_3^{2-}$

(3') [structure] $+ CO_2^{\cdot -} \longrightarrow CO + CO_3^{2-} + CO_2$

(4) $\quad CO_2^{\cdot -} + H_2O \rightleftharpoons HCO_2^{\cdot} + OH^-$

(5) $\quad HCO_2^{\cdot} + e^- \longrightarrow HCO_2^-$

(5') $\quad HCO_2^{\cdot} + CO_2^{\cdot -} \longrightarrow HCO_2^-$

**FIGURE 3.15** Reaction scheme of $CO_2$ reduction without specific interaction between electrode and the reactants, intermediate, and products. (Reproduced with permission from Gressin, J. et al. Electrochemical reduction of carbon-dioxide in low proton media. *Nouveau Journal De Chimie-New Journal of Chemistry*, 3(8–9), 545–554. Copyright 1979 American Chemical Society.)

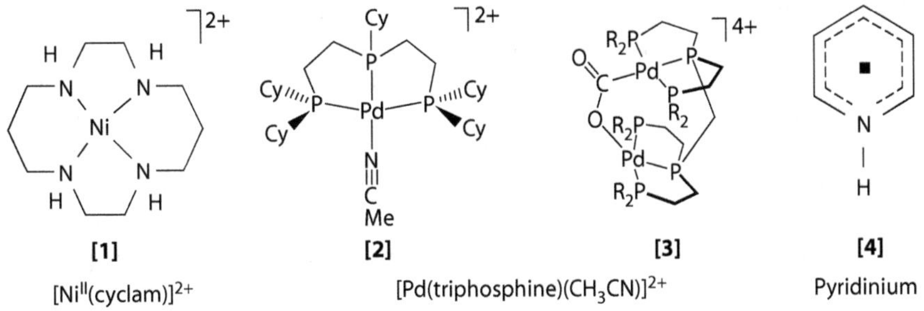

[1]

$[Ni^{II}(cyclam)]^{2+}$

[2]

[3]

$[Pd(triphosphine)(CH_3CN)]^{2+}$

[4]

Pyridinium

**FIGURE 3.16** Molecular catalysts of transition metal complexes and nonmetal nitrogen-based pyridimium.

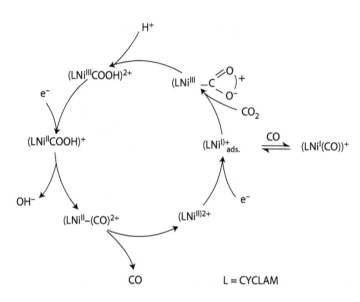

**FIGURE 3.17** Postulated mechanistic cycle for the electrocatalytic reduction of $CO_2$ into CO by Nicyclam$^{2+}$ in water. (Beley, M. et al. Electrocatalytic reduction of carbon dioxide by nickel cyclam2+ in water: Study of the factors affecting the efficiency and the selectivity of the process. *Journal of the American Chemical Society*. 1986; 108(24): 7461–7467. Reproduced by permission of the Royal Society of Chemistry.)

encircled nickel center to act as catalyst in the $CO_2$ electroreduction. The planar ligand geometry of the Ni(cyclam), allowing access to the metal center, has been suggested to be of significance with respect to the catalytic ability exhibited. The system favors the formation of the $Ni^I$ state [83,84]. [Ni(II)-cyclam] is reportedly a very selective electrocatalyst for $CO_2$ reduction to CO and was found to catalyze carbon dioxide reduction on mercury electrodes at ca. 0.5 V below the calculated thermodynamic value ($E^0 = -0.41$ V vs. NHE at pH 5) at a "remarkable" velocity of ca. 32 h$^{-1}$ with 95% CO formation CE. The catalyst was reported to cycle more than 1000 times with no notable deactivation. The system was found to be pH dependent, anion specific, prone to deactivation through CO buildup, resulting in Ni(cyclam) (CO) precipitation if the solution is not stirred and reported to require an Hg surface for appreciable rates and turnover [85, 86].

As being developed to date, the most efficient catalyst of modified Fe-porphyrin [87] with phenolic groups in all ortho and ortho' positions on the phenyl groups [88] as depicted in Figure 3.18 was reported to give CO in Faradaic yields >90% through $5 \times 10^7$ turnovers over 4 h at an overpotential of just 0.465 V in DMF with 0.2 M water added causing no observable degradation. The enhancement in activity was attributed to the higher local concentration of protons associated with the phenolic hydroxyl groups.

For [Pd(triphosphine)(CH$_3$CN)]$^{2+}$ catalysts (structure 2, Figure 3.16), the dissociation of a weakly coordinating solvent molecule of acetonitrile favors M–O bond formation during the electrocatalytic reduction cycle of $CO_2$ to CO. In addition,

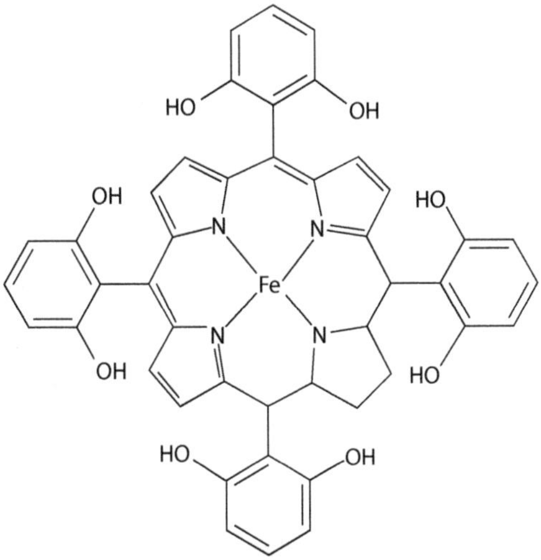

**FIGURE 3.18** Iron 5, 10, 15, 20-tetrakis(2′,6′-dihydroxyphenyl)-porphyrin, FeTDHPP. (Costentin, C., Robert, M. and Saveant, J.-M. Catalysis of the electrochemical reduction of carbon dioxide. *Chemical Society Reviews.* 2013; 42(6): 2423–2436. Reproduced by permission of the Royal Society of Chemistry.)

electron-donating substituents on the triphosphine ligand result in increased catalytic rates. The rate constant is between 5 and 300 $M^{-1}$ $s^{-1}$ [89]. The detail catalytic cycle is proposed as shown in Figure 3.19. As the Pd(II) complex gained an electron to form a Pd(I) intermediate, a reaction with $CO_2$ occurred to form a five-coordinate $CO_2$ adduct. Upon transfer of the second electron, the Pd(0) intermediate dissociated the solvent. Protonation of one of the oxygen atoms of coordinated $CO_2$ affords a metallocarboxylic acid intermediate, Pd–COOH. It is believed that the metallocarboxylic acid is protonated again to form a "dihydroxy carbene," and that CO is then formed by dehydration of the dihydroxy carbene. CO dissociation and solvent association regenerates the initial complex. In solutions of high acid concentration, the r.d.s. was found to be the reaction of the Pd(I) intermediate with $CO_2$, though, in solutions of low acid concentrations, the cleavage of the C–O bond to form carbon monoxide and water limits the rate of the catalytic cycle. These classes of Pd phosphine complexes have shown catalytic rates in the range of 10–300 $M^{-1}$ $s^{-1}$ and with >90% CEs for CO production. Overpotentials were in the range of 100–300 mV, yet turnover numbers were low (ca. 10–100) and the decomposition to Pd(I) dimers and hydrides eventually causes cessation of catalytic activity [85,90,91].

Molecular $CO_2$ reduction catalyst of binuclear palladium complex (structure 3, Figure 3.16), in which two or more independent Pd triphosphine units Pd(P₃) are incorporated and separated by a methylene spacer, binds $CO_2$ through two Pd sites, with one metal interacting with C, the other with O. This complex palladium catalysts show very high catalytic rates, increasing by three orders of magnitude, for $CO_2$ reduction (>104 $M^{-1}$ $s^{-1}$), but with a turnover number of ca. 10. Along with this increase in catalytic activity came an increase in the formation of Pd(I)–Pd(I) bonds,

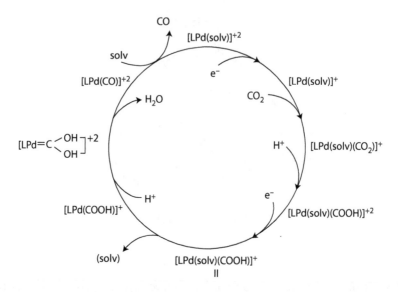

**FIGURE 3.19** Proposed catalytic cycle for the Pd$^{II}$ triphosphine complex catalyst. (Benson, E. E. et al. Electrocatalytic and homogeneous approaches to conversion of CO$_2$ to liquid fuels. *Chemical Society Reviews.* 2009; 38(1): 89–99. Reproduced by permission of The Royal Society of Chemistry.)

thus decreasing catalyst lifetimes. And when the bridge spacer is not methylene, the dendrimers of the Pd catalyst showed decreased activity and selectivity [91].

Studies on the CO$_2$ reduction to CO by the Re(bpy-R)(CO)$_3$X system (R = H, Me, tBu, X = halogen, OTf, or a neutral ligand with a noncoordinating anion) [93] indicated that the complex undergoes a two-electron reduction and loses the labile X ligand to form an anionic tricarbonyl species (A), reported to be the active form of the catalyst, best described as a Re$^0$(bpy-R)$^{-1}$ species [94]. The subsequent mechanistic steps are depicted in Figure 3.20. An intermediate of carboxylate species (B) is formed upon reaction of carbon dioxide with the anion (A). The carboxylate species is protonated to form a carboxylic acid species (C), which could then react with a second proton to liberate water and produce a tetracarbonyl cationic intermediate (D). Although the exact order of addition of protons, addition of electrons, and loss of product remain undetermined, the ligand loss and subsequent transfer of electron density to the metal center have been reported as essential steps in the process [93]. The intermediate species involved in catalysis will likely depend on reduction potentials and pKa's of the complexes compared with the applied potentials and pKa's of the Brönsted acids available in solution. TOFs of the bpy-tBu system greater than 250 s$^{-1}$ have been reported [93]. The increasing TOF, compared with that of relatively low value of 21.4 h$^{-1}$ for Re(bpy)(CO)$_3$Cl in DMF–H$_2$O (9:1) solution, forming CO at −1.49 V vs SCE with CEs at 98% and excellent selectivity over H$_2$ production [95], attributed to the inclusion of tertiary butyl groups at the 4,4′ positions of the bipyridyl ligand [93] and protic additives to the system [96]. To date, it is one of the best and most well-studied electrocatalyst system of CO$_2$, still underway almost 30 years on [92].

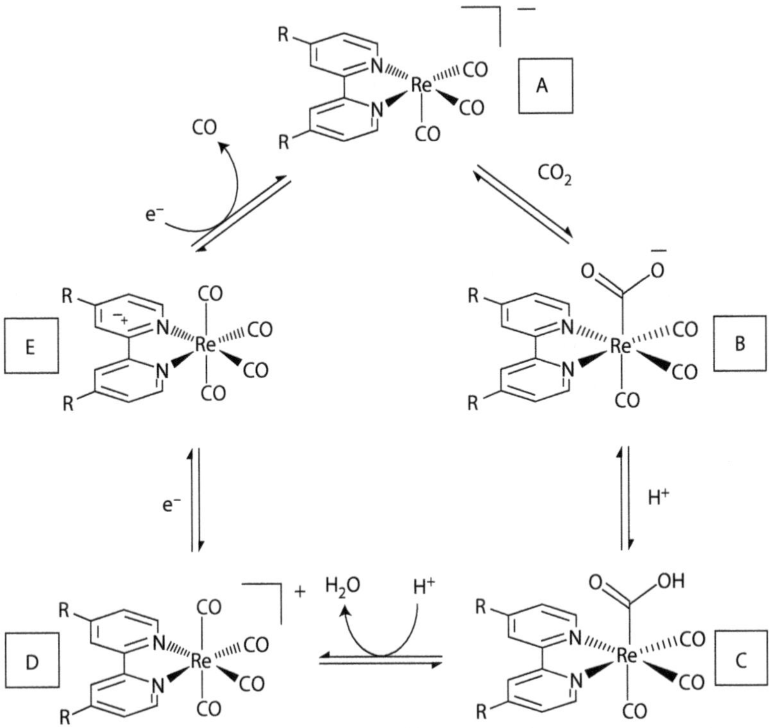

**FIGURE 3.20** Proposed mechanism for the catalytic reduction of CO₂ to CO by Re(bpy-R) (CO)₃X species. (Grice, K. A. et al. Carbon monoxide release catalysed by electron transfer: Electrochemical and spectroscopic investigations of [Re(bpy-R)(CO)4](OTf) complexes relevant to CO₂ reduction. *Dalton Transactions.* 2013; 42(23): 8498–8503. Reproduced by permission of The Royal Society of Chemistry.)

The bipyridine complexes of ruthenium, Ru(bipy)(CO)$_2^{2+}$, were found to electro-catalytically reduce $CO_2$ to CO and $HCOO^-$ at −1.40 V vs. SCE, when water was present [97], by a two-electron reduction process as shown in Figure 3.21 [85,97]. Following the production of CO, a five-coordinate neutral complex is formed. In the presence of $CO_2$, the complex forms an $\eta^1$-$CO_2$ adduct of Ru(0), which can also be formed by addition of two equivalents of $OH^-$ to Ru(bipy)(CO)$_2^{2+}$. Addition of a proton forms the LRu(CO)(COOH) species which under acidic conditions (pH 6.0) gains another proton to lose water and regenerate the catalyst. Under basic conditions (pH 9.5), the catalyst may undergo a two-electron reduction with the participation of a proton to create $HCOO^-$ and regenerate the five-coordinate Ru(0) complex. The insertion and two-electron attack rely on interactions between the reduced complexes and $CO_2$ as the electrophile. In this case, both are activated by the occupation of the $\pi^*$ orbitals of bipyridine. The ligand-based reduction exerts a strong influence on reactivity as electron density at the metal center and in the metal hydride bond is reported as increased making them more susceptible to $CO_2$ attack. The system is

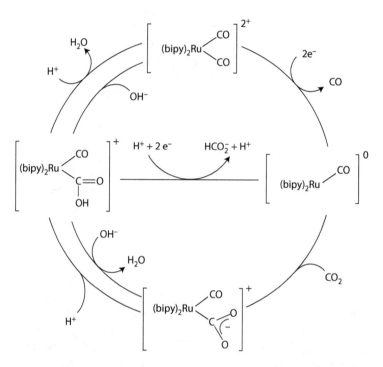

**FIGURE 3.21**   Proposed mechanism for the catalyst of bipyridine complexes of ruthenium. (Benson, E. E. et al. Electrocatalytic and homogeneous approaches to conversion of $CO_2$ to liquid fuels. *Chemical Society Reviews.* 2009; 38(1): 89–99. Reproduced by permission of The Royal Society of Chemistry.)

helpful to elucidate several of the key intermediates in the reduction of $CO_2$, though with low turnover numbers and low selectivity.

cis[Os(bpy)$_2$(CO)H][PF$_6$] reported by Bruce et al. [32] leads to associative activation of $CO_2$ in nonaqueous solution. At a platinum electrode, the reduction took place at −1.4 to −1.6 V (vs. a NaCl-saturated calomel electrode). The proposed mechanism is illustrated in Figure 3.22. The complex accepts two electrons to give the active species which reacts with $CO_2$, binding as an extra ligand, leading to coordination sphere expansion. There are different routes proposed where either the initial interaction with the catalyst results in the disproportionation or two of the direduced molecules may combine to give an uncharged complex and disproportionation products, CO and carbonate. In the experiments where water was present, for example, including trace water of less than 10 mM, they reported the incorporation of a proton into the intermediate direduced species, followed by combination to give formate or CO and hydroxide. The complex structure was investigated with the H found *cis* to the carbonyl being swapped for ligands of different sizes, H > CH$_3$ > CH$_2$–C$_6$H$_5$ > C$_6$H$_5$. It was seen that the rate constant for the reaction decreased with increasing size of this ligand. The *trans* structure was also trialed as a catalyst but the less sterically demanding configuration was less active. This highlighted the impact of relatively small differences in the molecular structure of the catalyst with a pronounced effect upon catalytic ability seen.

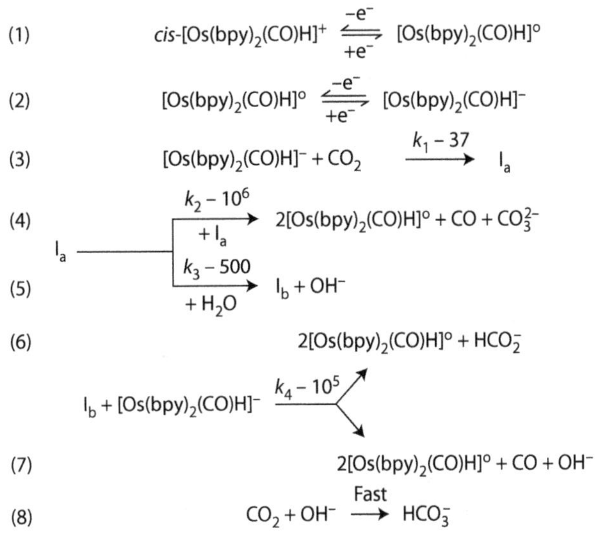

(1) $\quad$ cis-$[Os(bpy)_2(CO)H]^+ \underset{+e^-}{\overset{-e^-}{\rightleftharpoons}} [Os(bpy)_2(CO)H]^\circ$

(2) $\quad [Os(bpy)_2(CO)H]^\circ \underset{+e^-}{\overset{-e^-}{\rightleftharpoons}} [Os(bpy)_2(CO)H]^-$

(3) $\quad [Os(bpy)_2(CO)H]^- + CO_2 \xrightarrow{k_1 - 37} I_a$

(4) $\quad I_a \xrightarrow[+I_a]{k_2 - 10^6} 2[Os(bpy)_2(CO)H]^\circ + CO + CO_3^{2-}$

(5) $\quad \xrightarrow[+H_2O]{k_3 - 500} I_b + OH^-$

(6) $\quad 2[Os(bpy)_2(CO)H]^\circ + HCO_2^-$

$I_b + [Os(bpy)_2(CO)H]^- \xrightarrow{k_4 - 10^5}$

(7) $\quad 2[Os(bpy)_2(CO)H]^\circ + CO + OH^-$

(8) $\quad CO_2 + OH^- \xrightarrow{Fast} HCO_3^-$

**FIGURE 3.22** $\mathit{cis}$-$[Os(bpy)_2(CO)H][PF_6]$ catalyst proposed mechanism. (Reproduced with permission from Bruce, M. R. M. et al. Electrocatalytic reduction of carbon dioxide by associative activation. *Organometallics*. 7(1), 238–240. Copyright 1988 American Chemical Society.)

A dinuclear copper(I) complex incorporating pyridyl-like groups was reported to activate and convert $CO_2$ selectively into oxalate at only a very small applied potential of −0.03 V vs NHE, via a very different and effective catalytic cycle as seen in Figure 3.23. The electrocatalytic reduction of the copper(II) ion to copper(I) appears to be rate limiting, which occurs at a readily accessible potential, around 2 V less negative than that required for $CO_2$ reduction to $\cdot CO_2^-$ in acetonitrile on a glassy carbon electrode [98]. Two dinuclear direduced copper complexes are thought to form a tetranuclear complex with two bridging $CO_2$-derived oxalates which are subsequently released as $Li_2C_2O_4$ precipitate. The reduction of $CO_2$ in this system is found at a significantly less negative potential than that of dioxygen allowing the selective reduction of $CO_2$ from air. Unfortunately, this catalyst only completed six turnovers with 12 equivalents of oxalate formed in 7 h. So it cannot be considered suitable as a realistic catalyst, yet however is the most efficient route to oxalate in terms of overpotential reported to date [87].

Nonmetal nitrogen-based pyridinium cation (structure 4, Figure 3.16) catalyst [99,100] acts as one-electron shuttle to perform multiple-electron, multiple-proton reduction of $CO_2$ effectively in aqueous solution with Faradaic efficiencies of methanol observed ~30% at overpotentials of only −0.2 V on metal electrode. When this pyridinium ion catalyst was anchored to a p-GaP electrode, this photochemical system yielded nearly 100% Faradaic efficiency of methanol at potential 0.3 V below the thermodynamic potential of −0.52 V vs SCE for the reaction at pH 5.2 [101]. The detailed mechanism of pyridinium-catalyzed $CO_2$ reduction in aqueous solution was proposed [99] as given in Figure 3.24. The 6e-reduced product of methanol was

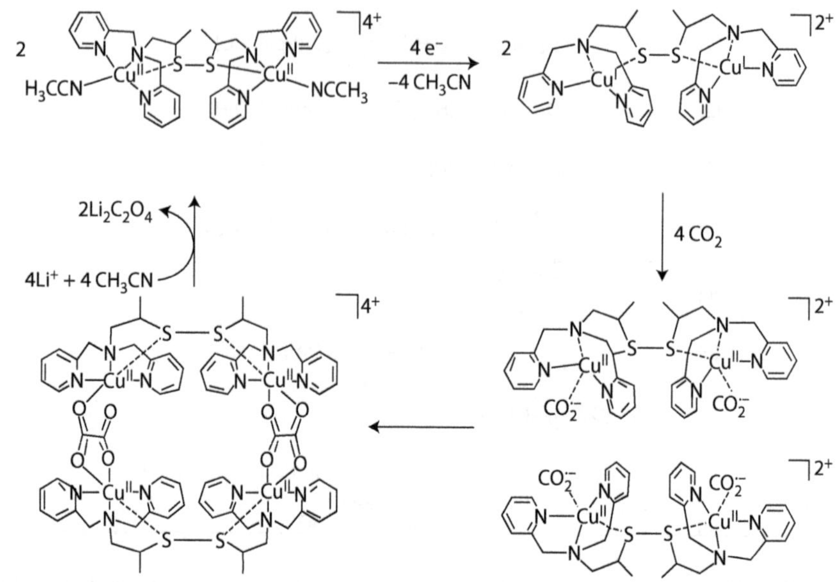

**FIGURE 3.23** Proposed mechanism of $CO_2$ reduction to oxalate by the dinuclear copper complex. (Costentin, C., Robert, M. and Saveant, J.-M. Catalysis of the electrochemical reduction of carbon dioxide. *Chemical Society Reviews*. 2013; 42(6): 2423–2436. Reproduced by permission of the Royal Society of Chemistry.)

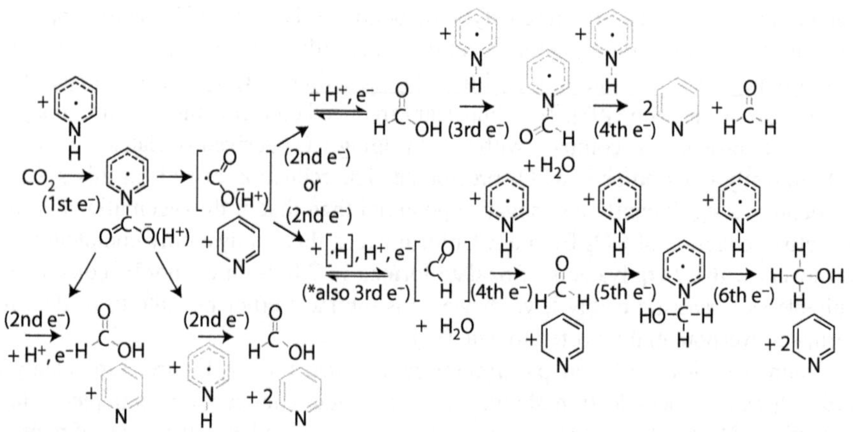

**FIGURE 3.24** Overall proposed mechanism for the pyridinium-catalyzed reduction of $CO_2$ to various products of formic acid, formaldehyde, and methanol. (Reproduced with permission from Barton Cole, E. et al. Using a one-electron shuttle for the multielectron reduction of $CO_2$ to methanol: Kinetic, mechanistic, and structural insights. *Journal of the American Chemical Society*. 132(33), 11539–11551. Copyright 2010 American Chemical Society.)

through a pathway of a series of single-electron transfer rather than the typical multielectron charge transfer. At metal electrodes, formic acid and formaldehyde were observed to be intermediate products with the pyridinium radical playing a role in the reduction of both intermediate products. The r.d.s. of the aqueous system was reportedly determined as being the initial step, formation of the pyridinum–$CO_2$ radical.

Large $NR_4^+$ cations reportedly shift $CO_2$ reduction to more negative potentials and decrease the associated rate constant as their size increases. This is hypothesized to be the blocking effect of the compact layer of adsorbed cations, perhaps linked to the catalysis by tetraalkylammonium cations on CdTe in DMF and acetonitrile [102,103]. The catalysis was most greatly enhanced by the shorter alkyl chain length with $NH_4^+$ optimal. The supposed mechanism through which the enhancement was obtained was given as [103]:

1. $NR_4^+ + e^- \rightarrow NR_4^{\cdot}(ads)$
2. $NR_4^{\cdot}(ads) + CO_2(ads) \rightarrow NR_4^+ + \cdot CO_2^-(ads)$
3. $\cdot CO_2^-(ads) + H^+ + e^- \rightarrow CO(ads) + OH^-$
4. $OH^- + H^+ \rightarrow H_2O$

Upon addition of crown ether catalysts with ring structures in the solution, the tetraalkylammonium species, best able to fit within the cavity, yielded higher catalytic activity. The additional lowering of the overpotential in this case was thought to be a result of a similar mechanism as mentioned above, with the crown ether allowing the ammonium ion to get closer to the electrode surface, which is of specific adsorption in the inner Helmholtz plane, rather than outer Helmholtz, increasing the rate of electron transfer to the $NR_4^+$ species [103].

### 3.4.4.3 Electrocatalytic Activity Degradation

The reduction of $CO_2$ at Cu is inefficient, occurring at high overpotentials toward most probable products of hydrocarbons, and the activity drops rapidly as well [4,104–107]. The proposed mechanisms for the activity degradation include adsorption of organic intermediate [105,106], black carbon deposition [105,108], poison of copper oxide patina [107], accumulation of an unknown low vapor pressure and soluble $CO_2$ reduction product [109], and contaminants $Fe^{2+}$ and $Zn^{2+}$ deposition [4]. However, the real reason behind this is still unclear. The formation mechanism [110] of hydrocarbons [111] is still of a great challenge, though a good start of simulating elements of the reaction mechanism on Cu using DFT [72,112,113]. It is worth noting that Cu electrodes containing a $Cu_2O$ surface layer demonstrated high activity toward $CH_3OH$ production [114,115]. Cu(I) species was considered to play a critical role in the activity of $CH_3OH$ formation. $CH_3OH$ yielded qualitatively follow Cu(I) concentrations.

## 3.5 PRODUCT DISTRIBUTION AND COMPETITIVE REDUCTION BETWEEN CO₂ AND SOLVENT

### 3.5.1 Product Distribution

There is a variety of possible product distributions as various pathway steps are involved in different experimental conditions, though a preference occurs for special occasion. It is noted that small changes can lead to competitive reactions and different routes. Those make generalizations based on individual studies very dubious and grouping difficult.

The reduction product at an electrode in aqueous electrolytes depends on anionic and cationic species of the electrolyte solution, the applied potential, and temperature as discussed. It is reported that CO and HCOO$^-$ were preferentially produced at less negative potentials of $>-0.4$ V vs. Ag/AgCl, whereas $CH_4$, $C_2H_4$, and $C_2H_5OH$ were major products at the potential $<-0.4$ V vs. Ag/AgCl. The origin of the selectivity was hypothesized to relate to the special affinity of hydrogen and/or proton to electrodes. For example, at more negative potential, protons are so abundant on the electrode surface that adsorbed hydrogen-containing carbon molecules are preferentially formed on an Ag electrode. As a result, hydrocarbonization reactions between $H_{ad}$ and $\cdot CO_2^-$ are possible, yielding $CH_4$, $C_2H_4$, and $C_2H_5OH$. On the contrary, at a less negative potential, desorption of $H_{ad}$ occurs, and adsorbed protons are not abundant enough to promote hydrocarbonization reactions. In this case, CO and HCOO$^-$ are preferentially produced.

### 3.5.2 Competitive Reduction between $CO_2$ and Solvent

Higher pH solution increases the presence of "H" on the electrode and the amount of carbon dioxide in solution; therefore, electrodes with low affinity for hydrogen adsorption such as aluminum will exhibit optimum $CO_2$ reduction kinetics at high pH but for others (e.g., palladium) the stronger M–H bonds will mean that H(ads) will dominate the electrode surface, inhibiting $CO_2$ reduction.

In general, to promote $CO_2$ reduction reaction while suppressing the competing HER, an electrocatalyst capable of mediating multiple electron and proton transfers at relatively low overpotentials is desirable [111,116–118].

In, Sn, Hg, and Pb electrodes, all have high overpotential for $H_2$ evolution, and negligible CO adsorption. In aqueous electrolyte, formic acid/formate is selectively produced with Faradaic efficiency between 70% and 100% [23,58,61,119–121].

d group metals (e.g., Pt, Pd, Ru, Fe, Cr, Mo, Ti, Nb, and Ni) have low activity to $CO_2$ reduction while $H_2$ evolution dominates. For example, on Ni electrodes, only CO, HCOOH, $CH_4$, and $C_2H_4$ with Faradaic efficiency less than 1% were detected in the aqueous electrolyte [120]. On Pd electrodes, CO and HCOOH are the main products from $CO_2$ reduction, with maximum Faradaic efficiency <30% for both products [31,120]. Under high pressure, for example, 30 atm, the Faradaic efficiency of CO and HCOOH became comparable with that of $H_2$ [122]. Interestingly, small yield of methanol was reported on Ru electrode [123,124]. Reduction of $CO_2$ on alloy Ru–Pd (1:1) mainly produced HCOOH with a maximum Faradaic efficiency of 90% [125]. $CO_2$ is reduced to adsorbed CO ($CO_{ad}$) on Pt catalysts, the activity of which remarkably depends on the symmetry of the single crystal surface. The following activity series are obtained for stepped surfaces: Pt(111) < Pt(100) < Pt(S) − [n(111) × (100)] < Pt(S) − [n(100) × (111)] < Pt(S) − [n(111) × (111)] < Pt(110). The initial rate of the $CO_2$ reduction gets higher with the increase of the step atom density [126]. Moreover, kinked step surfaces, which contain protruding atoms along the step lines, have higher activity for $CO_2$ reduction than the stepped surfaces [127].

The dominance of $H_2$ evolution could be shifted to $CO_2$ reduction on d group electrodes by increasing the pressure [120,122]. The formation of hydrocarbons on Cu electrode decreased upon increasing pressure [122,128], whereas lowering the temperature improved the hydrocarbons formation [60,105,120,129,130].

Figure 3.25 shows a result of controlled potential reduction of $CO_2$ at a Cu electrode in 0.1 M KHCO₃. The Faradaic yield of $HCOO^-$, which is thought to be the precursor of the final product of hydrocarbons/alcohols and not reduced at all at a Cu electrode, rises at −0.9 V, reaching maximum at −1.20 to −1.25 V, and drops with the increase of the cathodic potential, whereas $H_2$ is formed at lower potential with decreasing Faradaic yield along the line.

As shown in Table 3.6, KHCO₃, KCl, KClO₄, and K₂SO₄ solutions favor the $CO_2$ reduction. K₂HPO₄ solutions highly promote HER rather than $CO_2$ reduction at less negative potential. 0.5 M K₂HPO₄ gives much higher $H_2$ yields than 0.1 M K₂HPO₄. Such an enhancement of HER is attributed to lower pH value at the electrode/electrolyte interface, as mentioned previously. The pH would rise locally at the interface due to $OH^-$ generation in cathodic reactions in aqueous media. Nevertheless, the buffer action of $HPO_4^{2-}$ neutralizes the $OH^-$, keeping the pH at a lower value. KCl, KClO₄, and K₂SO₄ solutions do not have buffer ability, and thus the pH at the electrode/electrolyte interface rises. HER goes up with the concentration of KHCO₃ as indicated in Figure 3.26, owing to its buffer action.

In an ionic liquid catholyte of 1-ethyl-3-methyl imidazolium tetrafluoroborate (EMIM-BF4), the conversion of $CO_2$ to CO over Ag GDE (Figure 3.27) showed substantial improvement over both Faradaic efficiency (~100%) and the overpotential less negative than −0.2 V toward CO formation [29]. The ionic liquid was considered to react with $CO_2$ to form a complex such as $CO_2$–EMIM at potentials more negative

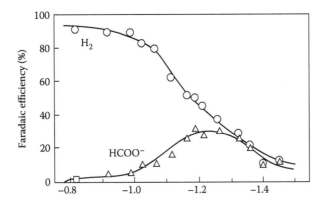

**FIGURE 3.25** Variation of the Faradaic efficiencies of different products in electrochemical reduction of $CO_2$ at a Cu electrode in controlled potential electrolysis, 0.1 mol kg⁻¹ KHCO₃, 1 atm $CO_2$, 18°C. (Hori, Y., Murata, A. and Takahashi, R. Formation of hydrocarbons in the electrochemical reduction of carbon dioxide at a copper electrode in aqueous solution. *Journal of the Chemical Society, Faraday Transactions 1: Physical Chemistry in Condensed Phases.* 1989; 85(8): 2309–2326. Reproduced by permission of the Royal Society of Chemistry.)

**TABLE 3.6**

**Faradaic Efficiencies of Products from the Electroreduction of $CO_2$ at a Cu Electrode at 5 mA cm$^{-2}$ in Various Solutions at 19°C**

| Solution | Concentration, M | pH[a] | Potential V vs. SHE | Faradaic Efficiency (%) | | | | | | | |
|---|---|---|---|---|---|---|---|---|---|---|---|
| | | | | $CH_4$ | $C_2H_4$ | EtOH | PrOH | CO | HCOO$^-$ | $H_2$ | Total |
| $KHCO_3$ | 0.1 | 6.8 | -1.41 | 29.4 | 30.1 | 6.9 | 3.0 | 2.0 | 9.7 | 10.9 | 92.0 |
| KCl | 0.1 | 5.9 | -1.44 | 11.5 | 47.8 | 21.9 | 3.6 | 2.5 | 6.6 | 5.9 | 99.8 |
| KCl | 0.5 | | -1.39 | 14.5 | 38.2 | b | b | 3.0 | 17.9 | 12.5 | 96.0 |
| $KClO_4$ | 0.1 | 5.9 | -1.40 | 10.2 | 48.1 | 15.5 | 4.2 | 2.4 | 8.9 | 6.7 | 96.0 |
| $K_2SO_4$ | 0.1 | 5.8 | -1.40 | 12.3 | 46.0 | 18.2 | 4.0 | 2.1 | 8.1 | 8.7 | 99.4 |
| $K_2HPO_4$ | 0.1 | 6.5 | -1.23 | 17.0 | 1.8 | 0.7 | tr | 1.3 | 5.3 | 72.4 | 98.5 |
| $K_2HPO_4$ | 0.5 | 7.0 | -1.17 | 6.6 | 1.0 | 0.6 | 0.0 | 1.0 | 4.2 | 83.3 | 96.7 |

*Source:* Hori, Y. et al. Formation of hydrocarbons in the electrochemical reduction of carbon dioxide at a copper electrode in aqueous solution. *Journal of the Chemical Society, Faraday Transactions 1: Physical Chemistry in Condensed Phases.* 1989; 85(8): 2309–2326. Reproduced by permission of The Royal Society of Chemistry.

[a] pH values were measured for bulk solution after electrolysis.

[b] Not analyzed.

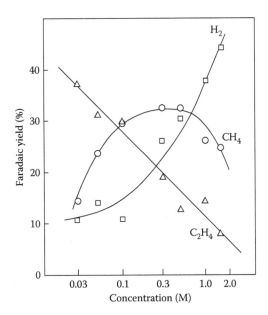

**FIGURE 3.26** Faradaic yields of the products in the $CO_2$ reduction at a Cu electrode at 19°C in $KHCO_3$ aqueous solutions of various concentrations. Current density: 5 mA cm⁻². (Hori, Y., Murata, A. and Takahashi, R. Formation of hydrocarbons in the electrochemical reduction of carbon dioxide at a copper electrode in aqueous solution. *Journal of the Chemical Society, Faraday Transactions 1: Physical Chemistry in Condensed Phases*. 1989; 85(8): 2309–2326. Reproduced by permission of the Royal Society of Chemistry.)

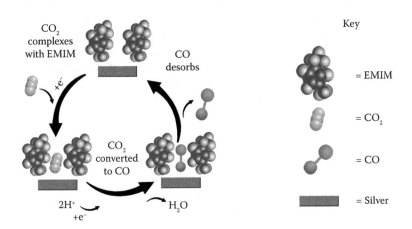

**FIGURE 3.27** Plausible $CO_2$ electrocatalytic reduction pathway in EMIM-BF₄ at Ag electrode. (Reproduced with permission from Rosen, B. A. et al. In situ spectroscopic examination of a low overpotential pathway for carbon dioxide conversion to carbon monoxide. *The Journal of Physical Chemistry C*. 116(29): 15307–15312. Copyright 2012 American Chemical Society.)

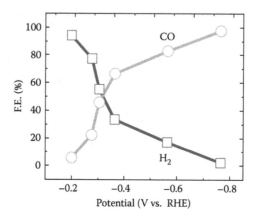

**FIGURE 3.28**　CO and $H_2$ Faradaic efficiency at different applied potentials for bulk $MoS_2$ in $CO_2$-saturated 96 mol% water and 4 mol% EMIM-$BF_4$ solution (pH = 4). (Reproduced by permission from Macmillan Publishers Ltd. *Nature Communications*, Asadi, M. et al. Robust carbon dioxide reduction on molybdenum disulphide edges. 5: 4470, copyright 2014.)

than −0.1 V with respect to SHE. This complex substantially lowered the barrier for formation of the intermediate ·$CO_2^-$ and suppressed $H_2$ evolution [131].

For $MoS_2$ catalysts in the electrolyte of 96 mol% water and 4 mol% EMIM-$BF_4$ at applied potentials between −0.2 and −0.764 V, both $CO_2$ reduction to CO and HER on the electrode surface are observed as shown in Figure 3.28 [132]. The measured FE of the CO and $H_2$ formation, ranging from 0% to ~100% in each component, strongly depends on the applied potential, probably attributed to the difference in the mechanisms of $CO_2$ reduction and HER. At low overpotential, $H_2$ is predominantly produced as the favorable thermodynamic potential for the $H_2$ evolution is low compared with $CO_2$ reduction. However, when the applied potential exceeds the onset potential of −0.164 V at this system for the $CO_2$ reduction, $CO_2$ reduction reaction is activated and dominated instead of HER, in which two $H^+$ from the electrolyte per one $CO_2$ molecule reduction are consumed for CO formation, coupled with the electron transfer on the catalyst surface [40,75,133,134].

## 3.6　SUMMARY AND CHALLENGES

Although the electrochemical conversion of $CO_2$ has great potential, significant technological advances are still needed for this process to become economically viable. There are three criteria [135]: (1) high energy efficiency, (2) high current density (i.e., high reaction rate or turnover number), and (3) long-term stability. So far, the development of electrochemical reduction of $CO_2$ system is slow due to the lack of scientific understanding of $CO_2$ reduction mechanism (C–O, C–C, and C–H bonds formation), pathways and intermediates to generate useful products, providing insights into the $CO_2$ reduction chemistry at the electrode/electrolyte interface and the degradation mechanism of Faradaic efficiency, although some attempts to understand the $CO_2$ electroreduction process through both experimental and theoretical modeling approaches. Such efforts in fundamental mechanistic studies will guide the development of new catalysts and the optimization of operating conditions.

High energetic efficiency can be achieved through a combination of high selectivity (Faradaic or CE) and low overpotentials (Figure 3.29).

Typically, the Faradaic efficiency for many products is >90% for formic acid and carbon monoxide, 65%–70% for methane and ethylene, respectively. High overpotentials are a major hindrance to improving energy efficiency. The reaction rate, as measured by the current density, is also an important parameter as it determines the reactor size and thus capital cost of the process. To date, researchers have reported moderate-to-high current densities (200–600 mA/cm²) using GDEs similar to those used in fuel cells.

The limiting step in the reduction of $CO_2$ is the formation of a $\cdot CO_2^-$ radical anion intermediate. (In this context, "$CO_2$-" does not necessarily denote a bare $\cdot CO_2^-$ anion; instead, it is whatever species forms when an electron is added to $CO_2$.) The equilibrium potential for $\cdot CO_2^-$ formation is very negative in water and in most common solvents [103,136] as discussed above, and this is the reason for the high overpotentials [11,29]. This potential can be improved by stabilizing the intermediate, which is one of the main functions of catalysts. Research has shown that the potential to form this radical anion can be improved by 0.3 V by adsorbing it on a catalyst surface [11]. Further improvements could be possible through optimization of the catalyst.

Another key limiting factor in $CO_2$ conversion is mass transfer of $CO_2$ to the cathode surface, especially given the low solubility of $CO_2$ in many electrolytes. As mentioned earlier, this has largely been overcome using GDEs, which create a three-phase interface between the gaseous reactants, the solid catalyst, and the electrolyte.

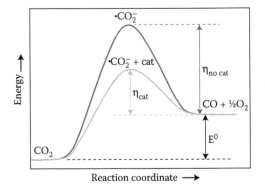

$$\varepsilon_{Faradaic} = \frac{n \cdot F \cdot n}{Q}$$

$$\varepsilon_{energetic} = \frac{E^0}{E^0 + \eta} \times \varepsilon_{Faradaic}$$

Where $\varepsilon_{energetic}$ = energetic efficiency
$\varepsilon_{Faradaic}$ = Faradaic efficiency
$n$ = number of electrons
$F$ = Faraday's constant
$n$ = number of moles of product
$Q$ = charge passed
$E^0$ = standard potential
$\eta$ = overpotential

**FIGURE 3.29** Qualitative reaction scheme for $CO_2$ conversion. Catalysts and electrolytes acting as cocatalysts can lower the energy of the intermediate and thus improve the energetic efficiency of the conversion. We use ε as the symbol for the Faradaic and energetic efficiencies to avoid confusion with the η symbol for the overpotential. (Reproduced with permission from Whipple, D. T. and Kenis, P. J. A. Prospects of $CO_2$ utilization via direct heterogeneous electrochemical reduction. *The Journal of Physical Chemistry Letters.* 1(24), 3451–3458. Copyright 2010 American Chemical Society.)

Thus, optimization of the electrode will be a key to improving current densities. The extensive work on GDE optimization for fuel cells over the past decades will greatly accelerate progress in this area.

Although remarkable achievements have been made in electrochemical reduction of $CO_2$ in the past two decades, there is limited scientific knowledge about $CO_2$ catalytic reduction. Due to the complexity of the reaction environment and multiple bond-forming and breaking processes, the detailed knowledge about the elementary pathways is obscured, which results in the difficulty in understanding the reaction kinetics and controlling the reaction chemistry. The determination of the transient catalytic intermediates is still not possible because of their very short lifetimes and low concentrations as well as difficulty in resolving intermediates spectroscopically. The characterization of surface intermediates under working conditions will require the development and application of sensitive ultrafast spectroscopic and pump-probe scattering methods. Coupling with the experimental probing, theoretical modeling, such as ab initio methods and DFT, is needed to calculate the overall reaction energies and activation barriers, which subsequently lead to construct overall potential energy surfaces. However, so far, simulating reaction conditions that explicitly treat the effects of surface coverage, electrolyte, electrochemical potential, and appropriate materials property, including the effects of particle size, surface orientation, and morphology, present significant challenges.

More work is needed to further understand the competition between $H_2$ evolution reaction and hydrogen protonation process. Multiphysical modeling is required to understand the ion (especially $H^+$) distribution across the buffer layer, the effect of electrolyte circulating rate, cell temperature and reaction gas pressure on the performance.

## REFERENCES

1   Yamamoto, T., Tryk, D. A., Fujishima, A. and Ohata, H. Production of syngas plus oxygen from $CO_2$ in a gas-diffusion electrode-based electrolytic cell. *Electrochimica Acta.* 2002; 47(20): 3327–3334.

2.  Seshan, K., Lercher, J., Paul, J. and Pradier, C. Carbon dioxide chemistry: Environmental issues. *The Royal Society of Chemistry.* 1994; 16.

3.  Coehn, A. and Jahn, S. Electrolytic reduction of carbon dioxide. *Berichte der Deutschen Chemischen Gesellschaft.* 1904; 37: 2836.

4.  Hori, Y., Konishi, H., Futamura, T., Murata, A., Koga, O., Sakurai, H. and Oguma, K. "Deactivation of copper electrode" in electrochemical reduction of $CO_2$. *Electrochimica Acta.* 2005; 50(27): 5354–5369.

5.  Ryu, J., Andersen, T. N. and Eyring, H. Electrode reduction kinetics of carbon dioxide in aqueous solution. *The Journal of Physical Chemistry.* 1972; 76(22): 3278–3286.

6.  Kapusta, S. and Hackerman, N. The electroreduction of carbon dioxide and formic acid on tin and indium electrodes. *Journal of the Electrochemical Society.* 1983; 130(3): 607–613.

7.  Udupa, K. S., Subramanian, G. S. and Udupa, H. V. K. The electrolytic reduction of carbon dioxide to formic acid. *Electrochimica Acta.* 1971; 16(9): 1593–1598.

8.  Sammells, A., Cook, R., Sullivan, B., Krist, K. and Guard, H. *Electrochemical and Electrocatalytic Reactions of Carbon Dioxide.* Elsevier, Amsterdam, 1993.

9. Paik, W., Andersen, T. N. and Eyring, H. Kinetic studies of the electrolytic reduction of carbon dioxide on the mercury electrode. *Electrochimica Acta.* 1969; 14(12): 1217–1232.

10. Kaneco, S., Iiba, K., Hiei, N.-h., Ohta, K., Mizuno, T. and Suzuki, T. Electrochemical reduction of carbon dioxide to ethylene with high Faradaic efficiency at a Cu electrode in CsOH/methanol. *Electrochimica Acta.* 1999; 44(26): 4701–4706.

11. Chaplin, R. P. S. and Wragg, A. A. Effects of process conditions and electrode material on reaction pathways for carbon dioxide electroreduction with particular reference to formate formation. *Journal of Applied Electrochemistry.* 2003; 33(12): 1107–1123.

12. DuBois, D. L. In electrochemical reactions of carbon dioxide. In: Scholz F and Pickett CJ, eds. *Encyclopedia of Electrochemistry.* Wiley-VCH, Weinheim, 2007, pp. 202–224.

13. Hori, Y., Murata, A. and Takahashi, R. Formation of hydrocarbons in the electrochemical reduction of carbon dioxide at a copper electrode in aqueous solution. *Journal of the Chemical Society, Faraday Transactions 1: Physical Chemistry in Condensed Phases.* 1989; 85(8): 2309–2326.

14. Jitaru, M., Lowy, D. A., Toma, M., Toma, B. C. and Oniciu, L. Electrochemical reduction of carbon dioxide on flat metallic cathodes. *Journal of Applied Electrochemistry.* 1997; 27(8): 875–889.

15. Scibioh, M. A. and Viswanathan, B. Electrochemical reduction of carbon dioxide: A status report. *Proceedings of the Indian National Science Academy.* 2004; 70A(3): 407–462.

16. Vassiliev, Y. B., Bagotsky, V. S., Osetrova, N. V., Khazova, O. A. and Mayorova, N. A. Electroreduction of carbon dioxide: Part I. The mechanism and kinetics of electroreduction of CO$_2$ in aqueous solutions on metals with high and moderate hydrogen overvoltages. *Journal of Electroanalytical Chemistry and Interfacial Electrochemistry.* 1985; 189(2): 271–294.

17. Teeter, T. E. and Van Rysselberghe, P. Reduction of carbon dioxide on mercury cathodes. *The Journal of Chemical Physics.* 1954; 22(4): 759–760.

18. Tryk, D. A. and Fujishima, A. Electrochemists enlisted in war on global warming: The carbon dioxide reduction battle. *The Electrochemical Society Interface.* 2001; 10(1): 32–36.

19. Schwarz, H. A. and Dodson, R. W. Reduction potentials of CO$_2^-$ and the alcohol radicals. *The Journal of Physical Chemistry.* 1989; 93(1): 409–414.

20. Surdhar, P. S., Mezyk, S. P. and Armstrong, D. A. Reduction potential of the carboxyl radical anion in aqueous solutions. *The Journal of Physical Chemistry.* 1989; 93(8): 3360–3363.

21. Lamy, E., Nadjo, L. and Saveant, J. M. Standard potential and kinetic parameters of the electrochemical reduction of carbon dioxide in dimethylformamide. *Journal of Electroanalytical Chemistry and Interfacial Electrochemistry.* 1977; 78(2): 403–407.

22. Pacansky, J., Wahlgren, U. and Bagus, P. S. SCF ab-initio ground state energy surfaces for CO$_2$ and CO$_2^-$. *The Journal of Chemical Physics.* 1975; 62(7): 2740–2744.

23. Kaneco, S., Iwao, R., Iiba, K., Ohta, K. and Mizuno, T. Electrochemical conversion of carbon dioxide to formic acid on Pb in KOH/methanol electrolyte at ambient temperature and pressure. *Energy.* 1998; 23(12): 1107–1112.

24. Li, H. and Oloman, C. The electro-reduction of carbon dioxide in a continuous reactor. *Journal of Applied Electrochemistry.* 2005; 35(10): 955–965.

25. Bard, A. J. and Faulkner, L. R. *Electrochemical Methods: Fundamentals and Applications.* 2nd ed. ed: John Wiley & Sons, Inc., New York, NY, p. 100. 2001.

26. Tao, G. Investigation of carbon dioxide electrolysis reaction kinetics in a solid oxide electrolyzer, 2003.

27. Cussler, E. L. *Diffusion: Mass Transfer in Fluid Systems.* Cambridge University Press, Cambridge, 2009.

28. Bockris, J. O. M. and Reddy, A. K. *Modern Electrochemistry: An Introduction to an Interdisciplinary Area*. Springer Science & Business Media, New York, NY, 2012.

29. Rosen, B. A., Salehi-Khojin, A., Thorson, M. R., Zhu, W., Whipple, D. T., Kenis, P. J. A. and Masel, R. I. Ionic liquid–mediated selective conversion of $CO_2$ to CO at low overpotentials. *Science*. 2011; 334(6056): 643–644.

30. Hori, Y., Murata, A., Kikuchi, K. and Suzuki, S. Electrochemical reduction of carbon dioxides to carbon monoxide at a gold electrode in aqueous potassium hydrogen carbonate. *Journal of the Chemical Society, Chemical Communications*. 1987; (10): 728–729.

31. Hori, Y., Wakebe, H., Tsukamoto, T. and Koga, O. Electrocatalytic process of CO selectivity in electrochemical reduction of $CO_2$ at metal electrodes in aqueous media. *Electrochimica Acta*. 1994; 39(11–12): 1833–1839.

32. Bruce, M. R. M., Megehee, E., Sullivan, B. P., Thorp, H., O'Toole, T. R., Downard, A. and Meyer, T. J. Electrocatalytic reduction of carbon dioxide by associative activation. *Organometallics*. 1988; 7(1): 238–240.

33. Li, C. W. and Kanan, M. W. $CO_2$ reduction at low overpotential on Cu electrodes resulting from the reduction of thick $Cu_2O$ films. *Journal of the American Chemical Society*. 2012; 134(17): 7231–7234.

34. Schrebler, R., Cury, P., Herrera, F., Gómez, H. and Córdova, R. Study of the electrochemical reduction of $CO_2$ on electrodeposited rhenium electrodes in methanol media. *Journal of Electroanalytical Chemistry*. 2001; 516(1–2): 23–30.

35. Frese, J. *Electrochemical Reduction of $CO_2$ at Solid Electrodes*. Elsevier, Amsterdam, 1993.

36. Noda, H., Ikeda, S., Yamamoto, A., Einaga, H. and Ito, K. Kinetics of electrochemical reduction of carbon dioxide on a gold electrode in phosphate buffer solutions. *Bulletin of the Chemical Society of Japan*. 1995; 68(7): 1889–1895.

37. Lu, Q., Rosen, J., Zhou, Y., Hutchings, G. S., Kimmel, Y. C., Chen, J. G. and Jiao, F. A selective and efficient electrocatalyst for carbon dioxide reduction. *Nature Communications*. 2014; 5: 3242.

38. Hori, Y. Electrochemical $CO_2$ reduction on metal electrodes. In: Vayenas C, White R and Gamboa-Aldeco M, eds. *Modern Aspects of Electrochemistry*. Springer, New York, 2008, pp. 89–189.

39. Gileadi, E. *Electrode Kinetics for Chemists, Chemical Engineers, and Materials Scientists*. Wiley-VCH Capstone, New York, NY, 1993.

40. Chen, Y., Li, C. W. and Kanan, M. W. Aqueous $CO_2$ reduction at very low overpotential on oxide-derived Au nanoparticles. *Journal of the American Chemical Society*. 2012; 134(49): 19969–19972.

41. Chen, Y. and Kanan, M. W. Tin oxide dependence of the $CO_2$ reduction efficiency on tin electrodes and enhanced activity for tin/tin oxide thin-film catalysts. *Journal of the American Chemical Society*. 2012; 134(4): 1986–1989.

42. Chen, Z., Cummins, D., Reinecke, B. N., Clark, E., Sunkara, M. K. and Jaramillo, T. F. Core–shell $MoO_3$–$MoS_2$ nanowires for hydrogen evolution: A functional design for electrocatalytic materials. *Nano Letters*. 2011; 11(10): 4168–4175.

43. Kibsgaard, J., Chen, Z., Reinecke, B. N. and Jaramillo, T. F. Engineering the surface structure of MoS2 to preferentially expose active edge sites for electrocatalysis. *Nature Materials*. 2012; 11(11): 963–969.

44. Kim, J. J., Summers, D. P. and Frese Jr, K. W. Reduction of $CO_2$ and CO to methane on Cu foil electrodes. *Journal of Electroanalytical Chemistry and Interfacial Electrochemistry*. 1988; 245(1–2): 223–244.

45. Atkins, P. and Paula, J. d. *Atkin's Physical Chemistry*. 8th ed. ed: Oxford University Press, Oxford, 2006.

46. Setterfield-Price, B. M. *Electrochemical Reduction of Carbon Dioxide*, 2013.

47. Einstein, A. A new determination of molecular dimensions. *Annals of Physics*. 1906; 19(2): 289–306.
48. Wilke, C. and Chang, P. Correlation of diffusion coefficients in dilute solutions. *AIChE Journal*. 1955; 1(2): 264–270.
49. Harte, C. R. and Baker, E. M. Absorption of carbon dioxide in aqueous sodium carbonate-bicarbonate solutions. *Industrial and Engineering Chemistry*. 1933; 25(10): 1128–1132.
50. Bachu, S. CO$_2$ storage in geological media: Role, means, status and barriers to deployment. *Progress in Energy and Combustion Science*. 2008; 34(2): 254–273.
51. Keene, F. R., Sullivan, B., Krist, K. and Guard, H. *Electrochemical and Electrocatalytic Reactions of Carbon Dioxide*. Elsevier, Amsterdam, 1993, pp. 1–18.
52. Palmer, D. A. and Van Eldik, R. The chemistry of metal carbonato and carbon dioxide complexes. *Chemical Reviews*. 1983; 83(6): 651–731.
53. Williams, R., Crandall, R. S. and Bloom, A. Use of carbon dioxide in energy storage. *Applied Physics Letters*. 1978; 33(5): 381–383.
54. Walker, A. C., Bray, U. B. and Johnston, J. Equilibrium in solutions of alkali carbonates. *Journal of the American Chemical Society*. 1927; 49(5): 1235–1256.
55. Perry, J. *Chemicals Engineers' Handbook*, 5th ed. ed: McGraw-Hill, New York, 1973.
56. Li, H. *Development of a Continuous Reactor for the Electro-Chemical Reduction of Carbon Dioxide*. University of British Columbia, British Columbia, Canada, 2006.
57. Gupta, N., Gattrell, M. and MacDougall, B. Calculation for the cathode surface concentrations in the electrochemical reduction of CO$_2$ in KHCO$_3$ solutions. *Journal of Applied Electrochemistry*. 2006; 36(2): 161–172.
58. Mahmood, M., Masheder, D. and Harty, C. Use of gas-diffusion electrodes for high-rate electrochemical reduction of carbon dioxide. I. Reduction at lead, indium-and tin-impregnated electrodes. *Journal of Applied Electrochemistry*. 1987; 17(6): 1159–1170.
59. Todoroki, M., Hara, K., Kudo, A. and Sakata, T. Electrochemical reduction of high pressure CO$_2$ at Pb, Hg and In electrodes in an aqueous KHCO$_3$ solution. *Journal of Electroanalytical Chemistry*. 1995; 394(1–2): 199–203.
60. Mizuno, T., Naitoh, A. and Ohta, K. Electrochemical reduction of CO$_2$ in methanol at −30°C. *Journal of Electroanalytical Chemistry*. 1995; 391(1–2): 199–201.
61. Mizuno, T., Ohta, K., Sasaki, A., Akai, T., Hirano, M. and Kawabe, A. Effect of temperature on electrochemical reduction of high-pressure CO$_2$ with In, Sn, and Pb electrodes. *Energy Sources*. 1995; 17(5): 503–508.
62. Köleli, F. and Balun, D. Reduction of CO$_2$ under high pressure and high temperature on Pb-granule electrodes in a fixed-bed reactor in aqueous medium. *Applied Catalysis A: General*. 2004; 274(1–2): 237–242.
63. Yano, H., Shirai, F., Nakayama, M. and Ogura, K. Efficient electrochemical conversion of CO$_2$ to CO, C$_2$H$_4$ and CH$_4$ at a three-phase interface on a Cu net electrode in acidic solution. *Journal of Electroanalytical Chemistry*. 2002; 519(1–2): 93–100.
64. Schizodimou, A. and Kyriacou, G. Acceleration of the reduction of carbon dioxide in the presence of multivalent cations. *Electrochimica Acta*. 2012; 78(0): 171–176.
65. Hori, Y. and Suzuki, S. Electrolytic reduction of carbon dioxide at mercury electrode in aqueous solution. *Bulletin of the Chemical Society of Japan*. 1982; 55(3): 660–665.
66. Studt, F., Sharafutdinov, I., Abild-Pedersen, F., Elkjær, C. F., Hummelshøj, J. S., Dahl, S., Chorkendorff, I. and Nørskov, J. K. Discovery of a Ni-Ga catalyst for carbon dioxide reduction to methanol. *Nature Chemistry*. 2014; 6(4): 320–324.
67. Behrens, M., Studt, F., Kasatkin, I., Kühl, S., Hävecker, M., Abild-Pedersen, F., Zander, S., Girgsdies, F., Kurr, P., Kniep, B.-L., Tovar, M., Fischer, R. W., Nørskov, J. K. and Schlögl, R. The active site of methanol synthesis over Cu/ZnO/Al$_2$O$_3$ industrial catalysts. *Science*. 2012; 336(6083): 893–897.

68. Herman, R. G., Klier, K., Simmons, G. W., Finn, B. P., Bulko, J. B. and Kobylinski, T. P. Catalytic synthesis of methanol from $COH_2$: I. Phase composition, electronic properties, and activities of the $Cu/ZnO/M_2O_3$ catalysts. *Journal of Catalysis*. 1979; 56(3): 407–429.

69. Bailey, S., Froment, G. F., Snoeck, J. W. and Waugh, K. C. A DRIFTS study of the morphology and surface adsorbate composition of an operating methanol synthesis catalyst. *Catalysis Letters*. 1995; 30(1–4): 99–111.

70. Sheffer, G. R. and King, T. S. Potassium's promotional effect of unsupported copper catalysts for methanol synthesis. *Journal of Catalysis*. 1989; 115(2): 376–387.

71. Cox, D. F. and Schulz, K. H. Methanol decomposition on single crystal $Cu_2O$. *Journal of Vacuum Science & Technology A*. 1990; 8(3): 2599–2604.

72. Peterson, A. A., Abild-Pedersen, F., Studt, F., Rossmeisl, J. and Norskov, J. K. How copper catalyzes the electroreduction of carbon dioxide into hydrocarbon fuels. *Energy & Environmental Science*. 2010; 3(9): 1311–1315.

73. Le, M. T. H. *Electrochemical Reduction of $CO_2$ to Methanol*. Louisiana State University, Louisiana, 2011.

74. Yang, Y., Evans, J., Rodriguez, J. A., White, M. G. and Liu, P. Fundamental studies of methanol synthesis from $CO_2$ hydrogenation on Cu (111), Cu clusters, and Cu/ZnO (0001 [combining macron]). *Physical Chemistry Chemical Physics*. 2010; 12(33): 9909–9917.

75. Gattrell, M., Gupta, N. and Co, A. A review of the aqueous electrochemical reduction of $CO_2$ to hydrocarbons at copper. *Journal of Electroanalytical Chemistry*. 2006; 594(1): 1–19.

76. Aylmer-Kelly, A. W. B., Bewick, A., Cantrill, P. R. and Tuxford, A. M. Studies of electrochemically generated reaction intermediates using modulated specular reflectance spectroscopy. *Faraday Discussions of the Chemical Society*. 1973; 56(0): 96–107.

77. Buxton, G. V. and Sellers, R. M. Acid dissociation constant of the carboxyl radical. Pulse radiolysis studies of aqueous solutions of formic acid and sodium formate. *Journal of the Chemical Society, Faraday Transactions 1: Physical Chemistry in Condensed Phases*. 1973; 69(0): 555–559.

78. Bagotzky, V. and Osetrova, N. Electrochemical reduction of carbon dioxide. *Russian Journal of Electrochemistry*. 1995; 31(5): 409–425.

79. Hammouche, M., Lexa, D., Momenteau, M. and Saveant, J. M. Chemical catalysis of electrochemical reactions. Homogeneous catalysis of the electrochemical reduction of carbon dioxide by iron("0") porphyrins. Role of the addition of magnesium cations. *Journal of the American Chemical Society*. 1991; 113(22): 8455–8466.

80. Gressin, J., Michelet, D., Nadjo, L. and Savéant, J. Electrochemical reduction of carbondioxide in low proton media. *Nouveau Journal De Chimie-New Journal of Chemistry*. 1979; 3(8–9): 545–554.

81. Amatore, C. and Saveant, J. M. Mechanism and kinetic characteristics of the electrochemical reduction of carbon dioxide in media of low proton availability. *Journal of the American Chemical Society*. 1981; 103(17): 5021–5023.

82. Gennaro, A., Isse, A. A., Severin, M.-G., Vianello, E., Bhugun, I. and Saveant, J.-M. Mechanism of the electrochemical reduction of carbon dioxide at inert electrodes in media of low proton availability. *Journal of the Chemical Society, Faraday Transactions*. 1996; 92(20): 3963–3968.

83. Beley, M., Collin, J. P., Ruppert, R. and Sauvage, J. P. Electrocatalytic reduction of carbon dioxide by nickel cyclam2+ in water: Study of the factors affecting the efficiency and the selectivity of the process. *Journal of the American Chemical Society*. 1986; 108(24): 7461–7467.

84. Shionoya, M., Kimura, E. and Iitaka, Y. Mono-, di- and tetrafluorinated cyclams. *Journal of the American Chemical Society*. 1990; 112(25): 9237–9245.

85. Benson, E. E., Kubiak, C. P., Sathrum, A. J. and Smieja, J. M. Electrocatalytic and homogeneous approaches to conversion of CO$_2$ to liquid fuels. *Chemical Society Reviews.* 2009; 38(1): 89–99.

86. Beley, M., Collin, J.-P., Ruppert, R. and Sauvage, J.-P. Nickel(II)-cyclam: An extremely selective electrocatalyst for reduction of CO$_2$ in water. *Journal of the Chemical Society, Chemical Communications.* 1984; (19): 1315–1316.

87. Costentin, C., Robert, M. and Saveant, J.-M. Catalysis of the electrochemical reduction of carbon dioxide. *Chemical Society Reviews.* 2013; 42(6): 2423–2436.

88. Costentin, C., Drouet, S., Robert, M. and Savéant, J.-M. A local proton source enhances CO$_2$ electroreduction to CO by a molecular Fe catalyst. *Science.* 2012; 338(6103): 90–94.

89. Rakowski Dubois, M. and Dubois, D. L. Development of molecular electrocatalysts for CO$_2$ reduction and H2 production/oxidation. *Accounts of Chemical Research.* 2009; 42(12): 1974–1982.

90. DuBois, D. L., Miedaner, A. and Haltiwanger, R. C. Electrochemical reduction of carbon dioxide catalyzed by [Pd(triphosphine)(solvent)](BF4)2 complexes: Synthetic and mechanistic studies. *Journal of the American Chemical Society.* 1991; 113(23): 8753–8764.

91. Raebiger, J. W., Turner, J. W., Noll, B. C., Curtis, C. J., Miedaner, A., Cox, B. and DuBois, D. L. Electrochemical reduction of CO$_2$ to CO catalyzed by a bimetallic palladium complex. *Organometallics.* 2006; 25(14): 3345–3351.

92. Grice, K. A., Gu, N. X., Sampson, M. D. and Kubiak, C. P. Carbon monoxide release catalysed by electron transfer: Electrochemical and spectroscopic investigations of [Re(bpy-R)(CO)4](OTf) complexes relevant to CO$_2$ reduction. *Dalton Transactions.* 2013; 42(23): 8498–8503.

93. Smieja, J. M. and Kubiak, C. P. Re(bipy-tBu)(CO)3Cl– improved catalytic activity for reduction of carbon dioxide: IR-spectroelectrochemical and mechanistic studies. *Inorganic Chemistry.* 2010; 49(20): 9283–9289.

94. Benson, E. E., Sampson, M. D., Grice, K. A., Smieja, J. M., Froehlich, J. D., Friebel, D., Keith, J. A., Carter, E. A., Nilsson, A. and Kubiak, C. P. The electronic states of rhenium bipyridyl electrocatalysts for CO$_2$ reduction as revealed by X-ray absorption spectroscopy and computational quantum chemistry. *Angewandte Chemie International Edition.* 2013; 52(18): 4841–4844.

95. Hawecker, J., Lehn, J.-M. and Ziessel, R. Electrocatalytic reduction of carbon dioxide mediated by Re(bipy)(CO)3Cl (bipy = 2,2[prime or minute]-bipyridine). *Journal of the Chemical Society, Chemical Communications.* 1984; (6): 328–330.

96. Smieja, J. M., Benson, E. E., Kumar, B., Grice, K. A., Seu, C. S., Miller, A. J. M., Mayer, J. M. and Kubiak, C. P. Kinetic and structural studies, origins of selectivity, and interfacial charge transfer in the artificial photosynthesis of CO. *Proceedings of the National Academy of Sciences.* 2012; 109(39): 15646–15650.

97. Ishida, H., Tanaka, K. and Tanaka, T. Electrochemical CO$_2$ reduction catalyzed by ruthenium complexes [Ru(bpy)2(CO)2]2+ and [Ru(bpy)2(CO)Cl]+. Effect of pH on the formation of CO and HCOO. *Organometallics.* 1987; 6(1): 181–186.

98. Angamuthu, R., Byers, P., Lutz, M., Spek, A. L. and Bouwman, E. Electrocatalytic CO$_2$ conversion to oxalate by a copper complex. *Science.* 2010; 327(5963): 313–315.

99. Barton Cole, E., Lakkaraju, P. S., Rampulla, D. M., Morris, A. J., Abelev, E. and Bocarsly, A. B. Using a one-electron shuttle for the multielectron reduction of CO$_2$ to methanol: Kinetic, mechanistic, and structural insights. *Journal of the American Chemical Society.* 2010; 132(33): 11539–11551.

100. Seshadri, G., Lin, C. and Bocarsly, A. B. A new homogeneous electrocatalyst for the reduction of carbon dioxide to methanol at low overpotential. *Journal of Electroanalytical Chemistry.* 1994; 372(1–2): 145–150.

101. Barton, E. E., Rampulla, D. M. and Bocarsly, A. B. Selective solar-driven reduction of $CO_2$ to methanol using a catalyzed p-GaP based photoelectrochemical cell. *Journal of the American Chemical Society.* 2008; 130(20): 6342–6344.

102. Taniguchi, I., Aurian-Blajeni, B. and Bockris, J. O. M. The mediation of the photoelectrochemical reduction of carbon dioxide by ammonium ions. *Journal of Electroanalytical Chemistry and Interfacial Electrochemistry.* 1984; 161(2): 385–388.

103. Bockris, J. M. and Wass, J. The photoelectrocatalytic reduction of carbon dioxide. *Journal of the Electrochemical Society.* 1989; 136(9): 2521–2528.

104. Kyriacou, G. and Anagnostopoulos, A. Electroreduction of $CO_2$ on differently prepared copper electrodes: The influence of electrode treatment on the current efficiences. *Journal of Electroanalytical Chemistry.* 1992; 322(1–2): 233–246.

105. DeWulf, D. W., Jin, T. and Bard, A. J. Electrochemical and surface studies of carbon dioxide reduction to methane and ethylene at copper electrodes in aqueous solutions. *Journal of the Electrochemical Society.* 1989; 136(6): 1686–1691.

106. Shiratsuchi, R., Aikoh, Y. and Nogami, G. Pulsed electroreduction of CO copper electrodes. *Journal of the Electrochemical Society.* 1993; 140(12): 3479–3482.

107. Smith, B., Irish, D., Kedzierzawski, P. and Augustynski, J. A surface enhanced Raman scattering study of the intermediate and poisoning species formed during the electrochemical reduction of $CO_2$ on copper. *Journal of the Electrochemical Society.* 1997; 144(12): 4288–4296.

108. Cook, R. L., MacDuff, R. C. and Sammells, A. F. On the electrochemical reduction of carbon dioxide at in situ electrodeposited copper. *Journal of the Electrochemical Society.* 1988; 135(6): 1320–1326.

109. Friebe, P., Bogdanoff, P., Alonso-Vante, N. and Tributsch, H. A real-time mass spectroscopy study of the (electro)chemical factors affecting $CO_2$ reduction at copper. *Journal of Catalysis.* 1997; 168(2): 374–385.

110. Grodkowski, J., Dhanasekaran, T., Neta, P., Hambright, P., Brunschwig, B. S., Shinozaki, K. and Fujita, E. Reduction of cobalt and iron phthalocyanines and the role of the reduced species in catalyzed photoreduction of CO2. *The Journal of Physical Chemistry A.* 2000; 104(48): 11332–11339.

111. Kuhl, K. P., Cave, E. R., Abram, D. N. and Jaramillo, T. F. New insights into the electrochemical reduction of carbon dioxide on metallic copper surfaces. *Energy & Environmental Science.* 2012; 5(5): 7050–7059.

112. Peterson, A. A. and Nørskov, J. K. Activity descriptors for $CO_2$ electroreduction to methane on transition-metal catalysts. *The Journal of Physical Chemistry Letters.* 2012; 3(2): 251–258.

113. Durand, W. J., Peterson, A. A., Studt, F., Abild-Pedersen, F. and Nørskov, J. K. Structure effects on the energetics of the electrochemical reduction of $CO_2$ by copper surfaces. *Surface Science.* 2011; 605(15–16): 1354–1359.

114. Frese, K. W. Electrochemical reduction of $CO_2$ at intentionally oxidized copper electrodes. *Journal of the Electrochemical Society.* 1991; 138(11): 3338–3344.

115. Le, M., Ren, M., Zhang, Z., Sprunger, P. T., Kurtz, R. L. and Flake, J. C. Electrochemical reduction of $CO_2$ to $CH_3OH$ at copper oxide surfaces. *Journal of the Electrochemical Society.* 2011; 158(5): E45–E49.

116. Cook, R. L., Macdugg, R. and Sammells, A. F. Gas-phase $CO_2$ reduction to hydrocarbons at metal/solid polymer electrolyte interface. *Journal of the Electrochemical Society.* 1990; 137(1): 187–189.

117. Whipple, D. T., Finke, E. C. and Kenis, P. J. Microfluidic reactor for the electrochemical reduction of carbon dioxide: The effect of pH. *Electrochemical and Solid-State Letters.* 2010; 13(9): B109–B111.

118. Spinner, N. S., Vega, J. A. and Mustain, W. E. Recent progress in the electrochemical conversion and utilization of $CO_2$. *Catalysis Science & Technology.* 2012; 2(1): 19–28.

119. Innocent, B., Liaigre, D., Pasquier, D., Ropital, F., Léger, J. M. and Kokoh, K. B. Electro-reduction of carbon dioxide to formate on lead electrode in aqueous medium. *Journal of Applied Electrochemistry.* 2009; 39(2): 227–232.

120. Azuma, M., Hashimoto, K., Hiramoto, M., Watanabe, M. and Sakata, T. Electrochemical reduction of carbon dioxide on various metal electrodes in low-temperature aqueous $KHCO_3$ media. *Journal of the Electrochemical Society.* 1990; 137(6): 1772–1778.

121. Köleli, F., Atilan, T., Palamut, N., Gizir, A. M., Aydin, R. and Hamann, C. H. Electrochemical reduction of $CO_2$ at Pb- and Sn-electrodes in a fixed-bed reactor in aqueous $K_2CO_3$ and $KHCO_3$ media. *Journal of Applied Electrochemistry.* 2003; 33(5): 447–450.

122. Hara, K., Kudo, A. and Sakata, T. Electrochemical reduction of carbon dioxide under high pressure on various electrodes in an aqueous electrolyte. *Journal of Electroanalytical Chemistry.* 1995; 391(1–2): 141–147.

123. Popić, J. P., Avramov-Ivić, M. L. and Vuković, N. B. Reduction of carbon dioxide on ruthenium oxide and modified ruthenium oxide electrodes in 0.5 M $NaHCO_3$. *Journal of Electroanalytical Chemistry.* 1997; 421(1–2): 105–110.

124. Frese, K. and Leach, S. Electrochemical reduction of carbon dioxide to methane, methanol, and CO on Ru electrodes. *Journal of the Electrochemical Society.* 1985; 132(1): 259–260.

125. Furuya, N., Yamazaki, T. and Shibata, M. High performance Ru Society of carbon dioxide to methane, metiffusion electrodes. *Journal of Electroanalytical Chemistry.* 1997; 431(1): 39–41.

126. Hoshi, N., Suzuki, T. and Hori, Y. Step density dependence of $CO_2$ reduction rate on Pt(S)-[n(111) × (111)] single crystal electrodes. *Electrochimica Acta.* 1996; 41(10): 1647–1653.

127. Hoshi, N., Suzuki, T. and Hori, Y. $CO_2$ reduction on Pt(S) – [n(111) × (111)] single crystal electrodes affected by the adsorption of sulfuric acid anion. *Journal of Electroanalytical Chemistry.* 1996; 416(1–2): 61–65.

128. Hara, K., Tsuneto, A., Kudo, A. and Sakata, T. Electrochemical reduction of $CO_2$ on a Cu electrode under high pressure factors that determine the product selectivity. *Journal of the Electrochemical Society.* 1994; 141(8): 2097–2103.

129. Kaneco, S., Hiei, N.-h., Xing, Y., Katsumata, H., Ohnishi, H., Suzuki, T. and Ohta, K. Electrochemical conversion of carbon dioxide to methane in aqueous $NaHCO_3$ solution at less than 273 K. *Electrochimica Acta.* 2002; 48(1): 51–55.

130. Kaneco, S., Katsumata, H., Suzuki, T., Ohta, K., Liu, C.-j., Mallinson, R. G. and Aresta, M. *Utilization of Greenhouse Gases: Electrochemical Reduction of CO2 on Cu Electrode in Methanol at Low Temperature.* American Chemical Society, Washington, DC, 2003.

131. Rosen, B. A., Haan, J. L., Mukherjee, P., Braunschweig, B., Zhu, W., Salehi-Khojin, A., Dlott, D. D. and Masel, R. I. In situ spectroscopic examination of a low overpotential pathway for carbon dioxide conversion to carbon monoxide. *The Journal of Physical Chemistry C.* 2012; 116(29): 15307–15312.

132. Asadi, M., Kumar, B., Behranginia, A., Rosen, B. A., Baskin, A., Repnin, N., Pisasale, D., Phillips, P., Zhu, W., Haasch, R., Klie, R. F., Kral, P., Abiade, J. and Salehi-Khojin, A. Robust carbon dioxide reduction on molybdenum disulphide edges. *Nature Communications.* 2014; 5: 4470.

133. Rosen, B. A., Zhu, W., Kaul, G., Salehi-Khojin, A. and Masel, R. I. Water enhancement of $CO_2$ conversion on silver in 1-ethyl-3-methylimidazolium tetrafluoroborate. *Journal of the Electrochemical Society.* 2013; 160(2): H138–H141.

134. Łukaszewski, M., Siwek, H. and Czerwiński, A. Electrosorption of carbon dioxide on platinum group metals and alloys—A review. *Journal of Solid State Electrochemistry.* 2009; 13(6): 813–827.

135. Whipple, D. T. and Kenis, P. J. A. Prospects of $CO_2$ utilization via direct heterogeneous electrochemical reduction. *The Journal of Physical Chemistry Letters*. 2010; 1(24): 3451–3458.

136. Chandrasekaran, K. and Bockris, L. O. M. In-situ spectroscopic investigation of adsorbed intermediate radicals in electrochemical reactions: $CO_2^-$ on platinum. *Surface Science*. 1987; 185(3): 495–514.

# 4 Catalysis of $CO_2$ Electroreduction

*Rongzhi Chen and Yuyu Liu*

## CONTENTS

## 4.1   INTRODUCTION

$CO_2$ conversion can be achieved by chemical methods [1–8], by photocatalytic and electrocatalytic reduction [9–17], and by a few other means [18–20]. However, at the present time, several disadvantages are associated with the practical application of these approaches, including (1) the high costs of $CO_2$ capture, separation, purification, and transportation to user sites; (2) the high energy requirements for $CO_2$ chemical/electrochemical conversion; (3) limitations in market size and investment incentives; (4) lack of industrial commitment to enhance $CO_2$-based chemicals; and (5) insufficient socioeconomic driving forces [21]. Despite such challenges, $CO_2$ capture, conversion, and utilization is still recognized as a feasible and promising cutting-edge area of exploration in energy and environmental research.

In recent years, $CO_2$ conversion using electrochemical catalysis approaches has attracted great attention for its several advantages: (1) the process is controllable by electrode potentials and reaction temperature; (2) the supporting electrolytes can be fully recycled so that the overall chemical consumption can be minimized to simply water or wastewater; (3) the electricity used to drive the process can be obtained without generating any new $CO_2$—sources include solar, wind, hydroelectric, geothermal, tidal, and thermoelectric processes; and (4) the electrochemical reaction systems are compact, modular, on-demand, and easy for scale-up applications. However, challenges remain, such as the slow kinetics of $CO_2$ electroreduction, even when electrocatalysts and high electrode reduction potential are applied; the low energy efficiency of the process, due to the parasitic or decomposition reaction of the solvent at high reduction potential; and high energy consumption. Researchers have recognized that the biggest challenge in $CO_2$ electroreduction is low performance of the electrocatalysts (i.e., low catalytic activity and insufficient stability).

## 4.2 CATALYST ACTIVITY AND STABILITY

Catalytic activity is normally evaluated by considering both the onset potential of reduction and the faradaic efficiency, while catalyst stability (or durability) is assessed according to variations in catalyst behavior with electrolysis time [22]. With respect to catalyst stability, the issue of deactivation has often been reported; the formation of poisonous intermediates and the deposition of inactive compositions on electrode surfaces are the main causes [22–28]. Hori et al. [26] put forth several possible factors, which can be summarized as (1) heavy metal impurities contained in reagent chemicals and introduced to the electrolyte solution; (2) very small amounts of organic substances possibly contained in water; and (3) intermediate poisoning species or products formed during $CO_2$ reduction and adsorbed on electrodes. Besides these, electrolysis mode and condition can also affect catalyst stability [29–31]. For example, deactivation of a Cu cathode was observed after only 3 h of electrolysis in a constant potential mode, while the electrocatalytic activity of Cu remained constant for 7 h if a superimposed potential method was applied [32].

Using the latter, the surface structure of the copper electrode was changed with the formation of cuprous oxide ($Cu_2O$), and the adsorption of amorphous graphite was prevented, leading to stable long-term electrolysis for $CH_4$ production [32]. The pulse electrolysis mode was also found to have a mitigating effect on electrode deactivation [33]. Changing the electrolytic conditions led to the deposition of poisoning species on a Cu electrode being highly suppressed, while the selectivity for $C_2H_4$ formation was enhanced. Details of the reduction mechanism on $Cu_2O$ remained unclear, and further study of the electrodeposition of $CO_2$ on the $Cu_2O$ cathode should be considered. An early study found that the deactivation of a Sn electrode was related to the formation of organometallic complexes on the electrode surface, which could accelerate the rate of hydrogen evolution [34]. Recently, Agarwal et al. investigated the long-term performance of Sn, together with other proprietary catalysts, in the electrochemical reduction of $CO_2$ (ERC) to HCOO/HCOOH at a gas/solid/liquid interface, using a flow-through reactor [35]. Although better durability was observed in Sn than in Cu, a color change appeared on the electrode surface, as well as slight deactivation. Wu et al. [36] observed the effects of the electrolyte on selectivity and activity with a Sn electrode. For a pure Sn electrocatalyst, a decrease in performance could be caused by several factors [37]: (1) cathodic degradation of the catalyst surface, (2) deposition of noncatalytic species from reaction intermediates in the reduction of the pollutant species, (3) deposition of noncatalytic metallic species from contaminants in the electrolyte [36], and (4) anodic degradation of the catalyst at sites where gas bubbles formed, preventing the cathodic polarization of the catalyst. Bujno et al. [38] conducted experiments in diluted solutions and confirmed that the Ni(I) complex catalysts present at the electrode surface were transformed into a catalytically inactive Ni(0) carbonyl deposit, blocking the electrode surface against further catalysis. Benson and Kubiak [39] investigated the deactivation pathway of the Lehn catalyst. One pathway was concluded to be the formation of thermodynamically stable and often catalytically inactive dimers [40]. Pugh et al. [41] found that the electrocatalytic activity of cis-[Ru(bpy)$_2$(CO)H]$^+$ decreased slowly over an extended period. Normally, active species (sites) in catalysts are always responsible for the catalytic activity and are indispensable for electrocatalytic $CO_2$ reduction [42–44]. Loss in catalytic activity is always associated with the disappearance of active sites. For example, during $CO_2$ reduction, Ru-based complex catalysts gradually lost their carbonyl-containing complexes, and inactive species, such as $[Ru(bpy)_2(CO_3)]$, were formed. The instability of the $[Cl(CO)_2\text{-}(bpy*^-)Ru\text{–}Ru(bpy)(CO)_2Cl]^-$ species was confirmed by a voluminous black precipitate, produced by exhaustive electrolysis at $-2.00$ V [45]. In a recent study of $CO_2$ reduction to CO at low overpotential in neutral aqueous solution by a Ni (cyclam) complex attached to poly (allylamine), Saravanakumar et al. [46] achieved a current efficiency of 92% during the initial 6 h of electrolysis, but this dropped to 88% at 12 h and then to 79% at 24 h. Hence, to mitigate the degradation of catalyst activity and stability, two major factors should be considered: (1) the effect of catalyst type, structure, and composite and (2) the effect of catalyst operating conditions.

ERC can proceed through two-, four-, six-, and eight-electron reduction pathways in gaseous, aqueous, and nonaqueous phases at both low and high temperatures. The major reduction products are carbon monoxide (CO), formic acid (HCOOH), or formate ($HCOO^-$) in basic solution, oxalic acid ($H_2C_2O_4$) or oxalate ($C_2O_4^{2-}$ in basic solution), formaldehyde ($CH_2O$), methanol ($CH_3OH$), methane ($CH_4$), ethylene ($CH_2CH_2$), ethanol ($CH_3CH_2OH$), as well as others. The thermodynamic electrochemical half-reactions of $CO_2$ reduction and their associated standard electrode potentials are listed in Table 4.1 [47]. Note that the reactions listed in Table 4.1 are thermodynamic, only indicating each reaction's tendency and possibility but giving no indication of the reaction's kinetics, such as rate and mechanism. In addition, the standard potentials listed in Table 4.1 are for aqueous solutions only; the potential values in nonaqueous solutions are different from those listed in Table 4.1 [48]. The kinetics of $CO_2$ electroreduction involve very complicated reaction mechanisms, and the reaction rates are very slow, even in the presence of electrocatalysts. In general, the catalysts currently being employed are still not active enough. Furthermore, in some cases, the product of the electroreduction is not a single species but a mixed product containing several component species (e.g., it could be a mixture of C, CO, HCOOH, $H_2C_2O_4$, $CH_2O$, $CH_3OH$, $CH_4$, $CH_2CH_2$, $CH_3CH_2OH$, and so on). The number of species and the amount of each species present are factors strongly dependent on the kind and selectivity of the electrocatalyst employed and what electrode potential is applied. This suggests that the currently employed electrocatalysts have insufficient catalytic selectivity and stability. In most cases, these catalysts can survive for fewer than 100 h [49–51], which is far below the requirements for practical use and technological commercialization. Therefore, unsatisfactory catalysis, including low catalytic activity, selectivity, and stability, is the biggest challenge in $CO_2$ electroreduction. In the past several decades, almost all the efforts in $CO_2$ electroreduction studies have been focused on research and development of electrocatalysts to overcome the above challenges [52,53]. Hence, while several review articles related to $CO_2$ reduction have been published [54], a comprehensive review specifically focusing on electrocatalysts for $CO_2$ electroreduction is definitely necessary to facilitate research and development in this area.

**TABLE 4.1**
**Selected Standard Potentials of $CO_2$ in Aqueous Solutions (V vs. SHE) at 1.0 atm and 25°C, Calculated According to the Standard Gibbs Energies of the Reactants in Reactions**

| Half-Electrochemical Thermodynamic Reactions | Electrode Potentials (V vs. SHE) under Standard Conditions |
|---|---|
| $CO_2$ (g) + $4H^+$ + $4e^-$ = C (s) + $2H_2O$ (l) | 0.210 |
| $CO_2$ (g) + $2H_2O$ (l) + $4e^-$ = C (s) + $4OH^-$ | −0.627 |
| $CO_2$ (g) + $2H^+$ + $2e^-$ = HCOOH (l) | −0.250 |
| $CO_2$ (g) + $2H_2O$ (l) + $2e^-$ = $HCOO^-$ (aq) + $OH^-$ | −1.078 |
| $CO_2$ (g) + $2H^+$ + $2e^-$ = CO (g) + $H_2O$ (l) | −0.106 |
| $CO_2$ (g) + $4H^+$ + $4e^-$ = $CH_2O$ (l) + $H_2O$ (l) | −0.070 |
| $CO_2$ (g) + $3H_2O$ (l) + $4e^-$ = $CH_2O$ (l) + $4OH^-$ | −0.898 |
| $CO_2$ (g) + $6H^+$ + $6e^-$ = $CH_3OH$ (l) + $6OH^-$ | 0.016 |
| $CO_2$ (g) + $5H_2O$ (l) + $6e^-$ = $CH_3OH$ (l) + $6OH^-$ | −0.812 |
| $CO_2$ (g) + $8H^+$ + $8e^-$ = $CH_4$ (g) + $2H_2O$ (l) | 0.169 |
| $CO_2$ (g) + $6H_2O$ (l) + $8e^-$ = $CH_4$ (l) + $8OH^-$ | −0.659 |
| $CO_2$ (g) + $2H^+$ + $2e^-$ = $H_2C_2O_4$ (aq) | −0.500 |
| $2CO_2$ (g) + $2e^-$ = $C_2O_4^{2-}$ (aq) | −0.590 |
| $2CO_2$ (g) + $12H^+$ + $12e^-$ = $CH_2CH_2$ (g) + $4H_2O$ (l) | 0.064 |
| $2CO_2$ (g) + $8H_2O$ (l) + $12e^-$ = $CH_2CH_2$ (g) + $12OH^-$ | −0.764 |
| $2CO_2$ (g) + $12H^+$ + $12e^-$ = $CH_3CH_2OH$ (l) + $3H_2O$ (l) | 0.084 |
| $2CO_2$ (g) + $9H_2O$ (l) + $12e^-$ = $CH_3CH_2OH$ (l) + $12OH^-$ | −0.744 |

*Source:* Qiao et al. *Chem Soc Rev*, 2014, 43, 631–675. Reproduced by permission of The Royal Society of Chemistry.

## 4.3 METAL CATALYSTS AND REACTION MECHANISMS

### 4.3.1 TITANIUM

Normally, titanium (Ti) on its own has no significant catalytic activity toward $CO_2$ electroreduction [55]. However, $TiO_2$ has shown some activity in both photocatalysis and electrocatalysis for $CO_2$ electroreduction [56]. When $TiO_2$ is used as an electrocatalyst, thin $TiO_2$ films or mixtures of $TiO_2$ and other metal oxides are usually deposited on a Ti electrode substrate to catalyze $CO_2$ electroreduction. For example, in the preparation of a $CO_2$ catalytic electrode, Monnier et al. [57,58] prepared $TiO_2$, $TiO_2$–Ru (or $RuO_2$), and $TiO_2$–Pt thin-film electrodes by thermal deposition on titanium rods. Bandi deposited metallic oxide mixtures (including $RuO_2$, $TiO_2$, $MoO_2$, $Co_3O_4$, and $Rh_2O_3$) on titanium foil [59]. Cueto and Hirata [60] prepared a $TiO_2$–indium-tin oxide (ITO) thin-film glass electrode via a fixed-potential bulk electrolysis process, in which 1-butyl-3-methylimidazolium tetrafluoroborate ($BMIm·BF_4$, an ionic liquid) was used as both solvent and supporting electrolyte. Several material characterization techniques were employed to analyze the coated electrode. X-ray diffraction results showed that $TiO_2$ films were suitable candidates for electrocatalytic processes since no phase changes could be observed, while a significant dissociation of $CO_2$ into chemisorbed $CO_3^{2-}$ was concluded from XPS measurements (Figure 4.1). Furthermore, observation using Auger electron spectroscopy indicated strong interactions between $TiO_2$ and $CO_2$ or $CO_3^{2-}$ and revealed a decrease in carbon content after the $CO_2$ electroreduction process. Recently, electrocatalytic synthesis of low-density polyethylene (PE) from $CO_2$ on a nanostructured $TiO_2$ (ns-$TiO_2$) film electrode was carried out by controlled potential electrolysis in a mixture of $H_2O$ and EMIm·$BF_4$ at room temperature and ambient pressure [61]. The ns-$TiO_2$ film appeared to be remarkably efficient and selective for the ERC when EMIm·$BF_4$ was the solvent [62], as EMIm·$BF_4$ maintained a high concentration of $CO_2$ at the electrode surface. According to the mechanism described in reactions (4.1) through (4.5), high pressure in the nanopores of the ns-$TiO_2$ film can lead to polymerization of $CH_2$ to form PE [61]:

$$CO_2 + Ti^{III} \rightarrow CO_2^{*-}{}_{(adsorbed\ on\ ns-TiO_2\ film)} + Ti^{IV} \tag{4.1}$$

$$Ti^{IV} + e^- \rightarrow Ti^{III} \tag{4.2}$$

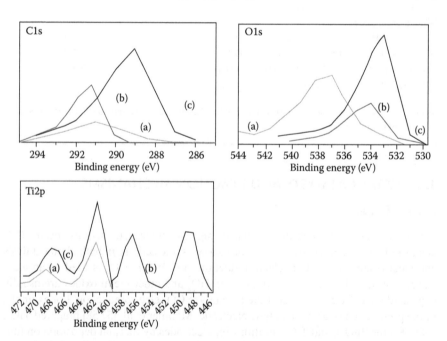

**FIGURE 4.1** XPS spectra of $TiO_2$/ITO thin-film C1s, Ti2p, and O1s chemical states (a) before ERC, (b) $CO_2$-saturated in [BMIm][BF4], and (c) after a 19 h $CO_2$ bulk electrolysis reduction reaction in [BMIm][BF4]. PQRE: Pt(s), Aux: Pt(s). (With kind permission from Springer Science + Business Media: *Journal of Sol–Gel Science and Technology*, Thin-film $TiO_2$ electrode surface characterization upon $CO_2$ reduction processes, 37(2), 2006, 105–109, Cueto, L. F., Hirata, G. A. and Sanchez, E. M.)

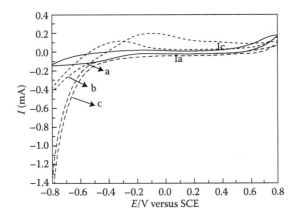

**FIGURE 4.2** Cyclic voltammograms of (a, b) RuO$_2$/TiO$_2$ NPs modified electrode in the presence of N$_2$ and CO$_2$, respectively and (c) RuO$_2$/TiO$_2$ NTs modified electrode in the presence of CO$_2$ in 0.5 M NaHCO$_3$, sweep rate: 10 mV s$^{-1}$. (Reprinted from *Electrochimica Acta*, 50(16–17), Qu, J. P. et al., Electrochemical reduction of CO$_2$ on RuO$_2$/TiO$_2$ nanotubes composite modified pt electrode, 3576–3580, Copyright 2005, with permission from Elsevier.)

$$CO_2{}^{*-}{}_{(ads)} + H^+ + e^- \rightarrow CO_{(ads)} + OH^- \qquad (4.3)$$

$$CO_{(ads)} + 4H^+ + 4e^- \rightarrow: CH_{2(ads)} + H_2O \qquad (4.4)$$

$$: CH_{2(ads)} \rightarrow [CH_2CH_2]_n \qquad (4.5)$$

Indeed, TiO$_2$ and carbon nanotubes (CNTs) have been explored as catalyst supports in the synthesis of nanostructured electrocatalysts for the selective reduction of CO$_2$. Due to the unique structures of these catalysts, some selectivity toward the desired products has been achieved. For example, Qu et al. [63] loaded RuO$_2$ onto TiO$_2$ nanotubes (NTs) or nanoparticles (NPs) to form RuO$_2$–TiO$_2$ (NTs) or RuO$_2$–TiO$_2$ (NPs), which were then coated onto a Pt electrode for the electrocatalytic reduction of CO$_2$. Figure 4.2 shows the voltammetric curves for these two electrodes in 0.5 M NaHCO$_3$ saturated with N$_2$ and CO$_2$, respectively. These modified electrodes exhibited two broad peaks (labeled Ia and Ic). Starting from negative values of the potential for the positive sweep in the presence of N$_2$, only processes connected with hydrogen adsorption and desorption can be seen at −0.2 V, which was not observed for the RuO$_2$ electrode and RuO$_2$–Ti electrode. This means that the NTs structure of the composite modified electrode for reduction of CO$_2$ exhibited a better electrocatalytic activity.

### 4.3.2 MOLYBDENUM, CHROMIUM, AND TUNGSTEN

Metallic electrodes of Cr, Mo, and W do not appear to show significant activity toward CO$_2$ electroreduction. For example, Noda et al. [55] tested electrodes of these metals at −1.6 V versus Ag/AgCl in KCl saturated 0.1 M KHCO₃ aqueous solution at 298°C, and did not observe significant catalytic activity for the ERC. However, in an earlier study [64], researchers using molybdenum metal electrodes in the electrolysis of $CO_2$-saturated 0.2 M $Na_2SO_4$ solution (pH 4.2) at 20°C in the potential range of −0.7 to −0.8 V versus saturated calomel electrode (SCE) observed a faradaic efficiency above 50%, with the main product being methanol. Even in 0.05 M $H_2SO_4$ at 20°C, methanol was obtained at −0.57 to −0.67 V versus SCE with a faradaic efficiency of up to 46%. Moreover, cyclic voltammetric (CV) measurements indicated that the corrosion of molybdenum metal to molybdenum dioxide ($MoO_2$) might be the source of electrons for the electroreduction of $CO_2$, As indicated in Figure 4.3, CO was formed at open circuit; $CO_2$ reduction (Mo + $2CO_2$ → $MoO_2$ + 2CO) and molybdenum oxidation (Mo + $2H_2O$ → $MoO_2$ + 4e⁻ + 4H+, E = 0.57 V vs. SCE at pH 4.2) occurred simultaneously.

In the late 1990s, Mo-containing catalysts were found to be useful in the reduction of chlorate, bromate, and iodide anions [64–69]. The electrochemistry of

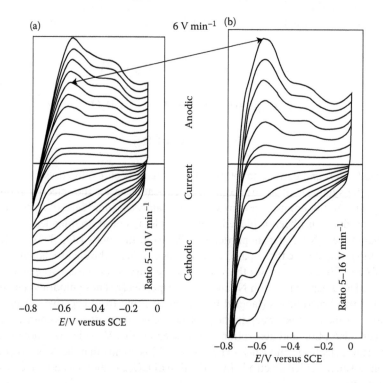

**FIGURE 4.3** Cyclic voltammetry for a Mo electrode in (a) $N_2$ and (b) $CO_2$-saturated 0.2 M $Na_2SO_4$ aqueous solution (pH 4.2). Electrodes (3.4 cm²) were pretreated with HCl and allowed to remain at open circuit for 1 h before scanning. (Reprinted from *Journal of Electroanalytical Chemistry*, 205(1–2), Summers, D. P., Leach, S. and Frese, K. W., The electrochemical reduction of aqueous carbon dioxide to methanol at molybdenum electrodes with low overpotentials, 219–232, Copyright 1986, with permission from Elsevier.)

molybdenum and molybdenum oxides has been described in detail [70]. Nakazawa et al. [71] employed two types of iron–sulfur clusters, [Fe$_4$S$_4$(SR)$_4$]$^{2-}$ (R = C$_6$H$_5$CH$_2$ and (CH$_3$)$_3$C) and [M$_2$Fe$_6$S$_8$(SCH$_2$CH$_3$)$_9$]$^{3-}$ (M = Mo or W), to catalyze the ERC. It was observed that the reduction potential was shifted by about 0.7 and 0.5 V, respectively, in the positive potential direction compared to what occurred without any catalyst, demonstrating the catalytic effect of these two catalysts. Bandi et al. [59] thermally decomposed a mixed metal oxide consisting of 25% RuO$_2$, 30% MoO$_2$, and 45% TiO$_2$ on Ti foil to catalyze CO$_2$ electroreduction to methanol, but no striking performance was observed. Regarding Cr-related catalysts, Ogura and Yoshida [72] employed Cr$^{(III)}$–TPPCl (TPP = 5,10,15,20-tetraphenylporphyrin) for CO$_2$ electroreduction in dimethylformamide (DMF) solutions and compared the results with those for Co$^{(II)}$–TPP, Ni(II)–TPP, Fe$^{(III)}$–TPPCl, and Fe$^{(II)}$–TPP. Cr complexes with 4-v-tpy and 6-v-tpy (v-tpy = vinyl-terpyridine) have also been reported for CO$_2$ electrocatalysts [73]. Sende et al. [74] electropolymerized several 4-v-tpy and 6-v-tpy complexes of transition metals (including Cr) onto glassy carbon electrodes (GCEs) for CO$_2$ electroreduction. They found that in CO$_2$-saturated 0.10 M aqueous NaClO$_4$ solution, their electropolymerized [Cr(4-v-tpy)$_2$]$^{2+}$ film was at −0.86 V versus Ag/AgCl, which was more positive than for Fe, Ni, Ru, and Os complexes in the potential range of −1.10 to −1.22 V versus Ag/AgCl; it also yielded a considerable amount of formaldehyde (CH$_2$O), up to 87% at −1.10 V (far higher than the 39% and 28% achieved with Co and Fe complexes, respectively). The turnover number (TON) for catalyzed CO$_2$ electroreduction was up to 6100 for the [Cr(4-v-tpy)$_2$]$^{2+}$ complex, about 50% lower than for both the [Co(4-v-tpy)$_2$]$^{2+}$ (11,000), and [Fe(4-v-tpy)$_2$]$^{2+}$ (15,000) complexes. In all cases, virtually the only reaction product detected was CH$_2$O, demonstrating the catalysts' high selectivity. It is worth mentioning that Sende et al. also explored one Cr salt (diaquabis(oxalato)-chromate(III) (K[Cr(III)(C$_2$O$_4$)$_2$(H$_2$O)$_2$]) as a homogeneous catalyst for the ERC as an Everitt's salt (ES), that is, KFe$^{II}$[Fe$^{II}$(CN)$_6$]-modified platinum gauze electrode. Regarding tungsten-related electrocatalysis for CO$_2$ electroreduction, Reda et al. [75] obtained a tungsten-containing formate dehydrogenase enzyme (FDH1) from *Syntrophobacter fumaroxidans* and adsorbed it on a freshly polished, pyrolytic graphite edge (PGE) electrode to catalyze CO$_2$ reduction. As either a homogeneous or a heterogeneous catalyst, FDH1 catalyzed the ERC in 20 mM Na$_2$CO$_3$ (pH 6.5) to produce only formate, with a reduction rate more than two orders of magnitude greater than that achieved with other known catalysts for the same reaction. Voltammograms of FDH1 catalyzing the interconversion of formate and CO$_2$ are presented in Figure 4.4. Both CO$_2$ reduction and formate oxidation are initiated at approximately −0.4 V, close to the reduction potential (see below), and then their rates increase sharply as the driving forces increase. The voltammetric wave shapes display a sigmoidal onset that changes into a linear dependency as the driving force (overpotential) is increased. This reflects the influence of the interfacial electron-transfer process that, even at the highest driving force applied, lags behind the rapid active-site turnover (24). CO$_2$ reduction reaches 0.08 mA cm$^{-2}$ at −0.8 V (pH 5.9), whereas formate oxidation increases to 0.5 mA cm$^{-2}$ at 0.2 V (pH 8.0) (Figure 4.4a and c). Figure 4.4b displays formate oxidation and CO$_2$ reduction in a single experiment. The electrocatalytic activity gradually decreases, because of enzyme desorption or denaturation, so that a potential at which all of the catalytic

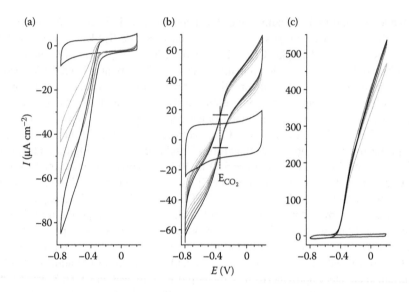

**FIGURE 4.4** Electrocatalytic voltammograms showing $CO_2$ reduction and formate oxidation by FDH1. Shown are the reduction of 10 mM $CO_2$ (pH 5.9) (a); electrocatalysis in 10 mM $CO_2$ and 10 mM formate (pH 6.4) (b), showing the points of intersection (marked with crosses) that define the reduction potential for the interconversion of $CO_2$ and formate; and the oxidation of 10 mM formate (pH 7.8) (c). The first voltammetric cycles are shown in black, subsequent cycles are in gray; and background cycles recorded in the absence of substrate are also shown (gray). (Reda, T., Plugge, C. M., Abram, N. J. and Hirst, J. Reversible interconversion of carbon dioxide and formate by an electroactive enzyme. *Proceedings of the National Academy of Sciences of the United States of America.* 105(31):10654–10658, Copyright 2008 National Academy of Sciences, U.S.A.)

voltammograms intersect is a potential of zero net catalytic current. Net formate oxidation occurs at all potentials above the intersection, and net $CO_2$ reduction occurs at all potentials below it.

### 4.3.3 RHENIUM AND MANGANESE

Normally, rhenium complexes have quite promising catalytic activity. In this chapter's context, tricarbonyl rhenium(I) complexes are a large group of catalysts for catalyzing the photochemical and ERC to CO [76–78]. Lehn et al. [77] studied the homogeneous catalysis of $CO_2$ electroreduction using $9.0 \times 10^{-4}$ M $Re^{(I)}(bpy)(CO)_3Cl$ (namely, Lehn catalyst; bpy = 2,2′-bipyridine) in 0.1 M $(CH_3CH_2)_4NCl$ DMF–$H_2O$ (9:1) solutions at 25°C on a GCE with a potential at −1.25 V versus normal hydrogen electrode (NHE), which was much less negative than −2.0 V in the absence of the catalyst. With the catalyst, the current efficiency reached 91%. The catalyst CV showed a reversible one-electron reduction process at −1.25 V versus NHE, forming $[Re^{(0)}(bpy)(CO)_3Cl]^-$. O'Toole et al. [79] electropolymerized another kind of tricarbonyl rhenium complex, $Re(CO)_3(vbpy)Cl$ (vbpy = 4-vinyl-4′-methyl-2,2′-bipyridine), on a Pt electrode to form a polymeric film that was used to heterogeneously catalyze the

ERC to CO. Results indicated that the heterogeneous catalyst could enhance the TON 20–30 times more than was observed in the homogeneous case with $Re(CO)_3(bpy)Cl$ catalyst. When the $Re(CO)_3(vbpy)Cl$ complex was coated on metallic Pt or on p-Si and polycrystalline thin films of p-WSe₂ semiconducting electrodes, the TONs were as high as ca. 600 and 450, respectively, for $CO_2$ electroreduction to CO [80]. Cosnier et al. [81,82] investigated the effects that electrode material, film thickness, and the structure of bipyridyl ligands had on the catalytic activity, stability, and current efficiency for $CO_2$ electroreduction catalyzed by electropolymerized *fac*-Re(L)(CO)₃Cl complexes (L = pyrrole-substituted bpy) on metallic Pt and carbon felt electrodes (Figure 4.5) [81,82]. The coatings obtained by electropolymerization of a monomer containing one pyrrole group (L1) seemed to be as stable as those prepared with monomers containing two pyrrole groups (L2), whereas poly-Re(L)(CO)₃Cl films (L3, L4, and L5) with lower reduction potentials were more stable but less active toward the ERC. O'Toole et al. [83] co-electropolymerized *cis*-[(bpy)₂Re(vpy)₂]²⁺ with *fac*-Re(CO)₃(vbpy)Cl or *fac*-[Re(CO)₃(vbpy)CH₃CN]⁺ on electrodes to form

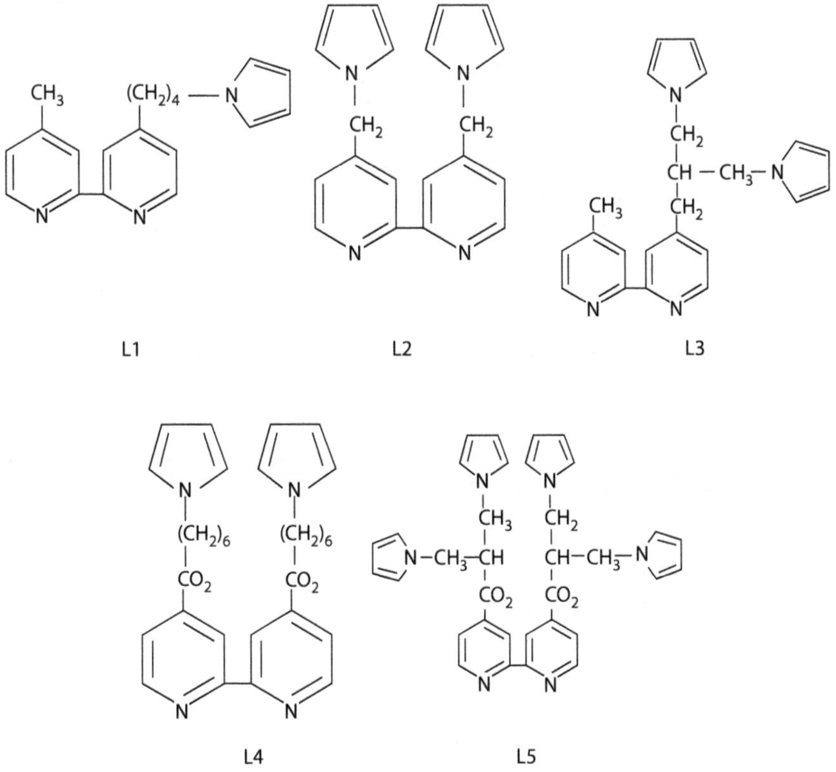

FIGURE 4.5 Structure of electropolymerized *fac*-Re(L)(CO)3Cl complexes. L = pyrrole-substituted bpy. (Reprinted from *Journal of Electroanalytical Chemistry*, 207(1–2), Cosnier, S., Deronzier, A. and Moutet, J. C., Electrochemical coating of a platinum-electrode by a poly(pyrrole) film containing the *fac*-Re(2,2'-bipyridine)(co)3cl system—Application to electrocatalytic reduction of $CO_2$, 315–321, Copyright 1986, with permission from Elsevier.)

thin polymeric films (heterogeneous catalysts). Metal sites on these films showed increased reactivity and stability toward $CO_2$ reduction compared to those catalyzed by $fac$-Re(CO)$_3$(bpy)Cl in solution. CO and oxalate (on the pure poly-Re(CO)$_3$(vbpy) Cl film) were found to be the main products. The results of electrochemical kinetic studies of poly-Re(CO)$_3$(vbpy)Cl showed that as the film thickness was increased, the film's rate-determining step could be changed from (1) the chemical reaction between reduced Re and $CO_2$ to (2) electron transport to the catalytic sites. Schrebler et al. [84] investigated the ERC on a Re film electrodeposited onto a polycrystalline Au support in a $CH_3OH$ solution with 0.1 M LiClO$_4$ under atmospheric pressure of $CO_2$. The $CO_2$ electroreduction displayed a Tafel slope of $-2RT/F$, suggesting that the first electronation of the $CO_2$ molecule to form $CO_2$*$^-$ was the rate-determining step. It was found that the product distribution was strongly dependent on the electrode potential at which the electrolysis was carried out, as well as on the hydrodynamic conditions. Under stirred conditions, the faradaic efficiency of CO production was 87% at $-1.35$ V, whereas under quiescent conditions, the faradaic efficiency of CO and $CH_4$ production was 57% and 10%, respectively. Schrebler et al. also prepared Re and Cu–Re microalloy polypyrrole (PPy) modified Au electrodes for the ERC in $CH_3OH$ solution with the same composition. Higher faradaic efficiencies for $CH_4$ were obtained at $-1.35$ V, with Au–PpyRe at 34% and Au–PpyCu–Re at 31%. Importantly, both Re and Cu–Re alloy could be highly dispersed on the PPy films, and the amount and selectivity of CO, $CH_4$, and $H_2$ were independent of the hydrodynamic conditions of the solution. The $fac$-Re(CO)$_3$(vbpy)Cl was also electropolymerized onto a mesoporous $TiO_2$ film coated on a $SnO_2$-doped glass electrode for $CO_2$ electroreduction [85]. The nanoporous nature of $TiO_2$ allowed an increase in the two-dimensional number of redox sites per surface area and hence achieved a significant enhancement in catalytic yield. In an effort to improve the catalytic activity of $CO_2$ electrocatalysts, Cheung et al. [86] recently electropolymerized a poly-Re(CO)$_3$(k$_2$-$N$,$N$-PPP)Cl film onto a GCE. Their results showed that the modified electrode also exhibited electrocatalytic activity for the reduction of $CO_2$ to CO. To understand the mechanism, Re(CO)$_3$LCl complexes with different bpy ligands that contain different substitutions on the benzene ring have been employed as example catalysts for $CO_2$ electroreduction. A systematic study of the ERC catalyzed by 1.0 mM Re(CO)$_3$LCl complexes (L = bpy, dcbpy, dmbpy, 4,40-di-$tert$-butyl-bpy, or 4,40-dimethoxy-bpy) in $CH_3CN$ + 0.1 M tetrabutylammonium (TBA) hydroxide on a glassy carbon working electrode revealed that the electron-donating/withdrawing substituents in the 4,40 positions of the bipyridine ligand had a significant effect on $CO_2$ electroreduction [87]. The catalytic activity was increased (with less negative reduction potentials) in the order $OCH_3$ < C(CH$_3$)$_3$ < $CH_3$ < H < COOH. When L = bpy, dcbpy, dmbpy, and 4,4'-di-$tert$-butyl-bpy, the Re complex gave the best catalytic activity, even 2.2 times higher than that of Lehn catalyst. Recently, $fac$-(5,5'-bisphenylethynyl-2,2'-bipyridyl)Re(CO)$_3$Cl was explored as a catalyst for $CO_2$ electroreduction. The results showed a 6.5-fold increase in the current density of $CO_2$ to CO at $-1.75$ V versus NHE compared to without $CO_2$, and the faradaic efficiency for CO production was around 45% [59]. In this study, the supporting electrolyte solution was $CH_3CN$ with 0.1 M (n-CH$_3$(CH$_2$)$_3$)4N·PF$_6$. Since the discovery of the Lehn catalyst ($fac$-Re(bpy)(CO)$_3$Cl) [61], the mechanism of $CO_2$ reduction has also been explored through chemical synthesis, electrochemical, and spectroscopic measurements [85–87]. Sullivan et al. [88] proposed catalytic pathways, including an initial one-electron reduction of the catalyst, which then catalyzed $CO_2$ reduction to form CO, as expressed in reactions (4.6) through (4.14):

$$fac - Re(bpy)(CO)_3Cl + e^- \leftrightarrow [Re(bpy)(CO)_3Cl]^{*-} \qquad (4.6)$$

$$[Re(bpy)(CO)_3Cl]^{*-} \rightarrow [Re(bpy)(CO)_3]^* + Cl^- \qquad (4.7)$$

$$[Re(bpy)(CO)_3]^* + S \rightarrow [Re(bpy)(CO)_3S]^* (S = \text{solvent molecule}) \qquad (4.8)$$

$$2[Re(bpy)(CO)_3]^* \rightarrow [fac - Re(bpy)(CO)_3]_2 \qquad (4.9)$$

$$[Re(bpy)(CO)_3]^* + CO_2 \rightarrow Re(bpy)(CO)_3CO_2 \qquad (4.10)$$

$$Re(bpy)(CO)_3CO_2 + CO_2 + 2e^- \rightarrow Re(bpy)(CO)_3 + CO_3^{2-} + CO \qquad (4.11)$$

$$[fac - Re(bpy)(CO)_3]_2 + 2e^- \leftrightarrow 2[Re(bpy)(CO)_3]^- \qquad (4.12)$$

$$[Re(bpy)(CO)_3]^- + CO_2 \rightarrow [Re(bpy)(CO)_3CO_2]^- \qquad (4.13)$$

$$[Re(bpy)(CO)_3CO_2]^- + A + e^-$$
$$\rightarrow [Re(bpy)(CO)_3]^- + CO + [AO]^- (A = \text{an oxide ion acceptor}) \qquad (4.14)$$

In their experimental observation of the above $CO_2$ electroreduction mechanism, Johnson et al. [89] reported the reaction products using an infrared spectroelectrochemical (IR-SEC) method, employing an optically transparent thin-layer electrochemical cell. The catalysts used were $[Re(CO)_3(bpy)P(OEt)_3]^+$, $[Re(CO)_3(bpy)CH_3CN]^+$, $Re(CO)_3(bpy)Cl$, and $Re(CO)_3(bpy)CF_3SO_3)$. They confirmed that the $[Re(CO)_3(bpy)Cl]^{*-}$ radical was only attacked by $CO_2$ to form $[Re(bpy)(CO)_3CO_2]^*$ after the dissociation of $Cl^-$, and that $[Re(CO)_3(dmbpy)Cl]^{*-}$ tended to form the $[Re(CO)_3(dmbpy)]^*$ radical for $CO_2$ reduction. Scheiring et al. [90] conducted a mechanism study using electron paramagnetic resonance (EPR) spectroscopy; the catalysts employed were paramagnetic $Re(CO)_3(bpy)X$ complexes $(X = Cl^-, CF_3SO_3^-, CH_3O^-, H^-$, tetrahydrofuran, $CH_3CN$, CO, $HCO_2^-$, $HCO_3^-$, and $CH_3C(O)^-)$. Furthermore, sum frequency generation spectroscopy and density functional theory (DFT) calculations indicated that $Re(CO)_3Cl(dcbpy)$ could bind to a rutile $TiO_2$ surface through the –COOH groups of dcbpy in bidentate or tridentate linkage motifs, and the Re atom was exposed to the solution in a configuration suitable for the catalysis of $CO_2$ electroreduction [91]. Table 4.2 summarizes all the mono-, bi-, and tridentate ligands employed in Re complexes. Manganese generally does not catalyze the

**TABLE 4.2**

**Tafel Parameters Obtained from the Tafel Plots (Figure 4.4)**

|                      | SnB                 | SnG                 | SnGDL               |
|----------------------|---------------------|---------------------|---------------------|
| $b_{c1}$ (mV)        | 116                 | 180                 | 185                 |
| $b_{c2}$ (mV)        | 430                 | 480                 | 458                 |
| $j_o$ (A cm$^{-2}$)  | $1.2 \times 10^{-6}$ | $2.8 \times 10^{-5}$ | $1.6 \times 10^{-4}$ |
| $E_{eq}$ (V)         | −0.10               | 0.030               | 0.053               |

*Source:* Reprinted from *Journal of Power Sources*, 223, Prakash, G. K. S., Viva, F. A. and Olah, G. A., Electrochemical reduction of CO$_2$ over Sn-Nafion (R) coated electrode for a fuel-cell-like device, 68–73, Copyright 2013, with permission from Elsevier.

*Note:* $b_{c1}$ represents the slope for the lower overpotential region and $b_{c2}$ for the high overpotential region.

ERC. However, some catalytically active Mn carbonyl complexes, that is, [Mn(bpy) (CO)$_3$]$^+$, [Mn(dmbpy)(CO)$_3$]$^+$ [92], and [Mn(bpyt-Bu)(CO)$_3$]$^+$ [93], recently were found to have some activity toward CO$_2$ reduction. X-ray crystallography of [Mn(bpy-tBu) (CO)$_3$]$^-$ showed a five-coordinate Mn center, similar to its rhenium analog and to the IR-SEC of Mn(bpy-tBu)(CO)$_3$Br [93].

### 4.3.4 IRON, COBALT, AND NICKEL

In general, the ERC on electrodes of group 8–10 metals, such as Fe, Co, and Ni, in aqueous solutions under ambient conditions produces H$_2$ and CO or other hydrocarbons. However, these metals have high activities for the hydrogenation of CO and/or CO$_2$ in heterogeneous catalytic reactions (e.g., the Fischer–Tropsch reaction). At 30 atm, the ERC on Fe electrodes at a constant current density of 120 mA cm$^{-2}$ produced HCOOH with a faradaic efficiency of approximately 60% [94]. Several factors that might have affected the reduction process were proposed, including the diffusion of molecular CO$_2$ to the electrode surface and the suppression of hydrogen evolution by CO adsorption on the electrode surface [95]. In the ERC at Ni electrodes in aqueous media, H$_2$ as well as some small hydrocarbons, such as CH$_4$, C$_2$H$_4$, and C$_2$H$_6$, was produced. Under high CO$_2$ pressure, the faradaic efficiency for CO$_2$ reduction on Ni electrodes could be increased by raising the CO$_2$ pressure, lowering the temperature, and polarizing the electrode potential at a more negative potential [96]. It was suggested that hydrocarbons were formed on Fe, Co, and Ni electrodes through pathways similar to the Fischer–Tropsch reaction of thermal catalysis. In CO$_2$ electroreduction catalyzed by Fe and Ni metals, CO adsorption on these two electrodes was observed by IR spectroscopy, and the results suggested that the catalytic activity was strongly related to the bonding between CO and the metal surface [97].

### 4.3.5  Pt Group Metals

Pt group metals are well-known catalysts for $CO_2$ electroreduction. $CO_2$, CO, and $H_2$ adsorption on Pt group metal-based electrode surfaces differs depending on the metal, leading to different catalytic activities/stabilities and product selectivities [98–103].

#### 4.3.5.1  Ruthenium

Normally, Ru metal shows high catalytic activity in the gas-phase conversion of $CO_2$ to $CH_4$ [103]. In the aqueous electrochemical process of $CO_2$ reduction, supported Ru sponge electrodes did not show high catalytic activity [100]. However, due to their high stability, Ru electrodes were used for long-term $CO_2$ reduction at a constant potential [99].

#### 4.3.5.2  Palladium

The electrocatalytic activity of Pd for $CO_2$ reduction was first investigated in a 1.0 M $NaHCO_3$ solution [104]. HCOOH and CO (main products) and small amounts of hydrocarbons (from methane to hexane) resulted from an electrolysis process catalyzed by a Pd electrode in $CO_2$-saturated $KHCO_3$ aqueous solution [105]. It was observed that the current yield of CO could be increased substantially by increasing the pressure; when the electrode potential was held at 1.8 V versus Ag/AgCl in 0.1 M $KHCO_3$, the yield went from 5.3% at 1.0 atm to 57.9% at 50 atm [106]. The evolution of $H_2$ can be suppressed by hydrogen absorption on the Pd surface, and this absorbed hydrogen can react with the adsorbed reaction intermediates to change the electrocatalytic activity [107]. Hydrogenated Cu-modified Pd electrodes also showed higher catalytic activities, producing HCOOH, $CH_4$, and $CH_3OH$ [108,109]. It was found that $CO_2$ electroreduction on Pd electrodes could occur at potentials higher than the reversible hydrogen potential, suggesting that adsorbed hydrogen atoms might take part in the slow stage of the electroreduction of $HCO_3^-$ ($CO_2$) to $HCOO^-$ [110]. Ohkawa et al. [111] studied the $CO_2$ electroreduction reaction on a Pd electrode in a nonaqueous $CH_3CN$ solution and compared the results to those obtained in an aqueous solution; they demonstrated that the concurrent desorption of hydrogen could lead to enhanced catalytic activity for $CO_2$ electroreduction.

#### 4.3.5.3  Platinum

$CO_2$ electroreduction on a Pt electrode surface was studied early on by Eggins and McNeill [112], in water, dimethyl sulfoxide (DMSO), $CH_3CN$, and propylene carbonate (PC) solutions, respectively. By applying differential electrochemical mass spectrometry, Brisard et al. [113] investigated the mechanism of $CO_2$ electroreduction catalyzed by polycrystalline Pt in acidic media. The results showed that the main product was methanol. A gas-diffusion electrode with Pt electrocatalyst was also employed for $CO_2$ reduction under high pressure (<50 atm); a faradaic efficiency of 46% was obtained at a current density of 900 mA $cm^{-2}$, and $CH_4$ and $CH_3CH_2OH$ were found to be the major products [114]. The faradaic efficiency for $CH_4$ formation was increased by increasing the $CO_2$ pressure, whereas the efficiency for hydrogen formation was decreased. In a water free electrolyte, $CO_2$ electroreduction on a Pt

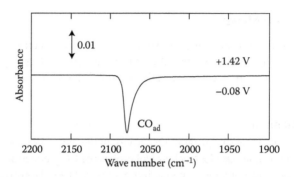

**FIGURE 4.6** FTIR spectrum of adsorbed CO on a Pt electrode formed during the ERC in the acetonitrile electrolyte with water concentration 10.5 mM. Potential with respect to Fc/Fc+: 11.42 and 20.08 V. (Reprinted with permission from Tomita, Y. et al. Electrochemical reduction of carbon dioxide at a platinum electrode in acetonitrile-water mixtures. *Journal of the Electrochemical Society.* 147(11), 4164–4167. Copyright 2000 American Chemical Society.)

electrode yielded oxalate as the main product. Since there was no water, $H_2$ evolution was not a concern. In this case, the $CO_2$ that diffused to the electrode might have been readily reduced to $CO_2^-$, with the $CO_2^-$ then reacting with $CO_2$ to form oxalate. By increasing the $H_2O$ concentration in the solution, $H_2$ evolution was enhanced, leading to the production of $HCOO^-$ and CO rather than oxalate, which was confirmed by *in situ* Fourier transform infrared spectroscopy (FTIR) reflection absorption spectroscopy (Figure 4.6) [115–117]. Regarding $CO_2$ electroreduction catalyzed by supported Pt NP catalysts, Centi et al. [118] employed carbon-supported Pt NPs (Pt/C) as the catalyst to convert $CO_2$ to long carbon-chain hydrocarbons (>C5) at room temperature and ambient atmospheric pressure in a continuous flow cell. Feng et al. [119] prepared a three-dimensional (3D) porous nanostructured electrode, composed of nanoporous CuPt composites, for $CO_2$ electroreduction in ionic liquid BMIM·BF$_4$. When Pt NPs were supported on either calcia-stabilized zirconia (Pt/CSZ) or $MnO_2$ (Pt/MnO$_2$) to form a high-temperature $CO_2$ reduction catalyst at 300–900°C, up to 100% selectivity for paraformaldehyde production was achieved [120]. More recently, Pt/C–TiO$_2$-and Pt-Pd/C–TiO$_2$-based nanocomposite cathodes were employed to catalyze the ERC to $CH_4$ and isopropanol [121]. In addition, electrochemical conversion of methanol and $CO_2$ to dimethyl carbonate (DMC) was realized at a graphite-Pt electrode in a dialkylimidazolium ionic liquid (1-benzyl-3-methylimidazolium chloride) methanol system without any other additives [122].

### 4.3.5.4 Single-Crystal Surfaces of Pt Group Metals

The surface structures of Pt single-crystal electrodes have long been found to have a significant influence on catalytic activity for $CO_2$ reduction [123,124]. A difference was observed between Pt(111) and Pt(110) single-crystal electrodes when researchers investigated the dynamic process of adsorbed CO formation from $CO_2$ and adsorbed hydrogen; the rate of CO formation on Pt(110) was more than 10 times higher than on Pt(111) [124]. Hoshi et al. [125] confirmed the order of activity for $CO_2$ reduction:

Pt(S)-[$n$(111) × (111)] > Pt(S)-[$n$(100) × (111)] > Pt(S)-[$n$(111) × (100)] (where S represents a single-crystal electrode) in 0.1 M $HClO_4$. They further found that $CO_2$ reduction rates on Pt(S)-[$n$(110)–(100)] electrodes ($n$ = 2, 9) were higher than those on Pt(110), and the rates on Pt(S)-[$n$(100) × (110)] ($n$ = 2, 3, 9) were over twice as high as on Pt(S)-[$n$(100) × (111)] [126]. In addition, Pt(210) (= Pt(S)-[2(100) × (110)]) yielded the highest rate of $CO_2$ reduction. This remarkably high activity might derive from the kink site characteristics of Pt(S)-[$n$(110) × (100)] and Pt(S)-[$n$(100) × (110)]. The order of activity for the stepped surfaces was Pt(110) > Pt(S)-[$n$(111) × (111)] > Pt(S)-[$n$(111) × (100)] > Pt(S)-[$n$(111) × (100)] > Pt(100) > Pt(111) [127]. The most active site in the stepped surfaces was derived from the pseudo-four-fold bridged site in Pt(S)-[$n$(111) × (111)]. Kinked stepped surfaces showed a higher activity than unkinked ones. The electrocatalytic activity of Pt single-crystal electrodes toward $CO_2$ reduction decreased in the order Pt(210) > Pt(310) > Pt(510) [128]. More recently, when evaluating 12 Bi–, Se–, Te–, and Sb–Pt(hkl) electrodes for the electrocatalysis of $CO_2$ reduction in both acid and neutral aqueous media, Sanchez-Sanchez et al. [129] found that only Bi–Pt(111), Te–Pt(111), and Sb–Pt(100) electrodes showed a visible current increase in the presence of $CO_2$, whereas Se–Pt(100) and Te–Pt(100) had lower catalytic activities than the corresponding unmodified Pt(100) electrode. The surface structure effects of Rh, Pd, and Ir single-crystal electrodes were also examined. It was found that the rate of $CO_2$ electroreduction on a Pd single-crystal electrode was strongly dependent on the crystal orientation. The rate of $CO_2$ reduction at −0.5 V versus reversible hydrogen electrode (RHE) on Pd(110) was two orders of magnitude higher than on Pt(110) [130].

### 4.3.5.5 Pt Group Metal Oxide (Mixture) Catalysts

In an early study on the electroreduction of $CO_2$ on various conductive oxide mixtures ($RuO_2$, $TiO_2$, $MOO_2$, $Co_3O_4$, and $Rh_2O_3$), two metal oxide electrodes—that is, $RuO_2$ (35, mole percentage) + $TiO_2$(65) and $RuO_2$(20) + $Co_3O_4$(10) + $SnO_2$(8) + $TiO_2$(62)—showed high current efficiencies for methanol production when the electrode potential was controlled near the equilibrium potential of hydrogen evolution in a solution of 0.2 M $Na_2SO_4$ (pH 4) saturated with $CO_2$. Later, Bandi and Kuhne investigated the electrocatalytic activities of mixed Ru/Ti oxide electrodes (titanium sheets); their results indicated that the overpotential for $H_2$ evolution increased with increasing $TiO_2$ content [131]. In a comparison study of $CO_2$ electroreduction in 0.5 M $NaHCO_3$ solution, three electrodes—Ru, Cu–Cd-modified Ru, and Cu–Cd-modified $RuO_x$ + $IrO_x$—were used for electrolysis for 8 h while the potential was held at −0.8 V versus SCE. Both methanol and acetone were produced [99]. A $RuO_2$-coated diamond film on a Si(111) wafer was also used as the electrode for the ERC [132]. The main reduction products obtained in acidic and neutral media were HCOOH and $CH_3OH$, with efficiencies of 40% and 7.7%, respectively. Qu et al. [63] loaded $RuO_2$–$TiO_2$ NTs and $RuO_2$–$TiO_2$ NPs, respectively, onto Pt electrodes for $CO_2$ electroreduction. Compared with electrodes coated with $RuO_2$ or with $RuO_2$–$TiO_2$ NPs, the electrodes coated with $RuO_2$–$TiO_2$ NTs had a higher electrocatalytic activity for the conversion of $CO_2$ to $CH_3OH$, with a current efficiency of up to 60.5%, suggesting that the NT structure might be important in achieving high efficiency and selectivity for $CO_2$ electroreduction.

### 4.3.6 Copper, Silver, Gold, Zinc, Cadmium, and Mercury

#### 4.3.6.1 Cu/Ag/Au/Zn/Cd/Hg Metal Electrodes

Several studies using metal electrodes for $CO_2$ electroreduction in aqueous $KHCO_3$ solution reported that $CH_4$ was predominantly the product at a Cu cathode, CO at Ag and Au cathodes, and $HCOO^-$ at a Cd cathode [133], as reviewed by Jitaru et al. [134] and Gattrell et al. [135]. It is suggested that the pathway by which methane, ethylene, carbon monoxide, and formic acid on Cu electrode are formed is given in Figure 4.7.

The metallic Cu electrodes thus far developed can be classified into several types: bulk Cu electrode, Cu electrodeposited GCE, *in situ* electrodeposited Cu electrode [136,137], and Cu-coated gas-diffusion electrode (GDE) [138]. Aside from low hydrocarbons such as $CH_4$, $C_2H_4$, CO, HCOOH, alcohols (methanol, $CH_3CH_2OH$, and $CH_3CH_2CH_2OH$), and esters, some relatively high hydrocarbons—such as paraffins and olefins containing up to six carbon atoms—can also be formed using Cu electrodes [139–143]. The faradaic efficiencies of these products were largely dependent on temperature, type, and concentration of electrolytes, electrode potential, pH, crystal surface, and even the purity of the cathode material (Cu). For example, in the electrolysis of $CO_2$-saturated 0.5 M $KHCO_3$ aqueous solution with a 99.999% Cu sheet cathode, Hori et al. [139] found that increasing the temperature (0–40°C) caused the faradaic efficiency of $CH_4$ production to drop rapidly from 65% to nearly zero, while that of $C_2H_4$ gradually increased by up to 20%. In addition, they carried out experiments using a range of electrolyte strengths, from 0.03 to 1.5 M [140]. Using voltammetric, coulometric, and chronopotentiometric measurements, they observed that CO was predominantly formed at potentials more positive than −1.2 V versus NHE, while hydrocarbons (e.g., $CH_4$ and $C_2H_4$) and alcohols (e.g., $CH_3CH_2OH$ and $CH_3CH_2CH_2OH$) were produced in

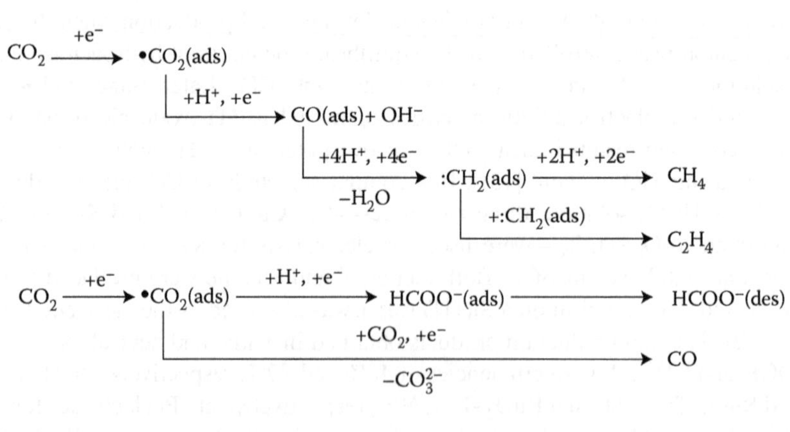

**FIGURE 4.7** Reaction mechanism of the ERC at Cu electrode in methanol. (Reprinted from *Electrochimica Acta*, 51(16), Kaneco, S., Katsumata, H., Suzuki, T. and Ohta, K., Electrochemical reduction of carbon dioxide to ethylene at a copper electrode in methanol using potassium hydroxide and rubidium hydroxide supporting electrolytes, 3316–3321, Copyright 2006, with permission from Elsevier.)

greater abundance below −1.3 V, where the faradaic efficiency of CO dropped. In fact, CO was a reaction intermediate that strongly adsorbed on the cathode, interfering with hydrogen formation. In KCl, $K_2SO_4$, $KClO_4$, and dilute $HCO3^-$ solutions, $C_2H_4$ and alcohols were found to be the main products, whereas $CH_4$ was preferentially produced in relatively concentrated $HCO3^-$ and $K_2HPO_4$ solutions. In nonaqueous solutions, the ERC on a 99.999% Cu wire in $CH_3OH$ at 20–25°C and 40 atm primarily yielded CO when TBA salts (TBA·$BF_4$ and TBA·$ClO_4$) were used as supporting electrolytes [143]. The formation of $HCOOCH_3$ became predominant when lithium salts ($LiBF_4$, $LiClO_4$, and $NH_4ClO_4$) were used. The current–potential curves for various concentrations of TBA·$BF_4$ showed a large cathodic current and a shoulder wave (Figure 4.8). Since the magnitude of the shoulder wave was dependent on the concentration of TBA·$BF_4$, this wave was attributed to the reduction of TBA+, presumably to TBA:

$$NR_4^+ + e^- \rightarrow NR_4 \qquad (4.15)$$

$$CO_2 + NR_4 \rightarrow CO_2^- + NR_4^+ \qquad (4.16)$$

At −2.3 to −4.0 V versus Ag/AgCl, the ERC at a 99.98% Cu electrode in a $CH_3OH$ solution containing 80 mM LiOH supporting salt at about −30°C, $CH_4$, $C_2H_4$, CO, and HCOOH were the main products [144]. The best current efficiency for $CH_4$ (the main product) was 63% at −4.0 V versus Ag/AgCl. When Kaneco et al. [145] studied $CO_2$ electroreduction at a 99.98% Cu electrode in a $CH_3OH$ solution with 80 mM CsOH supporting salt at about −30°C, $CH_4$, $C_2H_4$, $C_2H_6$, CO, and HCOOH were the

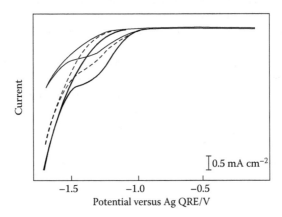

**FIGURE 4.8** Current–potential characteristics of the ERC in a $CO_2^+$ methanol medium with various concentrations of TBA·$BF_4$ as the supporting electrolyte. Concentrations of TBA·$BF_4$: 33 mM (dotted line); 66 mM (dashed line); and 0.1 M (solid line). (Reprinted from *Journal of Electroanalytical Chemistry*, 390(1–2), Saeki, T., Hashimoto, K., Kimura, N., Omata, K. and Fujishima, A., Electrochemical reduction of $CO_2$ with high-current density in a $CO_2$ plus methanol medium 2. Co formation promoted by tetrabutylammonium cation, 77–82, Copyright 1995, with permission from Elsevier.)

products. The maximum faradaic efficiency of $C_2H_4$ (the main product) was 32.3% at −3.5 V versus Ag/AgCl. The $C_2H_4/CH_4$ current efficiency ratio was in the range 2.9–7.9. It was thought that small cations, such as $Li^+$ and $Na^+$, might not easily adsorb on the electrode surface due to their strong hydration, while a less hydrated, bulky cation, such as $Cs^+$, might preferentially adsorb on the cathode, giving a less hydrated electrode surface. On such a surface, the conversion of the intermediate $Cu = CH_2$ to $CH_2CH_2$ might occur more easily than on a more hydrated one, producing more $CH_2CH_2$ in the presence of $Cs^+$. Under a $CO_2$ pressure of 10 atm at about −30°C in $CH_3OH$ with 0.5 M CsOH supporting salt, when the potential was increased from −3.5 to −2.0 V, the faradaic efficiencies of CO and $C_2H_4$ dropped slowly from 84% to 40% and 5% to 4%, respectively, but those of HCOOH, $CH_4$, and $H_2$ slowly increased [146]. Kaneco et al. [147] suspended 1 mm copper particles in $CH_3OH$ and conducted the ERC with Pb and Zn electrodes. Results showed that without the addition of the copper particles, only HCOOH and CO could be detected, but after their addition, hydrocarbons were formed. The faradaic efficiencies of $CH_4$ and $C_2H_4$ rose gradually as the amount of Cu particles in the solution was increased, while the currency efficiencies of HCOOH and CO decreased. When B-370 Cu (99.9% pure) was employed at the cathode instead of high-purity Cu [148], the faradaic efficiency of $CH_4$ was found to be lower. In Hori's study [139], which used pure Cu (99.999%) and different electrolytes—including KCl (0.1/0.5/1.5 M), $KHClO_4$ (0.1 M), $K_2SO_4$ (0.1 M), and $K_2HPO_4$ (0.1, 0.5 M)—the faradaic efficiency of $C_2H_4$ reached 38.2%–48.1%, while that of $CH_4$ was 11.5%–17.0%. Kim et al. [149] tried to enhance the formation rate of $CH_4$; the highest rates they obtained were $8 \times 10^{-5}$ mol $cm^{-2}$ $h^{-1}$ (22°C, 17 mA $cm^{-2}$) and $1.1 \times 10^{-5}$ mol $cm^{-2}$ $h^{-1}$ (0°C, 23 mA $cm^{-2}$) at −2.0 and −2.3 V versus SCE, respectively, on Cu foil in 0.5 M $KCO_3$ (pH 7.6). They found that the formation rate of $CH_4$ was even higher if the electrode surface was prepared by cleaning it using HCl rather than $HNO_3$ or by oxidation in air. Indeed, electrode surface conditions can have a major effect on an electrode's catalytic activity in $CO_2$ electroreduction. For example, some significant performance differences were found between rough and smooth electrodes as well as between thermally and nonthermally treated electrodes [25]. Ohta et al. [150] found that under ultrasonic irradiation, the production rates of $CH_4$, HCOOH, and CO were greatly affected. Cook et al. [151] employed an *in situ* electrodeposited Cu electrode (glass carbon substrate with *in situ* deposited Cu) in 0.5 M $KHCO_3$ aqueous solution. Although the main products were still $CH_4$ and $C_2H_4$, Cu purity and morphology were found to be crucial for promoting a high rate of $CO_2$ reduction. They achieved a faradaic efficiency of 71.3% for $C_2H_4$ and $CH_4$ production by employing a Cu-based GDE [137]. Recently, a Cu NP-covered electrode was reported to give better selectivity toward hydrocarbons than the surfaces of electropolished and argon-sputtered copper electrodes [152]. The copper NPs were formed in two steps. In the first step, the potential at the electropolished copper electrode was scanned between −0.6 and +1.15 V versus RHE at 20 mV $s^{-1}$ under $N_2$-saturated $KClO_4$. In the second step, the copper was redeposited on the electrode surface; this was performed under $CO_2^-$-saturated $KClO_4$ with a constant bias of −1.3 V versus RHE for 20 min. Scanning tunneling microscopy and scanning electron microscopy (SEM) was employed to compare various treated electrodes (Figure 4.9). A few layers of Cu NPs with sizes of 50–100 nm covered the Cu surface,

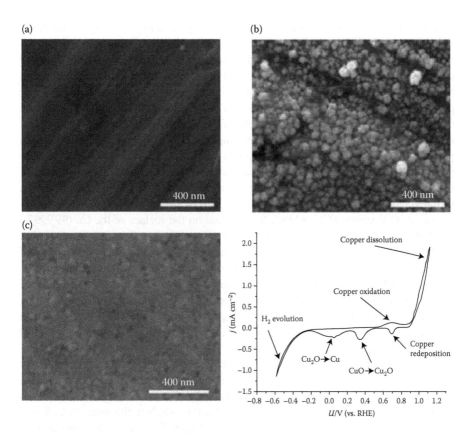

**FIGURE 4.9** SEM images for three types of surfaces: (a) electropolished, (b) copper NP covered, and (c) sputtered, and cyclic voltammograms (CVs) of the formation of copper NPs in 0.1 M KClO$_4$ purged with N$_2$ at pH 10.5 on the electropolished copper surface. (Wang et al. *Phys. Chem. Chem. Phys.*, 2012, 14, 76–81. Reproduced by permission of The Royal Society of Chemistry.)

creating a surface area two to three times greater than the geometric surface area of the Cu electrode. However, CV measurements in CO$_2$ (pH 6.0) showed that the current density of the NP-covered surface (at −0.75 V vs. RHE) was 10 times higher than that of the electropolished surface, indicating that surface morphology can contribute more to current density than just the effect of increased surface area.

The morphological effect was explained by the roughened surface having a greater abundance of undercoordinated sites; this was demonstrated by DFT calculations. In addition, electrochemical quartz crystal microbalance—a high-resolution (Bnanograms) mass sensing technique—has also been used to probe electroformed and electroreduced products on a copper electrode in aqueous solutions containing NaHCO$_3$ and Na$_2$CO$_3$ [153]. Li and Oloman [154] investigated the electroreduction of CO$_2$ in a laboratory bench-scale continuous reactor in which a flow-by 3D cathode of 30$^\#$ mesh tinned-copper was used as the cathode. With currents of 1–8 A, feeding gas-phase CO$_2$ concentrations of 16–100 vol%, and operating times of 10–180 min, a current efficiency of 86% was achieved for HCOO$^-$. The efficiency was dependent on current density and $CO_2$ pressure. In a study of copper-catalyzed $CO_2$ electroreduction, Kuhl et al. [155] reported an experimental methodology that allows for product identification and quantification with unprecedented sensitivity. Among all the possible products, $CH_4$ and $C_2H_4$ had the largest current efficiencies; the remaining products were oxygenates and other $C_2$ and $C_3$ species. The researchers offered some possible reaction pathways to account for the production of all the $C_2$ and $C_3$ species observed. Regarding the catalytic stability of $CO_2$ electroreduction on Cu electrodes, several other factors have been identified, including CO adsorption [156], electrode purity [148], the formation of carbon deposits [157], and the presence of other surface poisoning species [26,32]. The carbon deposited film seems to be a major factor in the irreversible degradation of the electrode surface. For example, when the ERC to both $CH_4$ and $C_2H_4$ was conducted in aqueous 0.5 M $KHCO_3$ solution at a constant potential of −2.00 V versus SCE, a black film formed on the surface of the Cu (99.999%) cathode [157]. It is important to ensure that the system is free of contaminants so that electrochemical and product measurements can truly be attributed to the activity of copper. SEM images of the copper electrode surface are shown after mechanical polishing and subsequent electropolishing (Figure 4.10), resulting in a smoother surface free of impurities. Electropolishing was performed in the copper ion diffusion limited regime of the CV (Figure 4.10c) before oxygen evolution begins and before surface roughening can occur. Figure 4.10d shows x-ray photoelectron spectroscopy (XPS) measurements before and after electrolysis. The peaks are attributable to only Cu, O, and C, revealing that no metal impurities are present within the detection limits of XPS. It indicated that this film was graphitic carbon formed by $CO_2$ reduction through $HCOO^-$. Graphitic carbon deposit was also found to cause a decline in Cu electrodes' catalytic activity [158,159]. To resolve this problem, the researchers developed a new method that could selectively convert $CO_2$ to $C_2H_4$ at the three-phase (gas/liquid/solid) interface on a $Cu^IBr$ confined Cu-mesh electrode in an aqueous solution of KBr. The conversion percentages of $CO_2$ and $H_2$ were found to be about 90% and 2%, respectively. It was suggested that the immobilized CuIBr, acting as a heterogeneous catalyst, offered some adsorption sites for reduction intermediates such as CO and carbene ($H_2C$:). The mechanism can be expressed as follows:

$$CO_2(g) + 2H^+ + 2e^- \rightarrow CO(g) + H_2O \tag{4.17}$$

$$CuBr(s) + CO(g) \rightarrow CuBr \cdots CO \tag{4.18}$$

$$CuBr \cdots CO + 4H^+ + 4e^- \rightarrow CuBr \cdots : CH_2 + H_2O \tag{4.19}$$

$$2CuBr \cdots : CH_2 \rightarrow CuBr \cdots : CH_2 = CH_2 + CuBr(s) \tag{4.20}$$

Similar to Pt group metals such as Rh [160], Pd [161], Ir [162], Pt [123], and Ag [163], single-crystal copper electrode surfaces can normally enhance the electrocatalytic activity in $CO_2$ electroreduction by introducing steps and kinks into atomically flat surfaces. For example, Hori et al. [164–166] investigated the ERC at various

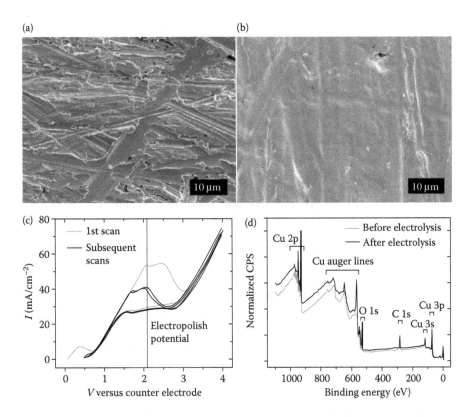

**FIGURE 4.10** Surface preparation of the Cu working electrode. (a) SEM of the sandpapered surface. (b) SEM of the surface after electropolishing. (c) Two electrode CV taken of the Cu electrode in 85% phosphoric acid indicating the potential where electropolishing occurred. (d) XPS of the surface before and after an electrolysis experiment. (Kuhl, K. P., Cave, E. R., Abram, D. N. and Jaramillo, T. F. New insights into the electrochemical reduction of carbon dioxide on metallic copper surfaces. *Energy & Environmental Science*. 2012; 5(5):7050–7059. Reproduced by permission of The Royal Society of Chemistry.)

types of copper single-crystal electrodes in 0.1 M KHCO$_3$ aqueous solution and found that the reaction selectivity could be greatly altered by changing the crystal orientation. The major product with electrodes based on (100) terrace surfaces (i.e., Cu(S)-[$n$(100) × (111)] and Cu(S)-[$n$(100) × (110)]) was C$_2$H$_4$. The formation of CH$_4$ was promoted at Cu(111) or by the introduction of (111) or (110) step atoms to the (100) basal plane. A Cu(S)-[$n$(111) × (111)] electrode yielded high amounts of C2+ substances (i.e., substances containing more than two carbon atoms), while a (110) electrode derived from Cu(S)-[$n$(111) × (111)] uniquely produced high yields of CH$_3$COOH, CH$_3$CHO, and C$_2$H$_5$OH. CO seems to be the key intermediate in CO$_2$ electroreduction to CH$_4$ and CH$_2$CH$_2$ on Cu electrodes [167,168]. Schouten et al. [168] observed two reaction pathways: a C1 pathway leading to CH$_4$ formation on single-crystal Cu(111) electrodes and a C2 pathway leading to ethylene formation on Cu(100) electrodes [169]. The authors also proposed a mechanism based on reactions as a function of potential, using online mass spectrometry combined with mechanisms suggested in the literature. Regarding theoretical studies of $CO_2$ electroreduction on Cu electrodes, some researchers have carried out studies to further our fundamental understanding of the effects of face-centered Cu facets—such as cubic fcc(111), fcc(100), and fcc(211)—on the energetics of $CO_2$ electroreduction. Durand et al. [170] reported that the intermediates in $CO_2$ reduction could be mostly stabilized by the (211) facet, followed by fcc(100) and fcc(111). This implied that the (211) facet should be the most active surface in producing $CH_4$, as well as the by-products $H_2$ and CO. HCOOH production might be mildly enhanced on the more closely packed surfaces (i.e., (111) and (100)). Their theoretical prediction of the trends for voltage requirements was consistent with experimental measurements. Liu et al. [171] studied the electroreduction of $CO_2$ to CO on Fe, Co, Ni, and Cu surfaces using a DFT method involving three reaction steps: adsorption of $CO_2$, decomposition of $CO_2$, and desorption of CO. Both the binding energies and the reaction energies were calculated. They found that the reaction energies and the total reaction energy barrier were strongly dependent on the type of electrode metal.

### 4.3.6.2 Ag and Au Electrodes

Similar to Cu, both Ag and Au electrodes also show considerable catalytic activity toward $CO_2$ electroreduction when appropriate electrolytes are employed [172]. In 0.1 $KHCO_3$ aqueous solution at 25°C, the faradaic efficiencies for CO production from $CO_2$ at Ag (99.98%) and Au (99.95%) electrodes at −1.6 V versus Ag/AgCl saturated with KCl were found to be 64.7% and 81.5%, respectively. A recent study reported that when Ag-coated nanoporous Cu composites were employed in $CO_2$ reduction in BMIm·$BF_4$, the electrosynthesis of DMC could be realized [142]. The highest yield of DMC was 80%; this was attributed to the electrodes' high surface area, open porosity, and high efficiency. A sputtering deposition technique was used to prepare an Au electrode for $CO_2$ reduction [173]. The results indicated that Ar pressure had an effect on the Au surface's geometrical structure and surface area, resulting in different $CO_2$ reduction potentials in both KCl and $KHCO_3$ solutions. A porous Au film electrode (a 200–260-nm Au film deposited on a porous hydrophilic polymer membrane) was also prepared by vapor deposition for the ERC to CO in 99.99% $KHCO_3$ aqueous solution [174]. Recently, ligand-protected $Au_{25}(SC_2H_4Ph)_{18}{}^{-}$ clusters were explored to promote the catalytic electroreduction of $CO_2$, and high catalytic activity in the conversion of $CO_2$ to CO was achieved [175].

### 4.3.6.3 Zn, Cd, and Hg Electrodes

Due to the high overpotentials of hydrogen evolution on Zn, Cd, and Hg electrodes, they were considered as suitable electrodes for $CO_2$ reduction. In 0.1 M $KHCO_3$ aqueous solution, the ERC on Zn, Hg, and Cd electrodes was found to be very selective for the formation of HCOOH and CO, with HCOOH seemingly the only product at the Hg cathode. On metallic Zn, Cd, and Hg electrodes, the faradaic efficiencies of HCOOH were measured to be 20%, 39%, and 94%, while those of CO were 39.6%, 14.4%, and undetectable, respectively. Shibata et al. [176] employed a Cd-loaded GDE to reduce $CO_2$ and nitrite ions with various catalysts and found that the maximum current efficiency of urea (($NH_2$)CO) formation was about 55% at −1.0 V versus SCE. In an organic solution such as DMSO, $CO_2$ electroreduction was also studied using both Au and Hg electrodes [177,178]. It should be mentioned that several decades ago, a mercury pool electrode was considered the best electrode for $CO_2$ reduction [179]. The results showed that in the neutral pH range, all the current was consumed in the production of HCOOH, while in acid solutions, both HCOOH and $H_2$ were produced. The mechanism was proposed to be as follows:

$$CO_2 + e^- \rightarrow CO_2^-(ads) \tag{4.21}$$

$$CO_2^-(ads) + H_2O \rightarrow HCO_3(ads) + OH^- \tag{4.22}$$

$$HCO_3(ads) + e^- \rightarrow HCOO^- \tag{4.23}$$

$$H^+ + e^- \rightarrow H\ ads \tag{4.24}$$

$$H^+ + CO_2 \rightarrow HCO_2 \tag{4.25}$$

$$H + H^+ + e^- \rightarrow H_2 \tag{4.26}$$

$$HCO_2 + H^+ + e^- \rightarrow HCOOH \tag{4.27}$$

With a Hg electrode, the formation of malate ($^-$OOCCH–(OH)CH$_2$COO$^-$) in the process of $CO_2$ reduction was also observed in aqueous solutions containing quaternary ammonium salts [180]. Since the observed coulombic yield was more than 100%, the overall reaction was believed to consist of not only reaction (4.50) but also possibly (4.51):

$$4CO_2 + 10H^+ + 12e^- \rightarrow \ ^-O\overset{\overset{\displaystyle O}{\|}}{C}\underset{\underset{\displaystyle OH}{|}}{C}HCH_2\overset{\overset{\displaystyle O}{\|}}{C}O^- + 3H_2O \tag{4.28}$$

$$4CO_2 + 9H_2O \rightarrow HO\overset{\overset{\displaystyle O}{\|}}{C}\underset{\underset{\displaystyle OH}{|}}{C}HCH_2\overset{\overset{\displaystyle O}{\|}}{C}OH + 6H_2O \tag{4.29}$$

However, with other supporting electrolytes, such as NaHCO$_3$, NaH$_2$PO$_4$–Na$_2$HPO$_4$, NaCl, NaClO$_4$, Na$_2$SO$_4$, LiHCO$_4$, and KHCO$_3$ and their mixtures, the favorable product was found to be HCOO$^-$, and the faradaic efficiency of HCOOH formation increased with increasing $CO_2$ pressure [181,182]. In 0.5 M KHCO$_3$ aqueous solution, the faradaic efficiency of HCOOH production was as high as 100% at a $CO_2$ pressure of 20 atm [182].

### 4.3.6.4   Cu and Au Oxide-Related Catalysts

#### 4.3.6.4.1   Cu Oxides

Cu oxides have been explored to catalyze $CO_2$ electroreduction. For example, Chang et al. [183] deposited as-prepared Cu$_2$O particles onto a carbon cloth electrode for

$CO_2$ reduction and carried out CV measurements; the results showed that $CH_3OH$ was the predominant product. Ohya et al. [184] prepared a CuO–Zn composite electrode by pressing a mixture of Zn particles (B7 mm) and CuO or $Cu_2O$ powder (B100 nm) for $CO_2$ reduction [184]. It was found that without the copper oxide particles; only HCOOH and CO were formed, whereas with the CuO–Zn composite electrode, hydrocarbons such as $CH_4$ and $C_2H_4$ could be obtained. The maximum formation efficiencies of $CH_4$ and $C_2H_4$ were 7.5% and 6.8%, respectively. Le et al. [29] examined the catalytic activity of an electrode electrodeposited as a cuprous oxide thin film and found that the faradaic efficiency for $CH_3OH$ production was 38%. They believed that Cu(I) species should play a critical role in selectivity for $CH_3OH$. More recently, Li and Kanan [43] prepared Cu as $Cu_2O$ layers for $CO_2$ reduction. The $Cu_2O$ layers formed at 500°C exhibited a large surface roughness, resulting in the electrochemically active surface area (ECSA) of a reduced electrode being 480 times larger than that of a polycrystalline Cu electrode. This improved ECSA resulted in a 0.5 V lower overpotential for $CO_2$ reduction than on a polycrystalline Cu electrode. An interesting result was that the SEM showed a dense array of rods with diameters of 100–1000 nm on the electrode surface. These rods were the outermost portion of a thick $Cu_2O$ layer coating the electrode. However, these rods were not necessary for efficient $CO_2$ reduction. It was observed that the Cu particles formed by reducing mm-thick (B3 mm) $Cu_2O$ films at the potentials at which $CO_2$ reduction occurred could catalyze the reduction of $CO_2$ to CO and HCOOH with high faradaic efficiencies and exceptionally low overpotentials. At high overpotentials, this Cu particle electrode could produce C2 hydrocarbons exclusively. To obtain insight into the mechanistic pathway(s) for $CO_2$ reduction, Tafel plots were analyzed. The plot for annealed Cu was found to be linear over the range of overpotentials from 0.05 to 0.3 V with a slope of 116 mV dec. This slope was consistent with a rate-determining initial electron transfer to $CO_2$ to form a surface-adsorbed $CO_2{}^*$ intermediate, which suggested that the Cu surfaces formed by reducing thick $Cu_2O$ layers could enable the formation of this $CO_2{}^*$ intermediate while suppressing $H_2O$ reduction. Regarding the understanding of fundamentals, Wu et al. [185] investigated the adsorption of $CO_2$, $H_2CO_3$, $HCO_3{}^-$, and $CO_3{}^{2-}$ on a $Cu_2O$ (111) surface by first-principles calculations based on DFT at the B3LYP hybrid functional level, in which the $Cu_2O$ (111) surface was modeled using an embedded cluster method. It was concluded that on the surface, $H_2CO_3$ was dissociated into an $H^+$ ion and an $HCO_3{}^-$ ion, which was the only activated $CO_2$ species on the surface.

### 4.3.6.4.2 Au Oxide Catalysts

Chen et al. [186] reduced Au oxide films to Au NPs on electrodes for $CO_2$ reduction. High selectivity from $CO_2$ to CO in water at overpotentials as low as 140 mV was observed. The high catalytic activity was thought to be due to the dramatically increased stabilization of the $CO_2{}^*$ intermediate on the surfaces of the oxide-derived Au electrodes. As shown in Figure 4.11, the reduction of thick Au oxide films results in the formation of Au NPs ("oxide-derived Au") that exhibit highly selective $CO_2$ reduction to CO in water at overpotentials as low as 140 mV and retain their activity for at least 8 h. Under identical conditions, polycrystalline Au electrodes and several other nanostructured Au electrodes prepared via alternative methods require at least

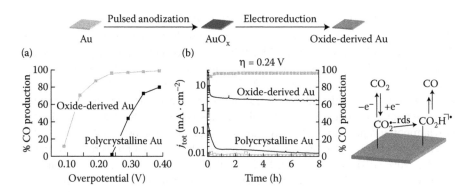

**FIGURE 4.11** (a) FEs for CO and $HCO_2^-$ production on oxide-derived Au and polycrystalline Au electrodes at various potentials between −0.2 and −0.5 V in 0.5 M $NaHCO_3$, and (b) $CO_2$ reduction activity of polycrystalline Au with that of oxide-derived Au. Total current density versus time (— left axis) and FE for CO production versus time (right axis) on oxide-derived Au (gray, ■) and polycrystalline Au (gray, □) in electrolyses at −0.35 V versus RHE. (Reprinted with permission from Chen, Y., Li, C. W. and Kanan, M. W. Aqueous $CO_2$ reduction at very low overpotential on oxide-derived Au nanoparticles. *Journal of the American Chemical Society*. 134(49), 19969–19972. Copyright 2012 American Chemical Society.)

200 mV of additional overpotential to attain comparable $CO_2$ reduction activity and rapidly lose their activity. Electrokinetic studies indicate that the improved catalysis is linked to dramatically increased stabilization of the $CO_2^{\bullet-}$ intermediate on the surfaces of the oxide-derived Au electrodes.

### 4.3.7 ALUMINUM, GALLIUM, INDIUM, AND THALLIUM

#### 4.3.7.1 Al Metal Electrodes and Al-Containing Catalysts

Al has been tried for the catalysis of $CO_2$ electroreduction, but unfortunately the catalytic activities have been extremely low [178]. For example, it was reported that in the ERC by a metallic Al electrode at 1.6 V versus Ag/AgCl in 0.1 M $KHCO_3$ aqueous solution, the faradaic efficiencies for $CH_4$, $C_2H_4$, and $C_2H_6$ production were as low as 0.58%, 0.04%, and 0.11%, respectively, versus 99% for $H_2$ evolution. In another experiment, at a constant potential of 2.2 V versus SCE in a $CO_2$-saturated 0.05 M $KHCO_3$ aqueous solution, the sum of the typical current efficiencies of $CH_4$, $C_2H_4$, $C_2H_6$, and HCOOH was less than 1%, compared with 95.7% for $H_2$ [178].

#### 4.3.7.2 Ga Electrodes and Ga-Containing Catalysts

In a photoelectrochemical cell, p-type Ga-containing semiconductors—that is, p-gallium phosphide (p-GaP) [187], p-galliumarsenide (p-GaAs) [188], and n-gallium arsenide (n-GaAs)—were explored as catalysts; high selectivity for $CH_3OH$ formation (a 6e reaction) was observed but only at some exceptionally high overpotentials. Recently, Barton et al. [189] reported the highly selective reduction of $CO_2$ to $CH_3OH$ at illuminated p-GaP photoelectrodes, with near 100% faradaic efficiency for underpotentials greater than 300 mV at 0.52 V versus SCE.

### 4.3.7.3    In Metal Electrodes and In-Containing Catalysts

Normally, $CO_2$ electroreduction at an indium metal electrode in aqueous media predominantly produces formate [34,133,190], but in nonaqueous solutions, the main product is CO [190]. Kapusta and Hackermant [34] reported $CO_2$ reduction to formate ions in a 0.5 M HCOOH + 0.5 M HCOONa solution with high current efficiency (about 95%), although the overall power efficiency was low due to the high overpotential of the reaction. Ikeda et al. [191] also observed the ERC in 0.1 M tetraethylammonium perchlorate (TEAP) aqueous electrolytes at potentials of 1.8–2.4 V versus Ag/AgCl; the current efficiencies for HCOOH formation were in the range of 80%–90% [191]. However, in 0.1 M TEAP/PC (nonaqueous electrolytes) at 2.0–2.2 V versus Ag/AgCl, the current efficiencies were up to 95% [190]. Todoroki et al. [182] achieved a faradaic efficiency of 100% under 60 atm of $CO_2$. They found that the efficiency of HCOOH formation became higher at high cathodic current densities (or more negative potentials), and that at less negative potentials, HCOOH formation could be suppressed and CO formation became relatively predominant. Mizuno et al. [192] also found that the faradaic efficiency for HCOOH was about 100% at 20–60°C, compared with 44.5% at 100°C. Recently, Narayanan et al. [193] studied the conversion of $CO_2$ to formate in an alkaline polymer electrolyte membrane cell in which In powder-coated porous carbon paper was the cathode. Three different aqueous solutions ($CO_2$-saturated deionized $H_2O$, 1 M $NaHCO_3$, and 1 M $Na_2CO_3$) were used as the cathode feed. The instantaneous faradaic efficiency of formate production achieved in $NaHCO_3$ solution was as high as 80% (although it decreased to B10% over a period of 1.0 h). Carbon mass transport was found to be the limiting factor for faradaic efficiency. However, this mass transport limitation could be mitigated using a high bicarbonate concentration or high carbon dioxide pressure. In their experiments, high faradaic efficiency could be maintained during continuous operation, even at moderately high current densities. An In-containing semiconductor (p-type indium phosphide, p-InP) electrode was used in the electrolysis of $CO_2$-saturated $Na_2SO_4$ aqueous solutions; the photocurrent densities for $CH_3OH$ formation were found to be 60–100 mA cm$^2$, and the current efficiencies were found to be 40%–80% [188]. Parkinson and Weaver [194] tried to fix $CO_2$ by combining a p-InP electrode with a biological catalyst (a FDH enzyme); they observed a 2e$^-$ reduction of $CO_2$ to HCOOH. Kaneco et al. [195] carried out a series of experiments on the photo ERC at a p-InP electrode in a methanol-based electrolyte (nonaqueous media). In $CO_2$-saturated 80 mM LiOH–methanol solution, they observed the maximum current efficiencies to be B40% for CO and B30% for HCOOH in the potential range of −2.2 to −2.5 V versus Ag/AgCl [195]. They also found that metal-modified InP electrodes could give different product selectivity [196]. For example, on Pb-, Ag-, Au-, and Cu-modified InP electrodes, the main reduction products of $CO_2$ were CO and HCOOH (the maximum current efficiency of CO was 80.4% on Ag-InP); in comparison, a Pd-modified electrode yielded only CO, and a Ni-modified electrode produced hydrocarbons with low faradaic efficiencies (0.7% for $CH_4$ and 0.2% for $CH_2CH_2$). Recently, they tried to catalyze $CO_2$ reduction by suspending Cu particles (1 mm diameter) in 0.10 M NaOH–methanol solution [197]. After the addition of Cu particles, the current efficiencies for methane and ethylene improved.

#### 4.3.7.4    Tl Metal Electrodes and Tl-Containing Catalysts

Tl metal electrodes in aqueous electrolytes have been found to favor the formation of formic acid, while in nonaqueous solutions, oxalic acid has been the dominant product [134,191]. Unfortunately, the literature contains only these two cited studies of Tl metal electrodes as cathode catalysts for $CO_2$ electroreduction.

### 4.3.8    TIN AND LEAD

#### 4.3.8.1    Sn Metal Electrodes and Sn-Containing Catalysts

Sn metal electrodes were reported to be most active toward $CO_2$ electroreduction in aqueous electrolytes, producing $HCOO^-$. However, in nonaqueous electrolytes, the predominant product was CO, with small amounts of formic acid, oxalic acid, and glyoxalic acid [134,191]. Some early studies indicated that Sn metal working electrodes could catalyze $CO_2$ electrochemical reduction in aqueous inorganic salt solutions to exclusively produce HCOOH with a current efficiency as high as B95% [34,133]. However, during the reduction reaction, organometallic complexes formed on the electrode surface, accelerating the rate of hydrogen evolution and leading to poor reaction efficiency [34]. Increasing the temperature also can cause a decrease in faradaic efficiency for formic acid production and an increase for hydrogen [192]. Li and Oloman [22,198] developed a scale-up reactor system for $CO_2$ electroreduction, in which a granulated tin cathode (99.9 wt% Sn) and a feed gas of 100% $CO_2$ were used. The results showed that the granulated tin cathode yielded better performance than the tinned-copper mesh cathode reported in their previous communications, in terms of both current efficiency and stability [153,198]. The formate current efficiencies were up to 91%. When tin was electrodeposited on a GDE in a zero-gap cell for $CO_2$ electroreduction, the electrode showed good stability [199]. Chen and Kanan [200] prepared a thin-film catalyst by simultaneously electrodepositing $Sn^0$ and $SnO_x$ onto a Ti electrode. Using an H-type cell reactor and $CO_2$ saturated aqueous $NaHCO_3$ solution, the $Sn^0/SnO_x$ catalyst exhibited up to eightfold higher partial current density and fourfold higher faradaic efficiency for $CO_2$ reduction, more than a Sn electrode coated with a native $SnO_x$ layer. They suggested that metal–metal oxide composite materials were promising catalysts for sustainable fuel synthesis. Recently, Prakash et al. [201] compared three kinds of Sn electrodes for $CO_2$ reduction in aqueous $NaHCO_3$ solution: a Sn-powder-decorated gas-diffusion layer (SnGDL) electrode, a Sn metal disk electrode (SnB), and a Sn powder-coated graphite (SnG) electrode (Figure 4.12). The exchange current densities ($j_0$, A $cm^{-2}$) of $CO_2$ reduction on these electrodes, determined by Tafel plots from current–voltage curves, showed a five-fold increase on the SnGDL electrode as compared to that on the SnG electrode (Table 4.2), although the SnG electrode showed a $j_0$ value two orders of magnitude higher than on the SnB electrode. The maximum current density obtained during electrolysis was 27 mA $cm^2$ at $-1.6$ V versus NHE, with 70% faradaic efficiency for the formation of formate, which probably is one of the highest values to be found in the literature on Sn electrodes at ambient pressure. The electrolyte also has an effect on $CO_2$ reduction on Sn-based electrodes. For example, Wu et al. [36] found that both $SO_4^{2-}$ and $Na^+$ gave higher faradaic and energy efficiencies (as high as

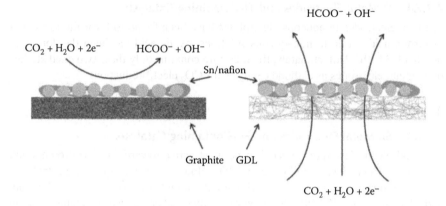

**FIGURE 4.12** ERC over Sn-Nafion coated electrode for a fuel-cell-like device. (Reprinted from *Journal of Power Sources*, 223, Prakash, G. K. S., Viva, F. A. and Olah, G. A., Electrochemical reduction of CO₂ over Sn-Nafion (R) coated electrode for a fuel-cell-like device, 68–73, Copyright 2013, with permission from Elsevier.)

B95% for 0.1 M $Na_2SO_4$ at a potential of −1.7 V vs. SCE), while $HCO_3^-$ and $K^+$ yielded a higher rate of HCOOH production (0.5 M $KHCO_3$ was found to be the optimal electrolyte for obtaining a high production rate of HCOOH, which reached over 3.8 µmol min⁻¹ cm⁻² at a potential of −2.0 V versus SCE while maintaining a faradaic efficiency of B 63%). Normally, the faradaic efficiency for $CO_2$ reduction rises with increasing electrolyte concentration.

### 4.3.8.2  Pb Metal Electrodes and Pb-Containing Catalysts

In general, lead, glassy carbon, mercury, platinum, and gold are popular electrode materials for $CO_2$ electroreduction [112]. The earliest $CO_2$ electroreduction was probably conducted on a lead cathode at −1.5 to −2.2 V versus NHE in quaternary ammonium salt aqueous solutions, with a carbonate/bicarbonate buffer being used to maintain the pH at 8.3 [202]. Pb metal electrodes in aqueous electrolytes favored the formation of formic acid, while in nonaqueous electrolytes, oxalic acid was the dominant product [191]. On a Pb electrode, the faradaic efficiency could be increased by increasing the $CO_2$ pressure and decreasing the temperature [34]. For example, $CO_2$ electroreduction on a Pb electrode in a $CO_2$-satuarated 0.05 M $KHCO_3$ solution at 0°C gave a higher current efficiency than at higher temperature [178]. At a high pressure of B50 atm and a high temperature of approximately 80°C, using Pb granule electrodes in a fixed-bed reactor fed with aqueous 0.2 M $K_2CO_3$ electrolyte solution, the only reaction product was found to be HCOOH, with a maximum faradaic efficiency of 94% at −1.8 V versus SCE [112]. If the reduction was conducted at ambient temperature and pressure, CO and methane were also produced [203]. The Tafel plots from the current–voltage curves for HCOOH and CO formation show a linearity in the entire potential range studied, indicating that the electroreduction of $CO_2$ to formic acid and CO was not limited by mass transfer. Eneau-Innocent et al. [204] employed techniques such as cyclic voltammetry, chronoamperometry, and *in situ* infrared reflectance spectroscopy to investigate the catalytic activity of lead electrodes toward $CO_2$ electrodimerization in 0.2MTEAP–PC solution. Cyclic voltammograms on the lead electrode in 0.2MTEAP–PrC were recorded with a potential sweep rate at 50 mV s⁻¹ and at 25°C in the absence (Figure 4.13a) and the presence (Figure 4.13b) of $CO_2$-saturated solution. As can be observed in Figure 4.13a, no reduction wave is obtained in the absence of $CO_2$. The voltammogram of Pb in Figure 4.13b was recorded after bubbling $CO_2$ during 15 min. During the negative-going scan, the cathodic current starts increasing at −2.05 V versus Ag/AgCl and reaches a maximum of −11.3 mA cm⁻² at −2.5 V versus Ag/AgCl. It is found that CO was not produced while oxalate was the main product.

The CV measurements for $CO_2$ reduction in aqueous medium on a lead plate in a filter-press cell showed that when the pH of the cathodic solution was 8, $CO_2$ existed predominantly in the form of $HCO_3^-$, and HCOOH was the exclusive product, with high faradaic yields of 65%–90% [205]. The main reaction can be expressed as follows:

$$HCO_3^- + H_2O + 2e^- \rightarrow HCOO^- + 2OH^- \tag{4.30}$$

Subramanian et al. [206] designed a flow type electrochemical membrane reactor to improve $CO_2$ mass transfer in electrocatalytic reduction in a potassium phosphate buffer solution. The anode and cathode chambers in this reactor were separated by a composite perfluoropolymer cation exchange membrane (Nafions 961 and Nafions 430). With a lead-coated cathode, a maximum current efficiency of 93% for formate formation was achieved.

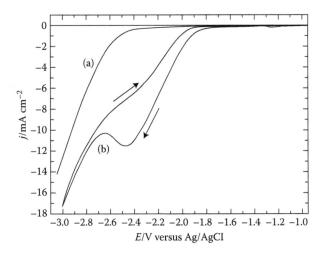

**FIGURE 4.13** Voltammograms of a Pb electrode in 0.2 M TEAP–PrC recorded at 50 mV s⁻¹ and 25°C in the absence (a) and the presence of a $CO_2$-saturated solution (b). (Reprinted from *Applied Catalysis B – Environmental*, 98(1–2), Eneau-Innocent, B., Pasquier, D., Ropital, F., Leger, J. M. and Kokoh, K. B., Electroreduction of carbon dioxide at a lead electrode in propylene carbonate: A spectroscopic study, 65–71, Copyright 2010, with permission from Elsevier.)

### 4.3.9 ALKALINE METALS AND ALKALINE EARTH METALS

Neither alkaline metals nor alkaline earth metals can be used as electrodes for $CO_2$ catalytic reduction catalysts due to their instability in electrochemical systems. However, their salts, commonly used as supporting electrolytes in electrochemical cells for $CO_2$ electroreduction, exhibit different effects on product selectivity, reduction reaction rate, and even catalyst stability. To study the effect of electrolytes on $CO_2$ electroreduction, Murata and Hori [207] tested several aqueous 0.1 M $MHCO_3$ solutions, where M = $Li^+$, $Na^+$, $K^+$, and $Cs^+$. They found that $H_2$ evolution prevailed over $CO_2$ reduction in the $Li^+$ electrolyte, whereas $CO_2$ reduction was favorable in $Na^+$, $K^+$, and $Cs^+$ solutions. In addition, they also observed that the magnitudes of the $C_2H_4/CH_4$ ratio were in the order $Li^+ > Na^+ > K^+ > Cs^+$, indicating that the current efficiency for the formation of $C_2H_4$ was apparently related to the size of the cation radius, which should have a strong effect on the outer Helmholtz plane potential of the electrode–electrolyte interface. Other salts have also been used as supporting electrolytes in $CH_3OH$ solution to study their effect on $CO_2$ electroreduction on a Cu electrode. They include Li salts ($LiBF_4$, $LiClO_4$, LiCl, LiBr, LiI, $LiClO_4$, and $CH_3COOLi$) [134,143], sodium salts ($CH_3COONa$, NaCl, NaBr, NaI, NaSCN, and $NaClO_4$) [208], potassium salts ($CH_3COOK$, KBr, KI, and KSCN) [209], cesium salts ($CH_3COOCs$, CsCl, CsBr, CsI, and CsSCN) [210], and alkali hydroxides such as LiOH [144], NaOH [211], KOH, RbOH [212], and CsOH [211]. Interested in the effect of multivalent supporting cations, Schizodimou and Kyriacou [213] investigated $CO_2$ electroreduction on a Cu(88)–Sn(6)–Pb(6) alloy cathode in 1.5 M HCl and found that the reduction rate slightly increased in the presence of divalent cations such as $Mg^{2+}$, $Ca^{2+}$, and $Ba^{2+}$. They also observed that the enhancement in the reduction rate was dependent on the charge number of the supporting electrolyte cation—the higher the charge number, the greater the rate enhancement, in the order $Na^+ < Mg^{2+} < Ca^{2+} < Ba^{2+} < Al^{3+} < Zr^{4+} < Nd^{3+} < La^{3+}$ (Figure 4.14).

## 4.4 METAL COMPLEX CATALYSTS AND REACTION MECHANISMS

### 4.4.1 FE/CO/NI COMPLEX CATALYSTS

A large number of studies have been conducted on $CO_2$ electroreduction catalyzed by iron, cobalt, and nickel complexes. Both the type of the metal and the structure of the ligands play important roles in the catalytic behavior of these complexes.

#### 4.4.1.1 Fe/Co/Ni Complexes of Porphyrin

In $CO_2$ electroreduction catalyzed by metal porphyrin (M–P)-modified electrodes (M = Fe(II, III), Co(II), Ni(II), and Cr(III); P = tetraphenylporphyrin) in DMF solutions, methanol production can be expressed as follows:

$$CO_2 + 6M - P + 6H^+ \rightarrow CH_3OH + 6M - P^+ + H_2O \qquad (4.31)$$

$$M - P^+ + e^- \rightarrow M - P \qquad (4.32)$$

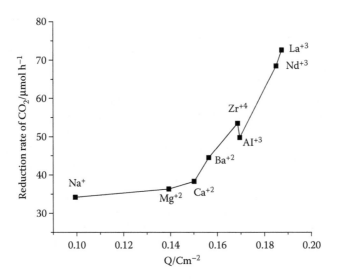

**FIGURE 4.14** Rate of the reduction versus surface charge of the cation at −0.65 V, in 1.5 mol L$^{-1}$ HCl containing various cations. Ionic strength = 2 mol L$^{-1}$. (Reprinted from *Electrochimica Acta*, 78, Schizodimou, A. and Kyriacou, G., Acceleration of the reduction of carbon dioxide in the presence of multivalent cations, 171–176, Copyright 2012, with permission from Elsevier.)

It has been reported that Fe(II) porphyrin yields efficient, CO-selective, and durable catalysts for $CO_2$ electroreduction when it is reduced to Fe(0)-porphyrin by two successive electrons during the reduction process. In an early study, Takahashi et al. [214] found that electrodes modified by Fe-*meso*-tetracarboxyphenyl porphyrin (Fe-mTCPP) and Fe-tetraphenylporphine sulfonate (Fe-sTPP) had insignificant catalytic activity toward $CO_2$ electroreduction in aqueous electrolytic solutions. However, in a DMF solution with tetraalkylammonium salts, three complexes of Fe-porphyrins, including Fe-tetraphenylporphyrin (FeTPP), showed electrocatalytic activity for $CO_2$ reduction, mainly producing CO, with a faradaic yield of over 94%. The addition of weak Bronsted acids to the solution, such as n-propanol, 2-pyrrolidone, or $CF_3CH_2OH$ [215,216], considerably improved the reduction process catalyzed by Fe(0)–TPP. The catalytic currents and the lifetime of the catalyst were both increased without significant hydrogen evolution. In contrast, the HCOOH yield decreased as the acidity of the acid synergist was increased, and finally became negligible when $CF_3CH_2OH$ was added. Most recently, Costentin et al. [217] found that modification of FeTPP through the introduction of phenolic groups in all ortho and ortho′ positions of phenyl groups to form iron 5,10,15,20-tetrakis (2060-dihydrolphenyl) porphyrin, that is, FeTDHPP, and iron 5,10,15,20-tetrakis (2060-dimethoxyphenyl) porphyrin, that is, FeTDDMPP—could considerably speed up the rate of $CO_2$ electroreduction to CO by an electrogenerated iron(0) complex on a GCE. The catalyst manifested a CO faradaic yield above 90% through 50 million turnovers over 4 h of electrolysis at low overpotential and 0.465 V, with no observed degradation. The reason for the enhanced activity appeared to be the high local concentration of protons associated with the phenolic hydroxyl substituents. The stoichiometric reactions involved in the catalytic reduction of $CO_2$ to CO by Fe(0)-porphyrins. Using CNTs for support, an FeTPPCl (FeP–CNT)-modified GCE for $CO_2$ electroreduction was studied using CV and $CO_2$ electrolysis in 0.1 M $NaHCO_3$ solution [218]. The FeP–CNT exhibited a less negative cathode potential and much higher reaction rate than pure FeP or CNT electrodes. By adding FDH, NADH, and methyl viologen to the electrolyte, HCOOH became the only product, and the concentrations of HCOOH formed at the cathode followed the order FeP–CNT > FeP > CNT. This improved catalytic activity was attributed to synergistic catalysis and direct electron transfer between Fe-porphyrin and CNTs, demonstrated using both electrochemical impedance spectroscopy and EPR analysis. The p–p interaction between the porphyrin ring and sidewalls of the CNTs reduced the electron density around the Fe nuclei in FeTPPCl, which expanded the macrocyclic conjugated structure of FeTPPCl and further increased the potential for $CO_2$ reduction [218]. Regarding cobalt complexes as catalysts, Co porphyrin catalyzed the electroreduction of $CO_2$, producing CO in high yield [219]. The electron-transfer rate may depend on the electrical contact between the catalyst and the gas-diffusion electrode and is not the main subject of this study. The protonation and C−OH cleavage steps in Figure 4.15 are at the heart of the catalytic function and must be fast and spontaneous; understanding of the scientific principles involved can lead to improved catalysts and reaction conditions. In the presence of aquopentacyanoferrate (II) or 2-hydroxyl-1-nitrosonaphthalene-3,6 disulfonatocobalt (II) as the homogeneous catalyst, the Pt plate electrodes modified by $Co^{II}TPP$, $Ni^{II}TPP$, $Fe^{III}TPPCl$, or $Fe^{II}TPP$ showed decreasing $CO_2$ reduction activity in an aqueous solution of methanol plus 0.1 M KCl (pH 3.5) [72], indicating that the $Co^{II}TPP$ complex, as a heterogeneous catalyst, certainly had higher catalytic activity. Using a GCE modified by $Co^{II}TPP$ or CoTPP-py-NHCO, the current efficiency of $CO_2$ electroreduction to CO in an aqueous phosphate buffer solution (a mixture of 1/15 M $NaH_2PO_4$ and 1/15 M $Na_2HPO_4$) was 92% at −1.1 V versus SCE.

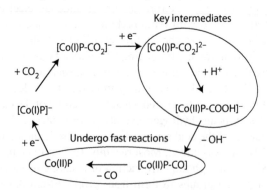

**FIGURE 4.15** Mechanism of $CO_2$ reduction with electron addition deduced from hybrid DFT plus dielectric continuum redox potential calculations. (Reprinted with permission from Leung, K., Nielsen, I. M. B., Sai, N., Medforth, C. and Shelnutt, J. A. Cobalt-porphyrin catalyzed electrochemical reduction of carbon dioxide in water. 2. Mechanism from first principles. *Journal of Physical Chemistry A*. 114(37), 10174–10184. Copyright 2010 American Chemical Society.)

A CoTPP-py-NHCO-modified GCE also showed high catalytic stability, with an overall TON exceeding 10 [7,88]. An amine cation radical method was explored to coordinate a CoII–TPP complex directly onto a GCE to catalyze $CO_2$ electroreduction, and the results were quite promising [220]. Compared to a GCE physically attached to Co(II)–TPP, this GCE chemically bonded to Co(II)–TPP yielded enhanced $CO_2$ electroreduction. In the exploration of new catalysts, 5,10,15,20-tetrakis (4-methoxyphenyl)porphyrinato cobalt(II) (CoTMPP) adsorbed on a nanoporous activated carbon fiber electrode was found to be an effective catalyst, yielding CO with current efficiencies of up to 70% [221]. The catalytic activity of a binuclear Co complex, Co$^{(I)}$TMPyP–MTPPS (TMPyP = a,b,g,d-tetrakis-(1-methylpyridinium-4-yl) porphyrin p-toluesulfonate; TPPS = tetraphenylporphine tetrasulfonic acid; M = metal), was also tested, and significant catalytic activity was observed [222]. The researchers believed that the catalytic activity should be strongly dependent on the central metal (M) of TPPS, which could serve as an electron mediator [222]. Imaoka et al. [223] prepared phenylazomethine dendrimers to bear CoTPP cores as homogeneous catalysts for $CO_2$ electroreduction on a GCE in a DMF solution containing a strong Lewis acid. They observed that the ligand had a strong steric effect on the catalytic activity of the CoTPP core. Recently, catalyst-modified diamond surfaces, dubbed "smart" electrodes, have been explored, with reports of good stability and electrocatalytic activity for $CO_2$ electroreduction to CO in $CH_3CN$ [224]. In this research, a catalytically active Co complex was covalently attached to a B-doped, p-type conductive diamond. In addition, some fluorinated derivatives of CoTPP (the phenyl rings of the CoTPP were substituted with electron withdrawing groups, such as F and $CF_3$) have also been explored as catalysts for $CO_2$ electroreduction, yielding positive results [225]. It should be noted that the catalytic behavior of a metal porphyrin for $CO_2$ electroreduction can be affected by many factors. Even the same catalyst may display different activity when used as a homogeneous catalyst rather than a heterogeneous catalyst [72,226]. $CO_2$ pressure and solution temperature also have a significant effect on $CO_2$ electroreduction. For example, when $CO_2$ pressure was increased from 1 to 20 atm, the current efficiencies of $CO_2$ electroreduction by Fe- and Co-*meso*-TPP-supported GDEs were increased by up to 97.45% and 84.6%, respectively [227]. According to the literature, the strength with which a catalyst bonds to the electrode surface has significant effects on the catalyst's activity [228], as has been demonstrated by photoelectron spectroscopy (PES) and scanning tunneling spectroscopy studies, as well as quantum-chemical calculations based on DFT.

### 4.4.1.2 Fe/Co/Ni Complexes of Phthalocyanine

Metal phthalocyanine (MPc) complexes seem to have less catalytic activity toward $CO_2$ electroreduction than metal porphyrin complexes. For example, iron phthalocyanine tetrasulfonate (FeTSPc), as distinct from CoTSP and NiTSP, did not show remarkable homogeneous catalytic activity for $CO_2$ electroreduction in Clark–Lubs buffer solution [229]. CoPc and NiPc complexes coated on a graphite electrode showed some catalytic activity for $CO_2$ electroreduction, with HCOOH being the predominant product in aqueous solutions (pH 3–7) [230–232]. The electrocatalytic reduction of $CO_2$ in an aqueous electrolyte catalyzed by a graphite electrode coated with a CoPc/poly-4-vinylpyridine (PVP) film was also studied, and enhanced catalytic activity was observed compared to what occurred with a pure CoPc catalyst [233]. The peripheral N atom on the Pc ring can be protonated by the addition of a proton following the first reduction. The second reduction generates the active species for the catalytic reduction of both proton (upward arrow) and $CO_2$ (downward arrow), generating $H_2$ and CO, respectively, at pH 4.4. When CoPc was incorporated into a PVP film coated on a graphite electrode, the polymer-incorporated CoPc showed high catalytic activity toward $CO_2$ electroreduction, with high selectivity from $CO_2$ to CO, where PVP functioned as a coordinative and weakly basic agent [234]. The coordination of PVP to CoPc was considered to have caused the increase in electron density on the central metal ion, facilitating the formation of the intermediate. Cobalt octabutoxyphthalocyanine $(CoPc(BuO)_8)$ coated on a graphite electrode was also explored as a catalyst for $CO_2$ reduction; the results showed that this catalyst had higher activity and selectivity from $CO_2$ to CO than nonsubstituted CoPc [235]. Under typical conditions at pH 4.4, the most active and selective $CO_2$ reduction was achieved at $-1.30$ V versus Ag/AgCl; the production selectivity of $CO/H_2$ was reported to be approximately 4.2. In addition, Zhang et al. [236] employed a rotating ring (platinum)-disk (graphite) electrode to analyze $CO_2$ electroreduction to CO, catalyzed by $N,N',N'',N'''$-tetramethyltetra-3,4-pyridoporphyrazinocobalt(II) coated on a graphite disk and protected by a Nafion film. In a microbial electrolysis cell, cobalt tetra-amino phthalocyanine (CoTAPc) coated on multiwalled CNTs was reported to produce high current efficiency for HCOOH production [237]. Besides phthalocyanine ligands, other kinds of hexaazamacrocycle ligands, such as phenanthroline and bipyridine, and their complexes with Co(II), Ni(II), and Cu(II) metal centers, have also been explored as $CO_2$ electroreduction catalysts [238]. For example, CV and UV–Vis spectroscopy have been employed to study the electroreduction of $CO_2$ using, as an electrocatalyst, hexaazamacrocycles derived from the condensation of 1,10-phenanthroline and its Co(II) complex, dissolved in DMF solution. The results showed that the ligand had no catalytic activity, whereas its cobalt complex showed electrocatalytic activity toward the reduction of $CO_2$, generating CO and HCOOH [239].

### 4.4.1.1.1   Fe/Co Complexes of Corroles

Some iron and cobalt complexes with corroles have also been explored as $CO_2$ electroreduction catalysts. For example, $Ph_3PCo^{III}(tpfc)$ (tpfc = 5,10,15-tris-(pentafluorophenyl)corrole), $ClFe^{IV}(tpfc)$, and $ClFe^{IV}(tdcc)$ (tdcc = 5,10,15-tris(2,6-dichlorophenyl)corrole) dissolved in $CO_2$-saturated $CH_3CN$ solution were tested, using CV, for their catalytic activity toward $CO_2$ electroreduction; the results indicated that the $Co^I$ and $Fe^I$ complexes were both effective catalyst centers [240].

### 4.4.1.1.2   Ni Complexes of Cyclams

Nickel complexes of cyclam (i.e., 1,4,8,11-tetraazatetradecane) and Ni(cyclam)$^{2+}$ have been recognized as highly selective catalysts for the electroreduction of $CO_2$ to CO in aqueous solution on a mercury cathode (as the working electrode) [44,241]. The stability of such catalysts was also found to be remarkable. Even after thousands of catalytic cycles, no significant deactivation was observed (the turnover frequency was $\sim 10^3$ mol of CO produced per mole of nickel complex in 1 h) [241]. A study of the reduction mechanism on mercury, using CV, polarography, and electrocapillarity, indicated that the adsorbed complex $Ni^{(I)}$-cyclam on Hg was the active catalyst [242]. Theoretical calculations performed by Sakaki [243], provided a reasonable description of the catalytic mechanism, based on the oxidation states of the adsorbed complexes ([Ni-cyclam]$^{2+}$ and [Ni-cyclam]$^{+}$), which were strongly dependent on the electrode potentials. Regarding the mechanism, Fujihira [242] observed that the desorption of $Ni^{(I)}$-cyclam gradually slowed down with the formation of nonadsorbable $Ni^{(I)}$-cyclam-CO by the reaction of $Ni^{(I)}$-cyclam in solution with electrocatalytically generated CO from $CO_2$. In the $CO_2$ reduction process, Ni(cyclam)$^{2+}$ was only weakly adsorbed over a limited potential range, and in quantities substantially less than one monolayer, whereas Ni(cyclam)$^{+}$ could be strongly adsorbed on the electrode surface over a wide potential range [244]. However, the high electrocatalytic activity of the Ni(cyclam)$^{2+}$ complex for the reduction of $CO_2$ at a static mercury electrode was severely diminished in the presence of CO when the mercury electrode was not stirred [245]. The cause of the decrease in activity was proposed to be an insoluble complex of Ni(O) and Ni(cyclam)CO, which formed during the reduction of $CO_2$. To explore this mechanism, *in situ* analyses of the products of $CO_2$ electroreduction catalyzed by Ni-cyclam were carried out using differential electrochemical mass spectroscopy during CV on an amalgamated-gold mesh electrode [246].

A binuclear nickel complex, $Ni_2$(biscyclam)$^{4+}$, was found to have similar catalytic activity to that of Ni(cyclam)$^{2+}$ for $CO_2$ electroreduction. In $Ni_2$(biscyclam)$^{4+}$, two Ni atoms are indirectly linked. When DMF with low water content was used as a solvent, high faradaic yields of $HCOO^-$ were observed (up to 75%) in addition to CO [247]. Methyl substitution of the amines on the cyclam ring was also explored as a ligand for the Ni complex when used as a $CO_2$ reduction electrocatalyst, and the result showed some catalytic effects. The $Ni^{(II)}$ complex of N-hydroxyethylazacyclam (i.e., 3-(20-hydroxyethyl)-1,3,5,8,12-penta-azacyclotetradecane nickel$^{(II)}$ perchlorate) appeared to be more active than unsubstituted Ni(cyclam) [248]. Recently, Schneider et al. [249] explored a series of materials that are structurally similar to [Ni(cyclam)]$^{2+}$ to test them as electrocatalysts for $CO_2$ reduction at a mercury pool working electrode in aqueous solution. [249]. Both [Ni(HTIM)]$^{2+}$ (HTIM = C-RRSS-2,3,9,10-tetramethyl-1,4,8,11-tetraazacyclotetradecane) and [Ni(DMC)]$^{2+}$ (DMC = C-*meso*-5,12-dimethyl-1,4,8,11-tetraazacyclotetradecane) showed better electrocatalytic activities than Ni(cyclam)$^{2+}$. Schneider et al. suggested that (1) the catalyst's geometry should be suitable for its adsorption onto the mercury electrode surface and (2) there should be electronic effects of methyl groups or cyclohexane rings on the cyclam backbone. Additional observations have also been made about the influence of methyl substitution on these catalysts' activities (Figure 4.16) [249,250]. Abba et al. found that the structural features of the cyclam and azacyclam framework played an important role in the enhanced catalytic efficiency of Ni-cyclam derivatives for $CO_2$ electroreduction. Even small deviations from such a geometrical arrangement caused the electrocatalytic effect to be drastically reduced or completely lost [251]. For $CO_2$ electroreduction catalyzed by Ni-cyclam complexes on nonmercury electrodes, a GCE was used as the electrode substrate on which a Ni-cyclam complex catalyst was coated, together with a Nafion film [252] or with a poly-(allylamine) (PALA) backbone, and some effective electrocatalytic activity was observed [46]. The orientation

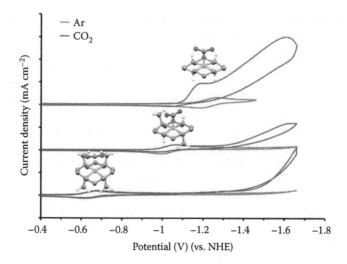

**FIGURE 4.16** Cyclic voltammograms of 1 mM Ni(cyclam)²⁺, Ni(DMC)²⁺, and Ni(TMC)²⁺ in a 0.08 M TBA hexafluorophosphate electrolyte (1:4 water/acetonitrile; GC electrode; 100 mV s⁻¹ scan rate). (Reprinted with permission from Froehlich, J. D. and Kubiak, C. P. Homogeneous $CO_2$ reduction by ni(cyclam) at a glassy carbon electrode. *Inorganic Chemistry.* 51(7), 3932–3934. Copyright 2012 American Chemical Society.)

of the nickel-cyclam complex on the electrode surface was found to be critical to the catalytic effect [252].

### 4.4.1.3 Ni/Co Complexes of Tetraazamacrocycles

Some nickel and cobalt complexes of tetraazamacrocycles, such as Co⁽ᴵ⁾-14-membered tetraazamacrocycles, can also catalyze the ERC to CO and $H_2$ in either $CH_3CN$ or water solution, with current efficiencies of greater than 90% [253]. In the reduction process, charge transfer from Co to bonded $CO_2$ was found to be an important factor in stabilizing the $CO_2$ adducts. The H-bonding interactions between the bound $CO_2$ and amine macrocycle N–H protons might serve to additionally stabilize the adducts in some cases, while steric repulsion by the macrocycle methyl groups could destabilize the adducts, depending on the nature of the individual complex catalyst. With a Ni complex, it was reported that both RRSS-NiIIHTIM(ClO₄)₂ (HTIM = 2,3,9,10-tetramethyl-1,4,8,11-tetraazacyclotetradecane) and NiIIDMC(ClO₄)₂ (DMC = C-*meso*-5,12-dimethyl-1,4,8,11-tetraazacyclotetradecane) showed electrocatalytic activity for $CO_2$ reduction. However, the latter was even more active than Ni(cyclam)²⁺ catalysts, whereas the former was not [254]. In addition, for Ni complex catalysts, the issue of catalyst poisoning during $CO_2$ electroreduction seems to be a concern. For example, this issue was reported when researchers employed isomers of a tetraazamacrocyclic Ni(II) complex in solutions saturated with argon, CO, and $CO_2$ [253–255]. Some Ni-macrocyclic complexes were found to have catalytic activity for $CO_2$ electroreduction from $CO_2$ to oxalate [256]. For example, with a homogeneous electron-transfer rate constant of approximately $10^5$ M⁻¹ s⁻¹, the complex Ni-Etn(Me/COOEt)-Etn was tested as a selective catalyst for $CO_2$ electroreduction to yield oxalate. In this test, electrolysis experiments were carried out using $1.5 \times 10^{-5}$ mol of the Ni complex as the catalyst in $CO_2$-saturated $CH_3CN$ + 0.25 M $Bu_4NClO_4$ solution. The overall reaction mechanism was interpreted in terms of an outer sphere electron-transfer reaction followed by dimerization of $CO_2$ radical anions. The structure of the Fe/Co/Ni complexes and their tetradentate ligands for $CO_2$ electrochemical reduction catalysts are summarized.

### 4.4.1.3.1 Fe/Co/Ni Complexes with Tridentate Ligands

Iron, cobalt, and nickel complexes with tridentate ligands such as 4'-vinyl-2,2':6',2"-terpyridine (v-tpy) can also be employed as catalysts for $CO_2$ electroreduction [74,249,257–259]. For example, an iron complex, $[Fe(4-v-tpy)_2]^{2+}$, electropolymerized onto an electrode yielded a TON in excess of 15,000. It was observed that heterogeneous catalysis with v-tpy (electropolymerized on the electrode) gave a higher catalytic activity than homogeneous catalysis (dissolved in solution) [259]. Heterogeneous catalysis by iron complexes of 4-v-tpy and 6-v-tpy, which were electropolymerized onto a GCE, was very selective, yielding only formaldehyde [74]. The geometric structure and degree of conjugation of the ligand, as well as the orientation and fixation of the macrocyclic metal complexes on the electrode surface, were found to be important for catalytic activity and stability in $CO_2$ electroreduction [260]. Furthermore, iron, cobalt, and nickel complexes of some diacetylpyridine-derived tridentate ligands also appear to be effective electrocatalysts for $CO_2$ reduction [261].

### 4.4.1.3.2 Fe/Co/Ni with Bidentate Ligands

Fe, Co, and Ni complexes with bidentate ligands such as $H_2$dophen [262], $H_2$salen [263], and salophen [264] have also been employed to electrocatalyze $CO_2$ reduction. In addition, polypyrrole Co(II) Schiff base complexes were electropolymerized on a polished GCE to catalyze $CO_2$ reduction [265]. However, the most studied systems seemed to be iron-based complexes. For example, electrolysis of $CO_2$ using $(Fe(dophen)Cl)_2 \cdot 2HCON(CH_3)_2$ and $Fe(dophen)-(N-MeIm)_2ClO_4$ as catalysts at $-2.0$ V versus a ferrocenium–ferrocene reference electrode gave a mixture of CO, $HCOO^-$ (major product) and $C_2O_4^{2-}$ [262]. In this catalysis process, the Fe(I) species seemed to play an important role. As well, the rate of $CO_2$ reduction was enhanced by adding $CF_3CH_2OH$ or $CH_3OH$, as a proton source, to the electrolyte. Both iron carbonyl and iron formato species were detected as intermediates by *in situ* FTIR measurements, where the formation of a Fe–Z1–$CO_2$ intermediate led to the production of CO. The formation of oxalate was attributed to the dimerization of two reduced $CO_2$ molecules. The overall reaction mechanism was proposed to be as follows:

$$Fe^{III}(dophen)Cl + e^- \leftrightarrow Fe^{II}(dophen) + Cl^- \tag{4.33}$$

$$Fe^{II}(dophen) + e^- \leftrightarrow [FeI(dophen)]^- \tag{4.34}$$

$$[Fe^I(dophen)]^- + e^- \leftrightarrow [Fe^I(dophen)]^{2-} \tag{4.35}$$

$$Fe^I + CO_2 \rightleftharpoons Fe^{II} - \overset{\displaystyle O}{\underset{\displaystyle O^-}{C}} \longleftrightarrow Fe^{II} - \overset{\displaystyle O^-}{\underset{\displaystyle O}{C}} \longleftrightarrow Fe^{III} = \overset{\displaystyle O^-}{\underset{\displaystyle O^-}{C}}$$

$$\big\Updownarrow\, 2HA$$

$$\left[ Fe(dophen) = C \overset{\displaystyle O\text{--}H\text{--}A}{\underset{\displaystyle O\text{--}H\text{--}A}{}} \right]^- \tag{4.36}$$

$$\left[ Fe(dophen) = C \overset{\displaystyle O\text{--}H\text{--}A}{\underset{\displaystyle O\text{--}H\text{--}A}{}} \right]^- \rightarrow [Fe(dophen)(CO)]^+ + 2A^- + H_2O \tag{4.37}$$

$$[Fe(dophen)(CO)]^+ + e^- + S \leftrightarrow Fe(dophen)S + CO \tag{4.38}$$

(S is a solvent molecule)

$$[Fe(dophen)(CO_2)]^- + S \leftrightarrow Fe(dophen)S + CO_2^{*-} \tag{4.39}$$

$$CO_2^{*-} \rightarrow C_2O_4^{2-} \tag{4.40}$$

The mechanism of HCOO_ formation was proposed to be as follows:

$$[Fe(dophen)]^- + AH \rightleftharpoons Fe(dophen)H + A^- \tag{4.41}$$

$$[Fe(dophen)H + CO_2 \rightleftharpoons [Fe(dophen)O\overset{\displaystyle O}{\overset{\|}{C}}H \tag{4.42}$$

$$[Fe(dophen)O\overset{\displaystyle O}{\overset{\|}{C}}H + e^- + S \rightarrow [Fe(dophen)S] + HCOO^- \tag{4.43}$$

In the above mechanisms expressed by reactions (4.21) through (4.27), HA is the proton source [262].

### 4.4.1.4    Other Fe/Co/Ni-Containing Catalysts

$Fe_4S_4$ cluster catalysts, namely cubane-type $Fe_4S_4$ clusters [266] are a unique type of iron complexes in which the "$Fe_4$ active site" structure plays a significant role in the electron-transfer reactions for $CO_2$ electrocatalytic reduction. In this regard, Tezuka et al. [71] reported the electroreduction of $CO_2$ catalyzed by two tetranuclear iron–sulfur clusters, $[Fe_4S_4(SCH_2C_6H_6)_4]^{2-}$ and $[Fe_4S_4(SC_6H_6)_4]^{2-}$, in DMF solution. Their results showed that without the catalysts, oxalate was predominantly formed in $CO_2$ reduction, together with small quantities of formate and CO, whereas formate was obtained preferentially in the presence of the catalyst clusters. It was found that $CO_2$ could be generated through electron transfer from the reduced clusters to $CO_2$ in the bulk solution. In another study of $CO_2$ electroreduction catalyzed by $[Fe_4S_4(SCH_2C_5H_6)_4]^{2-}$, it was found that the cubane structure of the cluster could be collapsed, generating two main products, $C_6H_5COO^-$ and $HCOO^-$ [267]. Adding $C_6H_5CH_2SH$ could preserve the cluster structure for a long time, producing more $C_6H_5COO^-$ than $HCOO^-$. Recently, Yuhas et al. [268] obtained enhanced electrocatalytic activity for $CO_2$ electroreduction catalyzed by both ternary $Ni–Fe_4S_4$- and $Co–Fe_4S_4$-based biomimetic chalcogels [268]. CNT supported metal catalysts also can be taken as promising candidate for $CO_2$ electroreduction. Unlike the Fe/Co/Ni metal electrodes or their complex catalysts mentioned above, Fe NPs supported on CNTs (Fe/CNTs) were found to show high electrocatalytic activity toward $CO_2$ reduction to small liquid fuels. Under conventional liquid-phase operations, $C_2H_4$ was found to be the main product of the electrocatalytic reduction of $CO_2$ at high operation potentials [29,269], whereas in the gas phase, isopropanol was the main product [270,271]. In this case, Fe/CNTs even showed better activity than Pt/CNTs but were unstable. The explanation for their degradation was thought to be that the electrolyte, especially K ions, reacted with iron particles, causing their dissolution and migration. In the case of Pt/CNTs, K ions could cover the Pt particles, causing deactivation. Preliminary tests indicated that Fe–Co/CNT catalysts were more stable. When CNTs were further doped with nitrogen to form N-doped CNTs (NCNTs), a supported $FeO_x$ catalyst ($FeO_x$/NCNTs) yielded enhanced electrocatalytic activity and selectivity compared to $FeO_x$ deposited on pristine or oxidized CNTs [272]. These iron NPs supported on NCNTs were able to selectively reduce $CO_2$ to isopropanol. To study the mechanism, the researchers employed microcalorimetry to determine the chemisorption sites of adsorbed $CO_2$; the results showed that NCNTs could cause the formation of small NPs on which there were reversible sites (120 kJ mol$^{-1}$). This approach had, in fact, already been utilized in the photoelectrocatalytic synthesis of solar fuels from $CO_2$ [271,273]. Furthermore, Bocarsly et al. [274] employed two N-containing heteroaromatics (imidazole and pyridine) as homogeneous "aromatic amine catalysts" in the photo ERC at some illuminated iron pyrite ($FeS_2$) electrodes. In their study of the catalysis mechanism of a series of imidazole derivatives, CV measurements were carried out (over a scan rate range of 5–200 mV s$^{-1}$), and they found that imidazole could reduce $CO_2$ to a mixture of CO and HCOOH at a moderate potential, while pyridine selectively produced formic acid.

### 4.4.1.5 Fe Complexes with CO and CN-Ligands

Rail and Berben [275] reported that under appropriate conditions, $Et_4N[Fe_4N(CO)_{12}]$ could be a catalyst for both the hydrogen evolution reaction and $CO_2$ electroreduction at 1.25 V versus SCE using a GCE. ES coated on a Pt plate electrode was also found to have activity for $CO_2$ electroreduction to methanol [276–278] in the presence of either pentacyanoferrate (II) ($Na_3 Fe^{II}(CN)_5H_2O$) or amminepentacyanoferrate (II) ($Na_3Fe^{II}(CN)_5NH_3$). The proposed reaction mechanism was that the ES (Prussian blue (PB) reduction product) was first reduced from PB ($KFe^{III}[Fe^{II}(CN)_6]$) on the cathode to start the reduction by transferring electrons to $CO_2$. In another study, it was found that $CO_2$ could also be reduced to $CH_3OH$ at an ES-mediated electrode in the presence of the 1,2-dihydroxybenzene-3,5-disulfonate (tiron) ferrate(III) complex and ethanol [279]. The proposed mechanism was as follows. First, a weak coordination bond was formed between the central metal and ethanol, followed by the insertion of $CO_2$ to form an intermediate Fe(III)–tiron–ethyl formate complex; this complex was then finally reduced by ES, with the consumption of protons in the solution, to $CH_3OH$ and the initial metal complex:

$$CO_2 + 6ES + 6H^+ \leftrightarrow CH_3OH + 6PB + 6K^+ + H_2O \tag{4.44}$$

$$6PB + 6K^+ + 6e^- = 6ES \tag{4.45}$$

Although catalyzed $CO_2$ electroreduction to $CH_3OH$ was also feasible with both quinone-derivative-coated Pt and stainless-steel electrodes [280], the current efficiency (>50%) was much lower than what the ES-coated Pt plate electrode yielded (>80%) [281]. The high performance of this ES catalyst was simultaneously confirmed for the photocatalytic conversion of CO and $CO_2$ [282], and the same catalyst was used to produce $CH_3OH$ fuel from $CO_2$ for a direct liquid fuel cell [283].

### 4.4.2  PB-Based Catalysts

Based on their observations regarding the catalytic activity of the Co-2-hydroxy-l-nitrosonaphthalene-3,6-disulfonate complex in the homogeneous catalysis of $CO_2$ electroreduction using an ES-mediated electrode in $CH_3OH$ [284], Ogura et al. [285] carried out a further study using a PB-polyaniline (PAn) dual-film coated Pt/Co-2-hydroxyl-nitrosonaphthalene-3,6-disulfonate complex electrode in aqueous solution [285]. The PB film was first electrodeposited on a Pt plate from an aqueous ferric ferricyanate solution; the PAn film was then deposited on a thin PB-coated electrode by repeated potential cycling in KCl solution (pH 1) containing 0.1 M $C_6H_5NH_2$. Lactic acid, ethanol, acetone, and methanol were detected at a low overpotential (−0.6 V vs. SCE) under ambient conditions.

Ogura et al. [286] then developed Fe-L (L = 4,5-dihydroxybenzene-1,3-disulfonate and 2-hydroxy-l-nitrosonaphthalene-3,6-disulfonate) complex-immobilized PAn/PB-modified electrodes. The results of $CO_2$ reduction at dual-film Pt electrodes modified with and without Fe(II) complexes confirmed that the Fe-4,5-dihydroxybenzene-1,3-disulfonate complex-immobilized PAn/PB-modified electrode yielded high catalytic activity and product selectivity for lactic acid. It was found that $CO_2$ could be reduced at the active centers existing in the coated film as well as at the electrode/solution interface. Interestingly, in $CO_2$ electroreduction catalyzed by Fe(II) complex-immobilized PAn/PB/Pt electrodes, formic acid was also one of the products, in addition to lactic acid [287]. The formation process for products such as lactic acid, formic acid, methanol, ethanol, and so on was examined using FTIR reflection spectra [288–290].

### 4.4.3  Multinuclear Nickel Complexes

Unlike iron and cobalt, nickel can form multinuclear complexes, which have also been explored as catalysts for $CO_2$ electroreduction [291]. Ni atoms were linked with each other either directly [291,292] or indirectly [247,293]. It has been found that in other multinuclear complexes, metal atoms are almost all indirectly linked [294–298]. Lee et al. [293] studied the electrocatalytic activity of a multi-nuclear nickel(II) complex, Ni$_3$(L)(ClO$_4$)$_6$ (L = 8,80,800-(2,20,200-nitrilotriethyl)-tris(1,3,6,8,10,13,15-heptaazatricyclo[11.3.1.13,15]octadecane)), toward CO$_2$ reduction and compared these results with those from catalysis by a mononuclear complex, such as [Ni(cyclam)]$^{2+}$, in CH$_3$CN–H$_2$O (9:1, v/v). Unfortunately, the catalytic effi-ciencies (TON) of [Ni$_3$(L)]$^{2+}$ were both lower than those of [Ni(cyclam)]$^{2+}$ and mono-metallic [Ni(3)]$^{2+}$. Simon-Manso et al. [299] synthesized several kinds of dinuclear nickel(0) complexes with dppa as a bridging ligand, expressed as Ni$_2$(m-dppa)$_2$(m-CNR)(CNR)$_2$ (R = Me(CH$_3$), n-Bu(CH$_3$CH$_2$CH$_2$CH$_2$), or 2,6-(CH$_3$)$_2$C$_6$H$_3$). The prod-ucts of CO$_2$ reduction electrocatalyzed by these complexes were mainly CO and CO$_3$$^{2-}$, with a small amount of HCOO$^-$ formed when residual water was present. The steric effect of such large multinuclear Ni complex molecules should be a major factor in the efficiency of CO$_2$ electroreduction [291]. Typical candidates are dinickel complexes [Ni$_2$L$_{2-6}$]4+ or pentaazamacrocycles with (CH$_3$OHCH$_2$)$_n$ bridges ($n$ = 2, 3, 4, 6) or a p-xylylenediamine linkage (L6). The redox potentials were remarkably constant, but the current peak separations increased, reflecting slower electron trans-fer due to the steric effect. The catalytic currents increased slightly as the linking chain length increased, due to improved stereochemical constraints.

### 4.4.4 Metal Complex Enzyme Catalysts

Shin et al. [300] reported using carbon monoxide dehydrogenase (CODH) from *Moorella thermoacetica* to catalyze CO$_2$ electroreduction to CO with a current effi-ciency of approximately 100% at −0.57 V versus NHE in a 0.1 M phosphate buffer solution (pH 6.3). CODH can in fact catalyze microbial interconversion between CO and CO$_2$ [301]. There are three types of CODH: Mo-CODH, in which the active site is a CuMo-pterin; Ni-CODH, in which the active site is a [Ni$_4$Fe-5S] cluster; and Ni-CODH/ACS, in which Ni-CODH is part of a larger complex structure for CO$_2$ reduction. Parkin et al. [302] studied rapid and efficient electrocatalytic CO$_2$/CO interconversion by *Carboxydothermus hydrogenoformans* CO dehydrogenase I (Ch Ni-CODH), which was attached onto PGE electrode as shown in Figure 4.17. This approach provided a way to study the electrocatalytic activity of enzymes under strict potential control; at the same time, it set a standard for future studies of CO and CO$_2$ electrochemical conversion.

### 4.4.5 Pt Group Metal Complex Catalysts

#### 4.4.5.1 Pt Group Metal Complexes with Tetradentate Ligands

Becker et al. [226] explored several Pd-tetradentate complexes, including Pd$^{II}$–por-phyrins (Pd$^{II}$–TPP) and Pd$^{II}$-2,3,7,8,12,13,17,18-octaethylporphyrin (Pd$^{II}$–OEP) com-plexes, as electrocatalysts for CO$_2$ reduction. CV results showed that only Pd$^{II}$–TPP and Pd$^{II}$–OEP displayed electrocatalytic activity toward CO$_2$ reduction, producing oxalic acid. Sende et al. [74] found that the reduction potential of CO$_2$ at electrodes modified with electropolymerized films of [Fe(4-v-tpy)$_2$]$^{2+}$, [Ru(4-v-tpy)$_2$]$^{2+}$, and [Os(4-v-tpy)$_2$]$^{2+}$ tended to become more negative from Fe to Os (−1.10, −1.20, and

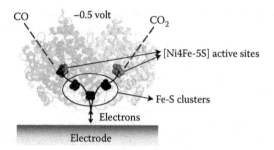

**FIGURE 4.17**   The Ni-containing carbon monoxide dehydrogenase I from *Carboxydothermus hydrogenoformans* adsorbed on a pyrolytic graphite "edge" electrode catalyzes rapid $CO_2$/CO interconversions at the thermodynamic potential. Structure of Ch CODH II showing the two subunits and arrangement of Fe-S clusters that relay electrons to and from the [Ni4Fe-5S] active sites. (Reprinted with permission from Parkin, A., Seravalli, J., Vincent, K. A., Ragsdale, S. W. and Armstrong, F. A. Rapid and efficient electrocatalytic $CO_2$/CO interconversions by carboxydothermus hydrogenoformans CO dehydrogenase I on an electrode. *Journal of the American Chemical Society.* 129(34), 10328–10329. Copyright 2007 American Chemical Society.)

–1.22 V, respectively), which is consistent with their positions in the periodic table (the first, second, and third rows).

### 4.4.5.2   Pt Group Metal Complexes of Polypyridine

Bolinger et al. [303] studied $CO_2$ electroreduction using [Ru(tpy)(dppene)Cl]$^+$ (dppene = *cis*-1,2-bis(diphenylphosphino)ethylene)   or   *cis*-[Rh(bpy)$_2$(TFMS)$_2$]+ (TFMS = trifluoromethanesulphonate anion) as the target catalysts. Two Ru complexes—[Ru(tpy)(bpy)(S)]$^{2+}$ (S = solvent) and Ru(tpy)(Mebin-py)(S)$^{2+}$ (Mebim-py = 3-methyl-1-pyridylbenzimidazol-2-ylidene)—and their catalytic activities in $CO_2$ electroreduction were recently reported [304]. In addition, Rh(III) complexes of tptz have also been explored and showed effective catalytic properties in $CO_2$ electroreduction [305]. Regarding transition metal polyphosphine complexes as electrocatalysts for $CO_2$ reduction, Slater and Wagenknecht [306] investigated Rh(diphos)$_2$Cl (diphos = 1,2-bis((diphenylphosphino) ethane2) in $CH_3CN$. DuBois and Miedaner [307] also observed the catalytic activity of M(PhP(CH$_2$CH$_2$PPh$_2$)$_2$) L(BF$_4$)$_2$ (M = Pd, Pt, Ni) for $CO_2$ reduction to CO in acidic $CH_3CN$ solutions. Pd complexes (L = CH$_3$CN, P(OMe)$_3$, PEt$_3$, P(CH$_2$OH)$_3$, or PPh$_3$) exhibited significant $CO_2$ catalytic activity, while Pt complexes (L = PEt$_3$) and Ni complexes (L = P(OMe)$_3$ and PEt$_3$) did not. In another report, DuBois et al. [42] confirmed the electrocatalytic activity of [Pd(tridentate)(CH$_3$CN)](BF$_4$)$_2$ complexes in acidic DMF or $CH_3CN$ solutions; they found that if one or more of the phosphorus atoms of the tridentate ligand were substituted with a nitrogen or sulfur heteroatom, the resulting complexes would not show significant catalytic activity toward $CO_2$ reduction. By comparing the rate constants of catalysts with different alkyl and aryl substituents on the terminal phosphorus atoms, they found that the reaction rate of Pd$^{(1)}$ intermediates with $CO_2$ was increased by increasing the electron-donating ability of the R groups, and that the steric interactions were of less importance.

Steffey et al. [299] synthesized and characterized many types of Pd complexes containing tridentate ligands with PXP (X = C, N, O, S, As) donor sets—for example, the complex Pd(PCP)(PEt$_3$)(BF$_4$) (where PCP is 2,6-bis((diphenylphosphino) methyl)phenyl)—and also evaluated their catalytic activities toward $CO_2$ electroreduction. They found that Pd(PCP)(CH$_3$CN)(BF$_4$) exhibited significant catalytic currents in the presence of acid and $CO_2$. The dependence of the catalytic current on $CO_2$ and acid concentrations was consistent with the formation of a hydroxycarbonyl intermediate, which decomposed in the presence of acid to form $H_2$ as the catalytic product. Raebiger et al. [297] explored a bimetallic Pd complex with bis(triphosphine) ligand (C$_6$H$_4$(P(CH$_2$CH$_2$P(C$_6$H$_{11}$)$_2$)$_2$)$_2$) and found that this complex could electrocatalyze the reduction of $CO_2$ to CO in acidic DMF solutions with a significantly higher TON than that achieved with other previous monometallic, bimetallic, and dendritic complexes of this catalyst class.

### 4.4.5.3 Pt Group Metal Complexes with Bidentate Ligands

Polypyridyl complexes of the second- and third-row transition metals have also been reported to act as efficient homogeneous catalysts for the electrocatalytic reduction of $CO_2$. Ishida et al. [308,309] conducted research on the controlled potential electrocatalysis of $CO_2$ reduction in a saturated $H_2O$–DMF (9:1, v/v) solution containing Ru(bpy)$_2$(CO)$_2$$^{2+}$ or Ru(bpy)$_2$(CO)Cl$^+$ at an electrode potential of 1.5 V versus SCE, and observed different product selectivity at different pH values. At pH 6.0, both CO and $H_2$ were the main products, whereas at pH 9.5, HCOO was produced. They considered [Ru(byp)$_2$(CO)COO]$^+$, that is, [Ru(byp)$_2$(CO)COO]$^0$, to be an important intermediate formed through the following reactions:

$$[Ru(bpy)_2(CO)_2]^{2+} + OH^- \rightarrow [Ru(byp)_2(CO)COOH]^+ \qquad (4.46)$$

$$[Ru(byp)_2(CO)COOH]^+ + OH^- \rightarrow [Ru(byp)_2(CO)COO]^+ + H_2O \qquad (4.47)$$

Moreover, the authors also found that the percentages of HCOO$^-$, CO, and $H_2$ produced during the reduction were largely dependent on the pK$_a$ [308]. For example, the current efficiency of HCOO formation was increased by increasing the pK$_a$ value and reached 84.3% in the presence of Me$_2$NHHCl. Pugh et al. [41] studied the catalytic activity of the complex cis-[Ru(bpy)$_2$(CO)H](PF$_6$) (i.e., cis-[Ru(bpy)$_2$(CO) H]$^+$) and, by means of FTIR spectroscopy, observed the cis-[Ru(bpy)$_2$(CO)H]$^+$, cis-[Ru(bpy)$_2$(CO)(OC(O)H)]$^+$, and cis-[Ru(bpy)$_2$(CO)(NCCH$_3$)]$^{2+}$ species in the solutions at the end of the electrolysis period. They suggested a different mechanism of HCOO$^-$ formation that included the formation of cis-[Ru(bpy)$_2$(CO)H]$^0$ and the insertion of $CO_2$.

Chardonnoblat et al. [310] prepared a "[Ru$^{II}$(bpy)(CO)$_2$]$_n$" polymeric film electrode by the electrochemical reduction of a mono- (bipyridine) complex, such as Ru$^{II}$(bpy)(CO)$_2$(Cl)$_2$ or Ru$^{II}$(bpy)-(CO)$_2$(CH$_3$CN)$_2$, for $CO_2$ electrocatalytic reduction. Exhaustive electrolysis at 1.55 V versus SCE produced CO with a current efficiency of 97% but yielded only 3% current efficiency for formate production. By comparing the electrochemical behaviors of isomers of Ru(bpy)(CO)$_2$Cl$_2$

(trans(Cl)-Ru(bpy)-(CO)$_2$Cl$_2$ and cis-(Cl)-Ru(bpy)(CO)$_2$Cl$_2$) and the behaviors of cis-(CO)-Ru(bpy)(CO)$_2$(C(O)OMe)Cl complexes, they found that for the cis-(Cl)-Ru(bpy)(CO)$_2$Cl$_2$ and cis-(CO)-Ru(bpy)(CO)$_2$(C(O)OMe)Cl complexes, catalyzed CO$_2$ electroreduction could lead to a Ru–Ru dimer when a chloride ion was lost, while the trans(Cl)-Ru(bpy)-(CO)$_2$Cl$_2$ complex could form a polymeric film of [Ru(bpy)(CO)$_2$]$_n$ [45]. This suggested that a choice could be made between homogeneous and heterogeneous systems using the same experimental setup simply by changing the stereochemistry of the precursor. Collombdunandsauthier et al. [311] also confirmed the catalytic activity of [Ru(II)(bpy)(CO)$_2$]$_n$ polymeric thin films. The (mononuclear) [(bpy)$_2$Ru(dmbbbpy)](PF$_6$)$_2$ and (dinuclear) [(bpy)$_2$Ru(dmbbbpy)Ru(bpy)$_2$](PF$_6$)$_4$ (dmbbbpy = 2,20-bis(1-methylbenzimidazol-2-yl)-4,40-bipyridine) was synthesized for CO$_2$ electroreduction by Ali et al. [312] In CO$_2$-saturated CH$_3$CN, CO$_2$ reduction catalyzed by [(bpy)$_2$Ru(dmbbbpy)](PF$_6$)$_2$ yielded a current efficiency of 89% for HCOOH in the presence of H$_2$O (~2.5%) and 64% for C$_2$O$_4^{2-}$ in the absence of H$_2$O; in a similar experiment, the current efficiencies for [(bpy)$_2$Ru(dmbbbpy)-Ru(bpy)$_2$](PF$_6$)$_4$ were 90% and 70%, respectively. Tanaka and Mizukawa [313] observed the highly selective formation of ketones (current efficiency of 20%) in the electrocatalytic reduction of CO$_2$ by Ru(bpy)(napy)$_2$(CO)$_2^{2+}$ (napy = 1,8-naphthyridine-KN) in the presence of (CH$_3$)$_4$NBF$_4$. Only CH$_3$C(O)CH$_3$ and CO$_3^{2-}$ were formed, according to the following reaction: $2CO_2 + 2(CH3)_4N^+ + 4e^- \rightarrow CH_3COCH_3 + CO_3^{2-} + 2(CH_3)_3N$. For structural and spectroscopic characterization, Tanaka et al. [314] employed Ru(II) complexes [Ru(bpy)$_2$(CO)L] (L = CH$_3$, C(O)H, and C(O)CH$_3$) as the model catalysts for a multistep CO$_2$ reduction study using IR, Raman, $^{13}$C-NMR, and single-crystal x-ray crystallography. To probe the mechanism, they further prepared a series of [Ru(bpy)$_2$(CO)L]$^{n+}$ (L = CO$_2$, C(O)OH, CO, CHO, CH$_2$OH, CH$_3$, and C(O)CH$_3$; $n = 0$–2) and determined their molecular structures by x-ray analyses [315]. They did so because they thought that these complexes might be the reaction intermediates in the multielectron reduction of CO$_2$ in protic media. They also found that Ru-(bpy)$_2$(2)$^{2+}$ (2 = 2-(2-pyridyl)-1-methylbenzimidazole) was inactive for CO$_2$ reduction and we have confirmed this (Figure 4.18a; Ru-(bpy)$_3^{2+}$ was also found to be inactive). However, we have found that the benzothiazole analog, Ru(bpy)$_2$(3)$^{2+}$ (3 = 2-(2-pyridyl)benzothiazole) is very active for CO$_2$ reduction (Figure 4.18b), with high activity at potentials as high as −1.0 V versus SCE. The magnitudes of the currents in the presence of CO$_2$ clearly indicate that there is electrocatalysis, rather than a simple irreversible chemical reaction between CO$_2$ and the reduced complex.

Not all ruthenium complexes of 2,20-bipyridine have been found to be electrocatalytically active toward CO$_2$ reduction [314]. Begum and Pickup [316] experimentally confirmed that Ru(2,20-bipyridine)$_2$[2-(2-pyridyl)benzothiazole]$^{2+}$ was a highly active catalyst whereas 1-methylbenzimidazole analog was not. Bolinger et al. [303] investigated the electrocatalytic reduction of CO$_2$ by 2,20-bipyridine complexes of both Rh and Ir (cis-[M(bpy)$_2$Cl$_2$]$^+$ or cis-[M(bpy)$_2$(TFMS)$_2$]$^+$, where TFMS is a trifluoromethanesulfonate anion). It was observed that cis-[Rh(bpy)$_2$X$_2$]$^+$ (X = Cl or TFMS) gave a current efficiency as high as 80% for CO$_2$ electroreduction to formate at a potential of −1.55 V versus SCE in CH$_3$CN solution. The proton source for the formate formation was considered to be [(n-Bu)$_4$N](PF$_4$) via the Hofmann degradation:

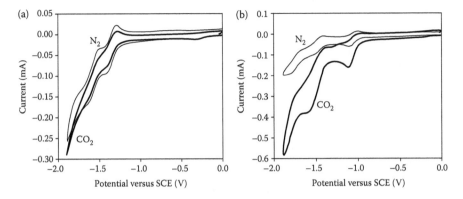

**FIGURE 4.18** Cyclic voltammetry (100 mV s$^{-1}$) of 1 mM Ru(bpy)$_2$(2)$^{2+}$ (a) and 1 mM Ru(bpy)$_2$(3)$^{2+}$ (b) in acetonitrile containing 0.1 M Et$_4$NClO$_4$ and 1% by volume H$_2$O under N$_2$ (light) and CO$_2$ (bold). (Reprinted from *Electrochemistry Communications*, 9(10), Begum, A. and Pickup, P. G., Electrocatalysis of CO$_2$ reduction by ruthenium benzothiazole and bithiazole complexes, 2525–2528, Copyright 2007, with permission from Elsevier.)

$$[Rh(bpy)_2] + CO^{2-} \rightarrow [Rh(bpy)2CO_2] \qquad (4.48)$$

$$[Rh(bpy)_2CO_2]^+(n-Bu)_4N^+$$
$$\rightarrow [Rh(bpy)_2CO_2H] + H_3CCH_2CHCH_2 + (n-Bu)_3N \qquad (4.49)$$

$$[Rh(bpy)_2CO_2H] \rightarrow [RhI(bpy)_2]^+ + HCOO^- \qquad (4.50)$$

The electrocatalytic activity of *cis*-[Os(bpy)$_2$(CO)H](PF$_4$) in the reduction of CO$_2$ was observed by Bruce et al. [317,318]. They found that under anhydrous conditions, CO was the dominant product. However, in the presence of water, 25% HCOO$^-$ could be formed. In the reduction process, [Os(byp)$_2$(CO)H] was considered an important intermediate that could be coordinated by CO$_2$ to produce CO and HCOO. Nallas and Brewer [296] explored a new family of catalysts for CO$_2$ electroreduction, namely, two mixed-metal trimetallic complexes: {[(bpy)$_2$Ru(BL)]$_2$IrCl$_2$}$^{5+}$ (BL = 2,3-bis(2-pyridyl)quinoxaline) and 3-bis(2-pyridyl)benzoquinoxaline. The two remote Ru centers served to tune the redox properties of the central catalytically active Ir$^{III}$(BL)$_2$Cl$_2$ core. These catalysts should represent a new class of systems in which the redox properties of catalytic sites can be altered through remote metal coordination and variation without changing the coordination environment of the catalytic iridium site. Some Pt group metal complexes with bidentate ligands, such as Ir complexes, have also been explored for CO$_2$ electrochemical reduction. For example, two [Ir$_2$(dimen)$_4$]Y$_2$ complexes (dimen = 1,8-diisocyanomenthane; Y = (PF$_6$) and [B(C$_6$H$_5$)$_2$]) were studied using infrared spectro-electrochemistry; the results indicated that [Ir$_2$(dimen)$_4$]$^{2+}$ first accepted two electrons to form [Ir$_2$(dimen)$_4$]$^0$, then reacted with CO$_2$ and H$_2$O to form two main products, formate and bicarbonate [319].

#### 4.4.5.4  Pt Group Metal Complexes with Monodentate Ligands

Hossain et al. reported the electrocatalytic reduction of $CO_2$ on either a GCE or a Pt electrode using a series of $PdL_2Cl_2$ complexes (L = substituted pyridine and pyrazole) as catalysts in $CH_3CN$ containing 0.1 M TEAP [320]. These catalysts (L = pyrazole, 4-methylpyridine, and 3-methylpyrazole) gave HCOOH yields of 10%–20%, while the current efficiency for hydrogen evolution was 31%–54%. $RhCl(CO)(PPh_3)_2$ and $IrCl(CO)(PPh_3)_2$ ($PPh_3$ = triphenylphosphine) showed different electrocatalytic behaviors in DMF solution. $IrCl(CO)(PPh_3)_2$ was found to be an efficient homogeneous catalyst for $CO_2$ electroreduction to CO and HCOOH [23]. In addition, the Pd–organophosphine dendrimer complex was tested as a $CO_2$ reduction catalyst [321].

#### 4.4.5.5  Other Catalysts Containing Pt Group Metals

Other complexes containing Pt group metals have also been explored as $CO_2$ reduction electrocatalysts [319,322–325]. A typical candidate was an Ir complex, for example, $Ir(Z_5\text{-}C_5Me_5))_2(Ir(Z_4\text{-}C_5Me_5)CH_2CN)(\grave{\imath}_3\text{-}S)_2$, as reported by Tanaka et al. [322]. They used it to catalyze $CO_2$ reduction to produce oxalate in $CO_2$-saturated $CH_3CN$ solution. Some water-stable iridium dihydride complexes supported by PCP-type pincer ligands in $CH_3CN$–water mixtures were also explored as catalysts for $CO_2$ reduction [325].

### 4.5  ORGANIC/BIO CATALYSTS AND REACTION MECHANISMS

Besides the metals and metal complexes mentioned above, organic molecules can also be mediators and catalysts for $CO_2$ electroreduction. Oh and Hu [326] have recently published a review of the literature on this subject.

#### 4.5.1  Conducting Polymer Electrodes

Conducting polymer electrodes have also been developed for the heterogeneous electrocatalysis of $CO_2$ reduction. For example, Koleli et al. [327] developed a polyaniline (PAn) electrode. In methanol solution, the maximum faradaic efficiencies were found to be 12% for formic acid and 78% for acetic acid. Aydin et al. [328] developed a polypyrrole (PPy) electrode. In the electrocatalytic reduction of $CO_2$ under high pressure in $CH_3OH$ at an overpotential value of 0.4 V versus Ag/AgCl, the maximum faradaic efficiencies at 20 bar were 1.9%, 40.5%, and 62.2% for HCHO, HCOOH, and $CH_3COOH$, respectively. Smith et al. [329] synthesized two organic polymers based on repeating benzimidazole and pyridine–bipyridine units, respectively. The pyridine-based polymer exhibited stable electrochemical behavior, while the bipyridine-based polymer gave large catalytic currents for $CO_2$ reduction in $CH_3CN$ containing 1% $H_2O$.

#### 4.5.2  Aromatic Amine Catalysts

Seshadri et al. [330] found that the pyridinium cation and its substituted derivatives could be effective and stable homogeneous electrocatalysts for the multiple-electron, multiple-proton reduction of $CO_2$ to methanol at low potentials, with faradaic yields of up to 30%. The selectivity for pyridinium-catalyzed methanol production was found to increase significantly under photoelectrocatalytic conditions.

Cole et al. [331] observed $CO_2$ reduction to HCOOH, HCHO, and $CH_3OH$ at 0.58 V versus SCE when using a Pt disk electrode in a 10 mM aqueous solution of pyridine (Py) at pH 5.3 as shown in Figure 4.19. In the proposed overall mechanism, with an inner-sphere-type electron transfer, the pyridinium radical was believed to play a role in the reduction. Ertem et al. [332] proposed that the pyridinium cation "PyH+" could undergo a one-electron reduction, forming hydrogen atoms adsorbed on the Pt surface, "Pt–H." This "Pt–H" was susceptible to electrophilic attack by $CO_2$, leading to a two-electron proton-coupled hydride transfer reaction:

$$CO_2 + Pt - H + PyH^+ + e^- \rightarrow Py + Pt + HCOOH \tag{4.51}$$

In a theoretical study of $CO_2$ electroreduction in the presence of a pyridinium cation and its substituted derivatives, Keith and Carter [333] employed first-principles quantum chemistry and the thermodynamic energies of various pyridine-derived intermediates, as well as energy barrier heights for key homogeneous reaction mechanisms. They predicted that the actual form of the co-catalyst was not the long-proposed pyridinyl radical in solution, but was more probably a surface-bound dihydropyridine species. Lim et al. [334] investigated the mechanism of homogeneous $CO_2$ reduction by pyridine (Py) in the Py–p–GaP system. Based on *ab initio* quantum-chemical calculations, they identified $PyCOOH^0$ as an important intermediate, whose formation was believed to be the rate-determining step for $CO_2$ reduction to $CH_3OH$. In the homogeneous phase, the formation of $PyCOOH^0$ proceeds by two potential routes. In route 1, Py is protonated to $PyH^+$, which is then reduced to $PyH^0$, which reduces $CO_2$ to form $PyCOOH^0$. In route 2, Py and $CO_2$ are combined to form $PyCO_2$, which is reduced to $PyCOO^-$ and, finally, protonated to $PyCOOH^0$. Bocarsly et al. [274] compared the catalytic activity of imidazole and pyridine in the photoelectrochemical reduction of $CO_2$ at illuminated iron pyrite (FeS$_2$) electrodes. The aqueous electrolyte was composed of 10 mM catalyst and 0.5 M KCl solution, and a 350–1350 nm light beam with an intensity of 890 mW cm$^2$ was used to illuminate the electrode surface. An equally viable alternative to the EC-based mechanism proposed above is a CE-type mechanism in which the NHC − $CO_2$ adduct is first formed via a purely solution chemical process and this species is then electrochemically reduced to generate formate as illustrated in Figure 4.20. The reaction is facile, occurring at room temperature. In this reaction, a chemical reduction step is not required to activate the imidazolium. Rather, a series of equilibria produce the 2-ylidene under exceptionally mild conditions, and the carbonylation of this species generates the electroactive intermediate leading to formate. Gennaro et al. [335] found that the anion radicals of certain aromatic esters (i.e., phenyl benzoate and methyl benzoate) and nitriles (i.e., benzonitrile) possessed remarkable catalytic activity toward $CO_2$ reduction to oxalate. The reduction pathway was simply described as follows: (1) A (aromatic ester or nitrile) was first reduced to A (an aromatic radical anion); (2) A$^-$ transferred the electron to $CO_2$, forming $CO_2^{*-}$; and (3) two $CO_2^{*-}$ dimerized to give oxalate.

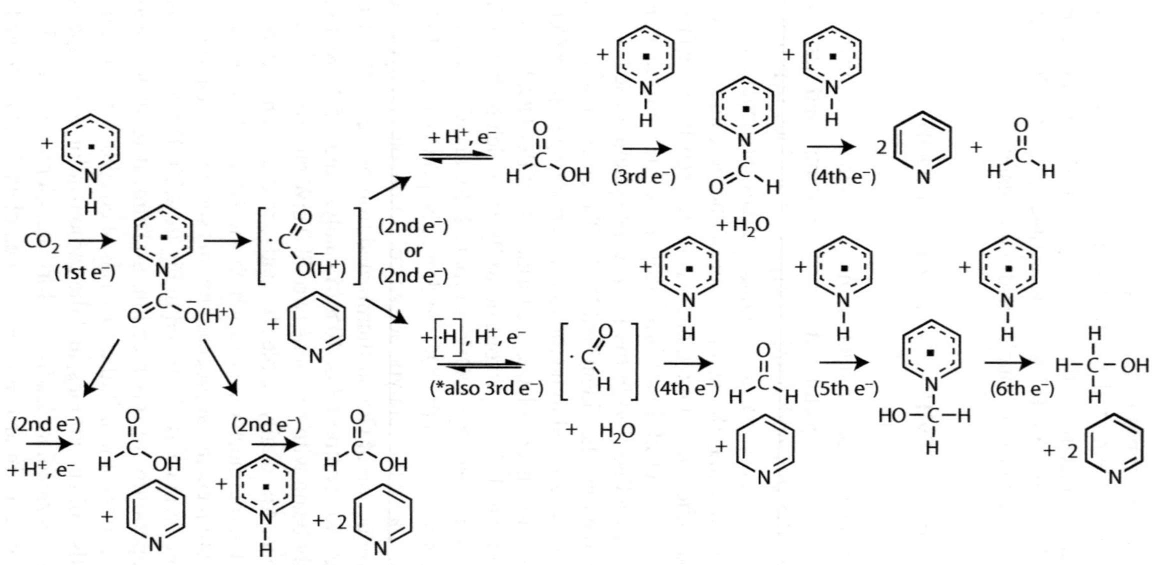

**FIGURE 4.19** Overall proposed mechanism for the pyridinium-catalyzed reduction of $CO_2$ to the various products of HCOOH, HCHO, and $CH_3OH$ at 0.58 V versus SCE when using a Pt disk electrode in a 10 mM aqueous solution of pyridine (Py) at pH 5.3. (Reprinted with permission from Cole, E. B. et al. Using a one-electron shuttle for the multielectron reduction of $CO_2$ to methanol: Kinetic, mechanistic, and structural insights. *Journal of the American Chemical Society.* 132(33), 11539–11551. Copyright 2010 American Chemical Society.)

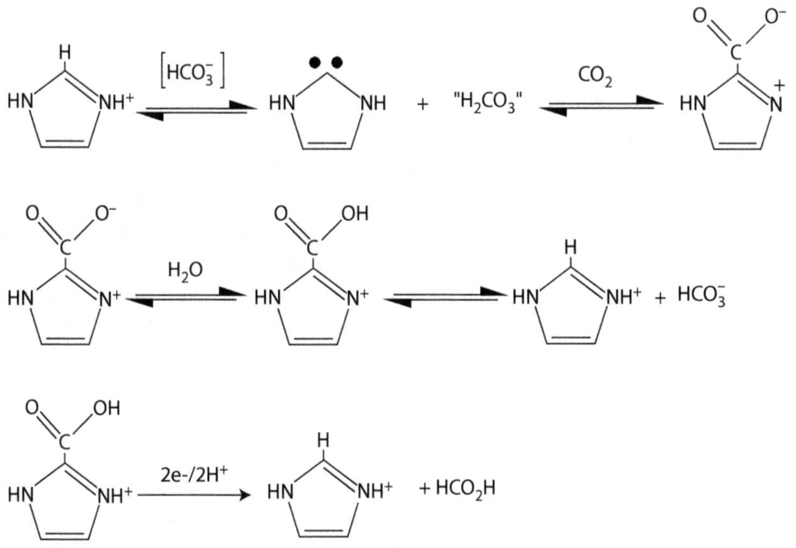

**FIGURE 4.20** Comparative mechanism of imidazole- and pyridine-catalyzed reduction of CO$_2$ at illuminated iron pyrite electrodes. (Reprinted with permission from Bocarsly, A. B. et al. Comparative study of imidazole and pyridine catalyzed reduction of carbon dioxide at illuminated iron pyrite electrodes. *ACS Catalysis.* 2(8), 1684–1692. Copyright 2012 American Chemical Society.)

### 4.5.3 IONIC LIQUID (CATALYSTS)

CO$_2$ has been dissolved in the ionic liquid EMIMBF$_3$Cl and electrochemically reduced at ambient pressure and room temperature [336]. The BF$_3$Clanion catalyzed the CO$_2$ reduction by forming a Lewis acid–base adduct, BF$_3$–CO$_2$. The B–Cl bond was relatively weak in BF$_3$Cl. Thermogravimetric analysis results showed that the decomposition of EMIMBF$_3$Cl required two steps. The first included the breaking of the B–Cl bond and then the release of BF3 at about 220°C; the second involved the breaking of other bonds which started at about 330°C.

### 4.5.4 ENZYME CATALYSTS

Two and a half decades ago, Sugimura et al. [337] used isocitrate dehydrogenase as an electrocatalyst for the fixation of CO$_2$ to isocitric acid (HOOCCH$_2$–CH(COOH) CH(OH)COOH) in oxoglutaric acid (HOOCCH$_2$CH$_2$–C(QO)COOH). The reaction occurred selectively, with current efficiencies approaching 100% at 0.95 V versus SCE in 0.2 M (HOCH$_2$)$_3$CNH$_2$ (tris buffer, pH 7). They also fixed CO$_2$ to yield pyruvic acid (CH$_3$C(QO)COOH) in acetyl-coenzyme A using pyruvate dehydrogenase complexes as electrocatalysts [338]. Solvent molecules seemed to be involved in the reactions. Kuwabat et al. [339] reported that at potentials between 0.7 and 0.9 V versus SCE, the electrolysis of CO$_2$-saturated phosphate buffer solutions (pH 7) that contained FDH and either methyl viologen (MV$^{2+}$)

or pyrroloquinolinequinone as an electron mediator yielded HCOO with current efficiencies of 90%. The FDH demonstrated considerable durability. They also found that the electrolysis of phosphate buffer solutions containing HCOONa in the presence of methanol dehydrogenase (MDH) and $MV^{2+}$ at 0.7 V versus SCE yielded HCHO when the enzyme concentration was low, whereas it produced both HCHO and $CH_3OH$ when the concentration became relatively high. The formation and accumulation of formaldehyde promoted methanol production [339]. Addo et al. [340] found that the addition of carbonic anhydrase could efficiently accelerate $CO_2$ reduction achieved by formate, aldehyde, and alcohol dehydrogenase, although the process could be catalyzed only by dehydrogenase. Hansen et al. [341] developed a model based on DFT calculations to describe trends in catalytic activity for $CO_2$ electroreduction to CO in terms of the adsorption energy of the reaction intermediates, CO and COOH. The model is applied to metal surfaces as well as the active site in the CODH enzymes and shows that the strong scaling between adsorbed CO and adsorbed COOH on metal surfaces is responsible for the persistent overpotential. The active site of the CODH enzyme is not subject to these scaling relations and optimizes the relative binding energies of these adsorbates, allowing for an essentially reversible process with a low overpotential (Figure 4.21). Enzymes that were able to catalytically transform small molecules (e.g., CO, formate, or protons) were a special category of electrocatalysts. Due to the presence of enzymes (catalysts), those previously irreversible processes could become electrochemically reversible [342]. In the active sites, some elements (e.g., Ni, Fe, Cu, Se, Mo, and W) have been found [343,344].

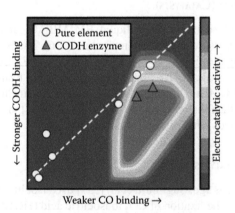

**FIGURE 4.21**   Kinetic volcano for CO evolution at a 0.35 V overpotential from the (211) step of transition metals. The transition metals fall along a trendline that does not pass over the top of the volcano. However, the noble metals are near the optimum along this trend line. The specific CO evolution current from ChCODH II and MbCODH enzyme models is comparable or better than that from the noble metals. (Reprinted with permission from Hansen, H. A., Varley, J. B., Peterson, A. A. and Norskov, J. K. Understanding trends in the electrocatalytic activity of metals and enzymes for $CO_2$ reduction to CO. *Journal of Physical Chemistry Letters*. 4(3), 388–392. Copyright 2013 American Chemical Society.)

### 4.5.5 Cu Organic Frameworks and Cu-Based Perovskite-Type Catalysts

Metal organic frameworks with crystalline ordered structures, extra-high porosity, high thermal stability, as well as adjustable chemical functionality have been pursued for many purposes, including gas-storage applications [345–350]. Recently, $Cu_3(BTC)_2$ (BTC = 1,3,5-benzenetricarboxylate), a Cu-based metal organic framework, was explored as an electrode for $CO_2$ reduction in a $CO_2$-saturated 0.01 M TBATFB–DMF solution (TBA TFB = tetrabutylammonium tetrafluoroborate) by Kumar et al. [31]. It was observed that this catalyst has high selectivity toward $CO_2$ reduction to oxalic acid, and the highly active site was believed to be a Cu(I) species. Cu-based Perovskite-type $A_{1.8}A'_{0.2}CuO_4$ (A = La, Pr, and Gd; A0 = Sr and Th) were also explored for catalyzing $CO_2$ electroreduction [351]. For example, when these kinds of catalysts were incorporated into a GDE, the cumulative faradaic efficiencies of $CH_3OH$, $CH_3CH_2OH$, and $CH_3CH_2CH_2OH$ reached 40% in $La_{1.8}Sr_{0.2}CuO_4$ GDE/0.5 M KOH aqueous solution under ambient conditions.

### 4.5.6 Other Catalysts

Dinuclear Cu complexes have also shown activity toward $CO_2$ electroreduction [294,295,298]. For example, Field et al. [294] synthesized one dinuclear Cu complex, $Cu_2(m$-$PPh_2bipy)_2(CH_3CN)_2(PF_6)_2$ (PPh₂bipy = 6-diphenylphosphino-2,20-bipyridyl), by treating $Cu(CH_3CN)_4PF_6$ with 6-diphenylphosphino-2,20-bipyridine in $CH_3CN$ solution, and used it as the catalyst for homogeneously catalyzing the ERC to selectively produce CO and $CO_3^{2-}$ in 0.1 M tetra-$N$-butylammonium perchlorate–$CH_3CN$ solution. Two sequential single-electron transfers to $[Cu_2(m$-$PPh_2bipy)_2$-$(CH_3CN)_2]^{2+}$ were observed at $E_{1/2(2+/+)} = -1.35$ V and $E_{1/2(+/0)} = -1.53$ V versus SCE, respectively. Thus, two possible routes were proposed, as follows:

$$[Cu_2(m-PPh_2bipy)_2(CH_3CN)_2]_{2+} + e^-$$

$$\rightarrow [Cu_2(m-PPh_2bipy)_2(CH_3CN)_2]^+ \qquad (4.52)$$

$$[Cu_2(m-PPh_2bipy)_2(CH_3CN)_2]^+ + e^-$$

$$\rightarrow [Cu_2(m-PPh_2bipy)_2(CH_3CN)_2]^0 \qquad (4.53)$$

$$[Cu_2(m-PPh_2bipy)_2(CH_3CN)_2]^0 + CO_2 \rightarrow product \qquad (4.54)$$

Indeed, the PPh₂bipy ligand offers the dual advantage of coordinated bipyridine and bridging phosphines. The Π*-unsaturation of the bipy component of the PPh₂bipy ligand could provide the ability to shuttle electrons in and out of a closed-shell $d^{10}$–$d^{10}$ binuclear complex. Kauffman et al. [175] recently electrocatalyzed the reduction of $CO_2$ to CO using ligand-protected $[Au_{25}(SC_2H_4Ph)_{18}]^-$ clusters in aqueous solution at −1.0 V versus RHE. The efficiency was approximately 100%, while the rate was 7–700 times higher than for larger Au catalysts and 10–100 times higher than for current state-of-the-art processes.

## REFERENCES

1. Meyer, T. J. Chemical approaches to artificial photosynthesis. *Accounts of Chemical Research.* 1989; 22(5):163–170.
2. Leitner, W. The coordination chemistry of carbon dioxide and its relevance for catalysis: A critical survey. *Coordination Chemistry Reviews.* 1996; 153:257–284.
3. Cheng, M., Lobkovsky, E. B. and Coates, G. W. Catalytic reactions involving c-1 feedstocks: New high-activity Zn(II)-based catalysts for the alternating copolymerization of carbon dioxide and epoxides. *Journal of the American Chemical Society.* 1998; 120(42):11018–11019.
4. Alstrum-Acevedo, J. H., Brennaman, M. K. and Meyer, T. J. Chemical approaches to artificial photosynthesis. 2. *Inorganic Chemistry.* 2005; 44(20):6802–6827.
5. Aresta, M. and Dibenedetto, A. Utilisation of $CO_2$ as a chemical feedstock: Opportunities and challenges. *Journal of the Chemical Society, Dalton Transactions.* 2007; (28):2975–2992.
6. Centi, G. and Perathoner, S. Opportunities and prospects in the chemical recycling of carbon dioxide to fuels. *Catalysis Today.* 2009; 148(3–4):191–205.
7. Omae, I. Recent developments in carbon dioxide utilization for the production of organic chemicals. *Coordination Chemistry Reviews.* 2012; 256(13–14):1384–1405.
8. Concepcion, J. J., House, R. L., Papanikolas, J. M. and Meyer, T. J. Chemical approaches to artificial photosynthesis. *Proceedings of the National Academy of Sciences of the United States of America.* 2012; 109(39):15560–15564.
9. Sakakura, T., Choi, J.-C. and Yasuda, H. Transformation of carbon dioxide. *Chemical Reviews.* 2007; 107(6):2365–2387.
10. Oloman, C. and Li, H. Electrochemical processing of carbon dioxide. *ChemSusChem.* 2008; 1(5):385–391.
11. Benson, E. E., Kubiak, C. P., Sathrum, A. J. and Smieja, J. M. Electrocatalytic and homogeneous approaches to conversion of $CO_2$ to liquid fuels. *Chemical Society Reviews.* 2009; 38(1):89–99.
12. Lee, J., Kwon, Y., Machunda, R. L. and Lee, H. J. Electrocatalytic recycling of $CO_2$ and small organic molecules. *Chemistry – An Asian Journal.* 2009; 4(10):1516–1523.
13. Windle, C. D. and Perutz, R. N. Advances in molecular photocatalytic and electrocatalytic $CO_2$ reduction. *Coordination Chemistry Reviews.* 2012; 256(21–22):2562–2570.
14. Finn, C., Schnittger, S., Yellowlees, L. J. and Love, J. B. Molecular approaches to the electrochemical reduction of carbon dioxide. *Chemical Communications.* 2012; 48(10):1392–1399.
15. Inglis, J. L., MacLean, B. J., Pryce, M. T. and Vos, J. G. Electrocatalytic pathways towards sustainable fuel production from water and $CO_2$. *Coordination Chemistry Reviews.* 2012; 256(21–22):2571–2600.
16. Schneider, J., Jia, H., Muckerman, J. T. and Fujita, E. Thermodynamics and kinetics of $CO_2$, CO, and h+ binding to the metal centre of $CO_2$ reduction catalysts. *Chemical Society Reviews.* 2012; 41(6):2036–2051.
17. Mori, K., Yamashita, H. and Anpo, M. Photocatalytic reduction of $CO_2$ with $H_2O$ on various titanium oxide photocatalysts. *RSC Advances.* 2012; 2(8):3165–3172.
18. Aresta, M. and Dibenedetto, A. *Carbon Dioxide Fixation into Organic Compounds.* Kluwer Academic Publishers, Boston, 2003, pp. 211–260.
19. Halmann, M. M. *Chemical Fixation of Carbon Dioxide – Methods for Recycling $CO_2$ into Useful Products.* CRC Press, Boca Raton, FL, 1993.

20. Sullivan, B. P., Krist, K. and Guard H. *Electrochemical and Electrocatalytic Reactions of Carbon Dioxide.* Elsevier, Amsterdam, 1993.

21. Song, C. S. Global challenges and strategies for control, conversion and utilization of $CO_2$ for sustainable development involving energy, catalysis, adsorption and chemical processing. *Catalysis Today.* 2006; 115(1–4):2–32.

22. Li, H. and Oloman, C. Development of a continuous reactor for the electro-reduction of carbon dioxide to formate—Part 2: Scale-up. *Journal of Applied Electrochemistry.* 2007; 37(10):1107–1117.

23. Szymaszek, A. and Pruchnik, F. P. Electrochemical reduction of carbon dioxide in the presence of $RhCl(CO)(PPh_3)_2$ and $IrCl(CO)(PPh_3)_2$. *Journal of Organometallic Chemistry.* 1989; 376(1):133–140.

24. Wasmus, S., Cattaneo, E. and Vielstich, W. Reduction of carbon dioxide to methane and ethene—An online MS study with rotating electrodes. *Electrochimica Acta.* 1990; 35(4):771–775.

25. Kyriacou, G. and Anagnostopoulos, A. Electroreduction of $CO_2$ on differently prepared copper electrodes. The influence of electrode treatment on the current efficiencies. *Journal of Electroanalytical Chemistry.* 1992; 322(1–2):233–246.

26. Hori, Y. et al. "Deactivation of copper electrode" in electrochemical reduction of $CO_2$. *Electrochimica Acta.* 2005; 50(27):5354–5369.

27. Smith, B. D., Irish, D. E., Kedzierzawski, P. and Augustynski, J. A surface enhanced raman scattering study of the intermediate and poisoning species formed during the electrochemical reduction of $CO_2$ on copper. *Journal of the Electrochemical Society.* 1997; 144(12):4288–4296.

28. Friebe, P., Bogdanoff, P., AlonsoVante, N. and Tributsch, H. A real-time mass spectroscopy study of the (electro)chemical factors affecting $CO_2$ reduction at copper. *Journal of Catalysis.* 1997; 168(2):374–385.

29. Le, M. et al. Electrochemical reduction of $CO_2$ to $CH_3OH$ at copper oxide surfaces. *Journal of the Electrochemical Society.* 2011; 158(5):E45–E49.

30. Burke, L. D. and Collins, J. A. Role of surface defects in the electrocatalytic behaviour of copper in base. *Journal of Applied Electrochemistry.* 1999; 29(12):1427–1438.

31. Kumar, R. S., Kumar, S. S. and Kulandainathan, M. A. Highly selective electrochemical reduction of carbon dioxide using cu based metal organic framework as an electrocatalyst. *Electrochemistry Communications.* 2012; 25:70–73.

32. Lee, J. and Tak, Y. Electrocatalytic activity of cu electrode in electroreduction of $CO_2$. *Electrochimica Acta.* 2001; 46(19):3015–3022.

33. Yano, J., Morita, T., Shimano, K., Nagami, Y. and Yamasaki, S. Selective ethylene formation by pulse-mode electrochemical reduction of carbon dioxide using copper and copper-oxide electrodes. *Journal of Solid State Electrochemistry.* 2007; 11(4):554–557.

34. Kapusta, S. and Hackerman, N. The electroreduction of carbon dioxide and formic-acid on tin and indium electrodes. *Journal of the Electrochemical Society.* 1983; 130(3):607–613.

35. Agarwal, A. S., Zhai, Y., Hill, D. and Sridhar, N. The electrochemical reduction of carbon dioxide to formate/formic acid: Engineering and economic feasibility. *ChemSusChem.* 2011; 4(9):1301–1310.

36. Wu, J., Risalvato, F. G., Ke, F.-S., Pellechia, P. J. and Zhou, X.-D. Electrochemical reduction of carbon dioxide. I. Effects of the electrolyte on the selectivity and activity with sn electrode. *Journal of the Electrochemical Society.* 2012; 159(7):F353–F359.

37. Chiacchiarelli, L. M., Zhai, Y., Frankel, G. S., Agarwal, A. S. and Sridhar, N. Cathodic degradation mechanisms of pure sn electrocatalyst in a nitrogen atmosphere. *Journal of Applied Electrochemistry.* 2012; 42(1):21–29.

38. Bujno, K., Bilewicz, R., Siegfried, L. and Kaden, T. Electroreduction of $CO_2$ catalyzed by Ni(II) tetraazamacrocyclic complexes—Reasons of poisoning of the catalytic surfaces. *Electrochimica Acta.* 1997; 42(8):1201–1206.

39. Benson, E. E. and Kubiak, C. P. Structural investigations into the deactivation pathway of the $CO_2$ reduction electrocatalyst Re(bpy)(CO)(3)Cl. *Chemical Communications*. 2012; 48(59):7374–7376.

40. Thoi, V. S. and Chang, C. J. Nickel n-heterocyclic carbene-pyridine complexes that exhibit selectivity for electrocatalytic reduction of carbon dioxide over water. *Chemical Communications*. 2011; 47(23):6578–6580.

41. Pugh, J. R., Bruce, M. R. M., Sullivan, B. P. and Meyer, T. J. Formation of a metal hydride bond and the insertion of $CO_2$—Key steps in the electrocatalytic reduction of carbon dioxide to formate anion. *Inorganic Chemistry*. 1991; 30(1):86–91.

42. Dubois, D. L., Miedaner, A. and Haltiwanger, R. C. Electrochemical reduction of $CO_2$ catalyzed by pd(triphosphine)(solvent) (bf4)$_2$ complexes—Synthetic and mechanistic studies. *Journal of the American Chemical Society*. 1991; 113(23):8753–8764.

43. Li, C. W. and Kanan, M. W. $CO_2$ reduction at low overpotential on cu electrodes resulting from the reduction of thick $Cu_2O$ films. *Journal of the American Chemical Society*. 2012; 134(17):7231–7234.

44. Beley, M., Collin, J. P., Ruppert, R. and Sauvage, J. P. Nickel(ii) cyclam—An extremely selective electrocatalyst for reduction of $CO_2$ in water. *Journal of the Chemical Society – Chemical Communications*. 1984;(19):1315–1316.

45. Chardon-Noblat, S., Deronzier, A., Ziessel, R. and Zsoldos, D. Selective synthesis and electrochemical behavior of *trans*(Cl)- and *cis*(Cl)-[Ru(bpy)(CO)$_2$Cl$_2$] complexes (bpy = 2,2′-bipyridine). Comparative studies of their electrocatalytic activity toward the reduction of carbon dioxide. *Inorganic Chemistry*. 1997; 36(23):5384–5389.

46. Saravanakumar, D., Song, J., Jung, N., Jirimali, H. and Shin, W. Reduction of $CO_2$ to CO at low overpotential in neutral aqueous solution by a ni(cyclam) complex attached to poly(allylamine). *ChemSusChem*. 2012; 5(4):634–636.

47. Bard, A. J., Parsons, R. and J. Jordan. *Standard Potentials in Aqueous Solutions*. CRC Press, Marcel Dekker, New York, 1985.

48. Lamy, E., Nadjo, L. and Saveant, J. M. Standard potential and kinetic-parameters of electrochemical reduction of carbon dioxide in dimethylformamide. *Journal of Electroanalytical Chemistry*. 1977; 78(2):403–407.

49. Hammouche, M., Lexa, D., Momenteau, M. and Saveant, J. M. Chemical catalysis of electrochemical reactions—Homogeneous catalysis of the electrochemical reduction of carbon dioxide by iron(0) porphyrins—Role of the addition of magnesium cations. *Journal of the American Chemical Society*. 1991; 113(22):8455–8466.

50. Bhugun, I., Lexa, D. and Saveant, J. M. Catalysis of the electrochemical reduction of carbon dioxide by iron(0) porphyrins. Synergistic effect of lewis acid cations. *Journal of Physical Chemistry*. 1996; 100(51):19981–19985.

51. Wong, K. Y., Chung, W. H. and Lau, C. P. The effect of weak bronsted acids on the electrocatalytic reduction of carbon dioxide by a rhenium tricarbonyl bipyridyl complex. *Journal of Electroanalytical Chemistry*. 1998; 453(1–2):161–169.

52. Saveant, J.-M. Molecular catalysis of electrochemical reactions. Mechanistic aspects. *Chemical Reviews*. 2008; 108(7):2348–2378.

53. Dubois, M. R. and Dubois, D. L. Development of molecular electrocatalysts for $CO_2$ reduction and H$_2$ production/oxidation. *Accounts of Chemical Research*. 2009; 42(12):1974–1982.

54. Costentin, C., Robert, M. and Saveant, J.-M. Catalysis of the electrochemical reduction of carbon dioxide. *Chemical Society Reviews*. 2013; 42(6):2423–2436.

55. Noda, H. et al. Electrochemical reduction of carbon dioxide at various metal-electrodes in aqueous potassium hydrogen carbonate solution. *Bulletin of the Chemical Society of Japan*. 1990; 63(9):2459–2462.

56. Fujishima, A., Zhang, X. and Tryk, D. A. Tio$_2$ photocatalysis and related surface phenomena. *Surface Science Reports*. 2008; 63(12):515–582.

57. Monnier, A., Augustynski, J. and Stalder, C. On the electrolytic reduction of carbon dioxide at $TiO_2$ and $TiO_2$-Ru cathodes. *Journal of Electroanalytical Chemistry.* 1980; 112(2):383–385.

58. Koudelka, M., Monnier, A. and Augustynski, J. Electrocatalysis of the cathodic reduction of carbon dioxide on platinized titanium-dioxide film electrodes. *Journal of the Electrochemical Society.* 1984; 131(4):745–750.

59. Bandi, A. Electrochemical reduction of carbon dioxide on conductive metallic oxides. *Journal of the Electrochemical Society.* 1990; 137(7):2157–2160.

60. Cueto, L. F., Hirata, G. A. and Sanchez, E. M. Thin-film $TiO_2$ electrode surface characterization upon $CO_2$ reduction processes. *Journal of Sol–Gel Science and Technology.* 2006; 37(2):105–109.

61. Chu, D. et al. Fixation of $CO_2$ by electrocatalytic reduction and electropolymerization in ionic liquid–$H_2O$ solution. *ChemSusChem.* 2008; 1(3):205–209.

62. Mizuno, T., Naitoh, A. and Ohta, K. Electrochemical reduction of $CO_2$ in methanol at −30-degrees-C. *Journal of Electroanalytical Chemistry.* 1995; 391(1–2):199–201.

63. Qu, J. P., Zhang, X. G., Wang, Y. G. and Xie, C. X. Electrochemical reduction of $CO_2$ on $RuO_2$/$TiO_2$ nanotubes composite modified pt electrode. *Electrochimica Acta.* 2005; 50(16–17):3576–3580.

64. Summers, D. P., Leach, S. and Frese, K. W. The electrochemical reduction of aqueous carbon dioxide to methanol at molybdenum electrodes with low overpotentials. *Journal of Electroanalytical Chemistry.* 1986; 205(1–2):219–232.

65. Wang, B. X. and Dong, S. J. Electrocatalytic properties of mixed-valence molybdenum oxide thin-film modified microelectrodes. *Journal of Electroanalytical Chemistry.* 1994; 379(1–2):207–214.

66. Bertotti, M. and Pletcher, D. The reduction of bromate at molybdenum oxide film cathodes. *Electroanalysis.* 1996; 8(12):1105–1111.

67. Bertotti, M. and Pletcher, D. Catalysis of the bromate reduction at a molybdenum oxide modified electrode. *Quimica Nova.* 1998; 21(2):167–171.

68. Kosminsky, L. and Bertotti, M. Studies on the catalytic reduction of iodate at glassy carbon electrodes modified by molybdenum oxides. *Journal of Electroanalytical Chemistry.* 1999; 471(1):37–41.

69. Kosminsky, L. and Bertotti, M. Determination of iodate in salt samples with amperometric detection at a molybdenum oxide modified electrode. *Electroanalysis.* 1999; 11(9):623–626.

70. Saji, V. S. and Lee, C.-W. Molybdenum, molybdenum oxides, and their electrochemistry. *ChemSusChem.* 2012; 5(7):1146–1161.

71. Nakazawa, M. et al. Electrochemical reduction of carbon dioxide using iron–sulfur clusters as catalyst precursors. *Bulletin of the Chemical Society of Japan.* 1986; 59(3):809–814.

72. Ogura, K. and Yoshida, I. Electrocatalytic reduction of $CO_2$ to methanol—Part 9. Mediation with metal porphyrins. *Journal of Molecular Catalysis.* 1988; 47(1):51–57.

73. Potts, K. T., Usifer, D. A., Guadalupe, A. and Abruna, H. D. 4-vinyl-2,2′-6′,2″, 6-vinyl-2,2′-6′,2″, and 4′-vinyl-2,2′-6′,2″-terpyridinyl ligands—Their synthesis and the electrochemistry of their transition-metal coordination-complexes. *Journal of the American Chemical Society.* 1987; 109(13):3961–3967.

74. Sende, J. A. R. et al. Electrocatalysis of $CO_2$ reduction in aqueous-media at electrodes modified with electropolymerized films of vinylterpyridine complexes of transition metals. *Inorganic Chemistry.* 1995; 34(12):3339–3348.

75. Reda, T., Plugge, C. M., Abram, N. J. and Hirst, J. Reversible interconversion of carbon dioxide and formate by an electroactive enzyme. *Proceedings of the National Academy of Sciences of the United States of America.* 2008; 105(31):10654–10658.

76. Hawecker, J., Lehn, J. M. and Ziessel, R. Efficient photochemical reduction of $CO_2$ to CO by visible-light irradiation of systems containing Re(bipy)(CO)3x or ru(bipy)$_3^{2+}$-$CO_2^+$ combinations as homogeneous catalysts. *Journal of the Chemical Society – Chemical Communications*. 1983; 9:536–538.

77. Hawecker, J., Lehn, J. M. and Ziessel, R. Electrocatalytic reduction of carbon dioxide mediated by Re(bipy)(CO)$_3$Cl(bipy = 2,2'-bipyridine). *Journal of the Chemical Society – Chemical Communications*. 1984; 6:328–330.

78. Hawecker, J., Lehn, J. M. and Ziessel, R. Photochemical and electrochemical reduction of carbon dioxide to carbon monoxide mediated by (2,2'-bipyridine) tricarbonylchlororhenium(I) and related complexes as homogeneous catalysts. *Helvetica Chimica Acta*. 1986; 69(8):1990–2012.

79. Otoole, T. R. et al. Electrocatalytic reduction of $CO_2$ at a chemically modified electrode. *Journal of the Chemical Society – Chemical Communications*. 1985; 20:1416–1417.

80. Cabrera, C. R. and Abruna, H. D. Electrocatalysis of $CO_2$ reduction at surface modified metallic and semiconducting electrodes. *Journal of Electroanalytical Chemistry*. 1986; 209(1):101–107.

81. Cosnier, S., Deronzier, A. and Moutet, J. C. Electrochemical coating of a platinum-electrode by a poly(pyrrole) film containing the *fac*-Re(2,2'-bipyridine)(co)3cl system—Application to electrocatalytic reduction of $CO_2$. *Journal of Electroanalytical Chemistry*. 1986; 207(1–2):315–321.

82. Cosnier, S., Deronzier, A. and Moutet, J. C. Electrocatalytic reduction of $CO_2$ on electrodes modified by *fac*-Re(2,2'-bipyridine)(CO)$_3$Cl complexes bonded to polypyrrole films. *Journal of Molecular Catalysis*. 1988; 45(3):381–391.

83. Otoole, T. R. et al. Electrocatalytic reduction of $CO_2$ by a complex of rhenium in thin polymeric films. *Journal of Electroanalytical Chemistry*. 1989; 259(1–2):217–239.

84. Schrebler, R., Cury, P., Herrera, F., Gomez, H. and Cordova, R. Study of the electrochemical reduction of $CO_2$ on electrodeposited rhenium electrodes in methanol media. *Journal of Electroanalytical Chemistry*. 2001; 516(1–2):23–30.

85. Cecchet, F., Alebbi, M., Bignozzi, C. A. and Paolucci, F. Efficiency enhancement of the electrocatalytic reduction of $CO_2$: *fac*-[Re(v-bpy)(CO)$_3$Cl] electropolymerized onto mesoporous $TiO_2$ electrodes. *Inorganica Chimica Acta*. 2006; 359(12):3871–3874.

86. Cheung, K.-C. et al. Electrocatalytic reduction of carbon dioxide by a polymeric film of rhenium tricarbonyl dipyridylamine. *Journal of Organometallic Chemistry*. 2009; 694(17):2842–2845.

87. Smieja, J. M. and Kubiak, C. P. Re(bipy-tbu)(CO)$_3$Cl-improved catalytic activity for reduction of carbon dioxide: IR-spectroelectrochemical and mechanistic studies. *Inorganic Chemistry*. 2010; 49(20):9283–9289.

88. Sullivan, B. P., Bolinger, C. M., Conrad, D., Vining, W. J. and Meyer, T. J. One-electron and 2-electron pathways in the electrocatalytic reduction of $CO_2$ by *fac*-Re(2,2'-bipyridine)(CO)$_3$Cl. *Journal of the Chemical Society – Chemical Communications*. 1985; 20:1414–1415.

89. Johnson, F. P. A., George, M. W., Hartl, F. and Turner, J. J. Electrocatalytic reduction of $CO_2$ using the complexes [Re(bpy)(CO)$_3$L]$^n$ ($n = +1$, L = P(OEt)$_3$, CH$_3$CN; $n = 0$, L = Cl$^-$, Otf$^-$; bpy = 2,2'-bipyridine; Otf$^-$ = CF$_3$SO$_3$) as catalyst precursors: Infrared spectroelectrochemical investigation. *Organometallics*. 1996; 15(15):3374–3387.

90. Scheiring, T., Klein, A. and Kaim, W. EPR study of paramagnetic rhenium (I) complexes (bpy($\cdot$-))Re(CO)$_3$X relevant to the mechanism of electrocatalytic $CO_2$ reduction. *Journal of the Chemical Society, Perkin Transactions*. 1997; 2:2569–2571.

91. Anfuso, C. L. et al. Covalent attachment of a rhenium bipyridyl $CO_2$ reduction catalyst to rutile $TiO_2$. *Journal of the American Chemical Society*. 2011; 133(18):6922–6925.

92. Bourrez, M., Molton, F., Chardon-Noblat, S. and Deronzier, A. Mn(bipyridyl)(co)(3) br: An abundant metal carbonyl complex as efficient electrocatalyst for $CO_2$ reduction. *Angewandte Chemie-International Edition*. 2011; 50(42):9903–9906.

93. Smieja, J. M. et al. Manganese as a substitute for rhenium in $CO_2$ reduction catalysts: The importance of acids. *Inorganic Chemistry*. 2013; 52(5):2484–2491.

94. Hara, K., Kudo, A. and Sakata, T. Electrochemical reduction of high-pressure carbon dioxide on Fe electrodes at large current density. *Journal of Electroanalytical Chemistry*. 1995; 386(1–2):257–260.

95. Koga, O. and Hori, Y. Reduction of adsorbed CO on a Ni electrode in connection with the electrochemical reduction of $CO_2$. *Electrochimica Acta*. 1993; 38(10):1391–1394.

96. Kudo, A., Nakagawa, S., Tsuneto, A. and Sakata, T. Electrochemical reduction of high-pressure $CO_2$ on Ni electrodes. *Journal of the Electrochemical Society*. 1993; 140(6):1541–1545.

97. Koga, O., Matsuo, T., Yamazaki, H. and Hori, Y. Infrared spectroscopic observation of intermediate species on Ni and Fe electrodes in the electrochemical reduction of $CO_2$ and CO to hydrocarbons. *Bulletin of the Chemical Society of Japan*. 1998; 71(2):315–320.

98. Hori, Y., Wakebe, H., Tsukamoto, T. and Koga, O. Electrocatalytic process of CO selectivity in electrochemical reduction of $CO_2$ at metal-electrodes in aqueous-media. *Electrochimica Acta*. 1994; 39(11–12):1833–1839.

99. Popic, J. P., Avramovlvic, M. L. and Vukovic, N. B. Reduction of carbon dioxide on ruthenium oxide and modified ruthenium oxide electrodes in 0.5 m $NaHCO_3$. *Journal of Electroanalytical Chemistry*. 1997; 421(1–2):105–110.

100. Frese, K. W. and Leach, S. Electrochemical reduction of carbon dioxide to methane, methanol, and co on ru electrodes. *Journal of the Electrochemical Society*. 1985; 132(1):259–260.

101. Lukaszewski, M. and Czerwinski, A. Electrochemical behavior of Pd–Rh alloys. *Journal of Solid State Electrochemistry*. 2007; 11(3):339–349.

102. Lukaszewski, M., Siwek, H. and Czerwinski, A. Electrosorption of carbon dioxide on platinum group metals and alloys—A review. *Journal of Solid State Electrochemistry*. 2009; 13(6):813–827.

103. McKee, D. W. Interaction of hydrogen and carbon monoxide on platinum group metals. *Journal of Catalysis*. 1967; 8(3):240–249.

104. Spichigerulmann, M. and Augustynski, J. Electrochemical reduction of bicarbonate ions at a bright palladium cathode. *Journal of the Chemical Society – Faraday Transactions I*. 1985; 81:713–716.

105. Azuma, M., Hashimoto, K., Watanabe, M. and Sakata, T. Electrochemical reduction of carbon dioxide to higher hydrocarbons in a $KHCO_3$ aqueous-solution. *Journal of Electroanalytical Chemistry*. 1990; 294(1–2):299–303.

106. Nakagawa, S., Kudo, A., Azuma, M. and Sakata, T. Effect of pressure on the electrochemical reduction of $CO_2$ on Group VIII metal electrodes. *Journal of Electroanalytical Chemistry*. 1991; 308(1–2):339–343.

107. Ohkawa, K., Hashimoto, K., Fujishima, A., Noguchi, Y. and Nakayama, S. Electrochemical reduction of carbon dioxide on hydrogen-storing materials – Part 1. The effect of hydrogen absorption on the electrochemical-behavior on palladium electrodes. *Journal of Electroanalytical Chemistry*. 1993; 345(1–2):445–456.

108. Ohkawa, K., Noguchi, Y., Nakayama, S., Hashimoto, K. and Fujishima, A. Electrochemical reduction of carbon dioxide on hydrogen-storing materials – Part II. Copper-modified palladium electrode. *Journal of Electroanalytical Chemistry*. 1993; 348(1–2):459–464.

109. Ohkawa, K., Noguchi, Y., Nakayama, S., Hashimoto, K. and Fujishima, A. Electrochemical reduction of carbon dioxide on hydrogen-storing materials – Part 3. The effect of the absorption of hydrogen on the palladium electrodes modified with copper. *Journal of Electroanalytical Chemistry*. 1994; 367(1–2):165–173.

110. Podlovchenko, B. I., Kolyadko, E. A. and Lu, S. G. Electroreduction of carbon dioxide on palladium electrodes at potentials higher than the reversible hydrogen potential. *Journal of Electroanalytical Chemistry.* 1994; 373(1–2):185–187.

111. Ohkawa, K., Noguchi, Y., Nakayama, S., Hashimoto, K. and Fujishima, A. Electrochemical reduction of carbon dioxide on hydrogen-storing materials: Part 4. Electrochemical behavior of the Pd electrode in aqueous and nonaqueous electrolyte. *Journal of Electroanalytical Chemistry.* 1994; 369(1–2):247–250.

112. Eggins, B. R. and McNeill, J. Voltammetry of carbon dioxide. 1. A general survey of voltammetry at different electrode materials in different solvents. *Journal of Electroanalytical Chemistry.* 1983; 148(1):17–24.

113. Brisard, G. M., Camargo, A. P. M., Nart, F. C. and Iwasita, T. On-line mass spectrometry investigation of the reduction of carbon dioxide in acidic media on polycrystalline Pt. *Electrochemistry Communications.* 2001; 3(11):603–607.

114. Hara, K., Kudo, A., Sakata, T. and Watanabe, M. High-efficiency electrochemical reduction of carbon dioxide under high-pressure on a gas-diffusion electrode containing Pt catalysts. *Journal of the Electrochemical Society.* 1995; 142(4):L57–L59.

115. Inui, T., Anpo, M., Izui, K., Yanagida, S. and Yamaguchi, Y. *Advances in Chemical Conversions for Mitigating Carbon Dioxide.* Delmon B, Yates JT, editors. Elsevier, Amsterdam, 1998.

116. Hori, Y. and Tomita, Y. Electrochemical reduction of carbon dioxide in acetonitrile-water mixtures at a platinum electrode. *Abstracts of Papers of the American Chemical Society.* 1998; 215:U405–U406.

117. Tomita, Y., Teruya, S., Koga, O. and Hori, Y. Electrochemical reduction of carbon dioxide at a platinum electrode in acetonitrile-water mixtures. *Journal of the Electrochemical Society.* 2000; 147(11):4164–4167.

118. Centi, G., Perathoner, S., Wine, G. and Gangeri, M. Electrocatalytic conversion of $CO_2$ to long carbon-chain hydrocarbons. *Green Chemistry.* 2007; 9(6):671–678.

119. Feng, Q., Liu, S., Wang, X. and Jin, G. Nanoporous copper incorporated platinum composites for electrocatalytic reduction of $CO_2$ in ionic liquid BMIMBF$_4$. *Applied Surface Science.* 2012; 258(12):5005–5009.

120. Hu, B., Stancovski, V., Morton, M. and Suib, S. L. Enhanced electrocatalytic reduction of $CO_2/H_2O$ to paraformaldehyde at Pt/metal oxide interfaces. *Applied Catalysis A – General.* 2010; 382(2):277–283.

121. de Tacconi, N. R. et al. Electrocatalytic reduction of carbon dioxide using Pt/C-TiO$_2$ nanocomposite cathode. *Electrochemical and Solid State Letters.* 2012; 15(1):B5–B8.

122. Yuan, X. et al. Electrochemical conversion of methanol and carbon dioxide to dimethyl carbonate at graphite-Pt electrode system. *Journal of the Electrochemical Society.* 2012; 159(12):E183–E186.

123. Nikolic, B. Z. et al. Electroreduction of carbon dioxide on platinum single-crystal electrodes—Electrochemical and *in situ* FTIR studies. *Journal of Electroanalytical Chemistry.* 1990; 295(1–2):415–423.

124. Hoshi, N., Mizumura, T. and Hori, Y. Significant difference of the reduction rates of carbon dioxide between Pt(111) and Pt(110) single-crystal electrodes. *Electrochimica Acta.* 1995; 40(7):883–887.

125. Hoshi, N., Suzuki, T. and Hori, Y. Catalytic activity of $CO_2$ reduction on Pt single-crystal electrodes: Pt(S)-[$n$(111) × (111)], Pt(S)-[$n$(111) × (100)], and Pt(S)-[$n$(100) × (111)]. *Journal of Physical Chemistry B.* 1997; 101(42):8520–8524.

126. Hoshi, N., Kawatani, S., Kudo, M. and Hori, Y. Significant enhancement of the electrochemical reduction of $CO_2$ at the kink sites on Pt(S)-[$n$(110) × (100)] and Pt(S)-[$n$(100) × (110)]. *Journal of Electroanalytical Chemistry.* 1999; 467(1–2):67–73.

127. Hoshi, N. and Hori, Y. Electrochemical reduction of carbon dioxide at a series of platinum single crystal electrodes. *Electrochimica Acta.* 2000; 45(25–26):4263–4270.

128. ChunJie, F., YbuJun, F., ChunHua, Z., QingWei, Z. and ShiGang, S. Studies of surface processes of electrocatalytic reduction of CO₂ on Pt(210), Pt(310) and Pt(510). *Science in China Series B – Chemistry.* 2007; 50(5):593–598.

129. Sanchez-Sanchez, C. M., Souza-Garcia, J., Herrero, E. and Aldaz, A. Electrocatalytic reduction of carbon dioxide on platinum single crystal electrodes modified with adsorbed adatoms. *Journal of Electroanalytical Chemistry.* 2012; 668:51–59.

130. Hoshi, N., Noma, M., Suzuki, T. and Hori, Y. Structural effect on the rate of CO₂ reduction on single crystal electrodes of palladium. *Journal of Electroanalytical Chemistry.* 1997; 421(1–2):15–18.

131. Bandi, A. and Kuhne, H. M. Electrochemical reduction of carbon dioxide in water—Analysis of reaction-mechanism on ruthenium–titanium-oxide. *Journal of the Electrochemical Society.* 1992; 139(6):1605–1610.

132. Spataru, N., Tokuhiro, K., Terashima, C., Rao, T. N. and Fujishima, A. Electrochemical reduction of carbon dioxide at ruthenium dioxide deposited on boron-doped diamond. *Journal of Applied Electrochemistry.* 2003; 33(12):1205–1210.

133. Hori, Y., Kikuchi, K. and Suzuki, S. Production of CO and CH₄ in electrochemical reduction of CO₂ at metal-electrodes in aqueous hydrogencarbonate solution. *Chemistry Letters.* 1985; 14(11):1695–1698.

134. Jitaru, M., Lowy, D. A., Toma, M., Toma, B. C. and Oniciu, L. Electrochemical reduction of carbon dioxide on flat metallic cathodes. *Journal of Applied Electrochemistry.* 1997; 27(8):875–889.

135. Gattrell, M., Gupta, N. and Co, A. A review of the aqueous electrochemical reduction of CO₂ to hydrocarbons at copper. *Journal of Electroanalytical Chemistry.* 2006; 594(1):1–19.

136. Cook, R. L., Macduff, R. C. and Sammells, A. F. Efficient high-rate carbon dioxide reduction to methane and ethylene at *in situ* electrodeposited copper electrode. *Journal of the Electrochemical Society.* 1987; 134(9):2375–2376.

137. Cook, R. L., Macduff, R. C. and Sammells, A. F. High-rate gas-phase CO₂ reduction to ethylene and methane using gas-diffusion electrodes. *Journal of the Electrochemical Society.* 1990; 137(2):607–608.

138. Hori, Y., Kikuchi, K., Murata, A. and Suzuki, S. Production of methane and ethylene in electrochemical reduction of carbon dioxide at copper electrode in aqueous hydrogen-carbonate solution. *Chemistry Letters.* 1986; 15(6):897–898.

139. Hori, Y., Murata, A., Takahashi, R. and Suzuki, S. Enhanced formation of ethylene and alcohols at ambient-temperature and pressure in electrochemical reduction of carbon dioxide at a copper electrode. *Journal of the Chemical Society – Chemical Communications.* 1988; 1:17–19.

140. Hori, Y., Murata, A. and Takahashi, R. Formation of hydrocarbons in the electrochemical reduction of carbon dioxide at a copper electrode in aqueous solution. *Journal of the Chemical Society – Faraday Transactions I.* 1989; 85:2309–2326.

141. Shibata, H., Moulijn, J. A. and Mul, G. Enabling electrocatalytic Fischer–Tropsch synthesis from carbon dioxide over copper-based electrodes. *Catalysis Letters.* 2008; 123(3–4):186–192.

142. Wang, X. Y. et al. Fixation of CO₂ by electrocatalytic reduction to synthesis of dimethyl carbonate in ionic liquid using effective silver-coated nanoporous copper composites. *Chinese Chemical Letters.* 2010; 21(8):987–990.

143. Saeki, T., Hashimoto, K., Kimura, N., Omata, K. and Fujishima, A. Electrochemical reduction of CO₂ with high-current density in a CO₂ plus methanol medium. II. CO formation promoted by tetrabutylammonium cation. *Journal of Electroanalytical Chemistry.* 1995; 390(1–2):77–82.

144. Kaneco, S., Iiba, K., Suzuki, S. K., Ohta, K. and Mizuno, T. Electrochemical reduction of carbon dioxide to hydrocarbons with high faradaic efficiency in LiOH/methanol. *Journal of Physical Chemistry B.* 1999; 103(35):7456–7460.

145. Kaneco, S. et al. Electrochemical reduction of carbon dioxide to ethylene with high faradaic efficiency at a Cu electrode in CsOH/methanol. *Electrochimica Acta.* 1999; 44(26):4701–4706.
146. Kaneco, S., Iiba, K., Katsumata, H., Suzuki, T. and Ohta, K. Electrochemical reduction of high pressure carbon dioxide at a Cu electrode in cold methanol with CsOH supporting salt. *Chemical Engineering Journal.* 2007; 128(1):47–50.
147. Kaneco, S., Ueno, Y., Katsumata, H., Suzuki, T. and Ohta, K. Electrochemical reduction of $CO_2$ in copper particle-suspended methanol. *Chemical Engineering Journal.* 2006; 119(2–3):107–112.
148. Cook, R. L., Macduff, R. C. and Sammells, A. F. Electrochemical reduction of carbon dioxide to methane at high-current densities. *Journal of the Electrochemical Society.* 1987; 134(7):1873–1874.
149. Kim, J. J., Summers, D. P. and Frese, K. W. Reduction of $CO_2$ and CO to methane on Cu foil electrodes. *Journal of Electroanalytical Chemistry.* 1988; 245(1–2):223–244.
150. Ohta, K., Suda, K., Kaneco, S. and Mizuno, T. Electrochemical reduction of carbon dioxide at Cu electrode under ultrasonic irradiation. *Journal of the Electrochemical Society.* 2000; 147(1):233–237.
151. Cook, R. L., Macduff, R. C. and Sammells, A. F. On the electrochemical reduction of carbon dioxide at *in situ* electrodeposited copper. *Journal of the Electrochemical Society.* 1988; 135(6):1320–1326.
152. Tang, W. et al. The importance of surface morphology in controlling the selectivity of polycrystalline copper for $CO_2$ electroreduction. *Physical Chemistry Chemical Physics.* 2012; 14(1):76–81.
153. Zhou, A. H. et al. Electrochemical quartz crystal microbalance probing the electro-formed and electro-reduced products on a copper electrode in aqueous solutions containing $NaHCO_3$ and $Na_2CO_3$. *Electrochimica Acta.* 2000; 45(24):3943–3950.
154. Li, H. and Oloman, C. The electro-reduction of carbon dioxide in a continuous reactor. *Journal of Applied Electrochemistry.* 2005; 35(10):955–965.
155. Kuhl, K. P., Cave, E. R., Abram, D. N. and Jaramillo, T. F. New insights into the electrochemical reduction of carbon dioxide on metallic copper surfaces. *Energy & Environmental Science.* 2012; 5(5):7050–7059.
156. De Jesus-Cardona, H., del Moral, C. and Cabrera, C. R. Voltammetric study of $CO_2$ reduction at Cu electrodes under different $KHCO_3$ concentrations, temperatures and $CO_2$ pressures. *Journal of Electroanalytical Chemistry.* 2001; 513(1):45–51.
157. Dewulf, D. W., Jin, T. and Bard, A. J. Electrochemical and surface studies of carbon dioxide reduction to methane and ethylene at copper electrodes in aqueous solutions. *Journal of the Electrochemical Society.* 1989; 136(6):1686–1691.
158. Yano, H., Tanaka, T., Nakayama, M. and Ogura, K. Selective electrochemical reduction of $CO_2$ to ethylene at a three-phase interface on copper(I) halide-confined Cu-mesh electrodes in acidic solutions of potassium halides. *Journal of Electroanalytical Chemistry.* 2004; 565(2):287–293.
159. Ogura, K., Yano, H. and Tanaka, T. Selective formation of ethylene from $CO_2$ by catalytic electrolysis at a three-phase interface. *Catalysis Today.* 2004; 98(4):515–521.
160. Hoshi, N., Ito, H., Suzuki, T. and Hori, Y. $CO_2$ reduction on Rh single-crystal electrodes and the structural effect. *Journal of Electroanalytical Chemistry.* 1995; 395(1–2):309–312.
161. Hoshi, N., Kuroda, M. and Hori, Y. Voltammograms of stepped and kinked stepped surfaces of palladium: Pd(S)-[$n$(111) × (100)] and Pd(S)-[$n$(100) × (110)]. *Journal of Electroanalytical Chemistry.* 2002; 521(1–2):155–160.
162. Hoshi, N., Uchida, T., Mizumura, T. and Hori, Y. Atomic arrangement dependence of reduction rates of carbon dioxide on iridium single-crystal electrodes. *Journal of Electroanalytical Chemistry.* 1995; 381(1–2):261–264.

163. Hoshi, N., Kato, M. and Hori, Y. Electrochemical reduction of $CO_2$ on single crystal electrodes of silver Ag(111), Ag(100) and Ag(110). *Journal of Electroanalytical Chemistry.* 1997; 440(1–2):283–286.

164. Hori, Y., Wakebe, H., Tsukamoto, T. and Koga, O. Adsorption of CO accompanied with simultaneous charge-transfer on copper single-crystal electrodes related with electrochemical reduction of $CO_2$ to hydrocarbons. *Surface Science.* 1995; 335(1–3):258–263.

165. Takahashi, I., Koga, O., Hoshi, N. and Hori, Y. Electrochemical reduction of $CO_2$ at copper single crystal Cu(S)-[$n$(111) × (111)] and Cu(S)-[$n$(110) × (100)] electrodes. *Journal of Electroanalytical Chemistry.* 2002; 533(1–2):135–143.

166. Hori, Y., Takahashi, I., Koga, O. and Hoshi, N. Electrochemical reduction of carbon dioxide at various series of copper single crystal electrodes. *Journal of Molecular Catalysis A – Chemical.* 2003; 199(1–2):39–47.

167. Goncalves, M. R. et al. Selective electrochemical conversion of $CO_2$ to $C_2$ hydrocarbons. *Energy Conversion and Management.* 2010; 51(1):30–32.

168. Schouten, K. J. P., Qin, Z., Gallent, E. P. and Koper, M. T. M. Two pathways for the formation of ethylene in CO reduction on single-crystal copper electrodes. *Journal of the American Chemical Society.* 2012; 134(24):9864–9867.

169. Schouten, K. J. P., Kwon, Y., van der Ham, C. J. M., Qin, Z. and Koper, M. T. M. A new mechanism for the selectivity to C1 and C2 species in the electrochemical reduction of carbon dioxide on copper electrodes. *Chemical Science.* 2011; 2(10):1902–1909.

170. Durand, W. J., Peterson, A. A., Studt, F., Abild-Pedersen, F. and Norskov, J. K. Structure effects on the energetics of the electrochemical reduction of $CO_2$ by copper surfaces. *Surface Science.* 2011; 605(15–16):1354–1359.

171. Liu, C., Cundari, T. R. and Wilson, A. K. $CO_2$ reduction on transition metal (Fe, Co, Ni, and Cu) surfaces: In comparison with homogeneous catalysis. *Journal of Physical Chemistry C.* 2012; 116(9):5681–5688.

172. Maeda, M., Kitaguchi, Y., Ikeda, S. and Ito, K. Reduction of carbon dioxide on partially-immersed Au plate electrode and Au-SPE electrode. *Journal of Electroanalytical Chemistry.* 1987; 238(1–2):247–258.

173. Ohmori, T., Nakayama, A., Mametsuka, H. and Suzuki, E. Influence of sputtering parameters on electrochemical $CO_2$ reduction in sputtered Au electrode. *Journal of Electroanalytical Chemistry.* 2001; 514(1–2):51–55.

174. Stevens, G. B., Reda, T. and Raguse, B. Energy storage by the electrochemical reduction of $CO_2$ to CO at a porous Au film. *Journal of Electroanalytical Chemistry.* 2002; 526(1–2):125–133.

175. Kauffman, D. R., Alfonso, D., Matranga, C., Qian, H. and Jin, R. Experimental and computational investigation of $Au_{25}$ clusters and $CO_2$: A unique interaction and enhanced electrocatalytic activity. *Journal of the American Chemical Society.* 2012; 134(24):10237–10243.

176. Shibata, M., Yoshida, K. and Furuya, N. Electrochemical synthesis of urea at gas-diffusion electrodes—IV. Simultaneous reduction of carbon dioxide and nitrate ions with various metal catalysts. *Journal of the Electrochemical Society.* 1998; 145(7):2348–2353.

177. Haynes, L. V. and Sawyer, D. T. Electrochemistry of carbon dioxide in dimethyl sulfoxide at gold and mercury electrodes. *Analytical Chemistry.* 1967; 39(3):332–338.

178. Azuma, M., Hashimoto, K., Hiramoto, M., Watanabe, M. and Sakata, T. Electrochemical reduction of carbon dioxide on various metal-electrodes in low-temperature aqueous $KHCO_3$ media. *Journal of the Electrochemical Society.* 1990; 137(6):1772–1778.

179. Paik, W., Andersen, T. N. and Eyring, H. Kinetic studies of electrolytic reduction of carbon dioxide on mercury electrode. *Electrochimica Acta.* 1969; 14(12):1217–1232.

180. Bewick, A. and Greener, G. P. Electroreduction of $CO_2$ to malate on a mercury cathode. *Tetrahedron Letters.* 1969; 10(53):4623–4626.

181. Hori, Y. and Suzuki, S. Electrolytic reduction of carbon dioxide at mercury-electrode in aqueous solution. *Bulletin of the Chemical Society of Japan.* 1982; 55(3):660–665.

182. Todoroki, M., Hara, K., Kudo, A. and Sakata, T. Electrochemical reduction of high-pressure $CO_2$ at Pb, Hg and In electrodes in an aqueous $KHCO_3$ solution. *Journal of Electroanalytical Chemistry.* 1995; 394(1–2):199–203.

183. Chang, T.-Y., Liang, R.-M., Wu, P.-W., Chen, J.-Y. and Hsieh, Y.-C. Electrochemical reduction of $CO_2$ by $Cu_2O$-catalyzed carbon clothes. *Materials Letters.* 2009; 63(12):1001–1003.

184. Ohya, S., Kaneco, S., Katsumata, H., Suzuki, T. and Ohta, K. Electrochemical reduction of $CO_2$ in methanol with aid of CuO and $Cu_2O$. *Catalysis Today.* 2009; 148(3–4):329–334.

185. Wu, H., Zhang, N., Cao, Z., Wang, H. and Hong, S. The adsorption of $CO_2$ $H_2CO_3$, $HCO_3^{3-}$ and $CO_3^{2-}$ on $Cu_2O(111)$ surface: First-principles study. *International Journal of Quantum Chemistry.* 2012; 112(12):2532–2540.

186. Chen, Y., Li, C. W. and Kanan, M. W. Aqueous $CO_2$ reduction at very low overpotential on oxide-derived Au nanoparticles. *Journal of the American Chemical Society.* 2012; 134(49):19969–19972.

187. Aurianblajeni, B., Halmann, M. and Manassen, J. Electrochemical measurements on the photo-electrochemical reduction of aqueous carbon dioxide on para-gallium phosphide and para-gallium arsenide semiconductor electrodes. *Solar Energy Materials.* 1983; 8(4):425–440.

188. Canfield, D. and Frese, K. W. Reduction of carbon dioxide to methanol on n-GaAs and p-GaAs and p-InP—Effect of crystal face, electrolyte and current density. *Journal of the Electrochemical Society.* 1983; 130(8):1772–1773.

189. Barton, E. E., Rampulla, D. M. and Bocarsly, A. B. Selective solar-driven reduction of $CO_2$ to methanol using a catalyzed p-gap based photoelectrochemical cell. *Journal of the American Chemical Society.* 2008; 130(20):6342–6344.

190. Ito, K., Ikeda, S., Yamauchi, N., Iida, T. and Takagi, T. Electrochemical reduction products of carbon dioxide at some metallic electrodes in nonaqueous electrolytes. *Bulletin of the Chemical Society of Japan.* 1985; 58(10):3027–3028.

191. Ikeda, S., Takagi, T. and Ito, K. Selective formation of formic-acid, oxalic-acid, and carbon-monoxide by electrochemical reduction of carbon dioxide. *Bulletin of the Chemical Society of Japan.* 1987; 60(7):2517–2522.

192. Mizuno, T. et al. Effect of temperature on electrochemical reduction of high-pressure $CO_2$ with In, Sn, and Pb electrodes. *Energy Sources.* 1995; 17(5):503–508.

193. Narayanan, S. R., Haines, B., Soler, J. and Valdez, T. I. Electrochemical conversion of carbon dioxide to formate in alkaline polymer electrolyte membrane cells. *Journal of the Electrochemical Society.* 2011; 158(2):A167–A173.

194. Parkinson, B. A. and Weaver, P. F. Photoelectrochemical pumping of enzymatic $CO_2$ reduction. *Nature.* 1984; 309(5964):148–149.

195. Kaneco, S., Katsumata, H., Suzuki, T. and Ohta, K. Photoelectrochemical reduction of carbon dioxide at p-type gallium arsenide and p-type indium phosphide electrodes in methanol. *Chemical Engineering Journal.* 2006; 116(3):227–231.

196. Kaneco, S., Katsumata, H., Suzuki, T. and Ohta, K. Photoelectrocatalytic reduction of $CO_2$ in LiOH/methanol at metal-modified p-InP electrodes. *Applied Catalysis B – Environmental.* 2006; 64(1–2):139–145.

197. Kaneco, S., Ueno, Y., Katsumata, H., Suzuki, T. and Ohta, K. Photoelectrochemical reduction of $CO_2$ at p-InP electrode in copper particle-suspended methanol. *Chemical Engineering Journal.* 2009; 148(1):57–62.

198. Li, H. and Oloman, C. Development of a continuous reactor for the electro-reduction of carbon dioxide to formate—Part 1: Process variables. *Journal of Applied Electrochemistry.* 2006; 36(10):1105–1115.

199. Machunda, R. L., Ju, H. and Lee, J. Electrocatalytic reduction of CO$_2$ gas at Sn based gas diffusion electrode. *Current Applied Physics*. 2011; 11(4):986–988.

200. Chen, Y. and Kanan, M. W. Tin oxide dependence of the CO$_2$ reduction efficiency on tin electrodes and enhanced activity for tin/tin oxide thin-film catalysts. *Journal of the American Chemical Society*. 2012; 134(4):1986–1989.

201. Prakash, G. K. S., Viva, F. A. and Olah, G. A. Electrochemical reduction of CO$_2$ over Sn-Nafion (R) coated electrode for a fuel-cell-like device. *Journal of Power Sources*. 2013; 223:68–73.

202. Bewick, A. and Greener, G. P. Electroreduction of CO$_2$ to glycollate on a lead cathode. *Tetrahedron Letters*. 1970; 11(5):391–394.

203. Kaneco, S., Iwao, R., Iiba, K., Ohta, K. and Mizuno, T. Electrochemical conversion of carbon dioxide to formic acid on Pb in KOH/methanol electrolyte at ambient temperature and pressure. *Energy*. 1998; 23(12):1107–1112.

204. Eneau-Innocent, B., Pasquier, D., Ropital, F., Leger, J. M. and Kokoh, K. B. Electroreduction of carbon dioxide at a lead electrode in propylene carbonate: A spectroscopic study. *Applied Catalysis B – Environmental*. 2010; 98(1–2):65–71.

205. Innocent, B. et al. Electro-reduction of carbon dioxide to formate on lead electrode in aqueous medium. *Journal of Applied Electrochemistry*. 2009; 39(2):227–232.

206. Subramanian, K., Asokan, K., Jeevarathinam, D. and Chandrasekaran, M. Electrochemical membrane reactor for the reduction of carbon dioxide to formate. *Journal of Applied Electrochemistry*. 2007; 37(2):255–260.

207. Murata, A. and Hori, Y. Product selectivity affected by cationic species in electrochemical reduction of CO$_2$ and CO at a Cu electrode. *Bulletin of the Chemical Society of Japan*. 1991; 64(1):123–127.

208. Kaneco, S., Katsumata, H., Suzuki, T. and Ohta, K. Electrochemical reduction of CO$_2$ to methane at the Cu electrode in methanol with sodium supporting salts and its comparison with other alkaline salts. *Energy & Fuels*. 2006; 20(1):409–414.

209. Kaneco, S., Iiba, K., Ohta, K. and Mizuno, T. Electrochemical reduction of carbon dioxide on copper in methanol with various potassium supporting electrolytes at low temperature. *Journal of Solid State Electrochemistry*. 1999; 3(7–8):424–428.

210. Kaneco, S., Iiba, K., Ohta, K. and Mizuno, T. Reduction of carbon dioxide to petrochemical intermediates. *Energy Sources*. 2000; 22(2):127–135.

211. Kaneco, S., Iiba, K., Katsumata, H., Suzuki, T. and Ohta, K. Effect of sodium cation on the electrochemical reduction of CO$_2$ at a copper electrode in methanol. *Journal of Solid State Electrochemistry*. 2007; 11(4):490–495.

212. Kaneco, S., Katsumata, H., Suzuki, T. and Ohta, K. Electrochemical reduction of carbon dioxide to ethylene at a copper electrode in methanol using potassium hydroxide and rubidium hydroxide supporting electrolytes. *Electrochimica Acta*. 2006; 51(16):3316–3321.

213. Schizodimou, A. and Kyriacou, G. Acceleration of the reduction of carbon dioxide in the presence of multivalent cations. *Electrochimica Acta*. 2012; 78:171–176.

214. Takahashi, K., Hiratsuka, K., Sasaki, H. and Toshima, S. Electrocatalytic behavior of metal porphyrins in the reduction of carbon dioxide. *Chemistry Letters*. 1979; 8(4):305–308.

215. Bhugun, I., Lexa, D. and Saveant, J. M. Ultraefficient selective homogeneous catalysis of the electrochemical reduction of carbon dioxide by an iron(0) porphyrin associated with a weak bronsted acid cocatalyst. *Journal of the American Chemical Society*. 1994; 116(11):5015–5016.

216. Bhugun, I., Lexa, D. and Saveant, J. M. Catalysis of the electrochemical reduction of carbon dioxide by iron(0) porphyrins: Synergystic effect of weak bronsted acids. *Journal of the American Chemical Society*. 1996; 118(7):1769–1776.

217. Costentin, C., Drouet, S., Robert, M. and Saveant, J.-M. A local proton source enhances CO$_2$ electroreduction to CO by a molecular Fe catalyst. *Science*. 2012; 338(6103):90–94.

218. Zhao, H.-Z., Chang, Y.-Y. and Liu, C. Electrodes modified with iron porphyrin and carbon nanotubes: Application to $CO_2$ reduction and mechanism of synergistic electrocatalysis. *Journal of Solid State Electrochemistry.* 2013; 17(6):1657–1664.
219. Leung, K., Nielsen, I. M. B., Sai, N., Medforth, C. and Shelnutt, J. A. Cobalt-porphyrin catalyzed electrochemical reduction of carbon dioxide in water. 2. Mechanism from first principles. *Journal of Physical Chemistry A.* 2010; 114(37):10174–10184.
220. Tanaka, H. and Aramata, A. Aminopyridyl cation radical method for bridging between metal complex and glassy carbon: Cobalt(II) tetraphenylporphyrin bonded on glassy carbon for enhancement of $CO_2$ electroreduction. *Journal of Electroanalytical Chemistry.* 1997; 437(1–2):29–35.
221. Magdesieva, T. V., Yamamoto, T., Tryk, D. A. and Fujishima, A. Electrochemical reduction of $CO_2$ with transition metal phthalocyanine and porphyrin complexes supported on activated carbon fibers. *Journal of the Electrochemical Society.* 2002; 149(6):D89–D95.
222. Enoki, O., Imaoka, T. and Yamamoto, K. Electrochemical reduction of carbon dioxide catalyzed by cofacial dinuclear metalloporphyrin. *Macromolecular Symposia.* 2003; 204:151–158.
223. Imaoka, T., Tanaka, R. and Yamamoto, K. Synergetic activation of carbon dioxide molecule using phenylazomethine dendrimers as a catalyst. *Journal of Polymer Science Part A – Polymer Chemistry.* 2006; 44(17):5229–5236.
224. Yao, S. A. et al. Covalent attachment of catalyst molecules to conductive diamond: $CO_2$ reduction using "smart" electrodes. *Journal of the American Chemical Society.* 2012; 134(38):15632–15635.
225. Behar, D. et al. Cobalt porphyrin catalyzed reduction of $CO_2$. Radiation chemical, photochemical, and electrochemical studies. *Journal of Physical Chemistry A.* 1998; 102(17):2870–2877.
226. Becker, J. Y., Vainas, B., Eger, R. and Kaufman, L. Electrocatalytic reduction of $CO_2$ to oxalate by Ag-II and Pd-II porphyrins. *Journal of the Chemical Society – Chemical Communications.* 1985; 21:1471–1472.
227. Sonoyama, N., Kirii, M. and Sakata, T. Electrochemical reduction of $CO_2$ at metal-porphyrin supported gas diffusion electrodes under high pressure $CO_2$. *Electrochemistry Communications.* 1999; 1(6):213–216.
228. Hieringer, W. et al. The surface trans effect: Influence of axial ligands on the surface chemical bonds of adsorbed metalloporphyrins. *Journal of the American Chemical Society.* 2011; 133(16):6206–6222.
229. Hiratsuka, K., Takahashi, K., Sasaki, H. and Toshima, S. Electrocatalytic behavior of tetrasulfonated metal phthalocyanines in reduction of carbon dioxide. *Chemistry Letters.* 1977; 10(10):1137–1140.
230. Meshitsu, S., Ichikawa, M. and Tamaru, K. Electrocatalysis by metal phthalocyanines in reduction of carbon dioxide. *Journal of the Chemical Society – Chemical Communications.* 1974; 5:158–159.
231. Lieber, C. M. and Lewis, N. S. Catalytic reduction of $CO_2$ at carbon electrodes modified with cobalt phthalocyanine. *Journal of the American Chemical Society.* 1984; 106(17):5033–5034.
232. Kapusta, S. and Hackerman, N. Carbon dioxide reduction at a metal phthalocyanine catalyzed carbon electrode. *Journal of the Electrochemical Society.* 1984; 131(7):1511–1514.
233. Yoshida, T. et al. Selective electrocatalysis for $CO_2$ reduction in the aqueous phase using cobalt phthalocyanine/poly-4-vinylpyridine modified electrodes. *Journal of Electroanalytical Chemistry.* 1995; 385(2):209–225.
234. Abe, T. et al. Factors affecting selective electrocatalytic $CO_2$ reduction with cobalt phthalocyanine incorporated in a polyvinylpyridine membrane coated on a graphite electrode. *Journal of Electroanalytical Chemistry.* 1996; 412(1–2):125–132.

235. Abe, T. et al. Electrocatalytic $CO_2$ reduction by cobalt octabutoxyphthalocyanine coated on graphite electrode. *Journal of Molecular Catalysis A – Chemical*. 1996; 112(1):55–61.
236. Zhang, J. J., Pietro, W. J. and Lever, A. B. P. Rotating ring-disk electrode analysis of $CO_2$ reduction electrocatalyzed by a cobalt tetramethylpyridoporphyrazine on the disk and detected as co on a platinum ring. *Journal of Electroanalytical Chemistry*. 1996; 403(1–2):93–100.
237. Zhao, H., Zhang, Y., Zhao, B., Chang, Y. and Li, Z. Electrochemical reduction of carbon dioxide in an MFC–MEC system with a layer-by-layer self-assembly carbon nanotube/cobalt phthalocyanine modified electrode. *Environmental Science & Technology*. 2012; 46(9):5198–5204.
238. Isaacs, M. et al. Electrocatalytic reduction of $CO_2$ by aza-macrocyclic complexes of Ni(ii), Co(ii), and Cu(ii). Theoretical contribution to probable mechanisms. *Inorganica Chimica Acta*. 2002; 339:224–232.
239. Isaacs, M. et al. Contribution of the ligand to the electroreduction of $CO_2$ catalyzed by a cobalt(II) macrocyclic complex. *Journal of Coordination Chemistry*. 2003; 56(14):1193–1201.
240. Grodkowski, J. et al. Reduction of cobalt and iron corroles and catalyzed reduction of $CO_2$. *Journal of Physical Chemistry A*. 2002; 106(18):4772–4778.
241. Beley, M., Collin, J. P., Ruppert, R. and Sauvage, J. P. Electrocatalytic reduction of $CO_2$ by Ni cyclam$^{2+}$ in water—Study of the factors affecting the efficiency and the selectivity of the process. *Journal of the American Chemical Society*. 1986; 108(24):7461–7467.
242. Fujihira, M., Hirata, Y. and Suga, K. Electrocatalytic reduction of $CO_2$ by nickel(II) cyclam—Study of the reduction-mechanism on mercury by cyclic voltammetry, polarography and electrocapillarity. *Journal of Electroanalytical Chemistry*. 1990; 292(1–2):199–215.
243. Sakaki, S. An ab initio MO/SD-CI study of model complexes of intermediates in electrochemical reduction of $CO_2$ catalyzed by NiCl$_2$(cyclam). *Journal of the American Chemical Society*. 1992; 114(6):2055–2062.
244. Balazs, G. B. and Anson, F. C. The adsorption of Ni(cyclam)$^+$ at mercury-electrodes and its relation to the electrocatalytic reduction of $CO_2$. *Journal of Electroanalytical Chemistry*. 1992; 322(1–2):325–345.
245. Balazs, G. B. and Anson, F. C. Effects of CO on the electrocatalytic activity of Ni (cyclam)(2+) toward the reduction of $CO_2$. *Journal of Electroanalytical Chemistry*. 1993; 361(1–2):149–157.
246. Hirata, Y., Suga, K. and Fujihira, M. In situ analysis of products in electrocatalytic reduction of $CO_2$ with Ni-cyclam by differential electrochemical mass-spectroscopy during cyclic voltammetry on an amalgamated-gold mesh electrode. *Chemistry Letters*. 1990; 19(7):1155–1158.
247. Collin, J. P., Jouaiti, A. and Sauvage, J. P. Electrocatalytic properties of Ni(cyclam)$^{2+}$ and Ni$_2$(biscyclam)$^{4+}$ with respect to $CO_2$ and $H_2O$ reduction. *Inorganic Chemistry*. 1988; 27(11):1986–1990.
248. Hay, R. W., Crayston, J. A., Cromie, T. J., Lightfoot, P. and deAlwis, D. C. L. The preparation, chemistry and crystal structure of the nickel(II) complex of N-hydroxyethylazacyclam 3-(2′-hydroxyethyl)-1,3,5,8,12-penta-azacyclotetradecane nickel(II) perchlorate. A new electrocatalyst for $CO_2$ reduction. *Polyhedron*. 1997; 16(20):3557–3563.
249. Schneider, J. et al. Nickel(II) macrocycles: Highly efficient electrocatalysts for the selective reduction of $CO_2$ to CO. *Energy & Environmental Science*. 2012; 5(11):9502–9510.
250. Froehlich, J. D. and Kubiak, C. P. Homogeneous $CO_2$ reduction by Ni(cyclam) at a glassy carbon electrode. *Inorganic Chemistry*. 2012; 51(7):3932–3934.
251. Abba, F. et al. Nickel(II) complexes of azacyclams—Oxidation and reduction behavior and catalytic effects in the electroreduction of carbon dioxide. *Inorganic Chemistry*. 1994; 33(7):1366–1375.

252. Jarzebinska, A. et al. Modified electrode surfaces for catalytic reduction of carbon dioxide. *Analytica Chimica Acta.* 1999; 396(1):1–12.

253. Fisher, B. and Eisenberg, R. Electrocatalytic reduction of carbon dioxide by using macrocycles of nickel and cobalt. *Journal of the American Chemical Society.* 1980; 102(24):7361–7363.

254. Fujita, E., Haff, J., Sanzenbacher, R. and Elias, H. High electrocatalytic activity of RRSS-[Ni(II)Htim)(ClO₄)₂ and [Ni(II)DMC (ClO₄)₂ for carbon dioxide reduction (HTIM= 2,3,9,10-tetramethyl-1,4,8,11-tetraazacyclotetradecane, DMC=C-*meso*-5,12-dimethyl-1,4,8,11-tetraazacyclotetradecane). *Inorganic Chemistry.* 1994; 33(21):4627–4628.

255. Bujno, K., Bilewicz, R., Siegfried, L. and Kaden, T. Electrochemical behaviour of isomers of a tetraazamacrocyclic Ni(II) complex in solutions saturated with argon, CO and CO₂. *Journal of Electroanalytical Chemistry.* 1996; 407(1–2):131–140.

256. Rudolph, M., Dautz, S. and Jager, E. G. Macrocyclic [N4²⁻] coordinated nickel complexes as catalysts for the formation of oxalate by electrochemical reduction of carbon dioxide. *Journal of the American Chemical Society.* 2000; 122(44):10821–10830.

257. Guadalupe, A. R. et al. Novel chemical pathways and charge-transport dynamics of electrodes modified with electropolymerized layers of [Co(v-terpy)2]2+. *Journal of the American Chemical Society.* 1988; 110(11):3462–3466.

258. Arana, C., Yan, S., Keshavarzk, M., Potts, K. T. and Abruna, H. D. Electrocatalytic reduction of carbon dioxide with iron, cobalt, and nickel-complexes of terdentate ligands. *Inorganic Chemistry.* 1992; 31(17):3680–3682.

259. Arana, C., Keshavarz, M., Potts, K. T. and Abruna, H. D. Electrocatalytic reduction of CO₂ and O₂ with electropolymerized films of vinyl-terpyridine complexes of Fe, Ni and Co. *Inorganica Chimica Acta.* 1994; 225(1–2):285–295.

260. Aga, H., Aramata, A. and Hisaeda, Y. The electroreduction of carbon dioxide by macrocyclic cobalt complexes chemically modified on a glassy carbon electrode. *Journal of Electroanalytical Chemistry.* 1997; 437(1–2):111–118.

261. Chiericato, G., Arana, C. R., Casado, C., Cuadrado, I. and Abruna, H. D. Electrocatalytic reduction of carbon dioxide mediated by transition metal complexes with terdentate ligands derived from diacetylpyridine. *Inorganica Chimica Acta.* 2000; 300:32–42.

262. Pun, S. N. et al. Iron(I) complexes of 2,9-bis(2-hydroxyphenyl)-1,10-phenanthroline (H₂dophen) as electrocatalysts for carbon dioxide reduction. X-ray crystal structures of [Fe(dophen)Cl]₂]•2HCON(CH₃)₂ and Fe(dophen)(N-Meim)₂ ClO₄ (N-Meim=1-methylimidazole). *Journal of the Chemical Society – Dalton Transactions.* 2002; 4:575–583.

263. Pearce, D. J. and Pletcher, D. A study of the mechanism for the electrocatalysis of carbon dioxide reduction by nickel and cobalt square-planar complexes in solution. *Journal of Electroanalytical Chemistry.* 1986; 197(1–2):317–330.

264. Isse, A. A., Gennaro, A., Vianello, E. and Floriani, C. Electrochemical reduction of carbon dioxide catalyzed by [CoI(salophen)Li]. *Journal of Molecular Catalysis.* 1991; 70(2):197–208.

265. Losada, J. et al. Electrocatalytic reduction of O₂ and CO₂ with electropolymerized films of polypyrrole cobalt(II) Schiff-base complexes. *Journal of Electroanalytical Chemistry.* 1995; 398(1–2):89–93.

266. Averill, B. A., Herskovi.T, Holm, R. H. and Ibers, J. A. Synthetic analogs of active-sites of iron–sulfur proteins. 2. Synthesis and structure of tetra[mercapto-Mu₃-sulfido-iron] clusters, [Fe₄S₄(SR)₄]². *Journal of the American Chemical Society.* 1973; 95(11):3523–3534.

267. Tezuka, M. et al. Electroreduction of carbon dioxide catalyzed by iron–sulfur clusters [Fe₄S₄(SR)₄]². *Journal of the American Chemical Society.* 1982; 104(24):6834–6836.

268. Yuhas, B. D., Prasittichai, C., Hupp, J. T. and Kanatzidis, M. G. Enhanced electrocatalytic reduction of CO₂ with ternary Ni-Fe₄S₄ and Co-Fe₄S₄-based biomimetic chalcogels. *Journal of the American Chemical Society.* 2011; 133(40):15854–15857.

269. Whipple, D. T. and Kenis, P. J. A. Prospects of $CO_2$ utilization via direct heterogeneous electrochemical reduction. *Journal of Physical Chemistry Letters.* 2010; 1(24):3451–3458.

270. Gangeri, M. et al. Fe and Pt carbon nanotubes for the electrocatalytic conversion of carbon dioxide to oxygenates. *Catalysis Today.* 2009; 143(1–2):57–63.

271. Ampelli, C., Centi, G., Passalacqua, R. and Perathoner, S. Synthesis of solar fuels by a novel photoelectrocatalytic approach. *Energy & Environmental Science.* 2010; 3(3):292–301.

272. Arrigo, R. et al. New insights from microcalorimetry on the FeOx/CNT-based electrocatalysts active in the conversion of $CO_2$ to fuels. *ChemSusChem.* 2012; 5(3):577–586.

273. Centi, G. and Perathoner, S. Towards solar fuels from water and $CO_2$. *ChemSusChem.* 2010; 3(2):195–208.

274. Bocarsly, A. B. et al. Comparative study of imidazole and pyridine catalyzed reduction of carbon dioxide at illuminated iron pyrite electrodes. *ACS Catalysis.* 2012; 2(8):1684–1692.

275. Rail, M. D. and Berben, L. A. Directing the reactivity of $[HFe_4N(CO)_{12}]^-$ toward $H^+$ or $CO_2$ reduction by understanding the electrocatalytic mechanism. *Journal of the American Chemical Society.* 2011; 133(46):18577–18579.

276. Ogura, K. and Yamasaki, S. Conversion of carbon monoxide into methanol at room temperature and atmospheric pressure. *Journal of the Chemical Society – Faraday Transactions I.* 1985; 81:267–271.

277. Ogura, K. and Kaneko, M. Reduction of CO to methanol by Everitts salt using pentacyanoferrate(II) or pentachlorochromate(III) and methanol as homogeneous catalysts. *Journal of Molecular Catalysis.* 1985; 31(1):49–56.

278. Ogura, K. and Takamagari, K. Electrocatalytic reduction of carbon dioxide to methanol. 2. Effects of metal complex and primary alcohol. *Journal of the Chemical Society – Dalton Transactions.* 1986; 8:1519–1523.

279. Ogura, K. and Yoshida, I. Electrocatalytic reduction of carbon dioxide to methanol in the presence of 1,2-dihydroxybenzene-3,5-disulphonatoferrate(III) and ethanol. *Journal of Molecular Catalysis.* 1986; 34(1):67–72.

280. Ogura, K. and Fujita, M. Electrocatalytic reduction of carbon dioxide to methanol. 7. With quinone derivatives immobilized on platinum and stainless steel. *Journal of Molecular Catalysis.* 1987; 41(3):303–311.

281. Ogura, K. and Uchida, H. Electrocatalytic reduction of $CO_2$ to methanol. 8. Photoassisted electrolysis and electrochemical photocell with n-$TiO_2$ anode. *Journal of Electroanalytical Chemistry.* 1987; 220(2):333–337.

282. Ogura, K. Catalytic conversion of carbon monoxide and carbon dioxide into methanol with photocells. *Journal of the Electrochemical Society.* 1987; 134(11):2749–2754.

283. Ogura, K., Migita, C. T. and Imura, H. Catalytic reduction of carbon dioxide with a hydrogen fuel cell. *Journal of the Electrochemical Society.* 1990; 137(6):1730–1732.

284. Ogura, K., Migita, C. T. and Wadaka, K. Homogeneous catalysis in the mediated electrochemical reduction of carbon dioxide. *Journal of Molecular Catalysis.* 1991; 67(2):161–173.

285. Ogura, K., Mine, K., Yano, J. and Sugihara, H. Electrocatalytic generation of $C_2$ and $C_3$ compounds from carbon dioxide on a cobalt complex-immobilized dual-film electrode. *Journal of the Chemical Society – Chemical Communications.* 1993; 1(1):20–21.

286. Ogura, K., Sugihara, H., Yano, J. and Higasa, M. Electrochemical reduction of carbon dioxide on dual-film electrodes modified with and without cobalt(II) and iron(II) complexes. *Journal of the Electrochemical Society.* 1994; 141(2):419–424.

287. Ogura, K., Nakayama, M. and Kusumoto, C. In situ fourier transform infrared spectroscopic studies on a metal complex-immobilized polyaniline Prussian blue modified electrode and the application to the electroreduction of $CO_2$. *Journal of the Electrochemical Society.* 1996; 143(11):3606–3615.

288. Ogura, K., Endo, N., Nakayama, M. and Ootsuka, H. Mediated activation and electroreduction of $CO_2$ on modified electrodes with conducting polymer and inorganic conductor films. *Journal of the Electrochemical Society.* 1995; 142(12):4026–4032.

289. Nakayama, M., Iino, M. and Ogura, K. In situ infrared spectroscopic investigations on the electrochemical properties of Prussian blue-polyaniline-modified electrodes with various anionic Fe(II) complexes working as a mediator for the electroreduction of $CO_2$. *Journal of Electroanalytical Chemistry.* 1997; 440(1–2):251–257.

290. Braunstein, P., Matt, D. and Nobel, D. Reactions of carbon dioxide with carbon carbon bond formation catalyzed by transition-metal complexes. *Chemical Reviews.* 1988; 88(5):747–764.

291. Ratliff, K. S., Lentz, R. E. and Kubiak, C. P. Carbon dioxide chemistry of the trinuclear complex [$Ni_3$(mu-3-CNMe)(mu-3-i)(dppm)3][PF6]—Electrocatalytic reduction of carbon dioxide. *Organometallics.* 1992; 11(6):1986–1988.

292. Simon-Manso, E. and Kubiak, C. P. Dinuclear nickel complexes as catalysts for electrochemical reduction of carbon dioxide. *Organometallics.* 2005; 24(1):96–102.

293. Lee, E. Y., Hong, D. W., Park, H. W. and Suh, M. P. Synthesis, properties, and reactions of trinuclear macrocyclic nickel(II) and nickel(I) complexes: Electrocatalytic reduction of $CO_2$ by nickel(II) complex. *European Journal of Inorganic Chemistry.* 2003; 2003(17):3242–3249.

294. Field, J. S., Haines, R. J., Parry, C. J. and Sookraj, S. H. Dicopper(I) complexes of the novel phosphorusbipyridyl ligand 6-diphenylphosphino-2,2'-bipyridyl. *Polyhedron.* 1993; 12(19):2425–2428.

295. Haines, R. J., Wittrig, R. E. and Kubiak, C. P. Electrocatalytic reduction of carbon dioxide by the binuclear copper complex [$Cu_2$(6-(diphenylphosphino)-2,2'-bipyridyl)$_2$(MeCN)$_2$][PF6]$_2$. *Inorganic Chemistry.* 1994; 33(21):4723–4728.

296. Nallas, G. N. A. and Brewer, K. J. Electrocatalytic reduction of carbon dioxide by mixed-metal trimetallic complexes of the form {[(bpy)$_2$Ru(BL)]$_2$IrCl$_2$}$^{5+}$ where bpy = 2,2'-bipyridine and BL = 2,3-bis(2-pyridyl)quinoxaline (dpq) or 2,3-bis(2-pyridyl)benzoquinoxaline (dpb). *Inorganica Chimica Acta.* 1996; 253(1):7–13.

297. Raebiger, J. W. et al. Electrochemical reduction of $CO_2$ to co catalyzed by a bimetallic palladium complex. *Organometallics.* 2006; 25(14):3345–3351.

298. Angamuthu, R., Byers, P., Lutz, M., Spek, A. L. and Bouwman, E. Electrocatalytic $CO_2$ conversion to oxalate by a copper complex. *Science.* 2010; 327(5963):313–315.

299. Steffey, B. D. et al. Synthesis and characterization of palladium complexes containing tridentate ligands with PXP (X = C, N, O, S, As) donor sets and their evaluation as electrochemical $CO_2$ reduction catalysts. *Organometallics.* 1994; 13(12):4844–4855.

300. Shin, W., Lee, S., Shin, J., Lee, S. and Kim, Y. Highly selective electrocatalytic conversion of $CO_2$ to CO at −0.57 V (NHE) by carbon monoxide dehydrogenase from *Moorella thermoacetica*. *Journal of the American Chemical Society.* 2003; 125(48):14688–14689.

301. Ragsdale, S. W. Life with carbon monoxide. *Critical Reviews in Biochemistry and Molecular Biology.* 2004; 39(3):165–195.

302. Parkin, A., Seravalli, J., Vincent, K. A., Ragsdale, S. W. and Armstrong, F. A. Rapid and efficient electrocatalytic $CO_2$/CO interconversions by carboxydothermus hydrogenoformans CO dehydrogenase I on an electrode. *Journal of the American Chemical Society.* 2007; 129(34):10328–10329.

303. Bolinger, C. M., Story, N., Sullivan, B. P. and Meyer, T. J. Electrocatalytic reduction of carbon dioxide by 2,2'-bipyridine complexes of rhodium and iridium. *Inorganic Chemistry.* 1988; 27(25):4582–4587.

304. Chen, Z. et al. Electrocatalytic reduction of $CO_2$ to CO by polypyridyl ruthenium complexes. *Chemical Communications.* 2011; 47(47):12607–12609.

305. Paul, P. Ruthenium, osmium and rhodium complexes of polypyridyl ligands: Metal-promoted activities, stereochemical aspects and electrochemical properties. *Proceedings of the Indian Academy of Sciences – Chemical Sciences.* 2002; 114(4):269–276.

306. Slater, S. and Wagenknecht, J. H. Electrochemical reduction of $CO_2$ catalyzed by Rh(diphos)$_2$Cl. *Journal of the American Chemical Society.* 1984; 106(18):5367–5368.

307. Dubois, D. L. and Miedaner, A. Mediated electrochemical reduction of $CO_2$—Preparation and comparison of an isoelectronic series of complexes. *Journal of the American Chemical Society.* 1987; 109(1):113–117.

308. Ishida, H., Tanaka, H., Tanaka, K. and Tanaka, T. Selective formation of $HCOO^-$ in the electrochemical $CO_2$ reduction catalyzed by [Ru(bpy)$_2$(CO)$_2$]$^{2+}$ (bpy = 2,2'-bipyridine). *Journal of the Chemical Society – Chemical Communications.* 1987; 2:131–132.

309. Ishida, H., Tanaka, K. and Tanaka, T. Electrochemical $CO_2$ reduction catalyzed by Ru(bpy)$_2$(CO)$_2$ 2+ and Ru(bpy)$_2$(CO)Cl + —The effect of Ph on the formation of CO and HCOO. *Organometallics.* 1987; 6(1):181–186.

310. Chardonnoblat, S., Collombdunandsauthier, M. N., Deronzier, A., Ziessel, R. and Zsoldos, D. Formation of polymeric [{Ru$^0$(bpy)(CO)$_2$}$_n$] films by electrochemical reduction of [Ru(bpy)$_2$(CO)$_2$](PF$_6$)$_2$—Its implication in $CO_2$ electrocatalytic reduction. *Inorganic Chemistry.* 1994; 33(19):4410–4412.

311. Collombdunandsauthier, M. N., Deronzier, A. and Ziessel, R. Electrocatalytic reduction of carbon dioxide with mono(bipyridine)carbonylruthenium complexes in solution or as polymeric thin films. *Inorganic Chemistry.* 1994; 33(13):2961–2967.

312. Ali, M. M. et al. Selective formation of $HCO_2^-$ and $C_2O_4^{2-}$ in electrochemical reduction of $CO_2$ catalyzed by mono- and di-nuclear ruthenium complexes. *Chemical Communications.* 1998; 2:249–250.

313. Tanaka, K. and Mizukawa, T. Selective formation of ketones by electrochemical reduction of $CO_2$ catalyzed by ruthenium complexes. *Applied Organometallic Chemistry.* 2000; 14(12):863–866.

314. Ooyama, D., Tomon, T., Tsuge, K. and Tanaka, K. Structural and spectroscopic characterization of ruthenium(II) complexes with methyl, formyl, and acetyl groups as model species in multi-step $CO_2$ reduction. *Journal of Organometallic Chemistry.* 2001; 619(1–2):299–304.

315. Tanaka, K. and Ooyama, D. Multi-electron reduction of $CO_2$ via Ru-$CO_2$, -C(O)OH, -CO, -CHO, and -CH$_2$OH species. *Coordination Chemistry Reviews.* 2002; 226(1–2):211–218.

316. Begum, A. and Pickup, P. G. Electrocatalysis of $CO_2$ reduction by ruthenium benzothiazole and bithiazole complexes. *Electrochemistry Communications.* 2007; 9(10):2525–2528.

317. Bruce, M. R. M. et al. Electrocatalytic reduction of $CO_2$ by associative activation. *Organometallics.* 1988; 7(1):238–240.

318. Bruce, M. R. M. et al. Electrocatalytic reduction of carbon dioxide based on 2,2'-bipyridyl complexes of osmium. *Inorganic Chemistry.* 1992; 31(23):4864–4873.

319. Cheng, S. C., Blaine, C. A., Hill, M. G. and Mann, K. R. Electrochemical and IR spectroelectrochemical studies of the electrocatalytic reduction of carbon dioxide by Ir-2(dimen)(4) (2+) (dimen equals 1,8-diisocyanomenthane). *Inorganic Chemistry.* 1996; 35(26):7704–7708.

320. Hossain, A., Nagaoka, T. and Ogura, K. Electrocatalytic reduction of carbon dioxide by substituted pyridine and pyrazole complexes of palladium. *Electrochimica Acta.* 1996; 41(17):2773–2780.

321. Miedaner, A., Curtis, C. J., Barkley, R. M. and Dubois, D. L. Electrochemical reduction of $CO_2$ catalyzed by small organophosphine dendrimers containing palladium. *Inorganic Chemistry.* 1994; 33(24):5482–5490.

322. Tanaka, K. et al. Catalytic generation of oxalate through a coupling reaction of two $CO_2$ molecules activated on [(Ir(eta(5)-c5me5))(2)(ir(eta(4)-c5me5)ch2cn)(mu(3)-s)(2)]. *Inorganic Chemistry.* 1998; 37(1):120–126.

323. Jessop, P. G., Ikariya, T. and Noyori, R. Homogeneous catalytic-hydrogenation of supercritical carbon dioxide. *Nature.* 1994; 368(6468):231–233.

324. Jessop, P. G., Joo, F. and Tai, C. C. Recent advances in the homogeneous hydrogenation of carbon dioxide. *Coordination Chemistry Reviews.* 2004; 248(21–24):2425–2442.

325. Kang, P. et al. Selective electrocatalytic reduction of $CO_2$ to formate by water-stable iridium dihydride pincer complexes. *Journal of the American Chemical Society.* 2012; 134(12):5500–5503.

326. Oh, Y. and Hu, X. Organic molecules as mediators and catalysts for photocatalytic and electrocatalytic $CO_2$ reduction. *Chemical Society Reviews.* 2013; 42(6):2253–2261.

327. Koleli, F., Ropke, T. and Hamann, C. H. The reduction of $CO_2$ on polyaniline electrode in a membrane cell. *Synthetic Metals.* 2004; 140(1):65–68.

328. Aydin, R. and Koleli, F. Electrocatalytic conversion of $CO_2$ on a polypyrrole electrode under high pressure in methanol. *Synthetic Metals.* 2004; 144(1):75–80.

329. Smith, R. D. L. and Pickup, P. G. Nitrogen-rich polymers for the electrocatalytic reduction of $CO_2$. *Electrochemistry Communications.* 2010; 12(12):1749–1751.

330. Seshadri, G., Lin, C. and Bocarsly, A. B. A new homogeneous electrocatalyst for the reduction of carbon dioxide to methanol at low overpotential. *Journal of Electroanalytical Chemistry.* 1994; 372(1–2):145–150.

331. Cole, E. B. et al. Using a one-electron shuttle for the multielectron reduction of $CO_2$ to methanol: Kinetic, mechanistic, and structural insights. *Journal of the American Chemical Society.* 2010; 132(33):11539–11551.

332. Ertem, M. Z., Konezny, S. J., Araujo, C. M. and Batista, V. S. Functional role of pyridinium during aqueous electrochemical reduction of $CO_2$ on Pt(111). *Journal of Physical Chemistry Letters.* 2013; 4(5):745–748.

333. Keith, J. A. and Carter, E. A. Electrochemical reactivities of pyridinium in solution: Consequences for $CO_2$ reduction mechanisms. *Chemical Science.* 2013; 4(4):1490–1496.

334. Lim, C. H., Holder, A. M., Musgrave, C. B. Mechanism of homogeneous reduction of $CO_2$ by pyridine: Proton relay in aqueous solvent and aromatic stabilization. *Journal of the American Chemical Society.* 2013;135(1):142–54

335. Gennaro, A., Isse, A. A., Saveant, J. M., Severin, M. G. and Vianello, E. Homogeneous electron transfer catalysis of the electrochemical reduction of carbon dioxide. Do aromatic anion radicals react in an outer-sphere manner? *Journal of the American Chemical Society.* 1996; 118(30):7190–7196.

336. Snuffin, L. L., Whaley, L. W. and Yu, L. Catalytic electrochemical reduction of $CO_2$ in ionic liquid $EMIMBF_3Cl$. *Journal of the Electrochemical Society.* 2011; 158(9):F155–F158.

337. Sugimura, K., Kuwabata, S. and Yoneyama, H. Electrochemical fixation of carbon dioxide in oxoglutaric acid using an enzyme as an electrocatalyst. *Journal of the American Chemical Society.* 1989; 111(6):2361–2362.

338. Kuwabata, S., Morishita, N. and Yoneyama, H. Electrochemical fixation of $CO_2$ in acetyl-coenzyme A to yield pyruvic-acid using pyruvate-dehydrogenase complexes as an electrocatalyst. *Chemistry Letters.* 1990; 19(7):1151–1154.

339. Kuwabata, S., Tsuda, R. and Yoneyama, H. Electrochemical conversion of carbon dioxide to methanol with the assistance of formate dehydrogenase and methanol dehydrogenase as biocatalysts. *Journal of the American Chemical Society.* 1994; 116(12):5437–5443.

340. Addo, P. K. et al. Methanol production via bioelectrocatalytic reduction of carbon dioxide: Role of carbonic anhydrase in improving electrode performance. *Electrochemical and Solid State Letters.* 2011; 14(4):E9–E13.

341. Hansen, H. A., Varley, J. B., Peterson, A. A. and Norskov, J. K. Understanding trends in the electrocatalytic activity of metals and enzymes for $CO_2$ reduction to CO. *Journal of Physical Chemistry Letters.* 2013; 4(3):388–392.

342. Armstrong, F. A. and Hirst, J. Reversibility and efficiency in electrocatalytic energy conversion and lessons from enzymes. *Proceedings of the National Academy of Sciences of the United States of America.* 2011; 108(34):14049–14054.

343. Moura, J. J. G., Brondino, C. D., Trincao, J. and Romao, M. J. Mo and W bis-MGD enzymes: Nitrate reductases and formate dehydrogenases. *Journal of Biological Inorganic Chemistry.* 2004; 9(7):791–799.

344. de Bok, F. A. M. et al. Two W-containing formate dehydrogenases ($CO_2$-reductases) involved in syntrophic propionate oxidation by *Syntrophobacter fumaroxidans*. *European Journal of Biochemistry.* 2003; 270(11):2476–2485.

345. Li, H., Eddaoudi, M., O'Keeffe, M. and Yaghi, O. M. Design and synthesis of an exceptionally stable and highly porous metal-organic framework. *Nature.* 1999; 402(6759):276–279.

346. Chui, S. S. Y., Lo, S. M. F., Charmant, J. P. H., Orpen, A. G. and Williams, I. D. A chemically functionalizable nanoporous material $[Cu_3(TMA)_2(H_2O)_3]_n$. *Science.* 1999; 283(5405):1148–1150.

347. Eddaoudi, M. et al. Systematic design of pore size and functionality in isoreticular mofs and their application in methane storage. *Science.* 2002; 295(5554):469–472.

348. Chae, H. K. et al. A route to high surface area, porosity and inclusion of large molecules in crystals. *Nature.* 2004; 427(6974):523–527.

349. Ye, J.-Y. and Liu, C.-J. $Cu_3(BTC)_2$: Co oxidation over MOF based catalysts. *Chemical Communications.* 2011; 47(7):2167–2169.

350. Kumar, R. S., Kumar, S. S. and Kulandainathan, M. A. Efficient electrosynthesis of highly active $Cu_3(BTC)_2$-MOF and its catalytic application to chemical reduction. *Microporous and Mesoporous Materials.* 2013; 168:57–64.

351. Schwartz, M. et al. Carbon dioxide reduction to alcohols using perovskite-type electrocatalysts. *Journal of the Electrochemical Society.* 1993; 140(3):614–618.

# 5 Electrochemical Methods for CO$_2$ Electroreduction

*Yu Chen and Dongmei Sun*

## CONTENTS

Electrochemical reduction of carbon dioxide (ERC) provides a means to convert greenhouse carbon dioxide gas to produce diverse attractive chemicals and fuels such as methanol, formic acid, methane, ethylene, carbon monoxide, and other hydrocarbons by using electricity as the source of energy [1] as shown in Equation 5.1. This is attractive because it can be carried out at room temperature and low pressure, and potentially reduces the dependence on foreign fuels as well as mitigates the concentration of CO$_2$ in the atmosphere.

$$CO_2 + xe^{-1} + xH^+ \xrightarrow{\ eV\ } CO, HCOOH, CH_4, (HCOOH)_2 \tag{5.1}$$

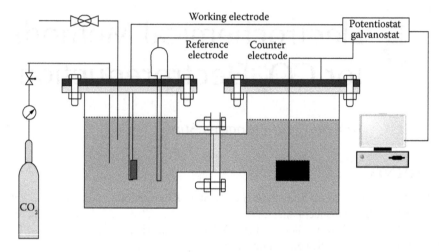

**FIGURE 5.1**  Schematic diagram of half-cell assembly. (Reproduced from Hossain, S. *Electrochemical Reduction of Carbon Dioxide to Hydrocarbons.* King Fahd University of Petroleum and Minerals, Saudi Arabia. 2011. With permission from ProQuest ILC.)

Nondestructive techniques, such as cyclic voltammetry (CV), are particularly useful for a first evaluation of the catalyst: determination of the potential at which the catalytic process can be run and of the catalytic efficiency as measured by the current density flowing through the electrode at this potential for a given concentration or partial pressure of the substrate. These techniques may then be used by means of a more detailed kinetic analysis to unravel the mechanism of the catalytic reaction and suggest improvements. Gauging of the selectivity of the catalytic reaction requires moving to preparative-scale electrolysis and determining the faradaic yields of each of the reaction products. Such sustained electrolyses are also necessary to estimate the stability of the catalyst by observing the variation of the preparative-scale current with time. Following the cyclic voltammetric response simultaneously is an additional way of observing the evolution of the catalyst in the course of electrolysis. Preparative-scale evaluation is thus required to establish the actual performances and viability of catalytic systems beyond the rapid test that CV allows.

Alternative methods for ERC investigation include steady-state polarization, linear sweep voltammetry, rotating electrode techniques, electrochemical impedance spectroscopy, and chronoamperometry (CA). Figure 5.1 schematically shows an assembly of a half-cell.

## 5.1  POTENTIOSTATIC OR GALVANOSTATIC STEADY-STATE POLARIZATION

### 5.1.1  Basic Concept

Electrode polarization is defined as the departure of an electrode potential from the equilibrium value upon passage of faradaic current. Current–potential curves under steady-state conditions are called polarization curves. For an ideal polarized

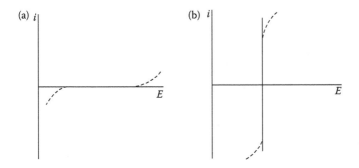

**FIGURE 5.2** Current–potential curves for ideal (a) polarizable and (b) nonpolarizable electrodes. Dashed lines show behavior of actual electrodes that approach the ideal behavior over limited ranges of current or potential. (Bard, A. J. and Faulkner, L. R.: *Electrochemical Methods Fundamentals and Applications.* 2nd edn. p. 22. 2001. Copyright Wiley-VCH Verlag GmbH & Co. KGaA. Reproduced with permission.)

electrode, an infinitesimal current induces a very large change in potential, characterized by a horizontal region of an *i–E* curve (Figure 5.2a). By contrast, for an ideal nonpolarizable electrode, where species is oxidized or reduced at its equilibrium potential value, the potential does not change at all upon passage of current, characterized by a vertical region of an *i–E* curve (Figure 5.2b) [3].

Steady-state polarization curves can be drawn by applying a series of potential/current in a certain step to an electrode, waiting for the response current (potentiostatic)/potential (galvanostatic) to stabilize, and then recording the corresponding data of *i* and *E*.

### 5.1.2 PRACTICAL APPLICATION FOR $CO_2$ ELECTROREDUCTION

Potentiostatic or galvanostatic steady-state polarization is investigated to estimate the efficiency and activity of various $CO_2$ reduction cathodes.

Figure 5.3 shows the simulated steady-state polarization curves in two different $CO_2$-saturated aqueous electrolytes of $KHCO_3$ and $NaClO_4$ both at 0.5 mol/L, where the hydrogen evolution reaction (HER) and CO evolution reaction (CER) were considered separately [4]. As observed, HER always occurs at a more positive potential compared to CER for both electrolytes and all possible reaction, indicating $H_2$ evolution is thermodynamically favored over CO evolution, in good agreement as reflected in the standard potentials. And, there appears obvious positive potential shift either for CO or $H_2$ evolution in $NaClO_4$ solution compared to $KHCO_3$ solution, as the pH value of aqueous $NaClO_4$ is ~3.91, lower than 7.27 of aqueous $KHCO_3$. However, the potential difference at equilibrium for $CO_2$ and $H_2O$ reduction is independent of pH as expected [4].

*In situ* electrodeposited inexpensive Bi catalyst can promote CO evolving (bismuth–carbon monoxide evolving catalyst [Bi-CMEC]) from $CO_2$ over electrochemical reduction with high activity and efficiency as shown in Figure 5.4 over potentiostatic steady-state polarization [5]. A significant lower overpotential compared to inexpensive cathode materials of Cu, Zn, Ni, and stainless steel (SS) is observed,

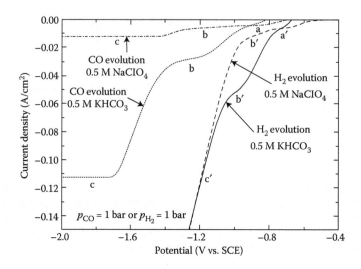

**FIGURE 5.3** Simulated steady-state current density as a function of cathode potential for HER (CER) when it is assumed equilibrated and CER (HER) is discarded ($p_{H_2} = 1$ bar when HER only; $p_{CO} = 1$ bar when CER only; thickness of the Nernst layer = 0.01 cm). (a) $3CO_2 + H_2O + 2e^- \rightleftharpoons CO(g) + 2HCO_3^-$, (b) $4HCO_3^- + 2e^- \rightleftharpoons CO(g) + 2H_2O + 3CO_3^{2-}$, (c) $CO_3^{2-} + 2H_2O + 2e^- \rightleftharpoons CO(g) + 4OH^-$, (a') $2CO_2 + 2H_2O + 2e^- \rightleftharpoons H_2(g) + 2HCO_3^-$, (b') $2HCO_3^- + 2e^- \rightleftharpoons H_2(g) + 2CO_3^{2-}$, and (c') $2H_2O + 2e^- \rightleftharpoons H_2(g) + 2OH^-$. Reactions (a), (b), and (c) and (a'), (b'), and (c') are written based on the predominate species, as functions of potential. (Delacourt, C., Ridgway, P.L. and Newman, J., *Journal of the Electrochemical Society.* 157(12), B1902–B1910, 2010. With permission from Electrochemical Society.)

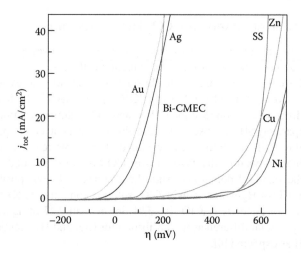

**FIGURE 5.4** Polarization curves recorded for several cathode materials for $CO_2$ reduction in $CO_2$-saturated MeCN with 300 mM [BMIM]OTf. (Reprinted with permission from Medina-Ramos, J., DiMeglio, J. L. and Rosenthal, J. Efficient reduction of $CO_2$ to CO with high current density using *in situ* or *ex situ* prepared bi-based materials. *Journal of the American Chemical Society.* 136(23): 8361–8367. Copyright 2014 American Chemical Society.)

while by contrast similar kinetics and a cathodic half reaction energy efficiency are obtained to that afforded by much more cost-prohibitive Au and Ag electrodes. In addition, the range of the rate of CO production by Bi-CMEC with a surface area of 1.0 cm² is at ~0.1–0.5 mmol/cm²/h for an applied overpotential of 250 mV.

The current–potential relations for the reduction of carbon dioxide to the formate anion in a neutral solution can be estimated through galvanostatic charging experiments as shown in the polarization curve of Figure 5.5 [6]. From that it is obvious that in region I at low overpotential with a Tafel slope of ~91 mV, the coverage by an adsorbed intermediate, which was suggested to be the formate radical of $HCO_{2ads}$, is large $(\theta \rightarrow 1)$, whereas in region II at high overpotential with a Tafel slope of ~240 mV, the coverage is small $(\theta \rightarrow 0)$. And for the concentration of carbon dioxide, the reaction order was found to be almost of zero and one in regions I and II, respectively. The reaction mechanism is proposed as follows [7]. In region I of the polarization curve, the rate-determining step is described by Equation 5.4, while in region II, it is the first step described by Equation 5.2.

$$CO_2 + e^- \rightarrow \cdot CO_{2\,ads}^- \tag{5.2}$$

$$\cdot CO_{2\,ads}^- + H_2O \xrightarrow{\text{fast}} HCO_{2ads} + OH^- \tag{5.3}$$

$$HCO_{2ads} + e^- \rightarrow HCO_2^- \tag{5.4}$$

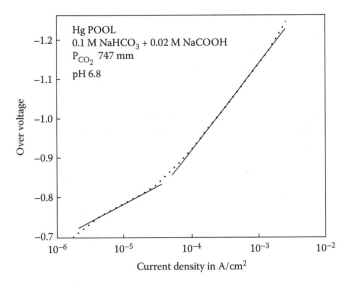

**FIGURE 5.5** Steady-state polarization curve for the reduction of carbon dioxide to the formate anion in a neutral solution. The overvoltage for the reduction of carbon dioxide is obtained by subtracting out the reversible potential which exists for the conditions of the experiment. The curve has been corrected for IR effects. (Reprinted from Russell, P. et al. The electrochemical reduction of carbon dioxide, formic acid, and formaldehyde. *Journal of the Electrochemical Society.* 1977; 124(9): 1329–1338, with permission from the Electrochemical Society.)

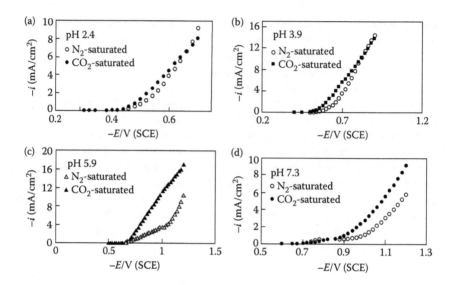

**FIGURE 5.6** Steady-state polarization curves in the absence and in the presence of dissolved $CO_2$ at several values of the pH: (a) pH = 2.4; (b) pH = 3.9; (c) pH = 5.9; and (d) pH = 7.3. (Reproduced from Spataru, N. et al., *Journal of applied Electrochemistry*, 33(12), 1205–1210, 2003. With permission from Kluwer.)

Water may be included in the first step, leading to the formation of $HCO_{2ads}$ [8]. For this case, the first step was the combination of Equations 5.2 and 5.3 [9].

The influence of the pH on the electrochemical reduction of $CO_2$ at the $RuO_2$-coated boron-doped diamond (BDD) electrodes was studied by potentiostatic steady-state polarization as shown in Figure 5.6 [10]. As observed in the pH and potential range investigated, the reduction of carbon dioxide is virtually always accompanied by hydrogen evolution during the whole process of interest. Prior to hydrogen evolution onset, there is no significant $CO_2$ reduction. By comparison, the best efficiency for $CO_2$ reduction can reach 80% for an applied potential of ∼−0.55 V at pH 3.9, probably suitable for practical purposes as the pH of a saturated carbonic acid aqueous solution is ~3.7 as well. At pH 5.9, wider potential range for efficient $CO_2$ reduction is achieved, allowing galvanostatic control during electrolysis, which is more convenient than potentiostatic control as usual. At neutral pH (Figure 5.6d), current densities for $CO_2$ reduction are reasonably high, and less affected by hydrogen evolution different from that in acidic media (Figure 5.6a and b) [10].

## 5.2 LINEAR POTENTIAL SWEEP VOTAMMETRY

### 5.2.1 BASIC CONCEPT

Linear sweep voltammetry (LSV) is developed to gain more information in a single experiment by linearly sweeping the potential with time at scan rates $v$ ranging from 10 mV/s to about 1000 V/s for conventional electrodes and up to $10^6$ V/s

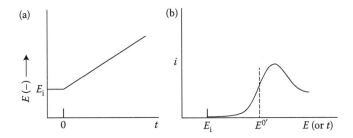

**FIGURE 5.7** (a) Potential scheme as a function of time for linear sweep voltammetry starting at $E_i$. (b) Resulting $i$–$E$ curve. (Bard, A. J. and Faulkner, L. R.: *Electrochemical Methods Fundamentals and Applications*. 2nd edn. p. 227. 2001. Copyright Wiley-VCH Verlag GmbH & Co. KGaA. Reproduced with permission.)

for ultramicroelectrode (UMEs), and recording the $i$–$E$ curve directly. A schematic waveform of the potential applied in a typical LSV is given in Figure 5.7a.

A typical LSV response curve for the reduction of redox active species is shown in Figure 5.7b. At initial electrode potential well positive of $E^{0'}$ for the reduction, only nonfaradaic current flows. When the potential continues to move negatively, the reduction begins at the vicinity of $E^{0'}$ and the current increases with the decrease of the growing potential. As the potential passes $E^{0'}$, the surface concentration of the active analyte drops nearly to zero, and the mass transfer to the electrode surface reaches a maximum rate, and then it declines as the depletion effect sets in, resulting in a peaked current–potential curve as depicted.

For a Nernstian wave, a convenient diagnostic between the peak potential, $E_p$ and half-peak potential, $E_{p/2}$ is expressed by Equation 5.5, Thus, $E_p$ is independent of

$$| E_p - E_{p/2} | = 2.20 \frac{RT}{nF} = \frac{56.5}{n} \text{ mV at 25°C} \tag{5.5}$$

the scan rate. The peak current $i_p$, described by the Randles–Sevčik Equation 5.7 as shown below in Section 5.3, is proportional to $v^{1/2}$, indicating a diffusion-controlled process. A convenient constant of $i_p/v^{1/2}C^*$ depends on $n^{3/2}$ and $D^{1/2}$ and can be used to estimate $n$ for an electrode reaction with a known $n$ value, if a value of $D$ can be estimated [3].

## 5.2.2 PRACTICAL APPLICATION FOR CO$_2$ ELECTROREDUCTION

Linear sweep voltammograms were recorded to determine a potential region of CO$_2$ reduction on a catalyst electrode.

Four post-transition-metal catalysts (Bi, Sn, Pb, and Sb), on glassy carbon or nickel electrodes, drive the formation of CO from CO$_2$ in the presence of [BMIM]OTf differently as shown in Figure 5.8 by LSVs [11]. The onset polarization potentials vary from ~−1.8 to −1.95 V. Of the four electrodeposited catalyst films, the Bi-based cathode shows the earliest onset potential of ~−1.8 V, followed by Sn, and Sb at ~−1.85 V

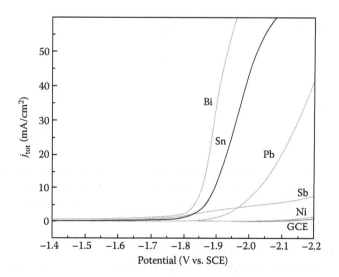

**FIGURE 5.8** Linear sweep voltammograms recorded for electrodeposited cathode materials in $CO_2$-saturated MeCN containing 100 mM [BMIM]OTf. (Reprinted with permission from Medina-Ramos, J. et al. Efficient conversion of $CO_2$ to CO using tin and other inexpensive and easily prepared post-transition metal catalysts. *Journal of the American Chemical Society*. 137(15), 5021–5027. Copyright 2015 American Chemical Society.)

and Pb at −1.95 V. Consequently, Sn and Bi catalysts proved to selectively promote rapid CO evolution with partial current densities of $j_{CO} = 5 - 8$ mA/cm² at applied overpotentials of $\eta < 250$ mV, whereas the ability of Pb and Sb catalysts to proficient carbon monoxide evolving was poor.

Reductive LSV scans at a reduced nano-$SnO_2$/carbon black coated glassy carbon electrode (Figure 5.9d) in 0.1 M $NaHCO_3$ indicate that the current density for $CO_2$ reduction can reach 6.2 mA/cm² at −1.8 V, which further increases to as high as 13.1 mA/cm² under the same conditions when these $SnO_2$ nanoparticles (NPs) were loaded onto graphene. By comparison, both a bare glassy carbon electrode (Figure 5.9a) and carbon black (Figure 5.9b) have negligible catalytic currents for $CO_2$ reduction, whereas electrodeposited ~200 nm tin NPs on glassy carbon provides obvious catalytic reduction current to a density of 3.6 mA/cm² at −1.8 V for $CO_2$ (Figure 5.9c). The specific current densities (with catalytic currents normalized to the mass of Sn catalysts) were calculated to be 266, 126, and 2 A/g for nano-$SnO_2$/graphene, nano-$SnO_2$/carbon black, and electrodeposited tin NPs, respectively [12].

The molybdenum(Mo)-terminated molybdenum disulfide ($MoS_2$) with abundant metallic-like d electrons in its edge states exhibits superior electrocatalytical activity to $CO_2$ reduction compared with the bulk Ag and Ag NPs (40 nm in average diameter) in $CO_2$-saturated 96 mol% water and 4 mol% EMIM-$BF_4$ solution (pH = 4) as demonstrated by linear sweep voltammetry shown in Figure 5.10a [13]. The $CO_2$ reduction to CO initiates at −0.164 V with faradaic efficiency (FE) at ~3%, suggesting a very low overpotential of 54 mV as the equilibrium potential is at −0.11 V in

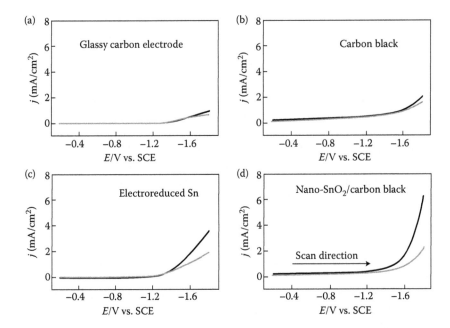

**FIGURE 5.9** (a) Single reductive linear sweep voltammetric scans at 50 mV/s under $N_2$ (gray line) and in a $CO_2$ (1 atm, black line)-saturated solution in aqueous 0.1 M $NaHCO_3$ at a bare glassy carbon electrode (a), glassy carbon electrode with carbon black (b), electrodeposited tin particles (c), and reduced nano-$SnO_2$/carbon black (d). (Reprinted with permission from Zhang, S., Kang, P. and Meyer, T. J. Nanostructured tin catalysts for selective electrochemical reduction of carbon dioxide to formate. *Journal of the American Chemical Society.* 136(5), 1734–1737. Copyright 2014 American Chemical Society.)

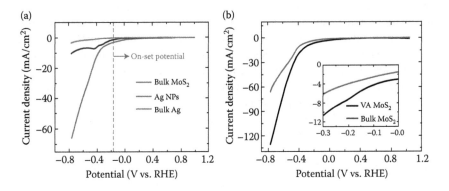

**FIGURE 5.10** (a) Linear sweep voltammetric curves for bulk $MoS_2$, Ag NPs and bulk Ag in $CO_2$-saturated 96 mol% water and 4 mol% EMIM-BF$_4$ solution (pH = 4). The vertical gray line indicates the low overpotential (~54 mV) for $CO_2$ reduction at bulk $MoS_2$. (b) $CO_2$ reduction performance of bulk $MoS_2$ and VA $MoS_2$. (Reprinted by permission from Macmillan Publishers Ltd. *Nature Communications*, Asadi, M. et al. Robust carbon dioxide reduction on molybdenum disulphide edges. 5: 4470, copyright 2014.)

the protic media [14,15]. When the potential increases to −0.764 V, $CO_2$ is selectively converted to CO (FE ~ 98%) at $MoS_2$ with a significantly high reduction current density of 65 mA/cm$^2$, whereas at Ag NPs catalyst with only a current density of 10 mA/cm$^2$ and 65% selectivity, and at bulk Ag with current density of 3 mA/cm$^2$ but for the $H_2$ formation [13]. This is in good agreement with that Mo-edged $MoS_2$ has significantly low work function of 3.9 eV compared to that of the bulk Ag (5.37 eV) and Ag NPs (5.38 eV) [16], indicating that the high current density and low over-potential (54 mV) are mainly due to the metallic character and the high d-electron density of the $MoS_2$ edges. Further study demonstrates that the EMIM$^+$ cation plays a crucial role to high $CO_2$ reduction performance by reducing the reaction barrier for electrons passing into $CO_2$ through *in situ* formation of [EMIM–$CO_2$]$^+$ complex, which could be adsorbed on the surface of negatively charged $MoS_2$ via coulombic and van der Waals coupling [14,16,17]. In addition, vertically aligned $MoS_2$ (VA $MoS_2$) catalyze $CO_2$ reduction similar to bulk $MoS_2$ in the same experimental condi-tion at low overpotential of 54 mV but with improved performance within complete applied potential range as shown in Figure 5.10b [13]. Two times higher $CO_2$ reduc-tion current density is observed in low potential region for VA $MoS_2$, whereas in high potential region at −0.764 V, a remarkably high current density of 130 mA/cm$^2$ was recorded compared with the bulk $MoS_2$, probably attributed to the high density of active sites preferably Mo atoms available [13].

## 5.3   CYCLIC VOLTAMMETRY

### 5.3.1   BASIC CONCEPT

CV is probably the most widely recognized and utilized electrochemical method due to its versatility, nondestructive nature, and rich information that can be extracted relatively quickly, with a wide range of applications in organic and inorganic chem-istry as well as materials development. The technique is often employed for initial investigations into prospective catalysts as it is quick and effective at providing useful data regarding electron transfers and allows identification of potentials required for various transformations. It is nondestructive as CV only involves the electrolysis of a small portion of the solution. The current is measured as a potential sweep is applied from an initial potential ($E_i$), generally where no net reaction, therefore minimal cur-rent, is expected and the composition of the analyte solution is not affected, to a cho-sen potential ($E_s$), relevant to the species being investigated. This overpotential can be oxidative or reductive depending upon the nature of the species under examina-tion but is scanned so that the concentration of the redox active substrate concentra-tion at the electrode surface is close to zero, passing through the standard potentials of redox couples studied, $E^0$, where $E_i < E^0 < E_s$. Once $E_s$ is reached, the potential sweep direction is reversed and returned to $E_i$, with any reverse reactions taking place observable in the current response obtained on the reverse sweep. A schematic of the potential waveform used in a typical CV is given in Figure 5.11.

The recorded current response is plotted against the potential to give a cyclic voltammogram, which shows current enhancements and peaks when electrochemi-cal oxidation and reduction take place. An example is given in Figure 5.12.

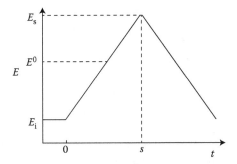

**FIGURE 5.11** Potential scheme as a function of time for CV. (Bard, A. J. and Faulkner, L. R.: *Electrochemical Methods: Fundamentals and Applications*. 2nd edn. p. 227. 2001. Copyright Wiley-VCH Verlag GmbH & Co. KGaA. Reproduced with permission.)

As seen in Figure 5.12, initially, at carefully selected potential of $E_i$, no current flows. However, when the potential is scanned closer to the standard potential of the electrochemical reaction, a current is produced which increases with potential to give a peak, where the maximum current ($I_{pa}$) is observed at the potential of $E_{pa}$ in a nonstirred homogeneous solution containing excess background electrolyte without convection and migration. Upon passing $E_{pa}$, the current response drops. The appearance of the peak is due to mass transport effects, where the movement of the substrate to the electrode surface is limited by the rate of diffusion which is slow compared with the electron-transfer step. Once high-enough overpotentials are applied for the redox process, the concentration of the substrate near the electrode surface drops to zero with the current response decaying as the depletion layer grows. The diffusion

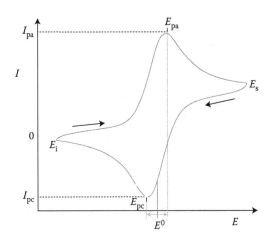

**FIGURE 5.12** Typical reversible cyclic voltammogram. Arrows indicate the direction of the potential sweep. (Bard, A. J. and Faulkner, L. R.: *Electrochemical Methods: Fundamentals and Applications*. 2nd edn. p. 227. 2001. Copyright Wiley-VCH Verlag GmbH & Co. KGaA. Reproduced with permission.)

layer thickness depends upon the experimental timescale but is usually of the order of μm to mm in width, and is the region of solution where the concentration varies from that of the bulk solution [18]. The voltammetric behavior prior to the diffusion limitation is described by the Nernst equation (5.6) [3]:

$$E = E^0 + \frac{RT}{nF} \ln\left(\frac{C_O}{C_R}\right)$$

(5.6)

This equation describes the concentration changes near the electrode, with $C_O/C_R$ being the concentration of oxidized/reduced redox species. $E$ is the applied potential, $E^0$ is the standard redox potential, and $R, T, n$, and $F$ are the ideal gas constant, temperature, number of electrons transferred per molecule, and Faraday's constant, respectively. The logarithmic relationship between the concentration change and potential is responsible for the initial rapid exponential growth in current. Upon reversal of the scan, for an electrochemically reversible system, the opposite processes occur with the product of the forward scan being reduced or oxidized back to the starting material as $E^0$ is reached and exceeded, giving a similar form of current response but of the opposite sign. For a completely reversible process, the ratio of the oxidative and reductive peak currents, $I_{pa}/-I_{pc}$, equals unity. The peak current for diffusion-limited processes is described by the Randles–Sevčik Equation 5.7 below, where the peak current ($I_p$) of a reversible system is dependent on the number of electrons transferred ($n$), the area of the electrode ($A$), the diffusion coefficient ($D$), the concentration of the redox species in the solution ($C$), and the scan rate ($v$) [3].

$$I_p = 0.4463 \left(\frac{n^3 F^3}{RT}\right)^{1/2} AD^{1/2} C v^{1/2}$$

(5.7)

When standard electrochemical units ($A$ in cm$^2$, $v$ in V/s, $D$ in cm$^2$/s, and $C$ in mol/cm$^3$) are used at 25°C, the Randles–Sevčik equation can be expressed as

$$I_p = 2.69 \times 10^5 n^{3/2} AD^{1/2} C v^{1/2}$$

(5.8)

The degree of reversibility can be assessed using CV through evaluation of the potentials as well as currents observed. If the peak separations of a redox couple increase with scan rate, this is indicative of quasi-reversible kinetics. If however the peaks associated with the forward and backward processes have no overlap on the potential axis or, as found for chemically irreversible processes there is a complete absence of a back-peak, the electron transfer is irreversible. The peak separation of a reversible one-electron process at 25°C is ~59 mV and independent of the scan rate. If $n$ electrons are transferred, the separation is described by Equation 5.9 [19]:

$$\Delta E = E_{pa} - E_{pc} = 2.218 \frac{RT}{nF}$$

(5.9)

The rate of electron transfer between the electrode and substrate, $k_f$, can be evaluated via voltammetry as it also shows a potential dependence, as described by the Butler–Volmer relation, Equation 5.10 [20]:

$$k_f = k^0 e^{[-(\alpha n F/RT)(E-E^0)]} \tag{5.10}$$

where $k^0$ is the standard heterogeneous rate constant and $\alpha$ is the transfer coefficient. The transfer coefficient represents the sensitivity of the transition state to the shift in potential upon moving from the electrode reactant to the product with possible values 0–1. In electron transfers without significant structural changes or alteration in solvation, this value is ~0.5 [20].

### 5.3.2 PRACTICAL APPLICATION FOR $CO_2$ ELECTROREDUCTION

Cyclic voltammograms are recorded at the Pt/C-based membrane electrode using a single cell equipped with a reference electrode of dynamic hydrogen electrode (DHE), to investigate the $CO_2$ reduction reaction as shown in Figure 5.13 [21]. Under the $CO_2$ atmosphere, there appear three characteristic peaks, designated as regions 1, 2, and 3, where region 1/2 located at 0.06–0.30 V versus DHE, the same as that under the $N_2$ atmosphere, corresponds to H-adsorption/desorption [22]. It is worth noting that the reduction current in the H-adsorption region under the $CO_2$ atmosphere is almost as high as that under the $N_2$ atmosphere, whereas the H-desorption current under the $CO_2$ atmosphere is much lower than that under the $N_2$ atmosphere. And a new peak in region 3 was observed under the $CO_2$ atmosphere. These results indicate that the $CO_2$ reduction occurs in region 1, competing with the H-adsorption reaction. And the product or intermediate of $CO_2$ reduction can be reoxidized in region 3 [21].

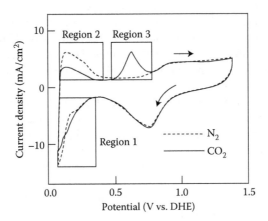

**FIGURE 5.13** Cyclic voltammograms obtained under $N_2$ (dotted line) and $CO_2$ (solid line) atmospheres for the Pt/C-based membrane electrode. Cell temperature: 40°C, scan rate: 10 mV/s. (Reprinted from *Journal of Power Sources*, 228(0), Shironita, S. et al., Feasibility investigation of methanol generation by $CO_2$ reduction using Pt/C-based membrane electrode assembly for a reversible fuel cell, 68–74, Copyright 2013, with permission from Elsevier.)

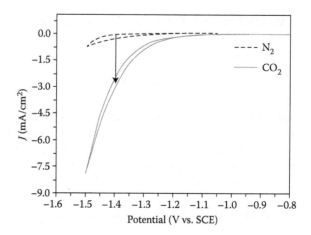

**FIGURE 5.14** Cyclic voltammograms of the $Cu_3Pt$ catalyst recorded in $N_2$- and $CO_2$-saturated 0.5 M $KHCO_3$ with a scan rate of 10 mV/s between −0.8 and −1.5 V (vs. SCE). (Guo, X. et al. Composition dependent activity of Cu–Pt nanocrystals for electrochemical reduction of $CO_2$. *Chemical Communications*. 2015; 51(7): 1345–1348. Reproduced by permission of The Royal Society of Chemistry.)

The $Cu_3Pt$ nanocrystal [23] ($Cu_3Pt$ NC with atomic ratio of Cu:Pt = 3:1) exhibited excellent catalytic activity toward $CO_2$ reduction as shown in Figure 5.14 over CV. It is observed that the reduction current in $CO_2$-saturated solution is much higher than that of a competing process of hydrogen evolution reaction in $N_2$-saturated solution, which strongly indicates that the reduction process is dominated by $CO_2$ reduction and the presence of $CO_2$ seems to inhibit the HER at potentials lower than −1.1 V. The current density of $CO_2$ reduction outcompeting HER could be simply calculated by subtracting the current density obtained at the dashed lines (HER current) from that recorded at the solid lines (the overall current of HER and $CO_2$ reduction), as inferred by the arrow lines in the figures. The onset potential is ~−0.972 V and the $CO_2$ reduction current density is ~0.598 mA/cm² at −1.3 V versus saturated calomel electrode (SCE).

Metal-free carbon nanofibers (CNFs) catalyst for $CO_2$ electroreduction was examined by CV in pure ionic liquid of EMIM-BF₄ by sweeping the applied voltage from 0.79 to −1.14 V versus SHE as shown in Figure 5.15 [24] compared to carbon film (C film) which neither possesses the overall porosity of CNFs nor the fractal-like corrugations characteristic of the individual fibers in CNFs. As observed in CV, in the presence of $CO_2$ for CNFs, a peak appears at −0.573 V, characteristic of the $CO_2$ reduction, similar to that for Ag-induced $CO_2$ reduction [16,17,25]. The calculated FE at −0.573 V is about 98% for CO formation by analyzing the collected products via gas chromatography. Compared with C film, the onset potential for $CO_2$ reduction in CNFs at about 0.23 V is more positive, indicating a smaller overpotential, which is similar to the smallest overpotential reported so far recorded by the Ag catalyst (0.17 V) [14]. Next, the current density at −0.573 V for CNFs is almost two times higher than that for the C film, indicating excellent reduction efficiency of CNFs catalysts, contributed to the high positive charges and spin density of the carbon

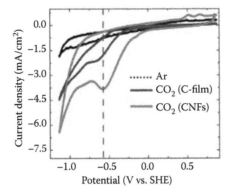

**FIGURE 5.15** CVs for $CO_2$ reduction in Ar-saturated and $CO_2$-saturated pure EMIM-BF₄ at 10 mV/s on carbon film electrode (black curve) and CNFs electrode (gray curve), respectively. The vertical dashed line represents the potential at which highest $CO_2$ reduction in the CNFs. (Reprinted by permission from Macmillan Publishers Ltd. *Nature Communications*, Kumar, B. et al. Renewable and metal-free carbon nanofibre catalysts for carbon dioxide reduction. 4, copyright 2013.)

atom around the electronegative nitrogen atom. The naturally oxidized carbon atom, acting as an effective active site, is incessantly renewable during the $CO_2$ electrocatalytic reduction process, resulting in stable current density even after 9 h [24].

The molecular electrocatalyst of $[Ni(cyclam)]^{2+}$ toward homogeneous reduction of $CO_2$ in acetonitrile (ACN) with or without water under $N_2$ and $CO_2$ is studied by CV as shown in Figure 5.16 [26]. As observed, in ACN without water, with the addition of $CO_2$, the Ni(II)/Ni(I) redox couple becomes irreversible and the reduction peak has a slight increase in current and is shifted positively by 30 mV. The positive shift is due to $CO_2$ binding to Ni(I). The irreversibility may arise from a fast, irreversible chemical step of an isomerization [27], following $CO_2$ binding. Interestingly, reversibility of the Ni(II)/Ni(I) couple is regained with scan rates above 3 V/s, assuming that with faster scan rates the chemical step of isomerization cannot proceed because the Ni(I) metal center is reoxidized to Ni(II) before the system can reach equilibrium. With a reversible couple, the $CO_2$ binding constant ($K_{CO_2}$) can be calculated from the shift in the Ni(II)/Ni(I) couple ($\Delta E$) under $N_2$ and $CO_2$ and Equation 5.11 [28]:

$$K_{CO_2}[CO_2] = e^{\Delta E(nF/RT)} - 1 \qquad (5.11)$$

Using this electrochemical determination method, $K_{CO_2}$ was found to be 6 $M^{-1}$. Information about the rate of the chemical step following $CO_2$ binding can be obtained as well. The half-life ($t_{1/2}$) is estimated to be ≈0.2 s under these conditions. By contrast, in ACN with water as the proton source, the reversibility is lost even at higher scan rates, suggesting that the chemical step responsible for irreversibility is most likely protonation of the $CO_2$ adduct. As shown in Figure 5.16b, two catalytic peaks develop. The first catalytic peak, a, simultaneously increases in

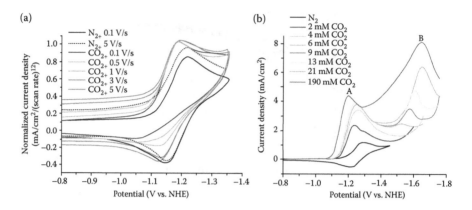

**FIGURE 5.16** (a) Cyclic voltammograms of 1 mM [Ni(cyclam)](PF$_6$)$_2$, 0.1 M TBAPF$_6$ in ACN. Current normalized by square root of scan rate. (b) Cyclic voltammograms of 5 mM [Ni(cyclam)](Cl)$_2$, 0.1 M TBAPF$_6$ in 1:4 water:ACN. Adding CO$_2$ with flow meter. The positive scan portion has been removed for clarity. Scan rate = 0.1 V/s. (Reprinted with permission from Froehlich, J. D. and Kubiak, C. P. The homogeneous reduction of CO$_2$ by [Ni(cyclam)]+: Increased catalytic rates with the addition of a CO scavenger. *Journal of the American Chemical Society*. 137(10), 3565–3573. Copyright 2014 American Chemical Society.)

current density and shifts positively in potential with increasing amounts of water, suggesting a proton-dependent electron transfer. The second reduction occurs only after protonation of the CO$_2$ adduct [29]. The current in peak b is assumed to be, in part, the reduction of a [Ni(cyclam)(CO)]$^+$ species [29] produced by reaction of [Ni-(cyclam)]$^+$ with CO. Additional contributions to the current in b could also be from CO$_2$ or proton reduction by a Ni(0) species or at the electrode. By varying the CO$_2$ concentration, the first catalytic peak, a, quickly reaches a plateau as the CO$_2$ concentration exceeds that of the catalyst, suggesting that CO$_2$ binding is not a rate-limiting step at high CO$_2$ concentrations. Peak b begins to appear at [CO$_2$] > 5 mM, which is probably related to CO$_2$ reduction by a species other than [Ni(cyclam)]$^+$. The catalytic current under excess CO$_2$ and proton source remains peak shaped, implying that substrate consumption is not the cause of the peak shape but is most likely due to an inhibition process such as catalyst degradation.

The CV of [Mn(mesbpy)-(CO)$_3$(MeCN)](OTf) in dry MeCN without added weak Brønsted acid of MeOH under N$_2$ consists of one couple of reversible reduction peaks at −1.55 V versus Fc$^+$/Fc with peak-to-peak separation of 39 mV (Figure 5.17a). This redox couple is best described as an electrical equivalent circuit (EEC) mechanism, where two one-electron reductions occur followed by the loss of a MeCN ligand [30]. The second of the two one-electron reductions occurs either at the same or at a lower potential than the first reduction [31]. This overall two-electron reduction leads to the anionic state, [Mn(mesbpy)(CO)$_3$]$^-$. No CV signal change was observed when the electrochemical solution was sparged with CO$_2$ in dry MeCN. However, once weak Brønsted acid of MeOH was added under CO$_2$, a current increase appeared at ~−2.0 V versus Fc$^+$/Fc, that is, ~400 mV after the two-electron reduction, solely due to the electrocatalytic reduction of CO$_2$ to CO. By contrast, no current increase

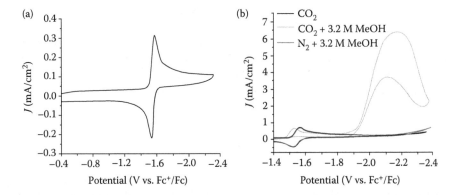

**FIGURE 5.17** (a) Cyclic voltammogram of 0.7 mM [Mn(mesbpy)-(CO)$_3$(MeCN)](OTf) in MeCN with 0.1 M TBAPF$_6$ under N$_2$ at 100 mV/s. (b) Cyclic voltammograms of 1 mM [Mn(mesbpy)(CO)$_3$(MeCN)](OTf) in 0.1 M TBAPF$_6$/MeCN under CO$_2$/N$_2$ with/without MeOH at 100 mV/s. Scan rate = 0.1 V/s. (Reprinted with permission from Sampson, M. D. et al. Manganese catalysts with bulky bipyridine ligands for the electrocatalytic reduction of carbon dioxide: Eliminating dimerization and altering catalysis. *Journal of the American Chemical Society.* 136(14), 5460–5471. Copyright 2014 American Chemical Society.)

was observed under N$_2$ with added weak acid, indicating that the current increase is not due to proton reduction (Figure 5.17b). Higher concentrations of weak Brønsted acid in CO$_2$ reduction electrocatalysis experiments resulted in increased current densities, before reaching a peak current density and leveling off or dropping with addition of more H$^+$ [30]. For a reversible electron-transfer reaction followed by a fast catalytic reaction ($E_R C_{cat}$ scheme), the peak catalytic current ($i_{cat}$) is given by Equation 5.12 [32]:

$$i_{cat} = n_{cat} FA[cat](Dk_{cat}[Q]^y)^{1/2} \qquad (5.12)$$

assuming that pseudo-first-order kinetics apply; that is, the reaction is first order in catalyst and that the concentrations of the substrates $Q$ are large in comparison to the concentration of the catalyst. In Equation 5.12, $n_{cat}$ is the number of electrons required for the catalytic reaction ($n_{cat} = 2$ for the reduction of CO$_2$ to CO), $F$ is Faraday's constant, $A$ is the surface area of the electrode, [cat] is the catalyst concentration, $D$ is the diffusion constant of the catalytically active species, $k_{cat}$ is the rate constant of the catalytic reaction, and [$Q$] is the substrate concentrations. At high [H$^+$], the electrocatalytic reduction of CO$_2$ is first order in the catalyst, first order in CO$_2$, independent of acid concentration, and at steady-state conditions [30]. Equation 5.13 describes the peak current of a complex with a reversible electron transfer and with no following reaction [33]:

$$i_p = 0.4463 n_p^{3/2} FA[cat]\left(\frac{F}{RT}\right)^{1/2} v^{1/2} D^{1/2} \qquad (5.13)$$

In Equation 5.13, $i_p$ is the peak current under $N_2$ with an amount of weak Brønsted acid corresponding to peak $i_{cat}$ conditions. $R$ is the universal gas constant, $T$ is the temperature, $n_p$ is the number of electrons in the reversible, noncatalytic reaction, and $v$ is the scan rate (0.1 V/s). Dividing Equation 5.12 by Equation 5.13 allows the determination of $i_{cat}/i_p$ and allows further calculation of the catalytic rate constant ($k_{cat}$) and the turnover frequency (TOF), as shown in Equation 5.14 [30]:

$$\text{TOF} = k_{cat}[Q] = \frac{Fvn_p^3}{RT}\left(\frac{0.4463}{n_{cat}}\right)^2\left(\frac{i_{cat}}{i_p}\right)^2 \tag{5.14}$$

assuming that the diffusion constant of the catalytically active species does not change significantly under $CO_2$ or $N_2$. Addition of MeOH ($pK_a$ of 29.0 in dimethyl sulfoxide (DMSO) [34]) resulted in a peak $i_{cat}/i_p = 30$ (7.6 mA/cm² peak current density) and TOF = 2000 s⁻¹ at 3.2 M MeOH. The pronounced deviation of the catalytic CVs from a steady-state "S-shaped" wave (Figure 5.17b) with peak maximum at ~-2.2 V versus Fc⁺/Fc and especially the peak in the return oxidation (~-2.1 V vs. Fc⁺/Fc) is quite unusual, likely arising from multiple factors. The main factor contributing to this odd current response is an overlapping bpy-based reduction at ~-2.3 V versus Fc⁺/Fc [30].

### 5.3.3 Calculation of Heterogeneous Rate Constant $k^0$

The standard rate constant for the heterogeneous electron transfer between the electrode and the redox species can be determined for voltammetric data using the dimensionless rate parameter $\psi$ defined by Equation 5.15 [3]:

$$\psi = \frac{(D_O/D_k)^{\alpha/2} k^0}{(\pi D_O fv)^{1/2}} \tag{5.15}$$

$f$ is a constant defined by $f = F/RT$. $D_O/D_R$ is the ratio of the diffusion coefficient of the starting material ($D_O$) and the diffusion coefficient of the reduced species formed by the electron transfer ($D_R$) to the power of $\alpha/2$. When the $D_R$ value is not expected to change compared to that of $D_O$ giving $D_O/D_R = 1$. This means that the transfer coefficient term, $\alpha$, does not need to be considered and the equation can be simplified to give Equation 5.16:

$$\psi = \frac{k^0}{(\pi D_O fv)^{1/2}} \tag{5.16}$$

By comparison of this with ideal data generated from numerical simulation of the electrochemical problem, $k^0$ can be determined for experimental data [3].

### 5.3.4 Homogeneous Electron-Transfer Rate Constant $k'$

The homogeneous rate constant can be determined if the reaction mechanism is known. For the $E_rC_i'$ mechanism, the plot as shown in Figure 5.18 can be employed, along with voltammetric data to gather $\lambda^{1/2}$ values.

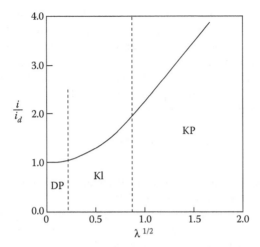

**FIGURE 5.18** Ratio of kinetic peak current for the $E_r C_i'$ reaction scheme to diffusion-controlled peak current as a function of $\lambda^{1/2}$. (Bard, A. J. and Faulkner, L. R.: *Electrochemical Methods: Fundamentals and Applications*. 2nd edn. p. 502. 2001. Copyright Wiley-VCH Verlag GmbH & Co. KGaA. Reproduced with permission.)

The kinetic parameter $\lambda$ is defined by Equation 5.17:

$$\lambda = \frac{k' C_Z^*}{v} \left( \frac{RT}{nF} \right) \qquad (5.17)$$

where $C_Z^*$ is the bulk concentration of the species present in excess which undergoes electron transfer from the redox species, such as [CO$_2$]; so with the experimental scan rates employed, a value of $k'$ can be established. In Figure 5.18, the DP zone is where there is pure diffusion control of the peak current. The peak current here is proportional to the square root of the scan rate and is reversible. The KP region is purely kinetically controlled with no diffusion contribution; therefore the peak current is independent of the scan rate. Complete kinetic control would be expected to result in the voltammogram shifting from peak shaped toward a sigmoidal form with a steady-state current being reached [3,35]. The KI region is the transitional region where both diffusion and kinetics play a part in the control of the limiting current and therefore the current ratio.

CV is not always particularly well suited for quantitative analysis of a system although it is easily employed for qualitative studies and the evaluation of potentials associated with electron transfer, etc.

## 5.4 ROTATING DISK ELECTRODE AND ROTATING RING-DISK ELECTRODE TECHNIQUES

### 5.4.1 Rotating Disk Electrode Technique

The rotating disk electrode (RDE) is a hydrodynamic technique where diffusion limitation plays a vital role, and forced convection is the dominant form of mass transport. Convection, which implies movement of the solution with respect to the electrode surface, is a much more effective form of transport than diffusion and when controlled allows for excellent reproducibility and increased sensitivity in electrochemical measurements. The RDE consists of a standard macroelectrode encased in insulating material that can be set to rotate about its own axis at selected specific frequencies, $f$ (revolutions per second), which can be easily transformed to angular velocity, $\omega$, through Equation 5.18:

$$\omega = 2\pi f \tag{5.18}$$

The convective flow of electrolyte to the electrode surface produced by the rotation is shown in the schematic Figure 5.19.

Rather than the diffusion-related peaks seen in CV, RDE gives a steady-state limiting current response dependent upon the fixed angular velocity employed as illustrated in the Levich equation [3] for the RDE (5.19):

$$I_l = 0.62nFAD^{2/3}v^{-1/6}C\omega^{1/2} \tag{5.19}$$

where $v$ is the kinematic viscosity of the electrolyte solution. The presence of the diffusion coefficient, $D$, in the equation describing a convection-controlled system may seem contradictory; however, although convection is dominant in the bulk solution, there is still a thin layer close to the electrode surface where diffusion plays an important role. The thickness of this layer ($\delta$) is dependent on rotational velocity as shown by Equation 5.20 [19]:

$$\delta = \frac{1.61D^{1/3}v^{1/6}}{\omega^{1/2}} \tag{5.20}$$

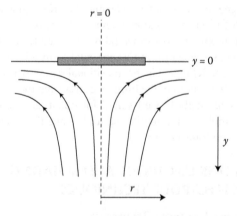

**FIGURE 5.19**  Schematic depicting RDE convective flow. (Bard, A. J. and Faulkner, L. R.: *Electrochemical Methods*: *Fundamentals and Applications*. 2nd edn. p. 337. 2001. Copyright Wiley-VCH Verlag GmbH & Co. KGaA. Reproduced with permission.)

### 5.4.2 ROTATING RING-DISK ELECTRODE TECHNIQUE

Generating a product or an intermediate at a disk electrode and collecting it at a ring electrode that concentrically surrounds the disk is an alternative to CV. The product or intermediate is generated by fixing the disk potential at an appropriate value and scanning the ring potential so as to obtain its oxidative or reductive signature as a steady-state current–potential curve. This technique has been extensively and successfully applied in the determination of the product, for example, CO, in the catalytic reduction of $CO_2$.

### 5.4.3 PRACTICAL APPLICATION FOR $CO_2$ ELECTROREDUCTION

The electrocatalytic activity of the complex $N,N',N'',N'$-tetramethyltetra-3,4-pyridoporphyrazinocobalt(II) ([Co(II)(Tmtppa)]$^{4+}$, Co(Tmtppa)) toward $CO_2$ reduction, is investigated using a rotating ring (platinum)-disk(EPG|Co(Tmtppa)|Nafion® electrode, consisting of an edged pyrolytic graphite (EPG) disk, modified with Co(Tmtppa) fixed by a Nafion film, and a polycrystalline platinum ring as shown in Figure 5.20 [36]. In a supporting electrolyte solution (0.1 M $NaClO_4$ + 0.1 M $CH_3COOH$ + $H_3BO_3$ + $H_3PO_4$ + NaOH buffer, pH 5.5) containing dissolved $CO_2$, when the rotating Co(Tmtppa)-modified EPG disk electrode is poised at a potential desirable for the electrocatalytic reduction of $CO_2$, the *in situ* formed product of CO will be thrown on to the platinum ring electrode and can be identified and detected with CV. As observed, a spark-like oxidation peak near 0.4 V appears in Figure 5.20b–d, indicating a surface process arising from the adsorbed CO, whereas hydrogen adsorption–desorption currents in the potential range 0 to −0.5 V gradually disappear, indicative of the strong competitive adsorption of CO at the hydrogen adsorption sites. A linear relationship between the ring current due to CO oxidation and $CO_2$ concentration in aqueous solution at certain range is observed as shown in Figure 5.20e, further confirming that the source of CO was in fact the $CO_2$ electrocatalytic reduction by immobilized Co(Tmtppa). When the dissolved $CO_2$ concentration exceeds $8 \times 10^{-3}$ M, the current reaches a plateau, probably due to the saturated adsorption on the ring electrode by the CO produced at the disk electrode. The CO oxidation reaction on the Pt ring electrode can be assigned to

$$[CO]_{ads} + H_2O \rightarrow CO_2 + 2H^+ + 2e^- \tag{5.21}$$

whereas the following shoulder near 0.5 V is recognized as splitting of the oxidation peak due to the effects of anion adsorption [37].

To ascertain the effect of mass transport limitations on the electrocatalytic reduction of $CO_2$ in 18 mol% EMIM-BF$_4$, electrochemical experiments using a rotating disk electrode (RDE) of Ag NPs-modified Au RDE polarized to −1.6 V versus AgQRE at the rotation speed range of 500–5000 rpm are carried out. As shown in Figure 5.21, in the absence of most mass transport effects, the turnover rate of the chemical reaction of interest can reach up to 60 turnovers per second and 62.1 mA/cm², based on the steady-state current measurements as a function of rotation speed [14].

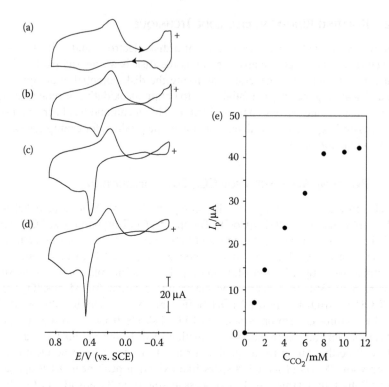

**FIGURE 5.20**  Ring electrode cyclic voltammograms of a rotating ring (platinum)-disk (EPG|Co(Tmtppa)|Nafion) electrode, recorded at a disk potential of −1.1 V in a supporting electrolyte solution (0.1 M NaClO$_4$ + 0.1 M CH$_3$COOH + H$_3$BO$_3$ + H$_3$PO$_4$ + NaOH buffer, pH 5.5) containing various concentrations of CO$_2$: (a) 0; (b) 2.0 × 10$^{-3}$ M; (c) 5.9 × 10$^{-3}$ M; and (d) 1.13 × 10$^{-2}$ M. Ring potential scan rate: 100 mV/s. Rotation rate: 900 rpm. (e) Ring CO oxidation peak currents near 0.4 V as a function of CO$_2$ concentration. (Reprinted from *Journal of Electroanalytical Chemistry*, 403(1–2), Zhang, J., Pietro, W. J. and Lever, A. B. P., Rotating ring-disk electrode analysis of CO$_2$ reduction electrocatalyzed by a cobalt tetramethylpyridoporphyrazine on the disk and detected as CO on a platinum ring, 93–100, Copyright 1996, with permission from Elsevier.)

## 5.5  ELECTROCHEMICAL IMPEDANCE SPECTROSCOPY

### 5.5.1  Basic Concept

Electrochemical impedance spectroscopy is a linear nondestructive technique, with high precision and a wide time span of 10$^4$–10$^{-6}$ s or large frequency range of 10$^{-4}$–10$^6$ Hz, due to the application of small perturbations to a system at equilibrium, which is powerful for the mechanism investigation of electrode reactions, the measurement of the dielectric and transport properties of materials, and the exploration of the properties of porous electrodes [3,38–41].

Generally, an electrode/electrolyte interface is considered simply an impedance to a small sinusoidal excitation and is described theoretically in terms of an EEC.

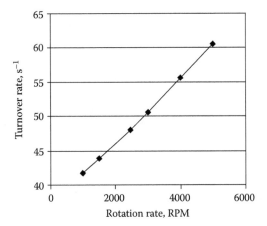

**FIGURE 5.21**  Turnover rate measured with a RDE at −0.9 V vs. normal hydrogen electrode (NHE) for $CO_2$ reduction on Ag NPs in 18 mol% EMIM-BF$_4$. (Reproduced from Rosen, B. A. et al. *Science*, 2011; 334(6056): 643–644. With permission from Science.)

By the electrical analogs, the electrochemical data are explained. For a fast charge-transfer reaction at a planar electrode,

$$O + ne \underset{k_b}{\overset{k_f}{\rightleftharpoons}} R \tag{5.22}$$

with both the oxidant $O$ and the reductant $R$ soluble, the electrified interface, illustrated in Figure 5.22a first proposed by Helmholtz [42], can be modeled by the Randles equivalent circuit [43] as shown in Figure 5.22b, which provides the most effective simulation of the impedance characteristics of the fast reaction. As observed in Figure 5.22, each component at the interface and in the solution during an electrochemical reaction is represented by comparison with a physical component. A double-layer capacitor, $C_d$, is in parallel with a pure polarization resistor, $R_p$, and a Warburg impedance, $Z_w$, then connected in series with a solution resistor, $R_s$. Among them, $R_p$ is a function of potential and becomes the charge-transfer resistance, $R_{ct}$, at $\eta = 0$, which manifests the kinetics of heterogeneous charge transfer. Warburg impedance has no simple electrical analog, related to diffusional mass transfer. The faradaic impedance, $Z_f$, broken down into two components of $R_p$ and the Warburg impedance, represents the effect of the heterogeneous electron-transfer process. The equivalent circuit is dependent on the type of a real electrochemical reaction involved at the interface, which could be much more complicated than the frequently used Randles circuit mentioned above. Here, the simplest electrode process, involving no adsorption of electroreactants, no multistep charge transfer, or no homogeneous chemistry, is discussed as in Figure 5.22. The positive charged oxidants accept electrons from the electrode at the interface and diffuse in, whereas the generated reductants flow out into the bulk of the solution.

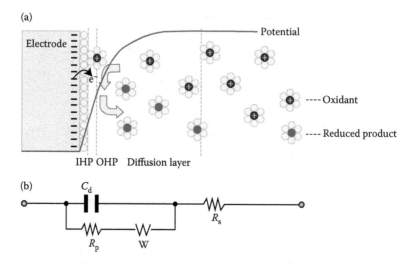

**FIGURE 5.22** Simple electrified interface with no specific adsorption, in which the vertical dotted lines in (a) are represented by the electronic components in (b). (a) IHP and OHP are the inner and outer Helmholtz planes, respectively. (Reprinted with permission from Park, S.-M. and Yoo, J.-S. Peer reviewed: Electrochemical impedance spectroscopy for better electrochemical measurements. *Analytical Chemistry.* 75(21), 455 A–461 A. Copyright 2003 American Chemical Society.)

The total impedance of an interface, $Z(\omega)$, is frequency-dependent and consists of two parts, a real number $Z'(\omega)$ and an imaginary number $Z''(\omega)$, expressed by $Z(\omega) = Z'(\omega) + jZ''(\omega)$ with the phase angle of $\phi = \tan^{-1}[Z''(\omega)/Z'(\omega)]$. Chemical information, such as system characteristics, $R_s$, $R_p$, and $C_d$ for an electrochemical reaction, can be extracted by treating the impedance data over Nyquist complex plane plots, in which $Z''(\omega)$ is plotted against $Z'(\omega)$ for different $\omega$ ($\omega = 2\pi f$ with $f$ being the frequency).

For the reaction (5.22), described by the Butler–Volmer equation, with the circuit elements shown in Figure 5.22b at the interface, the derived impedance expression can be simplified by considering the limiting behavior at high and low frequency:

1. At very high frequency, the contribution of the Warburg component is negligible in relation to $R_p$; the impedance is [41]

$$Z(\omega) = R_s + \frac{R_p}{1 + j\omega R_p C_d} = R_s + \frac{R_p}{1 + \omega^2 R_p^2 C_d^2} - \frac{j\omega R_p^2 C_d}{1 + \omega^2 R_p^2 C_d^2} = Z' + Z'' \quad (5.23)$$

2. At low frequency, the contribution of the Warburg component becomes important as the electron-transfer process at the interface may be limited by the mass transport of the electroactive species; the impedance is [41]

$$Z(\omega) = R_s + R_p\left[1 + \frac{\lambda}{\sqrt{2\omega}}\right] - R_p^2 \lambda_2 C_d - \frac{jR_p\lambda}{\sqrt{2\omega}} \quad (5.24)$$

where in $\lambda = (k_f/\sqrt{D_O}) + (k_b/\sqrt{D_R})$, $k_f$ and $k_b$ are the forward and backward electron-transfer rate constants, respectively, as shown in Equation 5.22. And, the frequency-dependent term of $\lambda/\sqrt{2\omega}$, appearing in both the real and the imaginary parts in Equation 5.24 is called the Warburg impedance.

Figure 5.23 shows the Nyquist plots for an electrochemical system, based on computer-generated data using Equation 5.24 at arbitrarily picked values of $R_p$, whereas the contribution of the Warburg impedance in stars is only for $R_p = 2$ k$\Omega$. As observed, the plot of $Z''(\omega)$ versus $Z'(\omega)$ is linear and has unit slope with the phase angle approaching 45° when the Warburg contribution becomes dominant in the low-frequency regime, characteristic of a diffusion-controlled electrode process. The intercept of the dashed line, representing Warburg components, is at $R_s + R_p - R_p^2\lambda^2C_d^2$ from which $\lambda = 38$ is obtained; therefore, $k_f = 4.1 \times 10^{-3}$ cm/s for a diffusion coefficient of $1.8 \times 10^{-6}$ cm²/s. $R_s$, 10 $\Omega$; $C_d$, 20 $\mu$F; $R_p$ value from inside to outside at 100 $\Omega$, 500 $\Omega$, 1 k$\Omega$, 2 k$\Omega$, respectively. Otherwise, with the obtained value of $\lambda$ and the calculated $k_f$ from $i_f = nFAk_fC_O$ ($C_O$, the bulk concentration of oxidant and $A$, the electrode area), the diffusion coefficient can be thereby calculated. According to Equation 5.23 as depicted in Figure 5.23, at high frequencies, $Z''(\omega)$ versus $Z'(\omega)$ gives a circular plot centered at $Z'(\omega) = R_s + R_p/2$ and $Z''(\omega) = 0$ and having a radius of $R_p/2$, where a maximum $Z''(\omega)$ is observed with the relationship of $R_p \times C_d = 1/\omega_{max} = 1/2\pi f_{max} = \tau_{rnx}$. Here, $\tau_{rnx}$ is the time constant of the electrochemical reaction, indicating how fast the reaction takes place. Also, from $R_p \times C_d$, $C_d$ can be obtained as $R_p$ is already known from the intercept on the $Z'(\omega)$ axis [41].

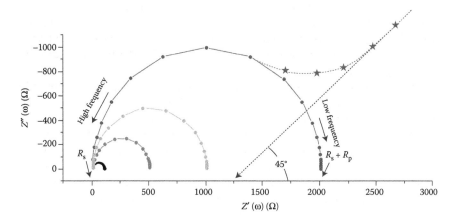

**FIGURE 5.23** Nyquist plots for an electrochemical system. Regions of mass transfer and kinetic control are found at low and high frequencies, respectively. (Reprinted with permission from Park, S.-M. and Yoo, J.-S. Peer Reviewed: Electrochemical impedance spectroscopy for better electrochemical measurements. *Analytical Chemistry*. 75(21), 455 A–461 A. Copyright 2003 American Chemical Society.)

## 5.5.2    PRACTICAL APPLICATION FOR $CO_2$ ELECTROREDUCTION

Electrochemical impedance spectroscopy was carried out to investigate the cell or electrode process for the electrochemical reduction of $CO_2$.

The electroreduction of $CO_2$ into CO and $CH_4$ in a proton conducting solid oxide electrolyzer with iron/iron oxide composite cathode using $BaCe_{0.5}Zr_{0.3}Y_{0.16}Zn_{0.04}O_{3-\delta}$ (BCZYZ) as an electrolyte was investigated by *in situ* AC impedance spectroscopy as shown in Figure 5.24 [44]. The $R_s$ is mainly from the BCZYZ electrolyte, which keeps almost constant and decreases slightly with increasing voltage, indicating increased electronic conduction as well as some oxygen-ion conduction at higher potentials. The $R_p$ is from electrode polarization, which displays sharp increases from 5.1 to 11.2 $\Omega$ cm$^2$ at low voltage from 0 to 0.5 V, decreases from 6.3 to 0.4 $\Omega$ cm$^2$ at high voltage from 1.0 to 2.0 V, and remains stable above 2.0 V, indicating activation of the redox processes dominate early stages at low voltage, whereas desirable electrode reactions with favorable thermodynamics and kinetics result in a large drop in $R_p$ at high voltage. The adsorption and diffusion processes on the electrode at low frequency below 10 kHz might be the rate-limiting step of the electrode process. The summit frequency of the process at low frequency changes from 0.5 to 19.9 Hz when the external load increases from 0.5 to 2.0 V, demonstrating an improvement of the low-frequency process at higher external loads.

## 5.6    CHRONOAMPEROMETRY

### 5.6.1    BASIC CONCEPT

CA involves the application of a fixed potential at a stationary electrode in an unstirred solution with the working electrode area kept small compared to the volume of solution under investigation. The current produced is measured as a function of time to give current transients or chronoamperograms. CA, like CV, usually involves an initial potential at which no reaction is anticipated therefore minimal current flows, $E_i$, which is then stepped to a set potential where the electron transfer of interest (either oxidation or reduction) takes place, $E_s$. The potential scheme is given in Figure 5.25.

The concentration of the substrate at the electrode surface during the application of $E_s$ is constantly approaching zero as it reacts to form a new species, creating a diffusion layer or depletion zone, and as such a concentration gradient to produce diffusion of the substrate from the bulk solution to the electrode surface and of the new species accumulating at the electrode toward the bulk phase. The diffusion layer increases with time as the electroactive substrate is depleted, resulting in lowering current responses [19]. The diffuse layer is the region over which the potential varies. This, as well as the electrical double layer, found in the solution immediately adjacent to the electrode, is depicted in Figure 5.26.

Figure 5.26 shows the potential change between the electrode surface ($\Phi^M$) and the bulk solution ($\Phi^S$) with distance from the electrode. Upon moving from the electrode surface, the first layer reached is the inner Helmholtz plane (IHP), which is defined as the locus of the electrical centers of the specifically adsorbed ions on the electrode surface. This specific adsorption is not necessarily charge dependent and

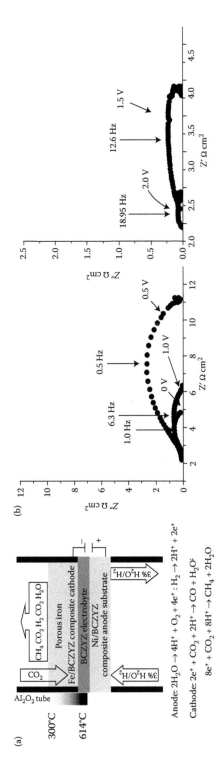

**FIGURE 5.24** (a) Schematic testing conditions of solid oxide electrolyzer. (b) AC impedance of solid oxide electrolyzer with different potentials. (Xie, K. et al. Electrochemical reduction of CO₂ in a proton conducting solid oxide electrolyser. *Journal of Materials Chemistry.* 2011; 21(1): 195–198. Reproduced by permission of The Royal Society of Chemistry.)

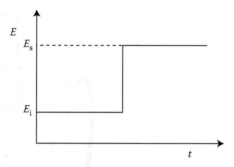

**FIGURE 5.25**   Potential scheme for CA. (Bard, A. J. and Faulkner, L. R.: *Electrochemical Methods: Fundamentals and Applications.* 2nd edn. p. 157. 2001. Copyright Wiley-VCH Verlag GmbH & Co. KGaA. Reproduced with permission.)

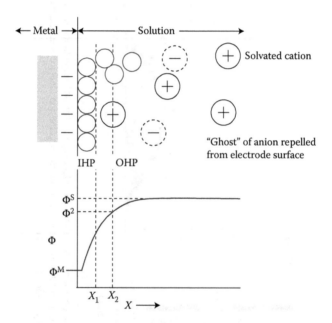

**FIGURE 5.26**   Phase boundary at the electrode–electrolyte interface showing the electrical double layer and diffuse layer. (Bard, A. J. and Faulkner, L. R.: *Electrochemical Methods: Fundamentals and Applications.* 2nd edn. p. 14. 2001. Copyright Wiley-VCH Verlag GmbH & Co. KGaA. Reproduced with permission.)

consists of nonsolvated species. The next layer is the outer Helmholtz plane (OHP), which marks the minimum distance of these nearest solvated ions. The ions here are predominantly of the opposite charge to that of the electrode and are responsible for the sharp drop in potential as the charge difference between the electrode and bulk solution is balanced. These two layers make up what is termed the double layer. Beyond this point, the potential drop is more gradual as the concentration of excess

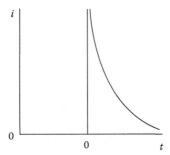

**FIGURE 5.27** Representative chronoamperogram of current flow versus time. (Bard, A. J. and Faulkner, L. R.: *Electrochemical Methods: Fundamentals and Applications.* 2nd edn. p. 157. 2001. Copyright Wiley-VCH Verlag GmbH & Co. KGaA. Reproduced with permission.)

counter ions decreases until the potential of bulk solution is matched marking the end of the diffuse layer [3].

A typical current transient can be seen in Figure 5.27.

At very short times, very high currents are usually produced due to the charging of the double layer (depicted in Figure 5.27). After this, the current drops as electrolysis occurs. The shape of the current transients is a result of the concentration gradients and Faraday's law (Equation 5.25) [45]:

$$Q = nFN \qquad (5.25)$$

where the charge transferred, $Q$, corresponding to the integral of the current transient, is proportional to the number of electrons transferred per molecule $n$, and the moles of reactant $N$. The current observed is proportional to the flux of the electroactive species ($J$) using Fick's diffusion laws, as can be seen from Equation 5.26:

$$I = nFAJ = nFAD\left(\frac{\partial C(x)}{\partial x}\right)_{x=0} \qquad (5.26)$$

At a planar electrode of working area, the Cottrell Equation 5.27 describes the instantaneous current [3]:

$$I = \frac{nFCD^{1/2}}{(\pi t)^{1/2}} \qquad (5.27)$$

where the symbols are defined as previously with the addition of $t$ which is time. This relationship allows reproducible quantitative evaluation of the currents and diffusion coefficient of a system at high overpotentials where the diffusion to the electrode is the limiting mode of mass transport. At very short times, deviation due to the charging and instrumental error is anticipated and over prolonged application of potential (>10–20 s) convection may become significant; however, Cottrell plots, where current is plotted as a function of $t^{-1/2}$, clearly show the period where the diffusion control is limiting with a linear region [3].

### 5.6.2 PRACTICAL APPLICATION FOR $CO_2$ ELECTROREDUCTION

By CA, electrolytic $CO_2$ reduction at a nanoporous silver (np-Ag) electrode in a $CO_2$-saturated 0.5 M $KHCO_3$ was investigated as shown in Figure 5.28 [46]. At −0.60 V with an overpotential of 490 mV relative to the $CO_2/CO$ equilibrium potential at −0.11 V versus reversible hydrogen electrode (RHE), a long-term stable current at ~18 mA/cm² throughout the electrocatalytic process was obtained with FE for CO calculated at ~92% (Figure 5.28a). The observed high initial current was thought from the reduction of a thin surface oxide layer formed in atmospheric air at about 0.5 nm thickness. By contrast, polycrystalline Ag displays a much lower current density of 0.47 mA/cm² and a poorer FE for CO at ~1.1% (Figure 5.28a) at the same condition. In addition, at more positive potential of −0.50 V with a lower overpotential of 390 mV, a smaller stable current of ~9.0 mA/cm² at an np-Ag electrode was observed with a CO efficiency decreased to ~90% (Figure 5.28b), whereas when the potential further positively increased to −0.40 V at 290 mV overpotential, the current dropped

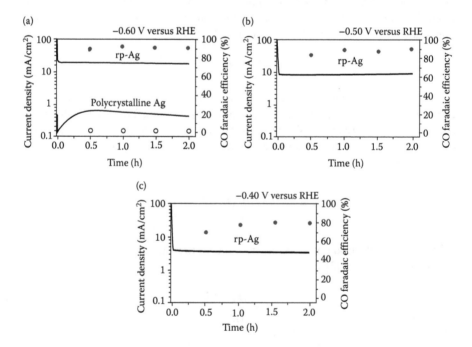

**FIGURE 5.28** $CO_2$ reduction activity of np-Ag and polycrystalline silver at (a) −0.60 V, np-Ag at (b) −0.50 V, and (c) −0.40 V versus RHE. Total current density versus time on (left axis) and CO FE versus time (right axis). (Reprinted by permission from Macmillan Publishers Ltd. *Nature Communications*. Lu, Q. et al. A selective and efficient electrocatalyst for carbon dioxide reduction. 5: 3242, copyright 2014.)

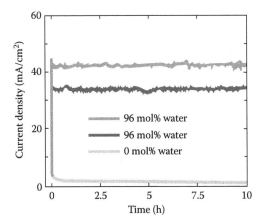

**FIGURE 5.29** Chronoamperometric curves of the bulk $MoS_2$ catalyst in $CO_2$-saturated EMIM-$BF_4$ solutions containing 96, 90, and 0 mol% water. (Reprinted by permission from Macmillan Publishers Ltd. *Nature Communications*. Asadi, M. et al. Robust carbon dioxide reduction on molybdenum disulphide edges. 5: 4470, copyright 2014.)

to ~3.3 mA/cm² with a CO efficiency only at ~79% (Figure 5.28c). Meanwhile, the rate of hydrogen evolution, occurred at a relatively small overpotential, increased correspondingly with the decrease of CO efficiency. The applied overpotential is not high enough to drive $CO_2$ reduction at a competitive rate [46].

The stability of the bulk $MoS_2$ catalyst for electrocatalytical reduction of $CO_2$ to CO is investigated at a prolonged period of 10 h in $CO_2$-saturated EMIM-$BF_4$ solutions containing 96, 90, and 0 mol% water by CA as shown in Figure 5.29 [13]. Stable steady-state current densities with negligible loss even after 10 h in all studied electrolytes are observed, indicating long-term stability and high efficiency of the $MoS_2$ catalyst. The more the water in the electrolyte, the higher the stable steady-state current densities, suggesting that water is involved in the $CO_2$ reduction reaction. The current density decreases rapidly at the initial period, attributed to the mass transport limitation from the bulk to the electrode surface without stirring. When the reaction rate of $CO_2$ electroreduction is governed by the reactant diffusion [3,24] after the reactant adjacent to the electrode surface is consumed immediately, a "steady-state current density" is observed. The obtained steady-state current density is much lower than that recorded in CV experiments at the same condition as there is an extra current due to interface charging involved in CV mode, but not for CA experiments [3,24].

### 5.6.3 CHRONOAMPEROMETRIC $k'$ DETERMINATION

In a similar manner to the voltammetric $k'$ determination, the homogeneous rate constant can be evaluated using chronoamperometric data and the working curve reproduced in Figure 5.30.

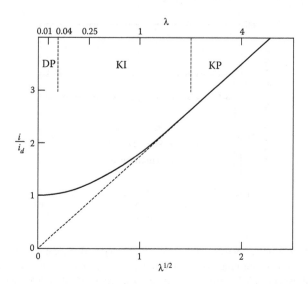

**FIGURE 5.30** Chronoamperometric working curve for the $E_rC_i'$ case for various values of $\lambda^{1/2}$. (Bard, A. J. and Faulkner, L. R.: *Electrochemical Methods: Fundamentals and Applications.* 2nd edn. p. 504. 2001. Copyright Wiley-VCH Verlag GmbH & Co. KGaA. Reproduced with permission.)

The $\lambda$ values are related to $k'$ for chronoamperometric experiments through Equation 5.28:

$$\lambda = K'C_z^*t \tag{5.28}$$

The plot is again split into the three zones described above for voltammetry but with the dashed line from the origin showing the KP region limiting line.

## 5.7 CHRONOPOTENTIOMETRY

Opposite to CA, for chronopotentiometry, the experiment is carried out by applying a constant current between stationary working and auxiliary electrodes with a galvanostat and recording the potential between the working and reference electrodes as the dependent variable as a function of time to give potential transients or chronopotentiograms, at similar condition such as small ratio of electrode area to solution volume and semi-infinite diffusion. The obvious disadvantage of controlled-current techniques is that double-layer charging effects occur throughout the experiment with larger values and no straightforward correction is available, though they can be of particular value for background process studying. Constant-current chronopotentiometry involves a set current $i$, where the electron transfer of interest (either oxidation or reduction) takes place at a constant rate. The current scheme is given in Figure 5.31a.

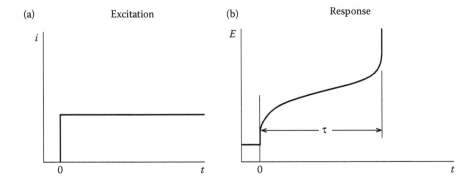

**FIGURE 5.31** (a) Current scheme as a function of time for chronopotentiometry. (b) A representative chronopotentiogram of potential change versus time. (Bard, A. J. and Faulkner, L. R.: *Electrochemical Methods: Fundamentals and Applications.* 2nd edn. p. 307. 2001. Copyright Wiley-VCH Verlag GmbH & Co. KGaA. Reproduced with permission.)

A typical potential transient can be seen in Figure 5.31b. When the steady current is applied, the electrode potential directly moves to values characteristic of the redox active couple. As the electrolysis continues, the potential varies since the concentration ratio of the oxidative/reductive species on the surface of the electrode changes with time. At the transition time, the concentration of active species drops to zero, then the flux of active species to the electrode surface is not large enough to accept or donate all electrons being forced across the electrode–solution interface, and the electrode potential rapidly shifts to a more extreme value where a new, second redox process can occurs. The shape and location of the resulting $E$–$t$ curve are governed by the reversibility, or the heterogeneous rate constant, of the redox process. The transition time, $\tau$, is defined as the time span from the beginning of the application of the constant current to the sharp decrease or increase in the electrode potential, which is the chronopotentiometric analog of the peak or limiting current, and related to the concentration and the diffusion coefficient.

At a planar working electrode in an unstirred solution, the Sand equation (5.29) describes the transition time, $\tau$:

$$\frac{i\tau^{1/2}}{C^*} = \frac{nFAD^{1/2}\pi^{1/2}}{2} = 85.5\,nD^{1/2}A\,\frac{mA-s^{1/2}}{mM}\text{(with } A \text{ in cm}^2\text{)} \qquad (5.29)$$

where the symbols either have their usual meanings or are defined as above. The $i\tau^{1/2}/c^*$ is called transition time constant, which is independent of $i$ or $c$ for a well-behaved system. The measured value of $\tau$ at known $i$ or $i\tau^{1/2}$ values at various currents can be used to determine $n$, $A$, $c$, or $D$. A lack of constancy in the parameter of $i\tau^{1/2}/c^*$ reflects a complicated electrode process, probably involving in a coupled homogeneous chemical reaction or adsorption, or due to double-layer charging or the onset of convection.

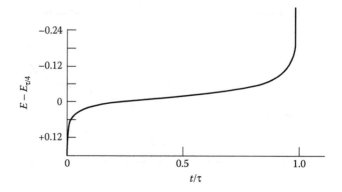

**FIGURE 5.32**  Theoretical chronopotentiogram for a Nernstian electrode process. (Bard, A. J. and Faulkner, L. R.: *Electrochemical Methods: Fundamentals and Applications.* 2nd edn. p. 310. 2001. Copyright Wiley-VCH Verlag GmbH & Co. KGaA. Reproduced with permission.)

For a Nernstian wave with rapid electron transfer, a convenient diagnostic between the electrode potential, $E$ and the quarter-wave potential of $E_{\tau/4}$ which is the chronopotentiometric equivalent of the voltammetric $E_{1/2}$, is expressed by Equation 5.30:

$$E = E_{\tau/4} + \frac{RT}{nF} \ln \frac{\tau^{1/2} - t^{1/2}}{t^{1/2}} \tag{5.30}$$

where $E_{\tau/4}$ is

$$E_{\tau/4} = E^{0'} - \frac{RT}{2nF} \ln \frac{D_O}{D_R} \tag{5.31}$$

For a reversible electrode process, the $E$–$t$ curve is a linear plot of $E$ versus log $(\tau^{1/2} - t^{1/2})/t^{1/2}$ with a slope of 59/$n$ mV at 25°C. The theoretical chronopotentiogram for a Nernstian electrode process is shown in Figure 5.32.

# REFERENCES

1. Jitaru, M., Lowy, D. A., Toma, M., Toma, B. C. and Oniciu, L. Electrochemical reduction of carbon dioxide on flat metallic cathodes. *Journal of Applied Electrochemistry.* 1997; 27(8): 875–889.
2. Hossain, S. *Electrochemical Reduction of Carbon Dioxide to Hydrocarbons.* King Fahd University of Petroleum and Minerals, Saudi Arabia. 2011.
3. Bard, A. J. and Faulkner, L. R. *Electrochemical Methods: Fundamentals and Applications.* 2nd edn. New York, NY: John Wiley & Sons, Inc. 2001.
4. Delacourt, C., Ridgway, P. L. and Newman, J. Mathematical modeling of $CO_2$ reduction to CO in aqueous electrolytes: I. Kinetic study on planar silver and gold electrodes. *Journal of the Electrochemical Society.* 2010; 157(12): B1902–B1910.

5. Medina-Ramos, J., DiMeglio, J. L. and Rosenthal, J. Efficient reduction of $CO_2$ to CO with high current density using *in situ* or *ex situ* prepared bi-based materials. *Journal of the American Chemical Society.* 2014; 136(23): 8361–8367.

6. Russell, P., Kovac, N., Srinivasan, S. and Steinberg, M. The electrochemical reduction of carbon dioxide, formic acid, and formaldehyde. *Journal of the Electrochemical Society.* 1977; 124(9): 1329–1338.

7. Paik, W., Andersen, T. N. and Eyring, H. Kinetic studies of the electrolytic reduction of carbon dioxide on the mercury electrode. *Electrochimica Acta.* 1969; 14(12): 1217–1232.

8. Haynes, L. V. and Sawyer, D. T. Electrochemistry of carbon dioxide in dimethyl sulfoxide at gold and mercury electrodes. *Analytical Chemistry.* 1967; 39(3): 332–338.

9. Ryu, J., Andersen, T. N. and Eyring, H. Electrode reduction kinetics of carbon dioxide in aqueous solution. *The Journal of Physical Chemistry.* 1972; 76(22): 3278–3286.

10. Spataru, N., Tokuhiro, K., Terashima, C., Rao, T. and Fujishima, A. Electrochemical reduction of carbon dioxide at ruthenium dioxide deposited on boron-doped diamond. *Journal of Applied Electrochemistry.* 2003; 33(12): 1205–1210.

11. Medina-Ramos, J., Pupillo, R. C., Keane, T. P., DiMeglio, J. L. and Rosenthal, J. Efficient conversion of $CO_2$ to CO using tin and other inexpensive and easily prepared post-transition metal catalysts. *Journal of the American Chemical Society.* 2015; 137(15): 5021–5027.

12. Zhang, S., Kang, P. and Meyer, T. J. Nanostructured tin catalysts for selective electrochemical reduction of carbon dioxide to formate. *Journal of the American Chemical Society.* 2014; 136(5): 1734–1737.

13. Asadi, M., Kumar, B., Behranginia, A., Rosen, B. A., Baskin, A., Repnin, N., Pisasale, D. et al. Robust carbon dioxide reduction on molybdenum disulphide edges. *Nature Communications.* 2014; 5: 4470.

14. Rosen, B. A., Salehi-Khojin, A., Thorson, M. R., Zhu, W., Whipple, D. T., Kenis, P. J. A. and Masel, R. I. Ionic liquid–mediated selective conversion of $CO_2$ to CO at low overpotentials. *Science.* 2011; 334(6056): 643–644.

15. Chen, Y., Li, C. W. and Kanan, M. W. Aqueous $CO_2$ reduction at very low overpotential on oxide-derived au nanoparticles. *Journal of the American Chemical Society.* 2012; 134(49): 19969–19972.

16. Salehi-Khojin, A., Jhong, H.-R. M., Rosen, B. A., Zhu, W., Ma, S., Kenis, P. J. A. and Masel, R. I. Nanoparticle silver catalysts that show enhanced activity for carbon dioxide electrolysis. *The Journal of Physical Chemistry C.* 2013; 117(4): 1627–1632.

17. Rosen, B. A., Haan, J. L., Mukherjee, P., Braunschweig, B., Zhu, W., Salehi-Khojin, A., Dlott, D. D. and Masel, R. I. *In situ* spectroscopic examination of a low overpotential pathway for carbon dioxide conversion to carbon monoxide. *The Journal of Physical Chemistry C.* 2012; 116(29): 15307–15312.

18. Kissinger, P. T. and Heineman, W. R. *Laboratory Techniques in Electroanalytical Chemistry, Revised and Expanded.* 2nd edn. New York: Marcel Dekker Inc. 1996.

19. Fischer, A. *Electrode Dynamics.* Oxford: Oxford University Press. 1996.

20. Girault, H. H. *Analytical and Physical Electrochemistry.* New York: Marcel Dekker, Inc. 2004.

21. Shironita, S., Karasuda, K., Sato, M. and Umeda, M. Feasibility investigation of methanol generation by $CO_2$ reduction using Pt/C-based membrane electrode assembly for a reversible fuel cell. *Journal of Power Sources.* 2013; 228(0): 68–74.

22. Conway, B. E., Angerstein-Kozlowska, H., Sharp, W. B. A. and Criddle, E. E. Ultrapurification of water for electrochemical and surface chemical work by catalytic pyrodistillation. *Analytical Chemistry.* 1973; 45(8): 1331–1336.

23. Guo, X., Zhang, Y., Deng, C., Li, X., Xue, Y., Yan, Y.-M. and Sun, K. Composition dependent activity of Cu–Pt nanocrystals for electrochemical reduction of $CO_2$. *Chemical Communications.* 2015; 51(7): 1345–1348.

24. Kumar, B., Asadi, M., Pisasale, D., Sinha-Ray, S., Rosen, B. A., Haasch, R., Abiade, J., Yarin, A. L. and Salehi-Khojin, A. Renewable and metal-free carbon nanofibre catalysts for carbon dioxide reduction. *Nature Communications*. 2013; 4: 2819.

25. Rosen, B. A., Zhu, W., Kaul, G., Salehi-Khojin, A. and Masel, R. I. Water enhancement of $CO_2$ conversion on silver in 1-ethyl-3-methylimidazolium tetrafluoroborate. *Journal of the Electrochemical Society*. 2013; 160(2): H138–H141.

26. Froehlich, J. D. and Kubiak, C. P. The homogeneous reduction of $CO_2$ by [Ni(cyclam)]+: Increased catalytic rates with the addition of a CO scavenger. *Journal of the American Chemical Society*. 2015; 137(10): 3565–3573.

27. Zilbermann, I., Winnik, M., Sagiv, D., Rotman, A., Cohen, H. and Meyerstein, D. Properties of monovalent nickel complexes with tetraaza-macrocyclic ligands in aqueous solutions. *Inorganica Chimica Acta*. 1995; 240(1–2): 503–514.

28. Gagne, R. R., Allison, J. L. and Ingle, D. M. Unusual structural and reactivity types for copper(I). Equilibrium constants for the binding of monodentate ligands to several four-coordinate copper(I) complexes. *Inorganic Chemistry*. 1979; 18(10): 2767–2774.

29. Beley, M., Collin, J. P., Ruppert, R. and Sauvage, J. P. Electrocatalytic reduction of carbon dioxide by nickel cyclam2+ in water: Study of the factors affecting the efficiency and the selectivity of the process. *Journal of the American Chemical Society*. 1986; 108(24): 7461–7467.

30. Sampson, M. D., Nguyen, A. D., Grice, K. A., Moore, C. E., Rheingold, A. L. and Kubiak, C. P. Manganese catalysts with bulky bipyridine ligands for the electrocatalytic reduction of carbon dioxide: Eliminating dimerization and altering catalysis. *Journal of the American Chemical Society*. 2014; 136(14): 5460–5471.

31. Tulyathan, B. and Geiger, W. E. Structural consequences of electron-transfer reactions. Part 12. Multi electron processes involving structural changes. The two-electron reduction of hexaosmium carbonyl cluster (Os6(CO)18). *Journal of the American Chemical Society*. 1985; 107(21): 5960–5967.

32. Savéant, J. M. and Vianello, E. Potential-sweep chronoamperometry theory of kinetic currents in the case of a first order chemical reaction preceding the electron-transfer process. *Electrochimica Acta*. 1963; 8(12): 905–923.

33. Bard, A. J. and Faulkner, L. R. *Electrochemical Methods*. New York: Wiley, 1980.

34. Olmstead, W. N., Margolin, Z. and Bordwell, F. G. Acidities of water and simple alcohols in dimethyl sulfoxide solution. *The Journal of Organic Chemistry*. 1980; 45(16): 3295–3299.

35. Costentin, C., Robert, M. and Saveant, J.-M. Catalysis of the electrochemical reduction of carbon dioxide. *Chemical Society Reviews*. 2013; 42(6): 2423–2436.

36. Zhang, J., Pietro, W. J. and Lever, A. B. P. Rotating ring-disk electrode analysis of $CO_2$ reduction electrocatalyzed by a cobalt tetramethylpyridoporphyrazine on the disk and detected as CO on a platinum ring. *Journal of Electroanalytical Chemistry*. 1996; 403(1–2): 93–100.

37. de Becdelièvre, A. M., de Becdelièvre, J. and Clavilier, J. Electrochemical oxidation of adsorbed carbon monoxide on platinum spherical single crystals: Effect of anion adsorption. *Journal of Electroanalytical Chemistry and Interfacial Electrochemistry*. 1990; 294(1–2): 97–110.

38. Macdonald, D. D. Reflections on the history of electrochemical impedance spectroscopy. *Electrochimica Acta*. 2006; 51(8): 1376–1388.

39. Smith, D. E. Data processing in electrochemistry. *Analytical Chemistry*. 1976; 48(2): 221A–240A.

40. Mcdonald, J. R., ed. *Impedance Spectroscopy: Emphasizing Solid Materials and Systems*. New York: Wiley/Interscience. 1987.

41. Park, S.-M. and Yoo, J.-S. Peer reviewed: Electrochemical impedance spectroscopy for better electrochemical measurements. *Analytical Chemistry*. 2003; 75(21): 455A–461A.

42. Parsons, R. The electrical double layer: Recent experimental and theoretical developments. *Chemical Reviews.* 1990; 90(5): 813–826.
43. Randles, J. E. B. A cathode ray polarograph. Part II. The current–voltage curves. *Transactions of the Faraday Society.* 1948; 44(0): 327–338.
44. Xie, K., Zhang, Y., Meng, G. and Irvine, J. T. S. Electrochemical reduction of $CO_2$ in a proton conducting solid oxide electrolyser. *Journal of Materials Chemistry.* 2011; 21(1): 195–198.
45. Monk, P. M. *Fundamentals of Electroanalytical Chemistry.* Chichester, England: John Wiley & Sons Ltd. 2001.
46. Lu, Q., Rosen, J., Zhou, Y., Hutchings, G. S., Kimmel, Y. C., Chen, J. G. and Jiao, F. A selective and efficient electrocatalyst for carbon dioxide reduction. *Nature Communications.* 2014; 5: 3242.

# 6 Mechanism of Catalytic and Electrocatalytic $CO_2$ Reduction to Fuels and Chemicals

*Neetu Kumari, M. Ali Haider,*
*and Suddhasatwa Basu*

## CONTENTS

## 6.1 INTRODUCTION

High carbon dioxide ($CO_2$) emission levels in the environment have necessitated efforts toward the development of technologies for $CO_2$ mitigation. Recent research work has highlighted the potential of $CO_2$ as a renewable resource for the production of fuels and chemicals which may help in reducing $CO_2$ concentration in the environment [1,2]. However, the thermodynamics of direct $CO_2$ reduction to a hydrocarbon product is not favorable [3]. $CO_2$ molecule is a stable linear structure (O=C=O), where the Gibbs free energy of its formation at standard condition is estimated to be $-396$ kJ/mol [2] (Figure 6.1). Therefore, the direct conversion of $CO_2$ to any other product molecule such as carbon monoxide, methane, or methanol will involve a substantial positive change in Gibbs free energy (Figure 6.1) at standard conditions and is unlikely to proceed [2,4]. In hydrogen-assisted conditions, the thermodynamics

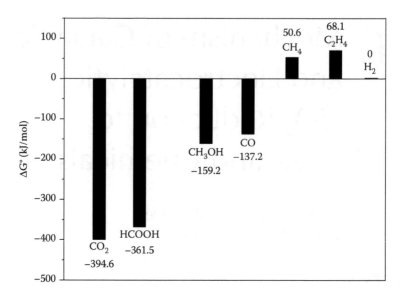

**FIGURE 6.1** Gibbs free energy of $CO_2$ compared to similar molecules and derived products at standard condition.

may become favorable. For example, the change in Gibbs energy ($\Delta G°$) for $CO_2$ conversions to methanol and methane at 298 K is estimated to be −9.2 and −130.8 kJ/mol [5] (Table 6.1), respectively. Interestingly, $\Delta G°$ as given by the Gibbs–Helmholtz equation ($\Delta G° = \Delta H° − T\Delta S°$) is dominated by the enthalpy change at standard conditions ($\Delta H°$) and remains less or not affected by the entropy contribution ($−T\Delta S°$) (Table 6.1). It should, however, be noted for reactions, where $\Delta G°$ is positive at standard conditions, such as conversion to carbon monoxide and formic acid, may become favorable with improved kinetics at higher temperatures. Endogenic reactions can

**TABLE 6.1**

**Enthalpy and Gibbs Free Energy Change for $CO_2$ Hydrogenations to Fuels and Chemicals at Standard Condition**

| Chemical Hydrogenation of $CO_2$ | $\Delta H°$ (kJ/mol) | $−T\Delta S°$ (kJ/mol) | $\Delta G°$ (kJ/mol) |
|---|---|---|---|
| $CO_2(g) + H_2(g) \rightarrow CO(g) + H_2O(g)$ | 41.2 | 22.6 | 18.6 |
| $CO_2(g) + H_2(g) \rightarrow CO(g) + H_2O(l)$ | −2.8 | 22.8 | 20.0 |
| $CO_2(g) + H_2(g) \rightarrow HCOOH(l)$ | −31.2 | 64.2 | 33.0 |
| $CO_2(g) + 3H_2(g) \rightarrow CH_3OH(l) + H_2O(l)$ | −131.3 | 122.1 | −9.2 |
| $CO_2(g) + 4H_2(g) \rightarrow CH_4(g) + 2H_2O(l)$ | −252.9 | 122.1 | −130.8 |
| $CO_2(g) + 6H_2(g) \rightarrow C_2H_4(g) + H_2O(l)$ | −307.2 | 210.6 | −96.5 |

*Source:* Reprinted from *Surf. Sci. Rep.*, 25(8), Freund H.-J., Roberts M.W., Surface chemistry of carbon dioxide, 225–273, Copyright 1996, with permission from Elsevier.

further be tuned for obtaining desired conversions by utilizing an electrocatalyst and thus changing the $\Delta G°$ of the reaction via an applied electrical potential, given by $\Delta G° = -nFE°$. Considering the overall energy cycle, the electrocatalytic conversions need to be implemented with one condition, that the required electricity has to be produced from a renewable source.

A variety of metal catalysts such as Cu, Pt, Ni, Ru, etc. have been experimented with for catalytic and electrocatalytic reactions involving CO$_2$ conversion to hydrocarbons and carbon monoxide (CO) [6–8]. Heterogeneous catalytic reactions are primarily focused on carbon dioxide reforming of methane (dry reforming) [9] assisted with several other reactions such as in oxyreforming [10], bi-reforming, and tri-reforming [11] reactions on Ni, Ru, Rh, Pt, and Pd-based metal catalyst. In an alternative route, CO$_2$ can be reduced to CO via a reverse water gas shift (RWGS) reaction and syngas thus produced can be converted into hydrocarbon fuels and chemicals via a Fischer–Tropsch process [12].

Electrocatalytic reactions for CO$_2$ reduction are primarily carried out at Cu electrodes at low temperatures, where hydrocarbons such as methane and methanol are produced [13–15]. Cu finds a unique position in the periodic table by producing hydrocarbons, whereas the rest of the metals either produce hydrogen or CO on electrolysis [16]. At high temperatures (>723 K), thermodynamics and kinetics for electrocatalytic CO$_2$ reduction are expected to be further improved [17]. A solid oxide fuel cell (SOFC) working in reverse direction can be utilized for co-electrolysis of CO$_2$ as a solid oxide electrolyzer for achieving high conversion. Electrode materials consisting of the cermet of Ni, Pt with yttria-stabilized zirconia (YSZ), or gadolinium-doped ceria (GDC), La$_{0.9}$Sr$_{0.1}$Ga$_{0.8}$Mn$_{0.2}$O$_{3-d}$ (LSGM) [18], La$_{0.6}$Sr$_{0.4}$Co$_{0.2}$Fe$_{0.8}$O$_{3-d}$ (LSCF) [19], and La$_{1-x}$Sr$_x$Cr$_{0.5}$Mn$_{0.5}$O$_{3-d}$ (LSCM) [20] are proposed in the literature for high-temperature CO$_2$ reduction reaction.

Hydrocarbon fuels such as methane, methanol, formic acid, and ethylene can be synthesized by direct hydrogenation of CO$_2$. Methane, the smallest hydrocarbon unit, is an important constituent of natural gas and is widely utilized as fuel in industries and in electric power generation. However, since the discovery of shale gas, methane production from CO$_2$ may not turn to be an economical option. In contrast, methanol production from CO$_2$ could be an attractive option and has been recently considered for commercialization [21]. As compared to CO hydrogenation to methanol ($\Delta H = -128.5$ kJ/mol), CO$_2$ hydrogenation to methanol could become energetically favorable ($\Delta H = -131$ kJ/mol) at standard conditions. Considerable progress has been made in the development of an effective catalyst for methanol production from CO$_2$ [6,21,22]. Methanol is one of the most important feedstock used in the chemical industry, with the worldwide demand of 50 million tons [23]. It is extensively used as a solvent in pharmaceutical industries, and as a precursor of various chemical intermediates mainly in the manufacturing of formaldehyde and formic acid. Similarly, CO$_2$ can be hydrogenated to other C$_1$-type molecules such as formic acid (HCOOH) which is used in the leather, textile, and rubber industries. Formic acid can be directly used as a fuel to generate electricity in fuel cells. Ethylene is produced by C–C coupling reaction from CO$_2$ and is utilized as the building block of polymers.

Irrespective of the catalyst, electrocatalyst, and reaction conditions, the mechanism of $CO_2$ reduction in the presence of hydrogen consists of similar intermediates. $CO_2$ on hydrogenation can produce a formate (HCOO) or a carboxyl (COOH) intermediate depending upon whether the carbon or the oxygen of $CO_2$ is hydrogenated. The two intermediates play a key role in determining the overall selectivity and yield of a desired product. Reaction energy diagrams are plotted via these key intermediates to decide upon the favored routes for $CO_2$ reduction to a target product. Understanding developed from energetics can be improvised by studying the kinetics of the intrinsic steps, which may further provide insights into the functioning of the catalyst surface. Possible mechanistic routes are outlined for four product molecules, which are methane, methanol, formic acid, and ethylene representing the key transformations in $CO_2$ conversions: CO bond dissociation, C–C coupling, C–H and O–H bond formations. The mechanistic discussions in this chapter are divided into three broad categories of reaction conditions:

1. Heterogeneous catalysis
2. Low-temperatures electrolysis
3. High-temperature electrolysis

The understanding thus developed will facilitate the design of novel catalyst materials for achieving better conversions of $CO_2$.

## 6.2 MECHANISM OF $CO_2$ REDUCTION TO DIFFERENT PRODUCTS

### 6.2.1 CARBON MONOXIDE

Carbon monoxide is an important constituent of syngas. There are two possible mechanisms through which $CO_2$ can be converted to CO. The first involves the $CO_2$ reduction at the catalyst surface in the presence of water via an RWGS reaction. The second type of reduction mechanism is proposed on reducible oxides having surface oxygen vacancies, where a redox mechanism is likely to occur.

In RWGS reaction, $CO_2$ may be hydrogenated to a COOH [9,24–27] or to a HCOO [28,29] intermediate. Carboxyl subsequently dissociates into CO and OH species on the surface. RWGS reaction with COOH-mediated species is proposed on Cu(111) surface using density functional theory (DFT) calculations [24]. Mechanisms of $CO_2$ hydrogenation and dissociation were studied on Cu(111) surface [30]. Theoretical calculations indicated that the hydrogenation of $CO_2$ to COOH is of significantly high activation barrier (~110 kJ/mol) and is likely to be the rate-determining step on the Cu surface [24]. Similarly, microkinetic results on Pt(111)-based catalysts also conform to the COOH-mediated formation of CO [25,26]. In this reaction as well, formation of COOH via $CO_2 + H \rightarrow COOH$ is proposed as the rate-limiting step on the Pt(111) surface. On the contrary, for $CO_2$ hydrogenation reactions on the $CeO_2$(110) surface, COOH dissociation to CO and OH was estimated to show a significantly high activation barrier [31] and is likely to be rate determining. Alternatively, in some studies, formate is proposed as an intermediate in RWGS reaction on Fe and W catalysts [28]. On Fe and W surfaces, DFT calculated activation energies showed a lower

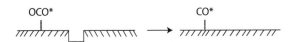

**FIGURE 6.2**   Schematic of CO$_2$ reduction to CO by redox mechanism.

barrier for HCOO formation as compared to the COOH, implying that on these two metals surfaces, CO$_2$ molecule reduces to CO via a formate-mediated mechanism. CO is produced by subsequent disproportionation of HCOO to CO and OH [32].

CO$_2$ could be reduced to CO on the catalyst having surface oxygen vacancies via a redox mechanism, where the oxygen atom of CO$_2$ is incorporated into the surface oxygen vacancy [33,34]. Schematics of the redox mechanism are shown in Figure 6.2. The activated CO$_2$ in the vicinity of oxygen vacancy dissociates into CO and atomic oxygen. Atomic oxygen heals the vacant site, leading to re-oxidation of the reduced surface into stoichiometric or partially reduced form of the metal oxide. Numerous studies have proposed a redox mechanism for CO oxidation to CO$_2$ [35–37]. Lewandowski et al. have studied CO oxidation to CO$_2$ on FeO$_2$/Pt(111) thin film by redox mechanism. Here, CO reacts with lattice oxygen of FeO$_2$ to form CO$_2$, leaving an oxygen vacancy on the surface. The oxygen vacancies are replenished by the reaction of oxygen gas molecules. In another study, CO oxidizes to CO$_2$ by the redox mechanism on the interface of the Au nano cluster supported on CeO$_2$(111) [38]. The oxygen atom at the interface of CeO$_2$(111) and Au is active to oxidize the CO molecule to CO$_2$. Similarly, CeO$_2$-based materials have high oxygen storage capacity, due to which CO can be oxidized by CeO$_2$ to CO$_2$, while CeO$_2$ is reduced to Ce$_2$O$_3$. Water present in the reaction mixture can re-oxidize the Ce$_2$O$_3$ back to CeO$_2$ (H$_2$O + Ce$_2$O$_3$ → H$_2$ + CeO$_2$) [39, 40] to complete the catalysis cycle. The reverse of this reaction is also possible to facilitate CO$_2$ reduction to CO on the CeO$_2$ surface. The CeO$_2$(110) surface has been studied by Cheng et al. to activate CO$_2$ into CO and O by performing DFT calculations [41]. One vacancy was generated in the $p(2 \times 2)$ supercell system of the CeO$_2$(110) surface. Oxygen from CO$_2$ can heal the oxygen vacancy site to regenerate the stoichiometric ceria surface. The authors have proposed the reaction path and possible intermediate species for the dissociation of CO$_2$. The electronic structure of surface atoms and adsorbate were analyzed and the results implied that the electrons were transferred from the reduced ceria surface to the adsorbed CO$_2$ molecule to dissociate the C–O bond. The overall activation barrier for CO$_2$ reduction to CO via this redox mechanism on ceria surface was calculated to be significantly higher (264 kJ/mol) and the reaction energy of this step was highly endothermic (~136 kJ/mol).

Once CO is produced either by the redox or RWGS reaction, it can possibly undergo for hydrogen-assisted dissociation to produce methane, which is discussed in detail in the next section of this chapter.

## 6.2.2   METHANE

Catalytic synthesis of methane by CO$_2$ hydrogenation, known as the Sabatier reaction, is an important process for chemical industries. The mechanism of chemical transformation of $CO_2$ to $CH_4$ on different metal and metal oxides constitutes a matter of detailed discussion. All suggested routes of chemical transformation of $CO_2$ to $CH_4$ are shown in Figure 6.3 [42–47]. The $CO_2$ methanation reaction has been theoretically studied on Ni(111) [42] and $CeO_2$(110) [31] via the dissociation of carboxyl to CO ($CO_2 \rightarrow COOH \rightarrow CO \rightarrow C + O \rightarrow CH_4$). In this reaction, dissociation of CO into atomic carbon and atomic oxygen has shown the highest intrinsic barrier as compared to the other elementary steps and is suggested to be a rate-limiting step. Nørskov et al. proposed an alternative route for CO dissociation in the electrochemical reduction model for $CO_2$ reduction to $CH_4$ by performing DFT calculation on a Pt(111) surface [46]. It was suggested that the $CO_2$ dissociation to CO may occur through the carboxyl intermediate, and the subsequent dissociation of CO is assisted by hydrogen, wherein the activation barriers are relatively lower (48.2 kJ/mol) as compared to the direct CO dissociation to carbon and atomic oxygen (286.6 kJ/mol on Ni(111)) [42]. In this proposed model, the hydrogen-assisted dissociation reaction could proceed in the sequence: $CO \rightarrow COH \rightarrow C \rightarrow CH_4$.

Similar to $CO_2$ hydrogenation to COOH (carboxyl) or HCOO (formate), CO hydrogenation to hydroxymethylidyne (COH) or formyl (HCO) forms the key in deciding upon the selectivity of a product. For example, few studies have suggested HCO as an active precursor for $CO_2$ hydrogenation to $CH_4$ on a Cu electrode [47]. The hydrogen atom attached to the carbon of CO assists in the activation of the C–O bond which on subsequent hydrogenation leads to the formation of $CH_4$. The sequence of elementary steps of this route is: $CO_2 \rightarrow COOH \rightarrow CO \rightarrow HCO \rightarrow HCOH \rightarrow H_2COH \rightarrow H_2C \rightarrow CH_4$. Asthagiri et al. have discussed on the rate-determining step and selective formation of $CH_4$ by the electrocatalytic reduction of $CO_2$ on Cu electrodes [48]. The authors have suggested that the rate-determining step of the mechanism is after the CO formation on the electrode surface. Once the CO is formed on the catalyst

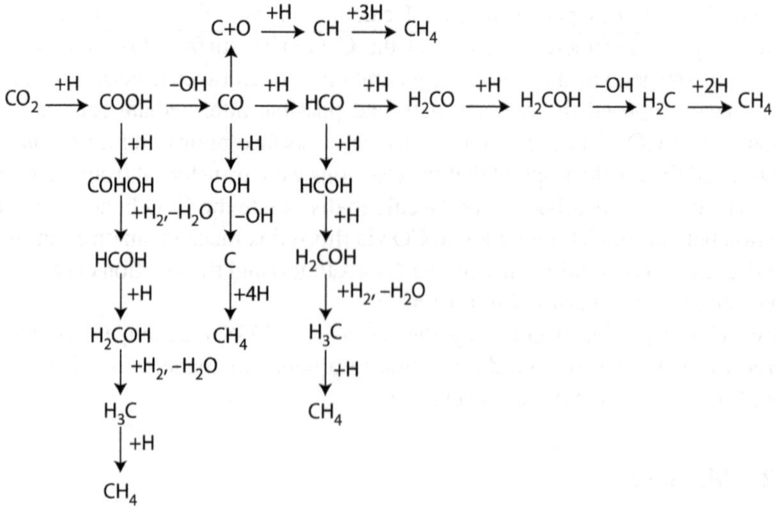

**FIGURE 6.3**   Reaction mechanisms of $CO_2$ reduction to $CH_4$.

surface, it may either hydrogenate to HCO or COH. The activation barrier for CO hydrogenation to COH was observed to be lower (by 17.4 kJ/mol) than the hydrogenation to HCO. Therefore, the path mediated by the COH intermediate is considered favorable for $CO_2$ hydrogenation to $CH_4$. A similar mechanism for $CH_4$ formation via the COH intermediate has also been proposed on Fe(100) surface [44,47].

It is not necessary that methane will be produced only via the formation of CO. Interestingly, $CO_2$ hydrogenation to methane on Ni(110) surface showed relatively higher barriers for the elementary steps of methane formation via the CO intermediate. Alternatively, a reaction mechanism via the formation of COHOH ($CO_2$ → COOH → COHOH → HCOH → $H_2$COH → $H_3$C → $CH_4$) is proposed (Figure 6.3), which was calculated to be energetically more favorable than the one proceeding through the formation of CO or HCOO [49].

## 6.2.3 ETHYLENE

High surface coverages of CO produced on $CO_2$ hydrogenation could lead to C–C coupling reactions via lateral interactions [45]. C–C coupling reactions followed by subsequent hydrogenation yield ethylene, as shown in Figure 6.4 [45,47]. CO dimers are suggested as a precursor for ethylene formation, which are calculated to be stable intermediates on a Cu(100) surface [47]. CO dimerization is followed by its successive hydrogenation to enediol, enediolate, and ethylene. The suggested rate-limiting step of this route is the coupling reaction between CO to form the dimer [47].

In an alternative route, it is proposed that $CH_2$ intermediates produced on CO hydrogenation can dimerize via the coupling reaction and successively hydrogenate to produce ethylene on a Cu electrode [48] (Figure 6.4). Hydrogenation of CO to HCO is the rate-determining step of this path. Several studies on catalytic as well as electrocatalytic hydrogenation of $CO_2$ have reported the formation of $C_2H_4$. At a significantly negative potential (−1.0 V vs. reversible hydrogen electrode [RHE]), Cu has been known as an active electrocatalyst for $CO_2$ hydrogenation to $C_2H_4$ [13,50,51].

## 6.2.4 FORMIC ACID

Formic acid is produced from $CO_2$ through two routes (Figure 6.5): the first route involves the formate (HCOO) intermediate and the second route involves the carboxyl intermediate (COOH). The likelihood of the formate or carboxyl-mediated mechanism depends on the nature of the surface morphology of the catalyst. For example, on a Cu(100) surface, $CO_2$ first hydrogenates to HCOO and then to formic acid (HCOOH) [52]. DFT calculations were performed for $CO_2$ hydrogenation on a

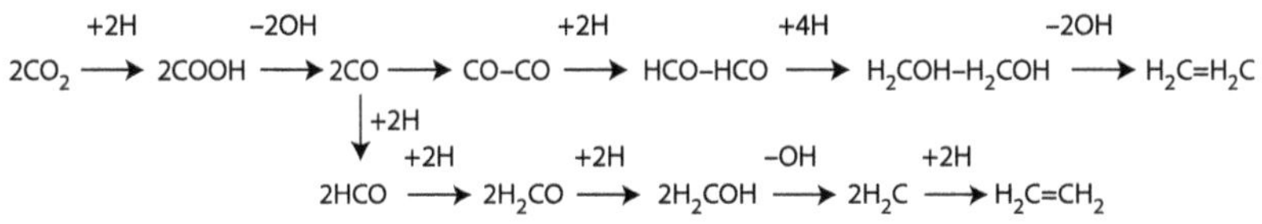

$$2CO_2 \xrightarrow{+2H} 2COOH \xrightarrow{-2OH} 2CO \longrightarrow CO\text{-}CO \xrightarrow{+2H} HCO\text{-}HCO \xrightarrow{+4H} H_2COH\text{-}H_2COH \xrightarrow{-2OH} H_2C\text{=}H_2C$$

$$\downarrow{+2H}$$

$$2HCO \xrightarrow{+2H} 2H_2CO \xrightarrow{+2H} 2H_2COH \xrightarrow{-OH} 2H_2C \xrightarrow{+2H} H_2C\text{=}CH_2$$

**FIGURE 6.4** Reaction mechanism for $CO_2$ reduction to $C_2H_4$.

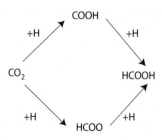

**FIGURE 6.5** Reaction mechanism for $CO_2$ reduction to HCOOH. (Reprinted with permission from Herron A., Scaranto J., Ferrin P., Li S., Mavrikakis M., Trends in formic acid decomposition on model transition metal surfaces: A density functional theory study, *ACS Catal.* 4(12), 4434 – 4445. Copyright 2014 American Chemical Society.)

Cu(111) surface [30]. The calculated activation energy for the formation of formate ($CO_2 + H \rightarrow HCOO$) is calculated to be slightly lower (by 8 kJ/mol) than the hydrogenation of formate to HCOOH. Therefore, on a Cu(111) surface, the rate-determining step was suggested to be the hydrogenation of HCOO to HCOOH. However, it needs to be further checked by careful experiments, considering that the activation barriers are of similar values for the two intrinsic steps. Similarly, DFT calculations on a Ni(111) surface showed that HCOO is more stable than COOH. Consequently, reaction mechanism mediated by the formate intermediate is more favorable on a Ni(111) surface.

Alternatively, the second route involves COOH intermediate formation. Hydrogenation of $CO_2$ to COOH ($CO_2 + H \rightarrow COOH$) is exothermic on the Cu(111) and $CeO_2$(110) surface whereas it is endothermic on Ag(111) and Pd(111) [53]. It was calculated on Cu(111) catalyst that the hydrogenation of COOH leads to the formation of COHOH with a lower activation energy as compared to the formation of HCOOH. Therefore, formic acid is unlikely to be produced via the carboxyl intermediate.

### 6.2.5 METHANOL

Mechanisms for the study of $CO_2$ hydrogenation to methanol have been studied extensively [30,31,44,48,54–57]. Suggested routes for $CO_2$ hydrogenation to methanol are illustrated in Figure 6.6. $CO_2$ could be hydrogenated either through formate [58–60] or carboxyl-mediated pathways to yield methanol [22]. Formic acid (HCOOH) produced by the hydrogenation of formate (HCOO) can subsequently hydrogenate to $H_2COOH$ and on further dissociation and hydrogenation lead to the formation of

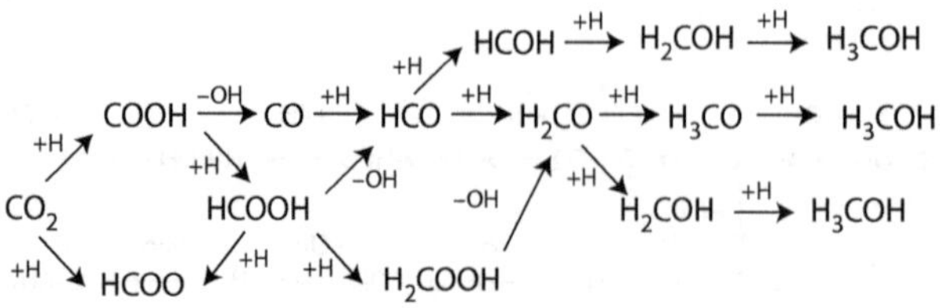

**FIGURE 6.6** Reaction mechanism for $CO_2$ hydrogenation to $CH_3OH$.

methanol [61]. The important intermediates and sequence of elementary steps are: $CO_2 \rightarrow HCOO \rightarrow HCOOH \rightarrow H_2COOH \rightarrow H_2COOH \rightarrow H_2CO \rightarrow H_3CO \rightarrow H_3COH$. On studying the micro kinetic model for this reaction on a Cu surface, hydrogenation of $H_3CO$ was observed to be the slowest step [61]. However, Zhao et al. have suggested that the formate species, produced by $CO_2$ hydrogenation are tightly bound to the catalyst surface and therefore it is difficult to further hydrogenate [30]. It may very well act as a spectator in the entire reaction. Furthermore, formic acid (HCOOH) is calculated to be weakly bound to the Cu(111) surface. It may desorb from the surface as a by-product or dissociate into the relatively stable species, HCOO.

These observations lead to the assertion that the hydrogenation of $CO_2$ to yield methanol may proceed via carboxyl (COOH) intermediates [30]. In this route, (COOH) produced by the hydrogenation of $CO_2$ can further hydrogenate to dihydroxycarbene (COHOH). Dissociation of COHOH can form hydroxymethylidene (COH) which on successive hydrogenation produces methanol. Sequences of this route are: $CO_2 \rightarrow COOH \rightarrow COHOH \rightarrow COH \rightarrow CHOH \rightarrow CH_2OH \rightarrow CH_3OH$ (Figure 6.6). COOH and COH are the key intermediates of this mechanism. The formation barrier for these two intermediates is significantly high at 112.9 and 97.4 kJ/mol. The other alternative route for methanol synthesis from $CO_2$ is the reduction of $CO_2$ to CO via RWGS reaction (via the formation of carboxyl) and then consecutive hydrogenation of CO to $CH_3OH$ [48,54,62]. In support of the carboxyl (COOH) acting as the precursor for methanol formation, the rate of conversion of $CO_2$ to methanol via COOH intermediate was observed to be faster than via HCOO, on a metal-doped Cu(111) surface [54]. On performing experiments, Graciani et al. have studied the simultaneous hydrogenation of $CO_2$ and CO on $Cu/CeO_{2-x}$ catalyst [62]. DFT calculations were performed at the interface of Cu and $CeO_2$ to study the thermochemistry of elementary steps involved in the hydrogenation process ($CO_2 \rightarrow COOH \rightarrow CO \rightarrow HCO \rightarrow H_2CO \rightarrow H_3CO \rightarrow H_3COH$) [62]. The activation barrier for $CO_2$ hydrogenation to COOH, and CO hydrogenation to HCO was calculated to be of similar in values, 45.9 and 58.5 kJ/mol, respectively, and comparable to the experimentally measured apparent activation energy of $CO_2$ hydrogenation on Cu. While the study suggests an important role of the interface of Cu and $CeO_2$ in reducing $CO_2$ to CO and then to methanol, it remains unclear whether the extended ceria surface may also activate $CO_2$ to produce methanol. In a DFT study on extended $CeO_2$ (110) surface, Kumari et al. have examined all suggested mechanisms for the transformation of $CO_2$ to methanol [31] on an extended $CeO_2$ (110) surface. Both carboxyl- and formate (HCOO)-mediated paths were considered for methanol synthesis. The formate intermediate was calculated to be tightly bound to the ceria surface and is unlikely to participate in the synthesis of methanol. On the contrary, the carboxyl-mediated route showed exothermic elementary steps all the way, except only to the slightly endothermic ($\Delta H = 5$ kJ/mol) dissociation of COOH into CO and OH as shown in Figure 6.7. The proposed route is similar to the one suggested for the $Cu/CeO_2$ interface. Dissociation of COOH was suggested to be rate determining as the calculated activation energy was significantly higher ($E_a = 126$ kJ/mol) [31]. Experimental studies on both catalytic and electrocatalytic environments are therefore required to further explore the possibility of methanol synthesis on an extended $CeO_2$ surface.

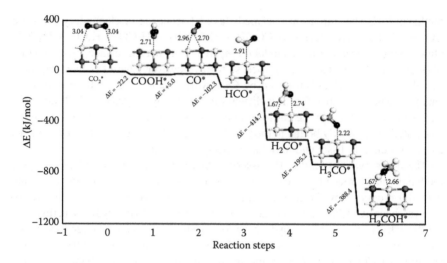

**FIGURE 6.7**   Energy diagram for $CO_2$ hydrogenation to $CH_3OH$. (Reprinted from *Electrochim. Acta*, 2, Kumari N., Sinha N., Haider M.A., Basu S., $CO_2$ reduction to methanol on $CeO_2$ (110) surface: A density functional theory study, Copyright 2015, with permission from Elsevier.)

In the next section, the importance of reaction conditions for both catalytic and electrocatalytic reaction is highlighted, whereas the selectivity to a desired product may get improved on tuning the reaction mechanism via a thoughtful choice of catalyst, support, and promoter for $CO_2$ hydrogenation reactions.

## 6.3   REACTION CONDITIONS

### 6.3.1   CHEMICAL REDUCTION

In chemical reduction of $CO_2$ to fuel and chemicals, mostly heterogeneous catalysts are employed. Heterogeneous catalysis on $CO_2$ reduction via hydrogenation is carried out at mid temperature range (483–543 K) and 2–5 MPa pressures [63] over a variety of catalysts which can be classified as: Cu-based catalyst, ceria-based catalysts, and miscellaneous supported metal catalysts as shown in Table 6.2. At thermodynamic equilibrium conditions, a maximum of 30% $CO_2$ conversion can be achieved with the methanol selectivity decreasing from 90% to 50% on increasing the temperature from 473 to 553 K [63]. Cu supported with ZnO or $ZrO_2$ is widely utilized for $CO_2$ hydrogenation to valuable products like methanol and CO [64]. ZnO helps in enhancing Cu dispersion on the surface and thus preventing further agglomeration [23]. Experimental studies on Cu/ZnO have shown up to 20% $CO_2$ conversion with product selectivity toward methanol and CO estimated to be 40% and 58%, respectively at 543 K. $CO_2$ hydrogenation to methanol is exothermic in nature, and its selectivity increases with the decrease in reaction temperature. On the contrary, CO selectivity increases with increase in temperature. At 483 K, the selectivity toward methanol and CO was observed to be 55% and 44%, respectively; however, with a relatively lower $CO_2$ conversion (~10%) [6]. This indicated

**TABLE 6.2**

**Catalyst Materials, Operating Temperature, Conversion, and Product Selectivity for Heterogeneous Catalytic Reduction of $CO_2$**

| Catalysts | Temperature (K) | $CO_2$ Conversion (%) | Selectivity (%) | | | Reference |
| | | | $CH_4$ | $CH_3OH$ | CO | |
|---|---|---|---|---|---|---|
| Cu/ZnO | 543 | 20.5 | 2.1 | 40.1 | 57.7 | [6] |
| Cu/ZnO | 483 | 9.7 | | 55.2 | 44.1 | [6] |
| Cu/ZrO$_2$ | 493 | 8 | | 49 | | [66] |
| Cu/ZrO$_2$/ZnO | 483 | 17 | | 55 | 44 | [6] |
| Cu/ZrO$_2$/ZnO | 543 | 28 | | 21 | 49 | [6] |
| Cu/Zn/Zr | 513 | 19 | | 50.3 | 49.7 | [63] |
| Cu/Ga$_2$O$_3$/ZnO | 523 | 6.0 | | 88 | 12 | [67] |
| Cu/Ga$_2$O$_3$/ZrO$_2$ | 473 | 13.7 | | 75.5 | 24.4 | [68] |
| Cu/ZnO/SiO$_2$ | 523 | 1.8 | | 99.5 | | [67] |
| Cu/ZnO/Ga$_2$O$_3$/SiO$_2$ | 523 | 3 | | 99 | | [67,69] |
| Cu/ZnO/Ga$_2$O$_3$/SiO$_2$ | 543 | 6 | | 76 | 24 | [67,69] |
| Cu/Zn/Al | 513 | 20.1 | | 55.9 | 44.1 | [63] |
| Cu/Zn/Al/Zr | 513 | 20.5 | | 61.0 | 39 | [63] |
| Ni/Al$_2$O$_3$ | 523 | 14.5 | 94 | 0 | 6 | [70] |
| Ni/Mo/Al$_2$O$_3$ | 523 | 12.9 | 95 | | 5 | [70] |
| Rh/Al$_2$O$_3$ | 443 | 10 | 99 | | | [71] |
| Ag/ZnO/ZrO$_2$ | 453 | 2 | | 97 | | [72] |
| Au/ZnO/ZrO$_2$ | 453 | 1.5 | | 100 | | [72] |
| Ni/CeO$_2$ | 503 | 2.5 | 91.2 | 0 | 8.8 | [7,73] |
| Pd/CeO$_2$ | 503 | 3.1 | 1.0 | 91.8 | 7.2 | [73] |
| Pt/CeO$_2$ | 503 | 8.1 | 17.9 | 68.2 | 13.9 | [7,73] |
| Ir/CeO$_2$ | 550 | | | | 97 | [7] |
| Ru/CeO$_2$ | 500 | 8.2 | 98.8 | 0 | 1.2 | [7] |
| Pd/ZnO/CNT | 523 | 6.3 | | 99.6 | | [74] |
| Pd/Ga$_2$O$_3$/SiO$_2$ | 523 | | | 68.9 | 27.6 | [75] |
| Pd/Mg/SiO$_2$ | 723 | 59 | 95 | | | [43] |

that at a higher temperature, RWGS reaction became favorable. The selectivity and yield of methanol, on a Cu/ZnO catalyst was observed to increase, if the hydrogenation reaction is carried out in the presence of CO. Here, CO acts as a reducing agent by creating oxygen vacancies on the metal oxide support and thus facilitating the hydrogenation reaction [6]. Cu/ZnO was observed to be active for methanol synthesis from CO and $H_2$. Water is produced as a by-product in the RWGS reaction, which may act as an inhibitor of active sites on the Cu/ZnO surface [6].

In place of ZnO, $ZrO_2$ could be a replacement to achieve significantly higher methanol selectivity with high catalyst stability. The activity of Cu/$ZrO_2$ interface has been studied by DFT calculations and kinetic Monte Carlo simulations. The calculated conversion of $CO_2$ and selectivity of methanol were 10% and 85%, respectively, [22,65] suggesting high activity of the Cu/$ZrO_2$ interface.

With significant $CO_2$ conversion (~8%), methanol selectivity up to 49% at 493 K is reported experimentally on the $Cu/ZrO_2$ catalyst (Table 6.2) [66]. The combination of $Cu/ZrO_2$ and ZnO may provide a highly active and selective catalyst for methanol synthesis. The $ZrO_2$-doped Cu/ZnO catalyst showed improved Cu dispersion leading to $CO_2$ conversion up to 17% at 483 K and 28% at 543 K, as compared to Cu/ZnO as shown in Table 6.2 [6]. Selectivity of methanol and CO on this catalyst was observed to be comparable to that on a Cu/ZnO catalyst. Besides the metal oxide composites, pure metal alloys of Cu/Zn/Zr are also active for $CO_2$ hydrogenation, where similar $CO_2$ conversion (~19%) and methanol selectivity (~50%) are reported (Table 6.2).

Addition of $Ga_2O_3$ to Cu-based catalyst improves the $Cu^+$ dispersion on the surface. Therefore, $Ga_2O_3$ is suggested to be a promoter to improve methanol selectivity [67,68]. For example, methanol selectivity on $Cu/Ga_2O_3/ZnO$ catalyst is 88%, albeit, with relatively low $CO_2$ conversion (~6%) [67]. Interestingly, $Ga_2O_3$ promoted $Cu/ZrO_2$ has shown higher $CO_2$ conversion (~14%), while retaining high methanol selectivity (~75.5) (Table 6.2). CO selectivity was observed to improve on a $Cu/Ga_2O_3/ZrO_2$ catalyst as compared to the $Cu/Ga_2O_3/ZnO$ catalyst as shown in Table 6.2. It could therefore be inferred that $ZrO_2$ as a support is relatively less active for CO or CO hydrogenation to methanol as compared to ZnO [76]. $Cu/Ga_2O_3/ZnO$ catalysts supported on silica have shown high selectivity for methanol (up to 99%); however, the conversion of $CO_2$ is limited only to 6% at 543 K (Table 6.2). The effect of silica may be hypothesized to inhibit the CO desorption reaction and to subsequently hydrogenate the CO to methanol [77].

Besides Ga and Zr, Al or $Al_2O_3$ are well-known promoters for Cu-based catalysts for $CO_2$ or CO hydrogenation reactions [70]. Al is shown to be effective in improving the performance of the catalyst in terms of its activity, selectivity, and stability [76]. Cu/Zn/Al alloy has been widely studied for methanol production from syngas [23,78,79]. Similarly, combinations like Cu/Zn/Al and Cu/Zn/Zr/Al catalysts have been studied for methanol synthesis [63]. The results suggested significant hydrogenation of $CO_2$ to methanol on the Cu/Zn/Zr/Al catalyst due to the promotional activity of Al and Zr, and better catalyst preparation methods [63]. Alumina-supported metal catalysts such as $Ni/Al_2O_3$ [70], $Ni/Mo/Al_2O_3$ [70], and $Rh/Al_2O_3$ [71] have shown high $CO_2$ conversions; however, the product is primarily methane with negligible methanol formation. It is expected that all these metal catalyst supported on alumina led to CO dissociation on the surface. Experimental studies have shown improvement in conversion with increasing partial pressure of hydrogen on the $Ru/Al_2O_3$ catalyst pointing toward the rate limitations offered by the hydrogenation steps. Conversion of $CO_2$ and selectivity of $CH_4$ on the mono-metallic (Ni) and bi-metallic Ni–Mo catalysts are similar in values, suggesting negligible effect of Mo promotion (Table 6.2) [70].

In the periodic table, besides Cu, elements of the same column, Au and Ag, are expected to play a similar role in $CO_2$ hydrogenation reactions. $ZnO/ZrO_2$-supported Au and Ag catalysts have been tested for methanol synthesis [72] and the results are reported in Table 6.2. While high selectivity (up to 100%) is reported toward methanol, the conversions are significantly low (up to 2%). The results suggested that both Au and Ag preferentially sit on the active sites for carbon hydrogenation and RWGS reaction and thus allow only methanol formation on relatively fewer sites which are available for the reaction.

Ceria ($CeO_2$)-supported noble metals (Ru, Ni, Pd, and Pt) catalyze $CO_2$ reduction to CO, $CH_3OH$, and $CH_4$ [6]. These metals surfaces act as a repository of hydrogen species to induce surface oxygen vacancy formation in ceria. The role of ceria support in the reduction reaction is important due to its high oxygen uptake and release capacity. $CO_2$ conversion on metal-supported ceria catalyst decreases in the order Ru > Pt > Pd > Ni (Table 6.2). Interestingly, similar trends have been reported for CO hydrogenation reactions on these metals [80]. The tendency to form methane is high on ceria-supported Ni and Ru catalyst, which is due to their known proclivity toward CO dissociation. For example, Ru catalyzes $CO_2$ hydrogenation to methane with selectivity up to 99% and conversion up to 7%. On the contrary, Pt and Pd tend to associatively adsorb CO and thus are favored for methanol production showing selectivity up to 68.2% and 91.8%, respectively, as given in Table 6.2. In a relatively recent study, Rodriguez et al. reported 200 fold enhancements in the turn over frequencies of $CO_2$ reduction to methanol on adding $CeO_2$ to the Cu catalyst [62]. The surface of Cu alone was showing significantly lower conversion of $CO_2$ to methanol. However, when coupled with a reducible oxide ($CeO_2$), the activity was observed to enhance by several orders. Cation atoms of ceria placed at the interface of $Cu(111)/CeO_2$ may show multiple oxidation states, for example, from +4 to +3, which enhances the metal support interaction, thus generating a high number of active sites for $CO_2$ hydrogenation.

The product selectivity of $CO_2$ hydrogenation to methanol may change with the change in support suggesting strong metal support interactions. For example, the Pd–ZnO catalyst, supported and promoted by carbon nanotubes (CNTs) has shown high selectivity (99.6%) for methanol at 523 K (Table 6.2) [74]. At similar temperature, $Ga_2O_3$ promoted Pd catalysts supported with $SiO_2$ are active toward methanol formation with selectivity of 69% as well as to CO formation via RWGS reaction with CO selectivity of 27%. On the contrary, when Pd is well dispersed with Mg on silica support, it produces methane, at a temperature of 723 K, with high selectivity (~95%) and conversion up to approximately 59% [43].

## 6.3.2 Electrocatalytic Reduction

### 6.3.2.1 Low-Temperature Reduction

In the heterogeneous catalyzed reduction of $CO_2$, measured conversion and selectivity are observed to be lower which may improve on electrochemical reduction. Electrocatalytic reduction of $CO_2$ produces a variety of useful chemicals such as methane, methanol, formic acid, and ethylene which are shown in Table 6.3 [81–84]. The position of the metal electrodes studied for low-temperature electrochemical $CO_2$ reduction is shown in the form of an abridged periodic table indicating the primary products of the reaction in Figure 6.8.

At room temperature, selective hydrogenation of $CO_2$ to ethylene was performed on a Cu mesh electrode [13]. Ethylene was produced with selectivity up to approximately 74% at a potential of −1.8 V and with a faradaic efficiency up to 69.5% [13]. At this potential, the hydrogen evolution reaction was suppressed and its faradaic efficiency was limited to only 12% (Table 6.3). In another study, Cu led to the formation of CO,

## TABLE 6.3

### Electrode Materials, Operating Conditions, and Faradaic Efficiency of Products for Low-Temperature Electrocatalytic Reduction of $CO_2$

| Cathode Catalyst | Potential (V) | Faradaic Efficiency (%) | | | | | | Reference |
|---|---|---|---|---|---|---|---|---|
| | | $CH_4$ | $C_2H_4$ | HCOOH | $CH_3OH$ | CO | $H_2$ | |
| Cu | −1.8[a] | 2.7 | 69.5 | 1.8 | 5.6 | 4.3 | 12.3 | [13] |
| Cu | −1.64[a] | 10.2 | 3.7 | 53.7 | | 20.1 | 2.5 | [8] |
| Cu | −0.6[b] | | | 35 | | 28 | | [14] |
| Cu(111) | −1.5[c] | 46.3 | 8.3 | 11.5 | | 6.4 | 16.3 | [85] |
| Cu(100) | −1.4[b] | 30.4 | 40.4 | 3.0 | | 0.9 | 6.8 | [85] |
| Cu(111) | −1.4[b] | 3.6 | 10 | | 0.1 | 4.6 | 54.3 | [15] |
| Cu/ZnO | −0.8[b] | 1.8 | 10.1 | | 2.8 | 5.4 | 45.1 | [15] |
| $Cu_2O/Zn$[d] | −3.0[a] | 7.5 | 6.8 | 25 | | 19 | 22 | [82] |
| CuO/Zn[d] | −3.0[a] | 1.0 | 5.0 | 25 | | 22 | 30 | [82] |
| Ag | −1.48[a] | | | 16.8 | | 75.6 | 3.9 | [8] |
| Au | −1.30[a] | | | 11.8 | | 64.7 | 15.4 | [8] |
| Ni | −1.59[a] | | | 31.3 | | 33.5 | 26.0 | [8] |
| Pd | −1.76[a] | | | 44.0 | | 35.2 | 13.8 | [8] |
| Pt | −1.48[a] | | | 50.4 | | 6.1 | 33.6 | [8] |
| Co | −1.54[a] | 3.0 | 0.4 | 21.9 | | 15.8 | 46.9 | [8] |
| Rh | −1.41[a] | 3.0 | | 19.5 | | 61.0 | 13.1 | [8] |
| Ir | −1.55[a] | | | 22.3 | | 17.5 | 48.3 | [8] |
| Fe | −1.63[a] | 2.0 | 0.1 | 28.6 | | 4.2 | 51.6 | [8] |
| $RuO_2/TiO_2$/ Pt NT | −0.8[c] | | | | 60.5 | | | [83] |
| Zn | −1.70[a] | | | 40.5 | | 48.7 | 2.8 | [8] |
| In | | | | 90.1 | | 3.8 | 5.6 | [8,86] |
| Sn | −1.39[a] | | | 92.3 | | 8.0 | 1.3 | [8] |
| Pb | −1.57[a] | | | 95.5 | | | 1.2 | [8] |

*Note:* With reference to [a]Ag/AgCl, [b]RHE, [c]SCE, [d]reaction temperature was 243 K and the rest of the experiments were performed at room temperature (298 K).

formic acid, methane, and ethylene at a potential of −1.64 V [8]. Formic acid is reported to be the primary product at this potential on the Cu electrode with a current efficiency of around 53.7% [8]. Water electrolysis to $H_2$ with a faradaic efficiency of 2.5% was observed to be significantly reduced [8]. Kanan et al. studied the electrochemical reduction of $CO_2$ on a Cu electrode prepared by the reduction of about 3 μm thick film of annealed $Cu_2O$ [14]. The Cu electrode catalyzes the $CO_2$ reduction to CO and to formic acid with a faradaic efficiency of 28% and 35%, respectively, at a relatively lower potential of −0.6 V [14]. These studies on Cu electrodes suggest that the product selectivity on Cu electrodes vary with the electrode potential. In addition, at the similar value of potential, there is a possibility to obtain different selectivity of the same product. This may be possible due to the change in the surface catalytic activity [16].

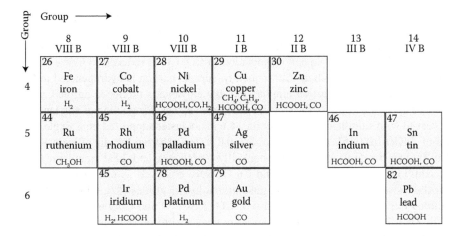

**FIGURE 6.8** Positions of metal electrodes, discussed in Table 6.3, in the periodic table showing major products obtained from $CO_2$ hydrogenation.

In order to explore the activity of the catalyst surface, surface science studies on Cu electrodes, Cu(111) and Cu(100), were carried out for the electrocatalytic reduction of $CO_2$ [15,85]. Hoshi et al. [85] observed the Cu(111) surface to be more active toward methane formation (faradaic efficiency of 46% at −1.5 V), whereas Cu(100) was favorable toward ethylene production (faradaic efficiency of 40% at −1.4 V) [85]. On the contrary, Andrew et al. have reported hydrogen as a primary product with the faradaic efficiency of 54% while hydrocarbons are produced with lesser faradaic efficiency on a Cu(111) surface (3.6% for methane and 10% for ethylene) [15].

A very little work has been reported on the use of Cu nanoparticles as a cathode catalyst and the reduction of $CO_2$ in gaseous form instead of absorbing in water. In most of the electrochemical cell, as mentioned above, the Cu plate has been used as a cathode instead of the Cu nanoparticle deposited on a high surface area substrate. Li et al. have clearly identified the disadvantages of $CO_2$ reduction in an aqueous medium, which include sluggish reduction kinetics, low selectivity, formation of various by-products, low solubility of $CO_2$ in aqueous electrolyte, deactivation, and low tolerance to impurities and contamination [87]. Garg and Basu have reported for the first time the liquid-phase as well as gas-phase electrochemical reduction of $CO_2$ on electrodes coated with Cu nanoparticles in an electrochemical reduction cell (ERC) with nafion 117 as electrolyte [88]. The product from ERC was analyzed in a mass spectrometer and high-performance liquid chromatograph and their variation with time is reported. The monitoring of product formation with time revealed that $CO_2$ is converted to CO which further hydrogenated to $CH_4$ and $CH_3OH$. Overall faradaic efficiency corresponding to $CO_2$ reduction was 39.6% whereas hydrogen production is suppressed to 4.5% till the ERC operation of about 40 min. Thereafter, the efficiency of ERC decreases and hydrogen evolution increases. Scanning electron microscope (SEM), energy dispersive x-ray spectroscopy (EDS), and x-ray diffraction (XRD) comparison of Cu nanoparticle coated electrode before and after the ERC operation reveals that Cu dissolution from the cathode may be the reason for the ERC deterioration.

Electrochemical reduction of $CO_2$ on a Cu(111) surface was explored by DFT studies [57]. Calculations on Cu(111) suggested that the presence of water in solvation layer facilitates the O–H formation which helps in the electrocatalytic conversion of $CO_2$ to COOH and subsequent dissociation to CO. CO is reduced to C via the formation of COH intermediate. Subsequently, surface carbon is hydrogenated to $CH_4$ [57]. The calculations further emphasized that ethylene is produced on the Cu(111) surface by the coupling of the $CH_2$ intermediate species. Production of ethylene was observed with significant selectivity on a Cu nanocluster supported on a single crystal of ZnO(1010) [15]. At a potential of −0.8 V, the reported faradic efficiency on a Cu/ZnO electrode was 1.8% (methane), 10% (ethylene), and 2.8% (methanol) [15]. In addition, oxides of Cu (CuO and $Cu_2O$) were observed as effective electrodes for the formation of saturated and unsaturated hydrocarbons [82]. Ohya et al. have studied the electrocatalytic reduction of $CO_2$ on $Cu_2O$ and CuO electrodes mixed with ZnO [82]. $Cu_2O$/ZnO electrodes were observed to produce hydrocarbons, for example, $CH_4$ and $C_2H_4$ with relatively higher faradaic efficiencies, 7.5% and 6.8%, respectively (Table 6.3) than CuO/ZnO [82]. On the contrary, faradaic efficiency of hydrogen formation is higher on CuO [82]. Cu with 2+ oxidation state showed less affinity for hydrogen adsorption, leading to desorption and $H_2$ evolution. Therefore, in the absence of any interaction with $H_2$, CO was observed as a main product on a CuO catalyst. In contrast, both CO and H were observed to be adsorbed on the $Cu_2O$ surface leading to subsequent hydrogenation and formation of hydrocarbon products.

The electrochemical reduction of $CO_2$ was performed on the transition metals belonging to group number 4–15. Ag and Au are known to produce CO as the main product of electrocatalytic reduction of $CO_2$. On an Ag electrode, CO yield is observed with a current efficiency of 75% and formic acid is observed with 17% faradaic efficiency, at a potential of around −1.5 V. Similarly, on an Au electrode, the current efficiency of CO and formic acid are around 65% and 12%, respectively, at −1.3 V. Metals belonging to group 10, Ni, Pd, and Pt, electrocatalyze the $CO_2$ primarily to CO and formic acid at relatively higher potential (−1.48 to −1.7 V) [8]. The faradaic efficiency of formic acid increases in the order of Ni > Pd > Pt. In group 9 metals (Co, Rh, and Ir), formic acid is produced with similar values of faradaic efficiency of approximately 20%; however, faradaic efficiency for CO was maximum, 61%, for Rh as compared to Co and Ir (Table 6.3). A high hydrogen evolution reaction was observed on Co and Ir electrodes with faradaic efficiency of 47% and 48%, respectively. Ru and Ti oxides are important for the synthesis of methanol from $CO_2$. $RuO_2$/$TiO_2$-promoted Pt electrode was studied for electrochemical reduction of $CO_2$ which led to the formation of methanol with a faradaic current efficiency of 60% [83]. Zn metal, belonging to group 12, showed high selectivity toward formic acid with a faradaic efficiency up to 40.5% and for CO faradaic efficiency was up to 48.7%, at a potential of −1.7 V. Similarly, group 13, In, and group 14, Sn and Pd, metals have shown the formation of formic acid as a primary product with reported faradaic efficiency of 90.1%, 92.3%, and 95.5%, respectively (Table 6.3).

## 6.3.2.2 High-Temperature Reduction

Catalyst poisoning by CO is a matter of concern in the low-temperature electrocatalytic reduction of $CO_2$. Low-temperature liquid-phase electrolysis operations also suffer a major drawback from corrosion and the loss of electrolyte [89]. To overcome these issues, high-temperature electrocatalytic reduction in a solid oxide electrolysis cell (SOEC) is suggested which may result in significantly lower Nernst potential and electrode overpotentials [90]. In a SOEC, the electrolyte conducts oxygen anions which is activated at high temperatures (>773 K). Materials which are commonly utilized as electrodes in a SOFC are being tested for $CO_2$ reduction and the results are shown in Table 6.4. Initial experiments of high-temperature electrolysis of $CO_2$ to produce oxygen were conducted at National Aeronautics and Space Administration (NASA), where a Pt electrode was utilized in combination with a YSZ electrolyte [91]. At the Pt electrode, $CO_2$ electrolysis was observed to be kinetically limited at higher potential (>1.0 V) and at relatively lower potential negligible $CO_2$ reduction was observed [91]. The electrode overpotential loss (>50% at 1.95 V) had a detrimental effect on the cell efficiency. In order to understand the limitations of electrode kinetics, the mechanism of the RWGS reaction was studied on Pt electrodes in a SOEC system. Experimentally, it was observed that the rate of CO formation increases with an increase in temperature. The favored mechanistic route of $CO_2$ hydrogenation was attributed to the formation of formate intermediates at the Pt electrode [92]. In another study, the NASA research group utilized a cermet of Pt/YSZ for $CO_2$ reduction to CO and $O_2$. A decrease in ohmic resistance by 2.5 times (Table 6.4) was observed at the anode, indicating better conductive interface and reduced delamination of Pt/YSZ [91]. However, the cathode reaction was observed to be sluggish. In addition, Pt being expensive may not be an economically feasible material for electrochemical $CO_2$ reduction [93]. Therefore, it is necessary to search for a cheaper electrode material and Ni is suggested as an alternative for Pt. Ni is widely used as Ni/YSZ cermet in the anode of a SOFC [93]. $CO_2$ electrolysis on the cermets of Ni/YSZ cathode was conducted for long durations (1283 h) and the effect of electrode passivation rate on electrolysis cell performance was analyzed [93]. At 1123 K temperature, the observed passivation rate was measured to be significantly lower and in the range of 0.22–0.44 mV/h, which signifies the long-term

## TABLE 6.4

### Electrode Materials, Operating Temperatures, and Inlet Gas Compositions for $CO_2$ Reduction in a SOEC

| Electrode Material | Operating Temperature (K) | Inlet Gas Composition | Reference |
|---|---|---|---|
| Pt | 1023–1123 | 100% $CO_2$ | [91] |
| Pt | 923–1023 | $CO_2$:$H_2$ = 1:1 | [92] |
| Pt/YSZ | 1023–1123 | 100% $CO_2$ | [91] |
| Ni/YSZ | 1023, 2730, 1123 | $CO_2$:CO = 1:1 | [93] |
| Ni/YSZ | 1173 | $CO_2$/CO = 9:1, 7:3, 1:1, 3:7 | [19] |
| LSCM/YSZ | 1173 | $CO_2$/CO = 9:1, 7:3, 1:1, 3:7 | [19] |
| LSCM/GDC | 1173 | $CO_2$/CO = 9:1, 7:3, 1:1, 3:7 | [19] |
| Cu/GDC | 1023 | CO/$CO_2$ = 1:1 | [94] |
| GDC | 923–1223 | CO/$CO_2$ = 0.05:1, 0.25:1, and 0.5:1 | [95] |

stability of the electrocatalyst. Interestingly, electrode passivation is not caused by coke deposition, for which Ni is notorious. Instead, the impurities of inlet gas mixture, like $H_2S$, were attributed for the deactivation of the cell by sulfur deposition. On the contrary, Irvine et al. utilized Ni/YSZ as the cathode material for $CO_2$ electrolysis and observed carbon formation on the electrode surface which led to an increase in the polarization resistance of the cathode [19]. A similar phenomenon was also observed on Ni-based cathode by other research groups [94]. Overall Ni-based electrodes may tend to show similar problems as observed in the case of Ni-based anodes for SOFC.

Alternatively, perovskite structured materials ($ABO_{3-x}$), having mixed ion and electron conducting properties have been utilized for $CO_2$ reduction at high temperature [19]. Perovskite material has a face centered crystalline structure in which "A" site metals (such as Ca, Sr, Ba, or La) are located at the corner sites of the unit cell. B site metals (like Co, Fe, or Mn) are located at the center which are octahedrally surrounded by oxide ions ($O^{2-}$) located at the face centered sites. Oxygen vacancies are created by the reduction of "B" site metal and doping of the "A" and "B" site metal ion with aliovalent cations. For example, on reducing the partial pressure of oxygen, $Mn^{3+}$ cation in LSCM reduces to $Mn^{2+}$, leaving an oxygen vacancy in the material. The reaction in Kröger–Vink notation can be represented as

$$4Mn_{Mn}^{x} + 2O_O^{X} \leftrightarrow 4Mn_{Mn}' + 2V_{\ddot{o}} + O_2$$

where $V_{\ddot{o}}$ denotes double charge oxygen vacancy in the electrode material.

Experiments for $CO_2$ reduction were performed on the LSCM/YSZ composite and electrode performance was studied by electrochemical impedance spectroscopy [93]. CO is co-fed with $CO_2$ to generate the reducing environment in the electrolysis system, thus facilitating the oxide anion transport via the formation of oxygen vacancies. The polarization resistance of the electrode was estimated to be significantly higher (1.1 $\Omega$ $cm^2$) and twice the value of the conventional Ni/YSZ electrode under similar reaction conditions (Table 6.4) [93]. These observations indicated that the LSCM/YSZ electrode had insufficient activity toward $CO_2$ electrolysis [19]. In order to improve the electrocatalytic activity of the YSZ-based composite electrode, GDC is utilized as an alternative in a SOFC where oxygen anions can be conducted at intermediate operating temperatures (773–973 K). Following up on the same principles, LSCM/GDC composite materials were tested for applications in a SOEC. The electrochemical performance measured in terms of ohmic loss and I–V characteristics of the LSCM/GDC electrode was observed to be better than the performance of the LSCM/YSZ electrode. The enhanced performance of the electrode could be attributed to the reduction of GDC [96] at high temperature, which helps in facilitating $CO_2$ reduction via the redox mechanism. The electrode further showed a stable performance for 200 h of operation and is a promising cathode material for $CO_2$ electrolysis [97]. Similar to the catalytic and low-temperature electrocatalytic reaction, ceria could play an important role in high-temperature electrochemical $CO_2$ reduction. Cheng et al. have utilized a combination of both Cu and GDC in the form of composite material as a cathode material for a SOEC. 100% current efficiency toward the formation of CO was observed at a potential of 1.5 V (Table 6.4). However, owing to poor dispersion of Cu on GDC, the area specific resistance of the electrode was measured to be 40% greater than the conventional Ni/YSZ electrode. It remained however unclear, whether doped ceria material (e.g., GDC) is also active in CO$_2$ reduction which could itself be a potential electrode material [95]. Adler et al. studied the electrochemical characterization of the GDC electrode in the different proportions of the CO$_2$/CO environment [95]. The area-specific resistance of this electrode was measured to be lower than Pt/YSZ as well as Ni/YSZ electrodes. Furthermore ceria-based materials showed better resistance to coke deposition [98] as compared to Ni-based electrodes. Oxygen vacancy concentrations in ceria could play an important role in improving the kinetics of CO$_2$ reduction via the redox mechanism. The electronic conductivity, oxygen anion transport, and vacancy concentrations can be further tuned by trying various aliovalent dopants such as Gd, Sm, or Pr [99]. Furthermore, surface oxygen vacancy concentrations of ceria are likely to improve its scavenging activity [100]. However, all of these mechanistic hypotheses need to be experimentally tested by fabricating electrodes of doped ceria materials and performing experiments on a SOEC in varying concentrations of the CO$_2$ environment, co-fed with CO or in a co-electrolysis mode with water.

## 6.4  CONCLUSIONS AND FUTURE DIRECTIONS

Hydrogenation of CO$_2$ on a heterogeneous catalyst or an electrocatalyst to a target chemical such as CO, CH$_4$, C$_2$H$_4$, HCOOH, and CH$_3$OH is discussed. The conversion of CO$_2$ to a desired product is carried out by key transformations on the catalyst surface which include CO bond dissociation, CO or CH$_2$ coupling, and C–H, O–H bond formations. The selectivity to a product primarily depends on the catalyst morphology, surface structure, and reaction conditions such as temperature, pressure, and applied potential. Nevertheless, the underlying mechanism could represent similar features irrespective of the applied reaction conditions. For example, carboxyl and formate species are suggested to form as intermediates in both catalytic and electrocatalytic reduction of CO$_2$, which led to the formation of the observed products. This chapter focused on explaining these key mechanistic steps and how they can be engineered by utilizing a different support or a promoter and/or changing the reaction conditions. The concepts introduced in this chapter cover the recent development in CO$_2$ hydrogenation reactions which may find applications in future design of a novel catalyst or an electrocatalyst.

Before we go in for the new design of a catalyst, we still need to thoroughly study and understand C–O bond cleavage and C–C and C–H bond formation, especially for C$_x$ ($x > 2$), in atomistic level on multisite metal catalyst surfaces promoted by second metals and hierarchal nanostructured catalyst support, and metal–organic framework hybrid catalysts. Although this may initially be investigated through DFT calculations to minimize experimental work, the real breakthrough to achieve high conversion and selectivity for CO$_2$ to C$_1$–C$_3$ hydrocarbons fuel lie in *in situ* studies, such as Fourier transform infrared spectroscopy (FTIR)-voltammeter or single crystal, atom chemical reaction-FTIR–XRD, etc. Nanotechnology provides the opportunity to control the surface facets, morphology, hierarchical structure, and multisite networks and may usher in advancement in designing a new catalyst, which may convert $CO_2$ to chemicals and fuels easily.

## REFERENCES

1. Centi G., Quadrelli E.A., Perathoner S., Catalysis for $CO_2$ conversion: A key technology for rapid introduction of renewable energy in the value chain of chemical industries, *Energy Environ. Sci.* 2013; 6(6): 1711–1731.
2. Jiang Z., Xiao T., Kuznetsov V.L., Edwards P.P., Turning carbon dioxide into fuel, *Philos. Trans. R. Soc. A.* 2010; 368(1923): 3343–3364.
3. Müller K., Mokrushina L., Arlt W., Thermodynamic constraints for the utilization of $CO_2$, *Chem. Ing. Tech.* 2014; 86(4): 497–503.
4. National Institute of Standards and Technology, Material measurement laboratory, U.S. Department of Commerce, http://webbook.nist.gov/chemistry/name-ser.html. Accessed on March 15, 2015.
5. Freund H.-J., Roberts M.W., Surface chemistry of carbon dioxide, *Surf. Sci. Rep.* 1996; 25(8): 225–273.
6. Yang C., Ma Z., Zhao N., Wei W., Methanol synthesis from $CO_2$-rich syngas over a $ZrO_2$ doped Cu/ZnO catalyst, *Catal. Today.* 2006; 115(1): 222–227.
7. De Leitenburge, Trovarelli A., Kaspar J., A temperature-programmed and transient kinetic study of $CO_2$ activation and methanation over $CeO_2$ supported noble metals, *J. Catal.* 1997; 166: 98–107.
8. Hara K., Kudo A., Sakata T., Electrochemical reduction of carbon dioxide under high pressure on various electrodes in an aqueous electrolyte, *J. Electroanal. Chem.* 1995; 391(1–2): 141–147.
9. Zhu Y.-A., Chen D., Zhou X.-G., Yuan W.-K., DFT studies of dry reforming of methane on Ni catalyst, *Catal. Today.* 2009; 148: 260–267.
10. Havran V., Dudukovic´ M.P., Lo C.S., Conversion of methane and carbon dioxide to higher value products, *Ind. Eng. Chem. Res.* 2011; 50(12): 7089–7100.
11. Mikkelsen M., Jørgensen M., Krebs F.C., The teraton challenge. A review of fixation and transformation of carbon dioxide, *Energy Environ. Sci.* 2010; 3(1): 43–81.
12. Khodakov A.Y., Chu W., Fongarland P., Advances in the development of novel cobalt Fischer–Tropsch catalysts for synthesis of long-chain hydrocarbons and clean fuels, *Chem. Rev.* 2007; 107(5): 1692–1744.
13. Ogura K., Yano H., Shira F., Catalytic reduction of $CO_2$ to ethylene by electrolysis at a three-phase interface, *J. Electrochem. Soc.* 2003; 150(9): D163–D168.
14. Li C.W., Kanan M.W., $CO_2$ reduction at low overpotential on Cu electrodes resulting from the reduction of thick $Cu_2O$ films, *J. Am. Chem. Soc.* 2012; 134(17): 7231–7234.
15. Andrews E., Ren M., Wang F., Zhang Z., Sprunger P., Kurtz R. et al., Electrochemical reduction of $CO_2$ at Cu nanocluster/(1010) ZnO electrodes. *J. Electrochem. Soc.* 2013; 160(11): H841–H846.
16. DeWulf D.W., Jin T., Bard A.J., Electrochemical and surface studies of carbon dioxide reduction to methane and ethylene at copper electrodes in aqueous solutions, *J. Electrochem. Soc.* 1989; 136(6): 1686–1691.
17. Ching-Shiun C., Wu-Hsun C., Study of iron-promoted Cu/$SiO_2$ catalyst on high temperature reverse water gas shift reaction, *Appl. Catal. A Gen.* 2004; 257(1): 97–106.
18. Chockalingam R., Ganguli A.K., Basu S., Praseodymium and gadolinium doped ceria as a cathode material for low temperature solid oxide fuel cells, *J. Power Sources.* 2014; 250(3): 80–89.

19. Yue X., Irvine J.T.S., Alternative cathode material for $CO_2$ reduction by high temperature solid oxide electrolysis cells, *J. Electrochem. Soc.* 2012; 159(8): F442–F448.

20. Bidrawn F., Kim G., Corre G., Irvine J.T.S., Vohs J.M., Efficient reduction of $CO_2$ in a solid oxide electrolyzer, *Electrochem. Solid-State Lett.* 2008; 11(9): B167–B170.

21. Goeppert A., Czaun M., Jones J.-P., Surya Prakash G.K., Olah G., Recycling of carbon dioxide to methanol and derived products—Closing the loop, *Chem. Soc. Rev.* 2014; 43(23): 7995–8048.

22. Tang Q.-L., Hong Q.-J., Liu Z.-P., $CO_2$ fixation into methanol at $Cu/ZrO_2$ interface from first principles kinetic Monte Carlo, *J. Catal.* 2009; 263(1): 114–122.

23. Behrens M., Studt F., Kasatkin I., Kühl S., Hävecker M., Abild-Pedersen F. et al., The active site of methanol synthesis over $Cu/ZnO/Al_2O_3$ industrial catalysts, *Science.* 2012; 336(6083): 893–897.

24. Gokhale A., Dumesic J., Mavrikakis M., On the mechanism of low-temperature water gas shift reaction on copper, *J. Am. Chem. Soc.* 2008; 130(4): 1402–1414.

25. Grabow L.C., Gokhale A.A., Evans S.T., Dumesic J.A., Mavrikakis M., Mechanism of the water gas shift reaction on PtRe: First principles and microkinetic modeling, *J. Phys. Chem. C.* 2008; 112(12): 4608–4617.

26. Stamatakis M., Chen Y., Vlachos D.G., First-principles-based kinetic Monte Carlo simulation of the structure sensitivity of the water–gas shift reaction on platinum surfaces, *J. Phys. Chem. C.* 2011; 115(5): 24750–24762.

27. Vesselli E., Rizzi M., De Rogatis L., Ding X., Baraldi A., Comelli G. et al., Hydrogen-assisted transformation of $CO_2$ on nickel: The role of formate and carbon monoxide, *J. Phys. Chem. Lett.* 2010; 1(1): 402–406.

28. Li H.-J., Ho J.-J., Density functional calculations on the hydrogenation of carbon dioxide on Fe(111) and W(111) surfaces, *J. Phys. Chem. C.* 2010; 114(2): 1194–1200.

29. Chen Y., Cheng J., Hu P., Wang H., Examining the redox and formate mechanisms for water–gas shift reaction on $Au/CeO_2$ using density functional theory, *Surf. Sci.* 2008; 602(17): 2828–2834.

30. Zhao Y.-F., Yang Y., Mims C., Peden C.H.F., Li J., Mei D., Insight into methanol synthesis from $CO_2$ hydrogenation on Cu(111): Complex reaction network and the effects of $H_2O$, *J. Catal.* 2011(2); 281: 199–211.

31. Kumari N., Sinha N., Haider M.A., Basu S., $CO_2$ reduction to methanol on $CeO_2$ (110) surface: A density functional theory study, *Electrochim. Acta.* 2015; 177(7): 21–29.

32. Daojian Cheng E.A., Negreiros F.R., Computation approaches to the chemical conversion of carbon dioxide, *Chem. Sus. Chem. Rev.* 2013; 6(6): 944–965.

33. Jin T., Okuhara T., Mains G.J., White J.M., Temperature-programmed desorption of CO and $CO_2$ from $Pt/CeO_2$, for lattice oxygen in CO oxidation, *J. Am. Chem. Soc.* 1987; 91(12): 3310–3315.

34. Doornkamp C., Ponec V., The universal character of the Mars and Van Krevelen mechanism, *J. Mol. Catal. A Chem.* 2000; 162(1–2): 19–32.

35. Agarwal J., Fujita E., Schaefer H.F., Muckerman J.T., Mechanisms for CO production from $CO_2$ using reduced rhenium tricarbonyl catalysts, *J. Am. Chem. Soc.* 2012; 134(11): 5180–5186.

36. Dow W., Huang T., Yttria-stabilized zirconia supported copper oxide catalyst: II. Effect of oxygen vacancy of support on catalytic activity for CO oxidation, *J. Catal.* 1996; 160(2): 171–182.

37. Gracia J.M., Prinsloo F.F., Niemantsverdriet J.W., Mars-van Krevelen-like mechanism of CO hydrogenation on an iron carbide surface, *Catal. Lett.* 2009; 133(3–4): 257–261.

38. Zhang L., Kim H.Y., Henkelman G., CO oxidation at the Au–Cu interface of bimetallic nanoclusters supported on $CeO_2$ (111), *J. Phys. Chem. Lett.* 2013; 4(17): 2943–2947.

39. Gorte R.J., Zhao S., Studies of the water–gas-shift reaction with ceria-supported precious metals, *Catal Today.* 2005; 104(1): 18–24.

40. Bunluesin T., Gorte R.J., Graham G.W., Studies of the water–gas-shift reaction on ceria-supported Pt, Pd, and Rh: Implications for oxygen-storage properties, *Appl. Catal. B Environ.* 1998; 15(1–2): 107–114.

41. Cheng Z., Sherman B.J., Lo C.S., Carbon dioxide activation and dissociation on ceria (110): A density functional theory study, *J. Chem. Phys.* 2013; 138(1): 014702–014713.

42. Choe S., Kang H., Kim S., Adsorbed carbon formation and carbon hydrogenation for $CO_2$ methanation on the Ni (111) surface: ASED-MO study, *Bull. Korean Chem. Soc.* 2005; 26(11): 1682–1688.

43. Park J.-N., McFarland E.W., A highly dispersed Pd–Mg/$SiO_2$ catalyst active for methanation of $CO_2$, *J. Catal.* 2009; 266(1): 92–97.

44. Bernstein N.J., Akhade S.A., Janik M.J., Density functional theory study of carbon dioxide electrochemical reduction on the Fe(100) surface, *Phys. Chem. Chem. Phys.* 2014; 16(27): 13708–13717.

45. Schouten K.J.P., Qin Z., Gallent E.P., Koper M.T.M., Two pathways for the formation of ethylene in CO reduction on single-crystal copper electrodes, *J. Am. Chem. Soc.* 2012; 134(24): 9864–9867.

46. Shi C., O'Grady C.P., Peterson A., Hansen H., Nørskov J.K., Modeling $CO_2$ reduction on Pt(111), *Phys. Chem. Chem. Phys.* 2013; 15(19): 7114–7122.

47. Schouten K.J.P., Kwon Y., van der Ham C.J.M., Qin Z., Koper M.T.M., A new mechanism for the selectivity to C1 and C2 species in the electrochemical reduction of carbon dioxide on copper electrodes, *Chem. Sci.* 2011; 2(10): 1902–1909.

48. Nie X., Esopi M.R., Janik M.J., Asthagiri A., Selectivity of $CO_2$ reduction on copper electrodes: The role of the kinetics of elementary steps, *Angew. Chem. Int. Ed.* 2013; 52(9): 2459–2462.

49. Bothra P., Periyasamy G., Pati S.K., Methane formation from the hydrogenation of carbon dioxide on Ni(110) surface—A density functional theoretical study, *Phys. Chem. Chem. Phys.* 2013; 15(15): 5701–5706.

50. Peterson A.A., Abild-Pedersen F., Studt F., Rossmeisl J., Norskov J.K., How copper catalyzes the electroreduction of carbon dioxide into hydrocarbon fuels, *Energy Environ. Sci.* 2010; 3(9): 1311–1315.

51. Kuhl K.P., Cave E.R., Abram D.N., Jaramillo T.F., New insights into the electrochemical reduction of carbon dioxide on metallic copper surfaces, *Energy Environ. Sci.* 2012; 5(5): 7050–7059.

52. Taylor P.A., Rasmussent P., Chorkendorff I., Is the observed hydrogenation of formate the rate-limiting step in methanol synthesis? *J. Chem. Soc. Faraday Trans.* 1995; 91(8): 1267–1269.

53. Herron A., Scaranto J., Ferrin P., Li S., Mavrikakis M., Trends in formic acid decomposition on model transition metal surfaces: A density functional theory study, *ACS Catal.* 2014; 4(12): 4434 – 4445.

54. Yang Y., White M.G., Liu P., A theoretical study of methanol synthesis from $CO_2$ hydrogenation on metal-doped Cu(111) surfaces, *J. Phys. Chem. C.* 2012; 116(1): 248–256.

55. Ye J., Liu C., Mei D., Ge Q., Methanol synthesis from $CO_2$ hydrogenation over a $Pd_4$/$In_2O_3$ model catalyst: A combined DFT and kinetic study, *J. Catal.* 2014; 317(1): 44–53.

56. Vidal A.B., Feria L., Evans J., Takahashi Y., Liu P., Nakamura K. et al., $CO_2$ activation and methanol synthesis on novel Au/TiC and Cu/TiC catalysts, *J. Phys. Chem. Lett.* 2012; 3(16): 2275–2280.

57. Nie X., Luo W., Janik M.J., Asthagiri A., Reaction mechanisms of $CO_2$ electrochemical reduction on Cu(1 1 1) determined with density functional theory, *J. Catal.* 2014; 312(3): 108–122.

58. Yoshihara J., Campbell C.T., Methanol synthesis and reverse water–gas shift kinetics over Cu(110) model catalysts: Structural sensitivity, *J. Catal.* 1996; 161(2): 776–782.

59. Rasmussen P.B., Holmblad P.M., Askgaard T., Ovesen C.V., Stoltze P., Norskov J.K. et al., Methanol synthesis on Cu(100) from a binary gas mixture of $CO_2$ and $H_2$, *Catal. Lett.* 1994; 26(3–4): 373–381.

60. Sio C., Sio C.Z., Fisher I., Bell T., In-situ infrared study of methanol synthesis from $H_2/CO_2$ over $Cu/SiO_2$ and $Cu/ZrO_2/SiO_2$, *J. Catal.* 1997; 237(1): 222–237.

61. Grabow L.C., Mavrikakis M., Mechanism of methanol synthesis on Cu through $CO_2$ and CO hydrogenation, *ACS Catal.* 2011; 1(4): 365–384.

62. Graciani J., Mudiyanselage K., Xu F., Baber E., Evans J., Senanayake S.D. et al., Highly active copper-ceria and copper-ceria-titania catalysts for methanol synthesis from $CO_2$, *Science.* 2014; 345(6196): 546–550.

63. An X., Li J., Zuo Y., Zhang Q., Wang D., Wang J., A Cu/Zn/Al/Zr fibrous catalyst that is an improved $CO_2$ hydrogenation to methanol catalyst, *Catal. Lett.* 2007; 118(3–4): 264–269.

64. Natesakhawat S., Ohodnicki P.R., Howard B.H., Lekse J.W., Baltrus J.P., Matranga C., Adsorption and deactivation characteristics of Cu/ZnO-based catalysts for methanol synthesis from carbon dioxide, *Top. Catal.* 2013; 56(18–20): 1752–1763.

65. Hong Q.J., Liu Z.P., Mechanism of $CO_2$ hydrogenation over $Cu/ZrO_2$(212) interface from first-principles kinetics Monte Carlo simulations, *Surf. Sci.* 2010; 604(21–22): 1869–1876.

66. Liu J., Shi J., He D., Zhang Q., Wu X., Liang Y. et al., Surface active structure of ultra-fine $Cu/ZrO_2$ catalysts used for the $CO_2 + H_2$ to methanol reaction, *Appl. Catal. A Gen.* 2001; 218(1–2): 113–119.

67. Toyir J., de la Piscina R.P., Fierro J.L.G., Homs N., Catalytic performance for $CO_2$ conversion to methanol of gallium-promoted copper-based catalysts: Influence of metallic precursors, *Appl. Catal. B.* 2001; 34(4): 255–266.

68. Liu X.-M., Lu G.Q., Yan Z.-F., Nanocrystalline zirconia as catalyst support in methanol synthesis, *Appl. Catal. A Gen.* 2005; 279(1–2): 241–245.

69. Toyir J., de la Piscina P.R., Fierro J.L.G., Homs N., Highly effective conversion of $CO_2$ to methanol over supported and promoted copper-based catalysts: Influence of support and promoter, *Appl. Catal. B.* 2001; 29(3): 207–215.

70. Aksoylu E.A., İnciİşli A., İlsen Önsan Z., Interaction between nickel and molybdenum in Ni–Mo/$Al_2O_3$ catalysts: I $CO_2$ methantion and SEM-TEM studies, *Appl. Catal. A.* 1998; 168(2): 387–397.

71. Solymosi F., Erdöhelyi A., Methanation of $CO_2$ on supported rhodium catalysts, *J. Catal.* 1981; 68(2): 1448–1449.

72. Słoczyński J., Grabowski R., Kozłowska A., Olszewski P., Stoch J., Skrzypek J. et al., Catalytic activity of the M/(3ZnO·$ZrO_2$) system (M = Cu, Ag, Au) in the hydrogenation of $CO_2$ to methanol, *Appl. Catal. A Gen.* 2004; 278(1): 11–23.

73. Tsubaki N., Fujimoto K., Promotional SMSI effect on supported palladium catalysts for methanol synthesis, *Top. Catal.* 2003; 22(3–4): 325–335.

74. Liang X.-L., Dong X., Lin G.-D., Zhang H.-B., Carbon nanotube-supported Pd–ZnO catalyst for hydrogenation of $CO_2$ to methanol, *Appl. Catal. B Environ.* 2009; 88(3–4): 315–322.

75. Collins S., Chiavassa D., Bonivardi A., Baltanás M., Hydrogen spillover in $Ga_2O_3$–Pd/$SiO_2$ catalysts for methanol synthesis from $CO_2/H_2$, *Catal. Lett.* 2005; 103(1–2): 83–88.

76. Liu X., Lu G., Yan Z., Beltramini J., Recent advances in catalysts for methanol synthesis via hydrogenation of CO and $CO_2$, *Ind. Eng. Chem. Res.* 2003; 42(25): 6518–6530.

77. Ichikawa M., Fukushima T., Infrared studies of metal additive effects on CO chemisorption modes on $SiO_2$-supported Rh-Mn, -Ti, and -Fe catalysts, *J. Phys. Chem.* 1985; 89(9): 1564–1567.

78. Studt F., Abild-Pedersen F., Wu Q., Jensen A.D., Temel B., Grunwaldt J.-D. et al., CO hydrogenation to methanol on Cu–Ni catalysts: Theory and experiment, *J. Catal.* 2012; 293(2): 51–60.

79. Zhang X.-R., Wang L.-C., Yao C.-Z., Cao Y., Dai W.-L., He H.-Y. et al., A highly efficient $Cu/ZnO/Al_2O_3$ catalyst via gel-coprecipitation of oxalate precursors for low-temperature steam reforming of methanol, *Catal. Lett.* 2005; 102(3–4): 183–190.

80. Wambach J., Baiker A., Wokaun A., $CO_2$ hydrogenation over metal/zirconia catalysts, *Phys. Chem. Chem. Phys.* 1999; 1(22): 5071–5080.

81. Yano H., Shirai F., Nakayama M., Ogura K., Electrochemical reduction of $CO_2$ at three-phase (gas|liquid|solid) and two-phase (liquid|solid) interfaces on Ag electrodes, *J. Electroanal. Chem.* 2002; 533(1–2): 113–118.

82. Ohya S., Kaneco S., Katsumata H., Suzuki T., Ohta K., Electrochemical reduction of $CO_2$ in methanol with aid of CuO and $Cu_2O$, *Catal. Today.* 2009; 148(3–4): 329–334.

83. Qu J., Zhang X., Wang Y., Xie C., Electrochemical reduction of $CO_2$ on $RuO_2/TiO_2$ nanotubes composite modified Pt electrode, *Electrochim. Acta.* 2005; 50(16–17): 3576–3580.

84. Yano H., Shirai F., Nakayama M., Ogura K., Efficient electrochemical conversion of $CO_2$ to CO, $C_2H_4$ and $CH_4$ at a three-phase interface on a Cu net electrode in acidic solution, *J. Electroanal. Chem.* 2002; 519(1–2): 93–100.

85. Hori Y., Takahashi I., Koga O., Hoshi N., Electrochemical reduction of carbon dioxide at various series of copper single crystal electrodes, *J. Mol. Catal. A Chem.* 2003; 199(1–2): 39–47.

86. Jitaru M., Lowy D.A., Toma M., Toma B.C., Oniciu L., Electrochemical reduction of carbon dioxide on flat metallic cathodes, *J. Appl. Electrochem.* 1997; 27(8): 875–889.

87. Li W., Electrocatalytic reduction of $CO_2$ to small organic molecule fuels on metal catalysts, In *Advances in $CO_2$ Conversion and Utilization*, Yan Hang Hu (Ed.), Vol. 1056, Chapter 5, American Chemical Society, 2010, pp. 5–55.

88. Garg G., Basu S., Studies on degradation of copper nano particles in cathode for $CO_2$ electrolysis to organic compounds, *Electrochim. Acta.* 2015; 177(7): 359–365.

89. Hori Y., Konishi H., Futamura T., Murata A., Koga O., Sakurai H. et al., "Deactivation of copper electrode" in electrochemical reduction of $CO_2$, *Electrochim. Acta.* 2005; 50(27): 5354–5369.

90. Laguna-Bercero M., Recent advances in high temperature electrolysis using solid oxide fuel cells: A review, *J. Power Sources.* 2012; 203(2): 4–16.

91. Tao C.L.C.G., Sridhar K.R., Study of carbon dioxide electrolysis at electrode/electrolyte interface: Part I. Pt/YSZ interface, *Solid State Ion.* 2004; 175(1–4): 615–619.

92. Pekridis G., Kalimeri K., Kaklidis N., Vakouftsi E., Iliopoulou E.F., Athanasiou C. et al., Study of the reverse water gas shift (RWGS) reaction over Pt in a solid oxide fuel cell (SOFC) operating under open and closed-circuit conditions, *Catal. Today.* 2007; 127(1–4): 337–346.

93. Ebbesen S.D., Mogensen M., Electrolysis of carbon dioxide in solid oxide electrolysis cells, *J Power Sources.* 2009; 193(1): 349–358.

94. Cheng C.Y., Kelsall G.H., Kleiminger L., Reduction of $CO_2$ to CO at Cu-ceria-gadolinia (CGO) cathode in solid oxide electrolyser, *J. Appl. Electrochem.* 2013; 43(11): 1131–1144.

95. Green R., Liu C., Adler S., Carbon dioxide reduction on gadolinia-doped ceria cathodes, *Solid State Ion.* 2008; 179(17–18): 647–660.

96. Chueh W.C., Falter C., Abbott M., Scipio D., Furler P., Haile S.M., High-flux solar-driven thermochemical dissociation of $CO_2$ and $H_2O$ using nonstoichiometric ceria, *Science.* 2010; 330(6012): 1797–1801.

97. Yue X., Irvine J.T.S., Impedance studies on LSCM/GDC cathode for high temperature CO$_2$ electrolysis, *Electrochem. Solid-State Lett.* 2012; 15(3): B31–B34.

98. Park S., Vohs J.M., Gorte R.J., Direct oxidation of hydrocarbons in a solid-oxide fuel cell, *Nature*. 2000; 404(6775): 265–267.

99. Bishop S.R., Stefanik T.S., Tuller H.L., Electrical conductivity and defect equilibria of Pr$_{0.1}$Ce$_{0.9}$O$_{(2-\delta)}$, *Phys. Chem. Chem. Phys.* 2011; 13(21): 10165–10173.

100. Trogadas P., Parrondo J., Ramani V., CeO$_2$ surface oxygen vacancy concentration governs in situ free radical scavenging efficacy in polymer electrolytes, *ACS Appl. Mater. Interfaces*. 2012; 4(10): 5098–5102.

# 7 Product Analysis for CO$_2$ Electroreduction

*Jingyu Tang, Xiaozhou Zou, and Feng Hong*

## CONTENTS

## 7.1 INTRODUCTION

CO$_2$ electroreduction involves multielectron processes which can form a large variety of products ranging from CO, CH$_4$ to higher hydrocarbons in the gas phase, and generate various oxygenates in the liquid phase such as alcohols, aldehydes, and carboxylic acids. Currently, controlling the CO$_2$ reduction pathways is still a challenging work. A thorough and accurate determination of products is of critical importance for the evaluation of electroreduction performance. Many groups have practically adopted the routine analysis of one or two products in either gas or liquid phase for evaluation of electrocatalytic efficiencies. However, to date, only very few studies have been focused on the analysis of reduction products, and more comprehensive and standard analysis methods are not yet reported in the literature. In this chapter, different analysis methods reported so far are summarized.

## 7.2 SAMPLE PREPARATION

The products of CO$_2$ electroreduction are affected by the electrode types, composition of electrolyte (aqueous or nonaqueous), and the overpotential. For example, in aqueous supporting electrolytes, metallic In, Sn, Hg, and Pb are selective for the production of formic acid; metallic Zn, Au, and Ag produce carbon monoxide; metallic Cu exhibits a high electrocatalytic activity for the formation of hydrocarbons, aldehydes, and alcohols, whereas metallic Al, Ga show low electrocatalytic activity in $CO_2$ electroreduction [1]. In nonaqueous supporting electrolytes on Pb, Tl, and Hg, the main product is oxalic acid; on Cu, Ag, Au, In, Zn, and Sn, carbon monoxide and carbonate ions are obtained, whereas Ni, Pd, and Pt are selective for CO formation; and Al, Ga form both CO and oxalic acid. This information will be very helpful for the product analysis for $CO_2$ electroreduction.

The products of $CO_2$ electroreduction can be divided into gas phase and liquid phase. The main products of the electroreduction of carbon dioxide are $CH_3CH_3$, CO, $CH_4$, $H_2$, $O_2$ in the gas phase, and oxygenates in the liquid phase such as alcohols, aldehydes, and carboxylic acids [2]. So the basic sample preparation is to collect the gas and the liquid products. Then, the next preparation step will be determined by the following analytic methods. For example, when the gas samples are analyzed by gas chromatography (GC) with flame ionization detector (FID), $O_2$ is usually needed to be removed.

## 7.3   CHEMICAL ANALYSIS

Many of the products in the liquid phase can be analyzed by chemical methods such as titration. Carbonate/bicarbonate concentration at the cathode side is usually determined by the sequential titration of the catholyte samples with hydrochloric acid, using phenolphthalein and methyl orange as indicators [3]. Carbonate is a stronger base than bicarbonate. At the beginning of titration, with phenolphthalein as the indicator, only carbonate reacts with hydrochloric forming bicarbonate. The red color of phenolphthalein will disappear when all carbonate turns into bicarbonate. The concentration of carbonate can be calculated according to the volume and concentration of the hydrochloric acid. Then methyl orange is added to the solution as an indicator for the titration of bicarbonate. The yellow solution turns into orange when the titration is just finished. The bicarbonate quantity obtained from the titration calculation consists of the original bicarbonate in the sample and the bicarbonate transformed from the titration of carbonate. As the quantity of carbonate has been estimated in the first step of titration, the bicarbonate concentration in the sample can be deduced sequentially.

Methanol/formaldehyde can be estimated by back titration with ferrous ammonium sulfate of the product from acid dichromate oxidation [2]. An alkaline permanganate oxidation technique can be used to quantify the formate in the cathodic liquid products [3]. Before titration, formic acid and formate are treated with an excess of sodium carbonate. The standard potassium permanganate solution is added to the hot formate solution until the clear liquid above the precipitate is just colored pink. In order to detect the exact end point in the presence of the brown precipitate, the pink solution is strongly acidified with dilute sulfuric acid, and also a known excess 0.1 N sodium oxalate solution is added, then the mixture is warmed until the precipitate has dissolved. The excess of oxalate is titrated with standard potassium permanganate solution. The quantity of formic acid and formate is calculated by using the conversion formula: 1 mL of 1 N KMnO$_4$ = 0.02301 g HCOOH.

## 7.4 INSTRUMENTAL ANALYSIS

### 7.4.1 GAS CHROMATOGRAPHY

GC is the most commonly used method for quantification of the gas species [4–9]. When the gas flows into the column, the column packing may adsorb CH$_3$CH$_3$, CO, CH$_4$, H$_2$, and other components differently, which cause different flow rates of different compounds under the impetus of the mobile phase. Then the compounds will be separated and detected when they get out of the column. The elution order and retention time are the main data for the identification of the products. The amount of the compounds can be calculated from the integral area of the chromatography peaks. It is reported that many types of column have been used for the separation of the CO$_2$ electroreduction products, such as carbon columns [10–13], molecular sieves [7,14], and Poraplot Q [6,8,9,15,16]. Thermal conductivity detector (TCD) and FID are the most commonly used detectors in GC. Both are sensitive for many compounds. FID is more sensitive than TCD for the analysis of low concentrations of CO and hydrocarbons. FID analysis is a destructive method. Therefore, the gas should first be analyzed by TCD for gas quantification followed by FID for the analysis of CO and alkanes [4]. GC calibration lines of some common products of CO$_2$ electroreduction, from which the linear correlativity between the peak area and the compound concentration can be used to quantify the products, always indicates high degree of accuracy of GC with a correlation index $R^2$ of almost close to 1.

However, the presence of CO$_2$ in the product stream can cause deactivation of certain types of columns, including the most commonly used molecular sieve column. Hong et al. [4] designed a better GC configuration to separate and detect all the gases, including CH$_4$, CH$_3$CH$_3$, CO, CO$_2$, O$_2$, N$_2$, and H$_2$ for product analysis of photocatalytic reduction of CO$_2$. The GC configuration can avoid regular column regeneration and achieve longer lifetime of the detector. Figure 7.1 shows the simplified configuration of their GC. Helium (99.9995%) was used as the carrier gas. The back channel of GC is equipped with two packed columns, Hayesep Q and Molsieve 5 Å, and three gas switch valves, V1–V3. During the analysis, 0.25 mL of gas sample in the sample loop of V1 was introduced to the Hayesep Q column where CO$_2$ was separated from the other gases due to its longer retention time. The other gases can be completely eluted from the first column and enter the Molsieve column before 2.5 min. At 2.5 min, Valve 2 is switched for CO$_2$ to bypass the Molsieve column and be detected by TCD. After TCD analysis, CO$_2$ is vented out by the switch of Valve 3 during 3–7 min. The rest of the gases after the Hayesep Q column were further separated by the Molsieve 5 Å column. At 10 min, V2 is switched back for these separated gases to be detected by TCD and then CH$_4$, CO, and CH$_3$CH$_3$ were further detected by FID with higher sensitivities. The role of the methanizer (nickel catalyst) is to convert CO to CH$_4$ for FID analysis, whereas alkanes remained unaffected. During this process, the GC oven was held at 60°C for 15 min and increased to 180°C for 5 min for the postrun.

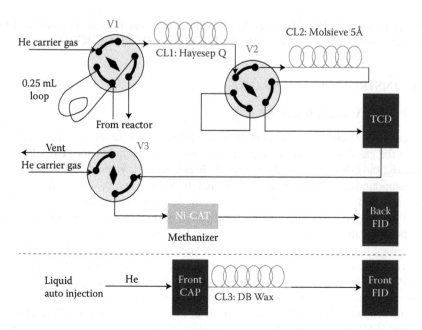

**FIGURE 7.1** Flowchart of gas chromatography for gas-phase product analysis and liquid-phase analysis of alcohols. V1–V3: gas switch valves, CL1–3: separation columns. (Reprinted by permission from Hong, J., Zhang, W., Ren, J., Xu, R., *Anal. Methods*, 5 (5), 1086–1097, 2013.)

GC with TCD/FID is also the main technique used for the analysis of different types of liquid oxygenates (mainly alcohols) [7–9,14,17]. Alcohols in the liquid phase are volatile compounds and can be easily vaporized into gases during the vaporization process. Then, gases such as $CH_4$, $CH_3CH_3$, CO, $CO_2$, $O_2$, $N_2$ described above will be separated and tested. For example, methanol can be analyzed by GC equipped with a DB-Wax column and a FID detector [18]. However, alcohol analysis is very sensitive to the alkaline conditions and organics in terms of peak areas [4]. Figure 7.2 illustrates the effect of organics and alkaline conditions to methanol, ethanol, and 1-propanol analysis.

However, carbonate, bicarbonate, and other salt compounds in the liquid sample are not easy to vaporize or easy to decompose when heated. GC is not suitable for the analysis of these compounds. Other methods such as high-performance liquid chromatography (HPLC), ion exchange chromatography (IEC), ultraviolet (UV)–visible spectroscopy (colorimetric assay), and nuclear magnetic resonance (NMR) are used to analyze the compounds in the liquid-phase products.

## 7.4.2 MASS SPECTROMETRY

Mass spectrometry (MS) is often equipped with GC to identify the compounds in the sample for its unique identification capability and high sensitivity [19–23]. During the GC–MS analysis, compounds are first separated by the GC column, and then the eluates are injected into MS for further analysis. Usually, nuclear mass ratio is the

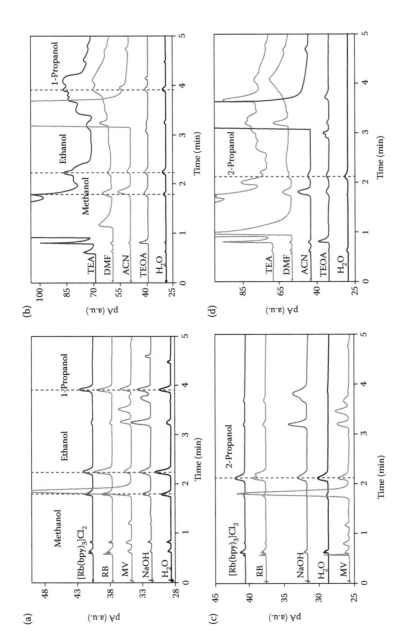

**FIGURE 7.2** GC peaks of methanol, ethanol, 1-propanol in (a) 0.1 M NaOH, 0.1 mM methyl viologen (MV), 0.1 mM rose bengal (RB), 0.1 mM [Ru(bpy)₃] Cl₂, and (b) 15 v% acetonitrile (ACN), 15 v% dimethylformamide (DMF), 15 v% triethylamine (TEA), 15 v% triethanolamine (TEOA); GC peaks of 2-propanol in (c) 0.1 M NaOH, 0.1 mM RB, 0.1 mM MV, 0.1 mM [Ru(bpy)₃]Cl₂; and (d) 15 v% ACN, 15 v% DMF, 15 v% TEA, 15 v% TEOA. (Reprinted by permission from Hong, J., Zhang, W., Ren, J., Xu, R., *Anal. Methods*, 5(5), 1086–1097, 2013.)

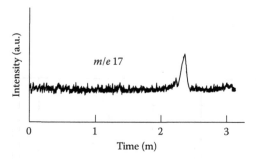

**FIGURE 7.3** GC–MS chromatogram at $m/e = 17$ attributable to $^{13}CH_4$ of the gas sample produced by the photocatalytic reaction using Pd (2%)-TiO$_2$ under a $^{13}CO_2$ atmosphere. (Reprinted by permission from Yui, T., Kan, A., Saitoh, C., Koike, K., Ibusuki, T., Ishitani, O., *ACS Appl. Mater. Interfaces*, 3 (7), 2594–2600, 2011.)

main data from the MS analysis. As the main products in electroreduction of $CO_2$ are some simple compounds such as $CH_4$, $CH_3CH_3$, CO, alcohols, aldehydes, and carboxylic acids, the nuclear mass data will be very helpful to identify the components of the products. For example, the $m/e = 17$ signal in a GC–MS chromatogram of the gas sample shown in Figure 7.3 is attributed to $^{13}CH_4$.

### 7.4.3 INFRARED SPECTROSCOPY

Infrared spectroscopy (IR) or diffuse reflectance infrared Fourier transform spectroscopy (DRIFT) has occasionally been employed to verify the consumption of $CO_2$ and the generation of CO [25–27]. IR or DRIFT analysis can give the functional group composition of the compounds separated by GC. Then the compounds separated by GC can be easily identified by the obtained IR/DRIFT data and the MS data. However, MS and IR/DRIFT are not suitable for the quantitative analysis of the components in the products of $CO_2$ electroreduction [4]. Figure 7.4 shows the *in situ* Fourier transform infrared spectroscopy (FT-IR) spectra using p-polarized light. Spectra were recorded at controlled potential for electrolyte solution saturated with $CO_2$ (Figure 7.4a) and in the absence of $CO_2$ (Figure 7.4b). Only the region between 1150 and 2000 cm$^{-1}$ is shown and remarkable differences are observed in this region. For the DMF solution saturated with $CO_2$ (Figure 7.4a), the most interesting features, with loss (i) and gain (ii) of absorbance, are: (i) 1759, 1670, 1508, 1410, 1262 cm$^{-1}$; (ii) 1732, 1639, 1604, 1478, 1370, 1330, 1300 cm$^{-1}$, and for the $CO_2$-free solution (Figure 7.4b): (i) 1662, 1508, 1473, 1406, 1391, 1362, 1223 cm$^{-1}$; (ii) 1739, 1566, and 1473 cm$^{-1}$ [28].

### 7.4.4 NUCLEAR MAGNETIC RESONANCE

NMR is also used to qualify the compounds in the sample [29,30]. It is usually equipped with GC or GC–MS. Almost all oxygenates in the products can be identified by this method. The compounds in the sample are first separated and then quantified by GC. MS is used to detect the molecular weight of the separated compounds.

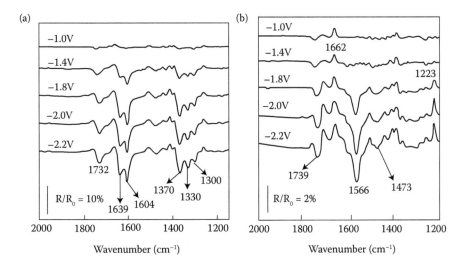

**FIGURE 7.4** *In situ* FT-IR external reflectance spectra on smooth gold with modulated potential. (IR region 1150–2000 cm$^{-1}$). P-polarized light. (a) DMF + TBAF (TBAF; 0.15 mol/dm$^3$) saturated with CO$_2$ and (b) without CO$_2$. Potentials vs. SCE. Temperature: 20°C. (Reprinted by permission from Pérez, E.R., Garcia, J.R., Cardoso, D.R., McGarvey, B.R., Batista, E.A., Rodrigues-Filho, U.P., Vielstich, W., Franco, D.W., *J. Electroanal. Chem.*, 578 (1), 87–94, 2005.)

$^1$H or $^{13}$C NMR analysis is then taken for structure analysis. Figure 7.5 shows $^1$H NMR spectra of formate produced from CO$_2$ reduction, from which $^1$H NMR signal of some compounds such as formate and water can be identified easily. Also like IR and MS, NMR is difficult for the quantification of the compounds. However, some work demonstrated the quantification of the products by NMR [31]. Formate products in aqueous solution were reported to be quantified by NMR with sodium 3-(trimethylsilyl) propionate 2,2,3,3-d (TSP) as an internal standard [30]. The chemical shift of the hydrogen atom in the CH group in formate (HCOO$^-$) is at about 8.443 ppm, whereas 0 ppm is assigned to the resonance of the CH$_3$ group of TSP. Formate concentration was determined with reference to the peak area of the CH proton in formate to that of the CH$_3$ protons in TSP. A calibration curve can be made by plotting the concentrations of standard formic acid solutions versus the concentrations measured by NMR. The formate concentration is then obtained by fitting the calibration curve. Figure 7.6 illustrates $^1$H NMR spectra of formate and CH$_3$ and CH$_2$ groups produced from CO$_2$ reduction at controlled potential, which will contribute to identify the components of the products for CO$_2$ electroreduction. $^{13}$C NMR also can be used to verify the origination of the products in the sample of CO$_2$ electroreduction. The disadvantage for the use of NMR is its high cost.

### 7.4.5 SPECTROPHOTOMETRIC METHOD

UV and visible spectrophotometry is widely used to quantify many compounds. Products such as formaldehyde and formate can be detected and quantified through

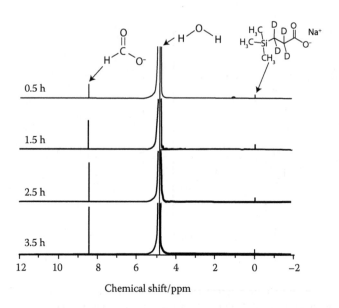

**FIGURE 7.5** $^1$H NMR spectra of formate produced from $CO_2$ reduction at $E = -2$ V vs. SCE in 0.5 M KHCO$_3$. (Reprinted by permission from Wu, J., Risalvato, F.G., Ke, F.-S., Pellechia, P., Zhou, X.-D., *J. Electroanal. Chem.*, 159 (7), F353–F359, 2012. The Electrochemical Society.)

the chromotropic acid test, which is a spectrophotometric method [33]. For the determination of formaldehyde, the sample is usually first diluted with sulfuric acid properly. Chromotropic acid is subsequently added and the solution is heated in a water bath at 60°C for 30 min. The absorbance of the solution at 484 and 578 nm is obtained and compared with those values obtained from a calibration curve using standard formaldehyde solutions. For the determination of formate, the electrolyzed solution is initially reduced with Mg/HCl and the same procedure is subsequently followed as described above. The amount of formate is determined by comparison of difference.

Formate production also can be measured by using UV–Visible spectrophotometry with the broad absorption region of 190–250 nm, as shown in Figure 7.7 [34]. Product samples are first acidified with sulfuric acid to neutralize bicarbonates/hydroxide/carbonates, and then boiled for 10 min to expel carbon dioxide from the solutions. The reference cell in the spectrophotometer is also filled with an appropriate concentration of sulfate as same as in the sample cell. The absorption at 230 nm is chosen for the formate concentration determination because of the saturation of the peak at 208 nm. The formate concentration is finally obtained from a calibration curve of standard formate solutions. The boiling step in this method is important because the UV absorption peaks for carbonates and carbon dioxide interfere with that of formate. With this method, cumulative efficiency of formate production at 40 Ma/cm$^2$ in a Nafion membrane cell with various cathode solutions was compared, as shown in Figure 7.8 [34].

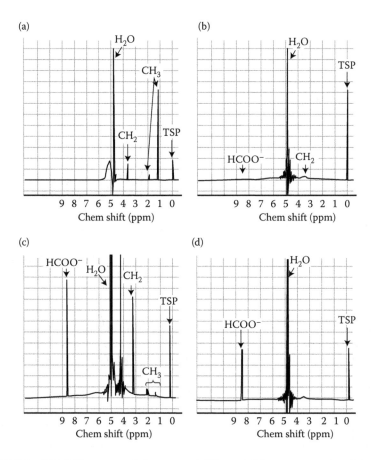

**FIGURE 7.6** $^1$H NMR spectra of formate and $CH_3$ and $CH_2$ groups produced from $CO_2$ reduction at controlled potential; (a) and (b): $E = -0.8$ and $-1.1$ V vs. Ag/AgCl in 0.5 M KHCO₃ for 90 min, Cu electrode surface obtained by pre-electroreduction in 1 M NaOH at $-3.0$ V for 10 min; (c) $E = -1.1$ V vs. Ag/AgCl in 0.5 M KHCO₃ for 90 min, Cu electrode surface obtained by pre-electroreduction in 1 M H₃PO₄ at $-3.0$ V for 10 min; (d) the standard $^1$H NMR spectra of HCOOH. (Reprinted by permission from Qiao, J., Jiang, P., Liu, J., Zhang, J., *Electrochem. Commun.*, 38, 8–11, 2014.)

### 7.4.6 HIGH-PERFORMANCE LIQUID CHROMATOGRAPHY

HPLC is a technique used to separate, identify, and quantify components in a mixture. This analytical method relies on a column filled with a solid adsorbent material, a mobile phase containing the sample mixture to reach the purpose of component separation. Pumps are used to pass the pressurized liquid mobile phase through columns. The components in the mixture interact differently with the adsorbent materials in the columns, leading to different flow rates for the various components. In the end, the components flow out from the column in order and can be detected by different detectors. The detectors must ensure the components from samples to give signals, which can be used for further analyses. Here, two typical types of detectors

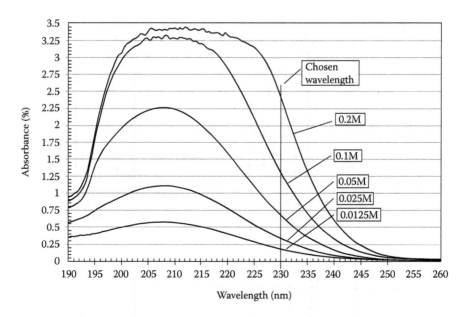

**FIGURE 7.7** UV–visible spectra for various concentrations of solutions of sodium formate acidified with sulfuric acid. (Reprinted by permission from Narayanan, S., Haines, B., Soler, J., Valdez, T., *J. Electroanal. Chem.*, 158 (2), A167–A173, 2011. The Electrochemical Society.)

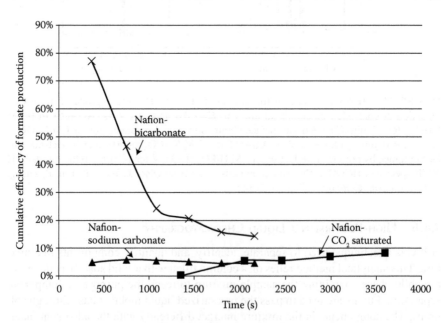

**FIGURE 7.8** Cumulative efficiency of formate production at 40 Ma/cm² in a Nafion membrane cell with various cathode solutions as indicated. (Reprinted by permission from Narayanan, S., Haines, B., Soler, J., Valdez, T., *J. Electroanal. Chem.*, 158 (2), A167–A173, 2011. The Electrochemical Society.)

are introduced as examples, the diode array detector (DAD) and reflective index detector (RID). A DAD is usually used to detect products based on their absorption of UV/visible light. A DAD detector provides multiple photodiode arrays that enable the detector to obtain information from a wide range of wave lengths. An RID works on the basis of the difference of refraction index between components from the sample and the mobile phase. Different products have different retention time, and the shift of product peak to a different retention time in HPLC always signifies different products. Both qualitative and quantitative information of the components can be received with the help of the obtained graphs.

HPLC has been used frequently to determine the liquid-phase products in the reactions of electrochemical reduction of carbon dioxide. HPLC allows separation of the compounds of the reaction mixture using different interactions with a liquid mobile phase and a solid stationary phase. For the analysis of the products, only a small volume of the sample is needed to be mixed with a solvent. The mixture is then forced to pass through the column under a high pressure which is given by pumps. The different components in the mixture move through the column with different rates, which depend on their own characteristics. Components such as formic acid, oxalic acid, and formaldehyde are often analyzed by HPLC [18,35–40]. Similarly with GC, the identification and quantization of the products by using HPLC also need calibration lines of standard compounds. Various organics and alkaline conditions may affect the acid analysis. Figure 7.9 illustrates HPLC peaks of formic acid and acetic acid obtained with different conditions.

Formic acid has been analyzed by using the following columns: Aminex ion exclusion column type HPX-87X [35], Shodex Ionpack KC-811 column [36,37], ODS-18 column [38,39], Atlantis dC18 column [39], and SupelcoGel C-610H column [40]. HPLC used for detection of formic acid is usually equipped with a UV detector [18,35,37]. For example, an Applied Chromatography Systems HPLC monitor fitted with a 254 nm filter [35] and a UV detector set at 220 nm [36,37] were used. The mobile phase should be chosen depending on the types of column used, for example, 0.0025 M sulfuric acid for an Aminex ion exclusion column type HPX-87X [35], 20% ACN in Milli-Q water as a solvent for Atlantis dC18 column [39].

Together with formic acid, oxalic acid can be analyzed at the same time, for example, an HPLC equipped with a Luna 5-μm C18 column and a UV detector operating at 208 nm, and eluted with 5% $H_3PO_4$ [41]. Formaldehyde is also analyzed by HPLC, and there can be a modification step before detection [18,42]. The liquid samples containing formaldehyde can be derived with 2,4-dinitrophenylhydrazine (DPNH) to form formaldehyde–DNPH samples, which can be measured by HPLC. An HPLC system was operated with Varian Inertsil 5-μ ODS-2 column (150 × 4.6 mm) and a UV detector.

### 7.4.7 ION CHROMATOGRAPHY

Ion chromatography (or IEC) separates ions and polar molecules based on their affinity to different types of ion exchanger. This method can be used for almost all kinds of charged molecules. The products of CO₂ electroreduction, for example, formic acid, formate, and oxalate, can also be detected with this method.

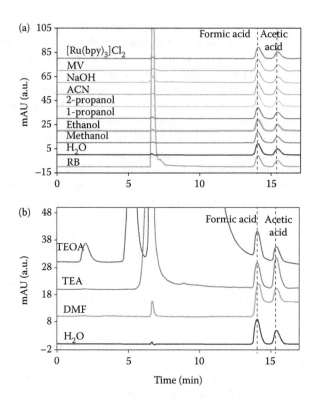

**FIGURE 7.9** HPLC peaks of formic acid and acetic acid in (a) 15 v% methanol, 15 v% ethanol, 15 v% 1-propanol, 15 v% 2-propanol, 15 v% ACN, 0.1 M NaOH, 0.1 mM MV, 0.1 mM [Ru(bpy)3]Cl$_2$, and 0.1 mM RB, and (b) 15 v% DMF, 15 v% TEA, and 15 v% TEOA. (Reprinted by permission from Hong, J., Zhang, W., Ren, J., Xu, R., *Anal. Methods*, 5 (5), 1086–1097, 2013.)

Formic acid in liquid samples was analyzed by this method with, for example, an IonPac AC20 column (4 × 250 mm) equipped with an ED40 conductance detector. There are other cases of formic acid analysis by ion chromatography ICS-900 Dionex using an AS23 analytical column and Na$_2$CO$_3$/NaHCO$_3$ as mobile phase [42–44].

Formate ion can be analyzed by the IC [45–47]. The ion exclusion chromatography system can be equipped with, for example, a Dionex AS15 column (2250 mm), an IonPac AG15 guard column (250 mm), and suppressed conductivity ICS-2000 detector [46]. Oxalate can be analyzed by the IC equipped with a UV detector operating at 210 nm and an Aminex HPX-87H ion exchange column eluted with 0.005 mol/dm$^3$ H$_2$SO$_4$, on which conversion of oxalate to oxalic acid takes place [48]. However, this system using Aminex HPX-87H and UV detector is often classified to HPLC.

## 7.5 ELECTROCHEMICAL ANALYSIS

Reaction rates for the processes of the reduction of carbon dioxide to liquid fuels, which is accomplished through photovoltaic or other electrochemical ways, can be estimated from the steady-state limiting current in cyclic voltammetry or by rotating disk voltammetry studies [49].

## 7.5.1 CYCLIC VOLTAMMETRY

Cyclic voltammetry is the most popular nondestructive technique applied to homogeneous catalysis systems [50]. It is widely used for first evaluation of the catalyst to determine potential of the catalytic process and the catalytic efficiency. Cyclic voltammetry consists of sweeping the potential back and forward at a fixed scan rate ($dE/dt$) while recording the current response. The obtained curve is known as cyclic voltammogram (CV), and its shape depends on the processes occurring at the electrode surfaces.

In a cyclic voltammetry experiment, the working electrode potential is ramped linearly versus time. During the anodic sweep toward positive potentials, oxidation processes are observed as a positive current, whereas in the cathodic sweep toward negative potentials, the reduction reactions are observed as a negative current. Because these processes are surface sensitive, the shape of the CV gives the information about the structure and the chemical nature of the electrode surface and the composition of the electrolyte. That is why CVs are considered as fingerprints of surfaces. Cyclic voltammetry is further used in a more detailed kinetic analysis to explain the mechanism of the catalytic reaction, and thus to be used to study electrochemical reactions.

Cyclic voltammetry can provide thermodynamic and kinetic information of the processes occurring at the electrode–electrolyte interface, and thus can also be used to determine CO product in electrochemical reactions. After adsorbing CO onto the surface, the potential can be swept toward positive values, and a positive current is observed due to the oxidation of CO.

## 7.5.2 ROTATING DISK VOLTAMMETRY AND ROTATING-RING DISK VOLTAMMETRY

Rotating disk electrode (RDE) voltammetry is a technique that has been mostly used in immobilized catalyst systems [50]. As shown in Figure 7.10a, the rapid rotation of the disk drives liquid to flow horizontally out and away from the center disk, thus serving as a motivation for replenished liquid to move axially to form an upward flow as shown by the arrows. The idea is to electrochemically generate a reactive species at the disk and then to monitor the species electrochemically as it is swept past the ring by laminar flow. Considerable information can be obtained about a redox system by rotating disk voltammetry. An initial voltammogram will locate redox couples in solution, giving information about redox potentials and reversibility.

A typical RDE and a rotating ring-disk electrode (RRDE) are both shown in Figure 7.10b. Compared with RDE, the disk of RRDE is encircled by a ring, which functions as a second working electrode. The idea is to generate a product or an intermediate at a disk electrode and collect it at a ring electrode that concentrically surrounds the disk. The product or intermediate is generated by fixing the disk potential at an appropriate value and scanning the ring potential so as to obtain

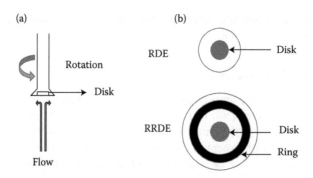

**FIGURE 7.10** A typical RDE and a rotating RRDE. (a) RDE with hydrodynamic flow pattern. (b) Bottom view of RDE and RRDE.

its oxidative or reductive signature as a steady-state current–potential curve. This technique is widely used in the study of reaction mechanisms as electroactive intermediates of coupled homogeneous chemical reactions can be monitored at the ring electrode.

RRDE can be used to determine the reduction product of $CO_2$ both in the liquid phase [51] and in the gas phase [52]. Formic acid can be detected by using the RRDE technique. The formic acid formed on a $SnO_2$ disk electrode surface as a reduction product of carbon dioxide can be detected at a Pt ring electrode, and its concentration is a parameter for the electrode. With the assumption that HCOOH molecules are all ionized, the detection limit of the present RRDE is expected to be quite close to that value of a pulsed electrochemical detector. Figure 7.11 shows the current–potential curves for a Pt (ring)–Pt (counter) in a 0.1 M $Na_2SO_4$ solution. Figure 7.12 gives an

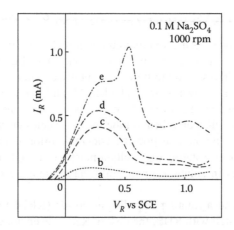

**FIGURE 7.11** Current–voltage curves for Pt-ring (anode) with and without HCOOH: (curve a) blank, (curves b–e) with 200, 500, 1000, and 5000 ppm of HCOOH. (Reprinted by permission from Aoki, A., Nogami, G., *J. Electroanal. Chem.*, 142 (2), 423–427, 1995. The Electrochemical Society.)

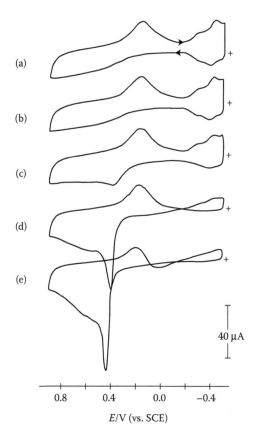

**FIGURE 7.12** Ring electrode cyclic votammograms of a rotating ring (platinum)-disk electrode, recorded at different disk electrode potentials: (a) open circuit; (b) –1.0 V; (c) –1.05 V; (d) –1.10 V; and (e) –1.15 V in a $3.2 \times 10^2$ M $CO_2$ solution (pH 5.5). Potential scan rate was 150 mV/s and rotation rate was 900 rpm. (Reprinted by permission from Zhang, J.J., Pietro, W.J., Lever, A.B.P., *J. Electroanal. Chem.*, 403 (1–2), 93–100, 1996.)

example of analysis of CO, which is a main product of $CO_2$ reduction catalyzed by Co(II) complexes, using the rotating ring (platinum)-disk electrode. The ring potential was scanned continuously in a certain range of voltage in order to define the activity of the platinum ring electrode for CO oxidation. In Figure 7.12a the disk electrode was open circuit and hence no CO is generated. A set of cyclic voltammograms (Figure 7.12b–e) were recorded as the disk electrode potential shifts more negatively into the region where $CO_2$ is reduced. When the disk electrode was polarized at –1.05 V, a small CO oxidation current near 0.4 V was observed (Figure 7.12c), which indicates that CO is produced at this potential, consistent with the appearance near –1.05 V of the tail where the catalytic current begins to arise in Figure 7.12b. When the potential is polarized more negatively (Figure 7.12d and e), the CO oxidation current grows rapidly, indicating greater production of CO at more negative disk potentials [52].

## REFERENCES

1.  Jitaru M, Lowy D, Toma M, Toma, B, Oniciu L. Electrochemical reduction of carbon dioxide on flat metallic cathodes. *Journal of Applied Electrochemistry* 1997, 27 (8), 875–889.
2.  Li H, Oloman C. The electro-reduction of carbon dioxide in a continuous reactor. *Journal of Applied Electrochemistry* 2005, 35 (10), 955–965.
3.  Welcher, F. J. A. Text-book of quantitative inorganic analysis including elementary instrumental analysis (Vogel, Arthur I.). *Journal of Chemical Education* 1963, 40 (1), A68.
4.  Hong J, Zhang W, Ren J, Xu R. Photocatalytic reduction of $CO_2$: A brief review on product analysis and systematic methods. *Analytical Methods* 2013, 5 (5), 1086–1097.
5.  Tomita Y, Teruya S, Koga O, Hori Y. Electrochemical reduction of carbon dioxide at a platinum electrode in acetonitrile-water mixtures. *Journal of the Electrochemical Society* 2000, 147 (11), 4164–4167.
6.  Azuma M, Hashimoto K, Hiramoto M, Watanabe M, Sakata T. Electrochemical reduction of carbon dioxide on various metal electrodes in low-temperature aqueous $KHCO_3$ media. *Journal of the Electrochemical Society* 1990, 137 (6), 1772–1778.
7.  Kaneco S, Iiba K, Katsumata H, Suzuki T, Ohta K. Effect of sodium cation on the electrochemical reduction of $CO_2$ at a copper electrode in methanol. *Journal of Solid State Electrochemistry* 2007, 11 (4), 490–495.
8.  Shiratsuchi R, Nogami G. Pulsed electroreduction of $CO_2$ on silver electrodes. *Journal of the Electrochemical Society* 1996, 143 (2), 582–586.
9.  Aydin R, Köleli F. Electrocatalytic conversion of $CO_2$ on a polypyrrole electrode under high pressure in methanol. *Synthetic Metals* 2004, 144 (1), 75–80.
10. Wang C, Thompson R. L, Baltrus J, Matranga C. Visible light photoreduction of $CO_2$ using CdSe/Pt/$TiO_2$ heterostructured catalysts. *The Journal of Physical Chemistry Letters* 2009, 1 (1), 48–53.
11. Moore G. F, Blakemore J. D, Milot R. L, Hull J. F, Song H.-E, Cai L, Schmuttenmaer C. A, Crabtree R. H, Brudvig G. W. A visible light water-splitting cell with a photoanode formed by codeposition of a high-potential porphyrin and an iridium water-oxidation catalyst. *Energy and Environmental Science* 2011, 4 (7), 2389–2392.
12. Roy L, Zimmerman P. M, Paul A. Changing lanes from concerted to stepwise hydrogenation: The reduction mechanism of frustrated Lewis acid–base pair trapped $CO_2$ to methanol by ammonia–borane. *Chemistry—A European Journal* 2011, 17 (2), 435–439.
13. Woolerton T. W, Sheard S, Pierce E, Ragsdale S. W, Armstrong F. A. $CO_2$ photoreduction at enzyme-modified metal oxide nanoparticles. *Energy and Environmental Science* 2011, 4 (7), 2393–2399.
14. Kaneco S, Iiba K, Suzuki S.-k, Ohta K, Mizuno T. Electrochemical reduction of carbon dioxide to hydrocarbons with high Faradaic efficiency in LiOH/methanol. *The Journal of Physical Chemistry B* 1999, 103 (35), 7456–7460.
15. Ohta K, Suda K, Kaneco S, Mizuno T. Electrochemical reduction of carbon dioxide at Cu electrode under ultrasonic irradiation. *Journal of the Electrochemical Society* 2000, 147 (1), 233–237.
16. Magdesieva T, Yamamoto T, Tryk D, Fujishima A. Electrochemical reduction of $CO_2$ with transition metal phthalocyanine and porphyrin complexes supported on activated carbon fibers. *Journal of the Electrochemical Society* 2002, 149 (6), D89–D95.
17. Kaneco S, Iiba K, Ohta K, Mizuno T. Electrochemical reduction of carbon dioxide on copper in methanol with various potassium supporting electrolytes at low temperature. *Journal of Solid State Electrochemistry* 1999, 3 (7–8), 424–428.
18. Li H, Oloman C. Development of a continuous reactor for the electro-reduction of carbon dioxide to formate—Part 2: Scale-up. *Journal of Applied Electrochemistry* 2007, 37 (10), 1107–1117.

19. Schrebler R, Cury P, Herrera F, Gomez H, Cordova R. Study of the electrochemical reduction of $CO_2$ on electrodeposited rhenium electrodes in methanol media. *Journal of Electroanalytical Chemistry* 2001, 516 (1), 23–30.

20. Thorson M. R, Siil K. I, Kenis P. J. Effect of cations on the electrochemical conversion of $CO_2$ to CO. *Journal of the Electrochemical Society* 2013, 160 (1), F69–F74.

21. Hori Y, Takahashi I, Koga O, Hoshi N. Electrochemical reduction of carbon dioxide at various series of copper single crystal electrodes. *Journal of Molecular Catalysis A: Chemical* 2003, 199 (1), 39–47.

22. Yang Z.-Y, Moure V. R, Dean D. R, Seefeldt L. C. Carbon dioxide reduction to methane and coupling with acetylene to form propylene catalyzed by remodeled nitrogenase. *Proceedings of the National Academy of Sciences USA* 2012, 109 (48), 19644–19648.

23. Hori Y, Takahashi I, Koga O, Hoshi N. Selective formation of C2 compounds from electrochemical reduction of $CO_2$ at a series of copper single crystal electrodes. *The Journal of Physical Chemistry B* 2002, 106 (1), 15–17.

24. Yui T, Kan A, Saitoh C, Koike K, Ibusuki T, Ishitani O. Photochemical reduction of $CO_2$ using $TiO_2$: Effects of organic adsorbates on $TiO_2$ and deposition of Pd onto $TiO_2$. *ACS Applied Materials & Interfaces* 2011, 3 (7), 2594–2600.

25. Simón-Manso E, Kubiak C. P. Dinuclear nickel complexes as catalysts for electrochemical reduction of carbon dioxide. *Organometallics* 2005, 24 (1), 96–102.

26. Endo N, Miho Y, Ogura K. Hydrogenation of $CO_2$ on the cathodized tungsten trioxide/polyaniline/polyvinylsulfate-modified electrode in aqueous solution. *Journal of Molecular Catalysis A: Chemical* 1997, 127 (1), 49–56.

27. Schrebler R, Cury P, Suarez C, Munoz E, Gomez H, Cordova R. Study of the electrochemical reduction of $CO_2$ on a polypyrrole electrode modified by rhenium and copper–rhenium microalloy in methanol media. *Journal of Electroanalytical Chemistry* 2002, 533 (1), 167–175.

28. Pérez E. R, Garcia J. R, Cardoso D. R, McGarvey B. R, Batista E. A, Rodrigues-Filho U. P, Vielstich W, Franco D. W. *In situ* FT-IR and ex situ EPR analysis for the study of the electroreduction of carbon dioxide in $N,N$-dimethylformamide on a gold interface. *Journal of Electroanalytical Chemistry* 2005, 578 (1), 87–94.

29. Wang C, Thompson R. L, Ohodnicki P, Baltrus J, Matranga C. Size-dependent photocatalytic reduction of $CO_2$ with PbS quantum dot sensitized $TiO_2$ heterostructured photocatalysts. *Journal of Materials Chemistry* 2011, 21 (35), 13452–13457.

30. Wu J, Risalvato F. G, Ke F.-S, Pellechia P, Zhou X.-D. Electrochemical reduction of carbon dioxide I. Effects of the electrolyte on the selectivity and activity with Sn electrode. *Journal of the Electrochemical Society* 2012, 159 (7), F353–F359.

31. Kuhl K. P, Cave E. R, Abram D. N, Jaramillo T. F. New insights into the electrochemical reduction of carbon dioxide on metallic copper surfaces. *Energy and Environmental Science* 2012, 5 (5), 7050–7059.

32. Qiao J, Jiang P, Liu J, Zhang J. Formation of Cu nanostructured electrode surfaces by an annealing–electroreduction procedure to achieve high-efficiency $CO_2$ electroreduction. *Electrochemistry Communications* 2014, 38, 8–11.

33. Ramos Sende J. A, Arana C. R, Hernandez L, Potts K. T, Keshevarz-K M, Abruna H. D. Electrocatalysis of $CO_2$ reduction in aqueous media at electrodes modified with electropolymerized films of vinylterpyridine complexes of transition metals. *Inorganic Chemistry* 1995, 34 (12), 3339–3348.

34. Narayanan S, Haines B, Soler J, Valdez T. Electrochemical conversion of carbon dioxide to formate in alkaline polymer electrolyte membrane cells. *Journal of the Electrochemical Society* 2011, 158 (2), A167–A173.

35. Mahmood M. N, Masheder D, Harty C. J. Use of gas-diffusion electrodes for high-rate electrochemical reduction of carbon-dioxide .1. Reduction at lead, indium-impregnated

and tin-impregnated electrodes. *Journal of Applied Electrochemistry* 1987, 17 (6), 1159–1170.

36. Hara K, Kudo A, Sakata T. Electrochemical reduction of carbon-dioxide under high-pressure on various electrodes in an aqueous-electrolyte. *Journal of Electroanalytical Chemistry* 1995, 391 (1–2), 141–147.

37. Sonoyama N, Kirii M, Sakata T. Electrochemical reduction of $CO_2$ at metal-porphyrin supported gas diffusion electrodes under high pressure $CO_2$. *Electrochemistry Communications* 1999, 1 (6), 213–216.

38. Koleli F, Balun D. Reduction of $CO_2$ under high pressure and high temperature on Pb-granule electrodes in a fixed-bed reactor in aqueous medium. *Applied Catalysis A-General* 2004, 274 (1–2), 237–242.

39. Zhang A. J, Zhang W. M, Lu J. X, Wallace G. G, Chen J. Electrocatalytic reduction of carbon dioxide by cobalt-phthalocyanine-incorporated polypyrrole. *Journal of Solid State Electrochemistry* 2009, 12 (8), E17–E19.

40. Prakash G. K. S, Viva F. A, Olah G. A. Electrochemical reduction of CO2 over Sn-Nafion (R) coated electrode for a fuel-cell-like device. *Journal of Power Sources* 2013, 223, 68–73.

41. Manfred Rudolph S. D. Ernst-Gottfried Jager Macrocyclic [N4$^{2-}$] coordinated nickel complexes as catalysts for the formation of oxalate by electrochemical reduction of carbon dioxide. *Journal of American Chemical Society* 2000, 122 (44), 10821–10830.

42. Peng Y. P, Yeh Y. T, Shah S. I, Huang C. P. Concurrent photoelectrochemical reduction of $CO_2$ and oxidation of methyl orange using nitrogen-doped $TiO_2$. *Applied Catalysis B-Environment* 2012, 123–124, 414–423.

43. Zhao H. Z, Chang Y. Y, Liu C. Electrodes modified with iron porphyrin and carbon nanotubes: application to $CO_2$ reduction and mechanism of synergistic electrocatalysis. *Journal of Solid State Electrochemistry* 2013, 17 (6), 1657–1664.

44. Zhao H. Z, Zhang Y, Zhao B, Chang Y. Y, Li Z. S. Electrochemical reduction of carbon dioxide in an MFC-MEC system with a layer-by-layer self-assembly carbon nanotube/cobalt phthalocyanine modified electrode. *Environmental Science and Technology* 2012, 46 (9), 5198–5204.

45. Hori Y, Kikuchi K, Suzuki S. Production of Co and Ch4 in electrochemical reduction of $CO_2$ at metal-electrodes in aqueous hydrogencarbonate solution. *Chemistry Letters* 1985, 14 (11), 1695–1698.

46. Reda T, Plugge C. M, Abram N. J, Hirst J. Reversible interconversion of carbon dioxide and formate by an electroactive enzyme. *Proceedings of National Academy of Sciences USA* 2008, 105 (31), 10654–10658.

47. Alvarez-Guerra M, Quintanilla S, Irabien A. Conversion of carbon dioxide into formate using a continuous electrochemical reduction process in a lead cathode. *Chemical Engineering Journal* 2012, 207, 278–284.

48. Kudo K, Ikoma F, Mori S, Komatsu K, Sugita N. Novel synthesis of oxalate from carbon-dioxide and carbon-monoxide in the presence of cesium carbonate. *Journal of Chemical Society, Chemical Communications* 1995, (6), 633–634.

49. Benson E. E, Kubiak C. P, Sathrum A. J, Smieja J. M. Electrocatalytic and homogeneous approaches to conversion of $CO_2$ to liquid fuels. *Chemical Society Review* 2009, 38 (1), 89–99.

50. Saveant J. M. Molecular catalysis of electrochemical reactions. Mechanistic aspects. *Chemical Reviews* 2008, 108 (7), 2348–2378.

51. Aoki A, Nogami G. Rotating ring-disk electrode study on the fixation mechanism of carbon dioxide. *Journal of Electrochemical Society* 1995, 142 (2), 423–427.

52. Zhang J. J, Pietro W. J, Lever A. B. P. Rotating ring-disk electrode analysis of $CO_2$ reduction electrocatalyzed by a cobalt tetramethylpyridoporphyrazine on the disk and detected as CO on a platinum ring. *Journal of Electroanalytical Chemistry* 1996, 403 (1–2), 93–100.

# 8 Modeling of Electrochemical $CO_2$ Reduction

*Yuanqing Wang, Makoto Hatakeyama, Koji Ogata, Fangming Jin, and Shinichiro Nakamura*

## CONTENTS

## 8.1 INTRODUCTION

### 8.1.1 EXPERIMENTAL RESULTS OF $CO_2$ REDUCTION

The basis and starting point for modeling are experiments. Some typical experimental results will be briefly introduced and the measured properties are shown in Table 8.1. Recently, Bocarlsy and coworkers proposed a promising way of reducing $CO_2$ into methanol in the presence of pyridine with low overpotential achieving nearly 100% faradaic efficiency [1–4]. The role of pyridine and detailed mechanism are still on debate [3]. However, the formation of a complex between $CO_2$ and pyridine derivatives, as the key step for reduction of $CO_2$, was considered by all the researchers [3–8]. Rosen et al. [9] reported a highly efficient method of reduction of $CO_2$ selectively into CO at overpotentials below 0.2 V catalyzed by an ionic liquid called "1-ethyl-3-methylimidazolium tetrafluoroborate (EMIM-BF$_4$)." The faradaic efficiencies of producing CO greater than 96% were obtained [9]. Other benefits of this technology are that ionic liquid itself is an electrolyte and $CO_2$ is remarkably soluble in the

**TABLE 8.1**

**Selected Results of Highly Efficient Electrochemical Reduction of $CO_2$**

| Electrode/ Electrocatalyst | Electrode Potential or Overpotential (V) | Products and Faradaic Efficiency (%) | Reference |
|---|---|---|---|
| Pd/pyridine and/or its derivatives (pH = 5.0) | Overpotentials below 0.2 V | Methanol (30%) | Seshadri et al. [1] |
| Pt, Ag/EMIM-BF$_4$ | Overpotentials below 0.2 V | CO (>96%) | Rosen et al. [9] |
| Pb/EMIM-Tf$_2$N | −2.25 V vs. Ag/AgNO$_3$ | CO (~45%) | Sun et al. [13] |
| Bi based/BMIM-BF$_4$ | −2.0 V vs. SCE | CO (~82%) | Medina-Ramos et al. [14] |
| Cu electrode | −1.39 V vs. SHE | HCOOH (15.4%), CO (1.5%), CH$_4$ (37.1%) | Hori et al. [15] |
| Modified Cu electrodes | Overpotentials as low as 0.19 V for CO and 0.25 V for HCOOH | CO (45%), HCOOH (33%) | Li et al. [16] |
| Au electrode | −1.14 V vs. SHE | HCOOH (0.4%), CO (81.2%) | Hori et al. [15] |
| Au nanoparticles | Overpotentials as low as 0.14 V | CO (>98%) | Chen et al. [17] |
| Nanoporous Ag | −0.60 V vs. RHE | CO (~92%) | Lu et al. [18] |

imidazolium-based ionic liquids. For instance, the solubility of $CO_2$ in the ionic liquid increases dramatically with increasing pressure, reaching 0.72 mole fraction $CO_2$ in BMIM-PF$_6$ at 93 bar [10]. An *in situ* spectroscopic observation of the reaction by sum frequency generation (SFG) showed that the adsorbed EMIM layer can react with $CO_2$ to form a complex such as $CO_2$–EMIM, which was then converted to CO [11]. It was found that the reaction rate varied remarkably with the metal electrode employed [11]. The addition of water to EMIM-BF$_4$ can surprisingly increase the efficiency of $CO_2$ conversion to CO, which was possibly due to the increased proton caused by the hydrolysis of tetrafluoroborate [12]. Other ionic liquids such as EMIM-Tf$_2$N [13] and BMIM-BF$_4$ [14] also showed selectivity to produce CO with high faradic efficiency.

Hori et al. reported that the distribution of products was governed by the metal electrode [15]. HCOO$^-$ was predominantly produced at Cd, In, Sn, and Pb cathodes. Au and Ag electrodes mainly produce CO, and Cu mainly gives CH$_4$. By employing a new experimental methodology, more intermediates and products from $CO_2$ on a copper electrode were detected [19]. The principal products are the hydrocarbons methane and ethylene, whereas the remaining 14 products are oxygenates, 11 of which are C2 and C3 products. Li et al. found that the Cu$_2$O-derived Cu electrode can catalyze the reduction of $CO_2$ to CO and HCOOH with high faradaic efficiencies at exceptionally low overpotentials (<100 mV) [16]. Nanomaterials-based electrode showed different catalytic activity with their bulk form in the reduction of $CO_2$. Chen et al. found that oxide-derived Au nanoparticles as the electrode could selectively reduce $CO_2$ into CO in water at overpotentials as low as 140 mV [17], although under identical conditions, polycrystalline Au electrodes require at least 200 mV of additional overpotential to attain comparable $CO_2$ reduction activity [17]. It was concluded that the Au nanoparticles can dramatically increase the stability of the $CO_2{}^{\cdot-}$ intermediate on the electrode surfaces from electrokinetic studies [17]. Similarly, nanoporous Ag can also obtain CO with a faradaic efficiency of about 92% [18].

## 8.1.2 CO₂ CHEMISTRY

The $CO_2$ molecule contains 22 electrons of which 16 electrons are in the valence shell. The atomic orbitals that are important to bonding are 2s and 2p of one carbon atom and two oxygen atoms, which lead to total 12 molecular orbitals. As shown in Walsh diagram of Figure 8.1a, in the linear structure, the combinations of 2s and $2p_z$ atomic orbitals lead to six molecule orbitals with $\sigma$ symmetry labeled as $3\sigma_g$, $4\sigma_g$, $5\sigma_g$ and $2\sigma_u$, $3\sigma_u$, $4\sigma_u$, respectively. The combinations of $2p_x$ and $2p_y$ atomic orbitals lead to three doubly degenerate $\pi$ orbitals labeled as $1\pi_g$, $1\pi_u$, and $2\pi_u$. The $1\pi_{gx}$ is the highest occupied molecule orbital (HOMO) and the $2\pi_u$ is the lowest unoccupied molecule orbital (LUMO). $CO_2$ molecule with a linear-shaped structure is relatively stable. When the structure of $CO_2$ bends from linear shape toward perpendicular shape, the degenerate $\pi$ orbitals all split, especially $2\pi_u$. One component of $2\pi_u$ becomes $2b_1$ with graduate energy change, whereas the energy of the other component that becomes $6a_1$ decreases sharply. Thus, the electron transfer (ET) to $CO_2$ will be favored by the bending structure with $6a_1$ orbital.

From a thermodynamic point of view, the reduction of $CO_2$ into reduced species such as CO and formate normally needs extra energy in room temperature. Meanwhile, an external hydrogen (or proton) is needed to reduce $CO_2$. Therefore, two factors, extra energy and hydrogen or proton, are required to fulfill the reduction. In the present context of electrochemistry, the energy source is electricity which can be derived from sustainable wind or hydrothermal energy or directly photoelectrochemistry [22]. On the other hand, proton can be from water in the aqueous solution. However, hydrogen evolution from water electrolysis can be a competitive process to harm the target reaction. Therefore, nonaqueous solution that shows higher solubility of $CO_2$ than water [9] is preferred due to its suppression to hydrogen evolution.

Another very important aspect of $CO_2$ is its aqueous chemistry. When $CO_2$ dissolves in water, it hydrates to yield $H_2CO_3$. Hydrated $CO_2$ then gives hydrated protons $H_3O^+$, bicarbonate ion $HCO_3{}^-$, and carbonate ion $CO_3{}^{2-}$. In equilibrium, $[H_2CO_3]$ is only about $10^{-3}$ as large as $[CO_2]$ and can often be omitted. So there are three possible electroactive species in the reduction of $CO_2$ in aqueous solution: the $CO_2$ molecule, $HCO_3{}^-$, and $CO_3{}^{2-}$. The ratios of carbonate species depend on acidity constants ($K_{a1}$ and $K_{a2}$) and pH value in the aqueous solution. As shown in Figure 8.1b, a Pourbaix diagram of $CO_2$ and its related substances can be drawn based on thermodynamic data [21]. Dissolved $CO_2$ dominates in the pH below 8; $HCO_3{}^-$ is stable in the pH range around 8–10; and $CO_3{}^{2-}$ above 10. In the lower pH range, the equilibrium potential of $CO_2$ reduction is more cathodic than that of the competitive hydrogen evolution reaction. In the pH range around 7–8, $CO_2$ reduction may proceed more effectively with the products of $HCOO^-$ instead of formic acid [21]. Although $CO_2$ is normally considered as an electroactive species in the reduction of $CO_2$, some authors suggested that $HCO_3{}^-$ could also be reduced [23]. However, Hori reported that the limiting

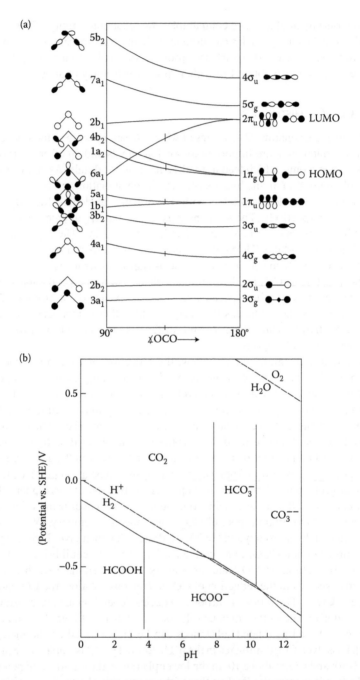

**FIGURE 8.1** General chemistry of $CO_2$. (a) Walsh diagram of $CO_2$ orbital energies in linear and bent geometries. (Reproduced from Freund, H. J. and Roberts, M. W. *Surface Science Reports* 1996; 25: 225–273.) (b) Pourbaix diagram of $CO_2$ and its related substances. (Reprinted from Hori, Y. and Suzuki, S. *Bulletin of the Chemical Society of Japan* 1982; 55: 660–665. With permission.)

current would be too low under usual experimental conditions, if $HCO_3^-$ were an electroactive species [24]. Besides, the electrochemical reduction of $HCO_3^-$ was not affected by the convective motion of the electrolyte [24]. These results are not consistent with the notion that $HCO_3^-$ is an electroactive species. Another possible function of $HCO_3^-$ is that it may behave as a proton source effectively [17].

### 8.1.3 PROPOSED MECHANISM

One ET to $CO_2$ forming $CO_2^{\cdot-}$ anion radical is widely considered as an initial step in the reduction of $CO_2$ [24]. As shown in Figure 8.2, the initial step

$$CO_2 + e^- = CO_{2ads}^{\cdot-} \tag{8.1}$$

is considered as the rate-determining step (RDS) on the copper electrode. Then two reaction pathways are available leading to CO and $HCOO^-$, respectively. The formation of $HCOO^-$ was probably more favored by a free $CO_2^{\cdot-}$ anion radical pathway on the Cd, Sn, In, Ti, and Hg electrode surface, [24] as shown in Equations 8.2 and 8.3.

$$CO_{2ads}^{\cdot-} + H_2O = HCOO_{ads}^{\cdot} + OH^- \tag{8.2}$$

$$HCOO_{ads}^{\cdot} + e^- = HCOO^- \tag{8.3}$$

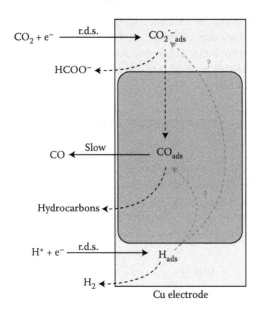

**FIGURE 8.2** The main reaction pathways at the electrode surface, with adsorbed CO as the key intermediate leading to CO or higher reduced products. (Reprinted from *Journal of Electroanalytical Chemistry*, 594, Gattrell, M., Gupta, N. and Co, A., A review of the aqueous electrochemical reduction of $CO_2$ to hydrocarbons at copper, 1–19, Copyright 2006, with permission from Elsevier.)

On the Cu surface, both $CO_2$ reduction and hydrogen evolution occur. The adsorbed hydrogen could be a potential proton source for hydrogenation of $CO_2$. The adsorbed $CO_2^{\cdot-}$ anion radical subsequently reacts with the adsorbed hydrogen to form adsorbed CO as shown in Equation 8.4, which is a slow process.

$$CO_{2ads}^{\cdot-} + H_{ads} + H^+ = CO_{ads} + H_2O \qquad (8.4)$$

The desorption of adsorbed CO can lead to CO gas. Alternatively, adsorbed CO continues to react with hydrogen (proton and electron) to produce hydrocarbons including C2 and C3 molecules. In the following sections, we will introduce and review the recent theoretical calculations of reduction of $CO_2$ in the heterogeneous and homogeneous phase, respectively. In the heterogeneous reactions, the metal electrodes were often introduced into the model as a periodic system. Electronic structure calculations were often performed using density functional theory (DFT) in a plane wave pseudopotential implementation. In the homogeneous reactions, the electrode effect was not considered in most cases.

## 8.2 HETEROGENEOUS CATALYSIS

### 8.2.1 COPPER SURFACE

#### 8.2.1.1 Computational Hydrogen Electrode Model

The computational hydrogen electrode (CHE), which was described by Nørskov et al. [26,27], was adopted to calculate the free energy change of each reaction step that involves an electrochemical proton–electron transfer as a function of applied electrode potential. In this method, a reversible hydrogen electrode (RHE) is defined as the following equation:

$$H^+ + e^- = \frac{1}{2}H_2 \qquad (8.5)$$

Zero voltage is defined in its equilibrium at all values of pH, at all temperatures, and with $H_2$ at 101325 Pa pressure. Therefore, at zero voltage, the chemical potential of one proton and one electron is equal to half of chemical potential of gaseous hydrogen. At specific applied potential ($U$), the chemical potential ($\mu$) of one proton and one electron can be calculated as follows, where e is the elementary charge (positive value).

$$\mu(H^+) + \mu(e^-) = \frac{1}{2}\mu(H_{2(g)}) - eU \qquad (8.6)$$

Then the free energy change ($\Delta G$) of one electrochemical reaction of Equation 8.7, where * denotes the surface bound species, can be calculated as follows (Equation 8.8):

$$A* + H^+ + e^- = AH* \qquad (8.7)$$

$$\Delta G = \mu(AH^*) - \mu(A^*) - \left[\frac{1}{2}\mu(H_{2(g)}) - eU\right] \tag{8.8}$$

### 8.2.1.2 Calculations of C1 Products

As shown in Figure 8.3, Peterson et al. calculated the reaction pathways of reduction of $CO_2$ into hydrocarbons in the electrochemical system [27]. They proposed that methane is formed by a CHO intermediate through the following path:

$$CO_2 \rightarrow COOH \rightarrow CO \rightarrow \mathbf{CHO} \rightarrow CH_2O \rightarrow CH_3O \rightarrow CH_4 + O$$

$$\rightarrow CH_4 + OH \rightarrow CH_4 + H_2O \tag{8.9}$$

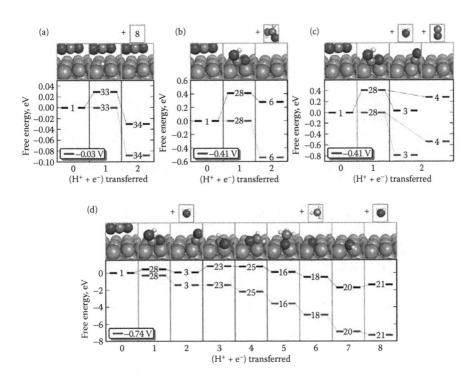

**FIGURE 8.3** Calculated free energy diagrams for the lowest-energy pathways to (a) $H_2$, (b) HCOOH, (c) CO, and (d) $CH_4$. In each diagram, the black bars (higher) pathway represents the free energy at 0 V vs. RHE and the gray bars (lower) pathway represents the free energy at the indicated potential. The numbers in the middle of the bars indicate different intermediates. The shadow ball represents oxygen atom, the gray ball represents carbon atom, and the white ball represents hydrogen atom. (Peterson, A. A., Abild-Pedersen, F., Studt, F., Rossmeisl, J. and Norskov, J. K. How copper catalyzes the electroreduction of carbon dioxide into hydrocarbon fuels. *Energy & Environmental Science* 2010; 3: 1311–1315. Reproduced by permission of The Royal Society of Chemistry.)

Figure 8.3 shows that the first pathway to react in the reactions is the production of $H_2$. It is predicted to occur at about $-0.03$ V vs. RHE at which all the steps are exergonic (see Figure 8.3a). The productions of HCOOH and CO then follow at about $-0.41$ V (Figure 8.3b and c). When CO is adsorbed, the subsequent reduction of CO to $CH_4$ occurs at about $-0.74$ V. The sequence of formation of products with applied potential follows the same order as in the experiments reported by Hori et al [28]. As shown in Figure 8.3d, the key step in the formation of $CH_4$ is the hydrogenation of the adsorbed CO to adsorbed CHO (species 23). According to this pathway, the selectivity of $CH_4$ over $CH_3OH$ was determined by the step of $CH_3O$ that can lead to $CH_4$ or $CH_3OH$. However, the experimental fact that the $CH_3OH$ is over $CH_4$ by direct reduction of $CH_2O$ did not support the proposed pathway [29]. Therefore, Nie et al. calculated two parallel pathways through CHO and COH intermediate, based on the CHE model and explicit water [30]. They found that the calculated reaction free energies are not enough to determine the reaction pathways in this system. A transferable method of determining activation barriers for electrochemical reactions was used [30].

$$E_{act}(U) = E_{act}^{o}(U^{o}) + \beta'(U - U^{o}) \qquad (8.10)$$

where $E_{act}^{o}$ and $E_{act}$ are activation barriers, and $U^{o}$ is the equilibrium potential for the reductive adsorption of a proton, at which the chemical potential of the adsorbed H* species is equal to the chemical potential of a proton–electron pair. $\beta'$ denotes an effective reaction symmetry factor, which is approximated to be 0.5 followed by corrections. After comparison of the positions of transition states, a more preferred pathway to produce $CH_4$ is proposed with COH as the intermediate [30].

$$CO_2 \rightarrow COOH \rightarrow CO \rightarrow \mathbf{COH} \rightarrow C \rightarrow CH \rightarrow CH_2 \rightarrow CH_3 \rightarrow CH_4 \quad (8.11)$$

### 8.2.1.3   Calculations of C2 Products

In addition to C1 products, the mechanism of C2 production and higher carbon number products is also of great importance, because it is related to C–C bond formation, which may have significance in providing insights into producing fuels. Schouten et al. investigated the electroreduction of various intermediates and suggested that the dimerization of CO is the rate-determining step for the C2 pathway, followed by the formation of a surface-bonded enediol or enediolate, or an oxametallacycle which all lead toward ethylene [29]. Further experiments confirmed that the formation of ethylene consisting of two reaction pathways is preferably through CO dimerization with low overpotentials at Cu(100) facet [31]. Figure 8.4 presents lowest-energy pathways from CO to $C_2H_4$, MeCHO and EtOH using CHE model. As shown in Figure 8.4, an applied potential of $-0.40$ V vs. RHE makes all the steps thermodynamically viable, which is close to the experimental value [25]. The first proton–electron transfer is the potential-determining step (PDS) [32], as shown in Equation 8.12.

$$* + 2CO_{(g)} + (H^{+} + e^{-}) \rightarrow *CO - COH \qquad (8.12)$$

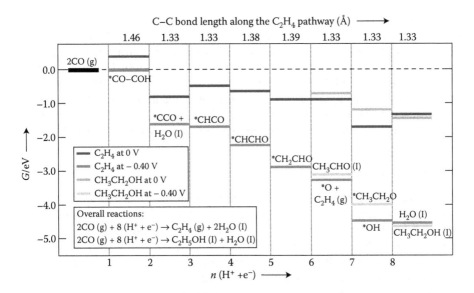

**FIGURE 8.4** Lowest-energy pathways for the electroreduction of CO to $C_2H_4$, MeCHO/ EtOH at 0 V (shadow, silver) and −0.40 V (gray, cloud). C–C bond lengths are given in the upper x-axis for the pathway to $C_2H_4$. (Calle-Vallejo, F. and Koper, M. T. M.: Theoretical considerations on the electroreduction of CO to C2 species on Cu(100) electrodes. *Angewandte Chemie International Edition*. 2013. 52. 7282–7285. Copyright Wiley-VCH Verlag GmbH & Co. KGaA. Reproduced with permission.)

Two CO molecules of the gas phase are adhered and coupled on the Cu(100) surface. The oxygen atom of CO is hydrogenated through one proton–electron transfer. The C–C bond length of *CO–COH is calculated at 1.46 Å which is in typical of C–C bond. The second proton–electron transfer leads to intermediate *CCO with a C–C bond length of 1.33 Å, which is in typical of C = C bond that is preserved in the subsequent steps. It can also be found that the path through intermediate *CH₂CHO splits leading to $C_2H_4$ and EtOH, respectively. Specifically, the first proton–electron step was proposed to consist of three sequential steps in which proton transfer (PT) and ET are decoupled.

Adsorption:

$$* + CO_{(g)} \rightarrow *CO \tag{8.13}$$

C–C coupling:

$$*CO + CO_{(g)} + e^- \rightarrow *C_2O_2^- \tag{8.14}$$

Hydrogenation:

$$*C_2O_2^- + H^+ \rightarrow *CO - COH \tag{8.15}$$

The rate-determining step was calculated, with the calculation of the transition state, to be the coupling step (Equation 8.14). This is before the PT step (Equation 8.15). These findings could explain an important experimental observation [25,29] that the formation of $C_2H_4$ is independent of the pH value on Cu(100). However, Montoya et al. provided another DFT calculations on the Cu(211) surface and found that kinetic barrier to C–C coupling decreases significantly with the degree of hydrogenation of adsorbates [34]. That is, by comparison with dimerization of adsorbed CO, the C–C coupling is more kinetically favorable through *CO and *CHO or two *CHO.

### 8.2.2   AU SURFACE

For a long time, gold had been considered to be an inert material and cannot be a catalyst. However, it was found by Haruta et al. [35,36], in the late 1980s, that gold exhibited surprisingly high catalytic activity for CO oxidation when formed as nanoparticles. These findings motivated the intense research of gold catalysis in the next two decades [37]. Recently, gold nanoparticles and nanowires have also been proved to be effective and efficient for electrochemical $CO_2$ reduction [17,38,39]. DFT calculations suggest that the edge sites of gold nanoparticles are more selective for $CO_2$ compared with corner sites [38] due to the optimum binding strengths of COOH and CO [39]. Kauffman et al. [40] have studied the interaction between $CO_2$ and ligand-protected $Au_{25}$ cluster by experiments and DFT methods. The $Au_{25}$ catalyst was found to produce CO with approximately 100% faradaic efficiency at −1.0 V vs. RHE and a rate 7–700 times higher than 2–5 nm Au nanoparticles and bulk Au. $Au_{25}$ cluster consists of core atoms and shell atoms. DFT modeling indicates that the spectroscopic and electrochemical changes, probably caused from a reversible $Au_{25}$–$CO_2$ interaction, may be a $CO_2$-induced redistribution of charge within the cluster. The enhanced electrocatalytic activity was proposed to be due to the anionic charge of $Au_{25}$, which promotes $CO_2$ adsorption, and its unique reactive site, which promotes $C = O$ bond activation and $H_{ads}$ formation.

### 8.3   HOMOGENEOUS CATALYSIS

The experimental data of reduction of $CO_2$ in the pyridine system are reported by Bocarsly and coworkers [1,2,22,41,42]. They proposed a reaction mechanism (see Equation 8.16) that aqueous pyridine (Py) accepts a proton to form aqueous PyH+, which is in turn reduced into the aqueous pyridinyl radical (PyH•). Thus, formed PyH· then reacts with $CO_2$ to form a pyridinium radical–$CO_2$ complex, which is estimated to be the rate-determining step with $16.5 \pm 2.4$ kcal/mol activation barrier [42]. Lastly, the produced PyCOOH• is reduced into methanol through multiple proton and electron transfer. The whole process was considered to occur in the homogeneous phase [41].

$$Py + H^+ \longrightarrow PyH^+ \xrightarrow{+e^-} PyH^• \xrightarrow{+CO_2, \text{rate-determining step}}$$

$$PyCOOH^• \xrightarrow{nH^+ + ne^-} \text{methanol} \qquad (8.16)$$

Lim et al. calculated the homogeneous PyCOOH• formation using *ab initio* calculations in which the reaction proceeds through a proton relay of H$_2$O molecules [8]. The solvent is considered by using the mixed models: implicit/explicit and only-implicit solvation, respectively [8]. In the absence of proton relay, the activation barrier is quite high (about 46 kcal/mol), whereas in the presence of proton relay it decreases to the range between 13.6 and 18.5 kcal/mol. The values are close to the estimated experimental value. As shown in Figure 8.5, along with the intrinsic reaction coordinate (IRC) of the reaction step, the nitrogen in PyH• first attacks C of CO$_2$ to form C–N bond (Figure 8.5a). The bending of linear O=C=O structure is followed by one PT from PyH• to one neighboring H$_2$O to form H$_3$O$^+$ (Figure 8.5b). The PT then proceeds through a water network. In the end, PyCOOH• is formed receiving one proton from the last H$_3$O$^+$ in the network (Figure 8.5f). With an aid of the water proton-shuttling network, the strain in the transition state is reduced and the produced PyCOOH• is stabilized. Furthermore, it was found that ET precedes PT in the PyCOOH• formation process, which avoids the formation of high-energy radicals: Py$^-$ and CO$_2^-$. This calculation is the first application of proton relay in the reduction of CO$_2$, though the proton-shuttling mechanism has already been proposed in other processes [43].

Keith et al., however, questioned the mechanism proposed by Bocarsly et al. about the existence of PyH• radical by the calculation results of acidity constant (p$K_a$ value) and reduction potential [5,6]. The methodology of the calculations is briefly

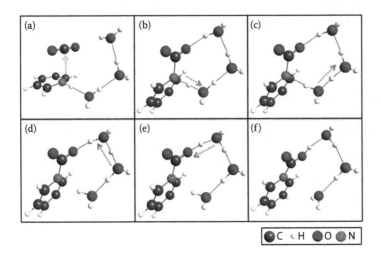

**FIGURE 8.5**  Structures along the IRC for the PyH• + CO$_2$ + 3H$_2$O reaction step of indirect PT from PyH• to CO$_2$ via a proton relay consisting of a chain of three H$_2$O molecules. (a) Reactants, (b) TS for PyCOO$^-$ formation by ET followed by PT from PyH• to a H$_2$O, (c) and (d) PT from a H$_3$O$^+$ to a neighboring H$_2$O, (e) PT from H$_3$O$^+$ to PyCOO$^-$, and (f) the trans PyCOOH· product. The dashed gray arrows indicate the direction of PT and the solid silver arrow the nucleophilic attack on the C of CO$_2$. (Reprinted with permission from Lim, C. H., Holder, A. M. and Musgrave, C. B. Mechanism of homogeneous reduction of CO$_2$ by pyridine: Proton relay in aqueous solvent and aromatic stabilization. *Journal of the American Chemical Society* 135: 142–154. Copyright 2013 American Chemical Society.)

introduced here. Taking reaction 8.17 as an example, $pK_a$ value is calculated using the following Equation 8.18:

$$Py + H^+ \leftrightarrow PyH^+ \tag{8.17}$$

$$pK_a = -\log K_a = \frac{(G_{298}(Py) + G_{298}(\text{proton}) - G_{298}(PyH^+))}{(2.303 \cdot RT)} \tag{8.18}$$

Here, $R$ is the gas constant and $T$ is the absolute temperature, which is set at 298 K in this case. Absolute free energies, $G_{298}$, is calculated using Equation 8.19:

$$G_{298} = G_g + \Delta G_s \tag{8.19}$$

where $G_g$ is the gas-phase Gibbs free energy at 1 atm and 298 K, and $\Delta G_s$ is calculated from a solvation model using either implicit or mixed implicit/explicit solvation models, if an explicit water was shown to improve solvation energies [5,44]. The strategy for calculating $\Delta G_s$ using a mixed implicit/explicit solvation model is shown in Figure 8.6. As shown in Figure 8.6, by adding explicit waters to $X^{m\pm}$ to form $[X(H_2O)_n]^{m\pm}$ in the gas phase and aqueous phase, respectively, a thermodynamic cycle can be drawn. The value of $\Delta G_s$ is, therefore, calculated based on this thermodynamic cycle as the following equation:

$$\Delta G_s(X^{m\pm}) = \Delta G_g^o(\text{bind}) + \Delta G_s([X(H_2O)_n]^{m\pm}) - \Delta G_s((H_2O)_n) - \Delta G_{aq}^* \tag{8.20}$$

where the first term in the right-hand side of the equation is the gas-phase free energy change due to binding water molecules to $X^{m\pm}$, the second and third terms are the solvation energy of $(H_2O)_n$ and $[X(H_2O)_n]^{m\pm}$, and the fourth term is considered as 0. Note that the second and third terms in Equation 8.20 contain not only free energies due to solvation but also additional energy contributions due to standard state changes as well as calculation methods used. The detailed calculation techniques were explained in Reference 44.

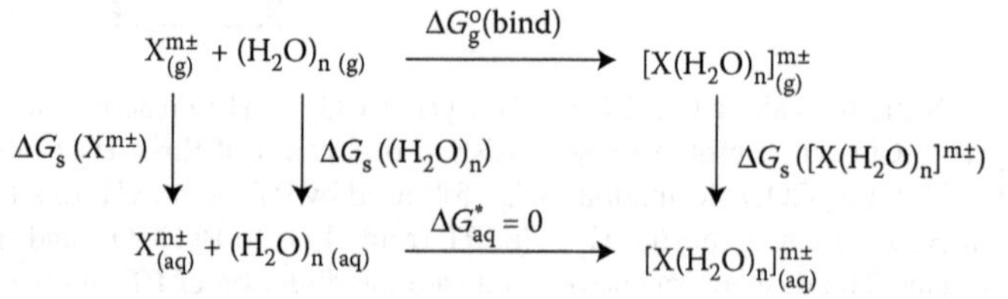

**FIGURE 8.6** Thermodynamic cycle to compute $\Delta G_s$ for a general species within a mixed implicit–explicit solvation approach. (Reprinted from Keith, J. A. and Carter, E. A. *Journal of Chemical Theory and Computation* 2012; 8: 3187–3206.)

For a general multiple proton-coupled ET reaction (PCET, as shown in Equation 8.21), the reduction half potential is calculated using Equation 8.22 [45]:

$$A + n_e e^- + n_H H^+ \rightarrow AH_{n_H}^{(n_H - n_e)} \qquad (8.21)$$

$$E_{1/2} = -\left(\frac{\Delta G_{1/2}^0}{n_e F}\right) - \frac{RT \cdot n_H \cdot \ln 10}{n_e} pH - E_{1/2}^{0,\text{ref}} \qquad (8.22)$$

where $F$ is Faraday's constant, $n_e$ is the number of electrons, $n_H$ is the number of protons, $E_{1/2}^{0,\text{ref}}$ is a reference potential (usually at 4.44 V for normal hydrogen electrode), and $\Delta G_{1/2}^0$ is aqueous-phase reaction free energies and determined from Equation 8.23.

$$\Delta G_{1/2}^0 = G^0\left(AH_{n_H}^{(n_H - n_e)}\right) - G^0(A) - n_H G^0(H^+) \qquad (8.23)$$

The calculated $pK_a$ of PyH$^\bullet$ is a very high value, which is about 27 [5]. Thus, the deprotonation of PyH$^\bullet$ should not be facile. It means that the rate-determining step of the reaction from PyH$^\bullet$ and $CO_2$ to PyCOOH$^\bullet$ is not realistic in the homogeneous solution. Another puzzle is the large discrepancy between observed reduction potential of −0.58 V vs. saturated calomel electrode (SCE) and the calculated value of −1.47 V of the homogeneous reduction of PyH$^+$ [5]. It leads to the doubt whether PyH$^\bullet$ itself is really formed in the homogeneous solution. Lim et al. argued that the calculation assumed a homogeneous environment; however, it was not the case for actual reduction of PyH$^+$ that in fact was largely dependent on electrode surface in experiment [8]. In other words, the surface effect may cause the discrepancy between experiment and calculation. Some experiments suggested that a surface-mediated reduction of PyH$^+$ into PyH$^\bullet$, which then desorbs from the electrode surface to the homogeneous solution [8,46], whereas electron paramagnetic resonance (EPR) and ultraviolet–visible (UV/vis) spectroscopy showed no evidence of the formation of PyH$^\bullet$ [3].

The debate on the existence of the PyH$^\bullet$ formation indicated that there are probably other electroactive species involved in the reduction of $CO_2$. Keith et al. proposed 4-4′-bipyridine (BPy) as a solid candidate to catalyze $CO_2$. The assumption was supported by BPy and its reduction derivatives, which show low calculated acidity constants. Also, the calculated redox potential is closer to the experimental one [5]. Later, they proposed another possible catalyst, which is a surface-bound dihydropyridine species formed from two electrons reduction [6]. In contrast to the commonly adopted PCET mechanism, Ertem et al. recently provided an alternative proton-coupled hydride transfer (PCHT) mechanism [7]. As shown in Figure 8.7, starting with the protonation of pyridine to pyridinium (step i), which then assists to produce hydrogen atoms adsorbed on the Pt(111) surface (step ii), the formed hydrogen adsorbates are thus able to attack the carbon atom of $CO_2$ with another proton transferred from pyridinium to the oxygen atom of $CO_2$ simultaneously (step iii). The calculated redox potential of step ii is estimated at about −0.72 V vs. SCE, which is in agreement with the observed value of −0.58 V in the Pt electrode [42]. Besides,

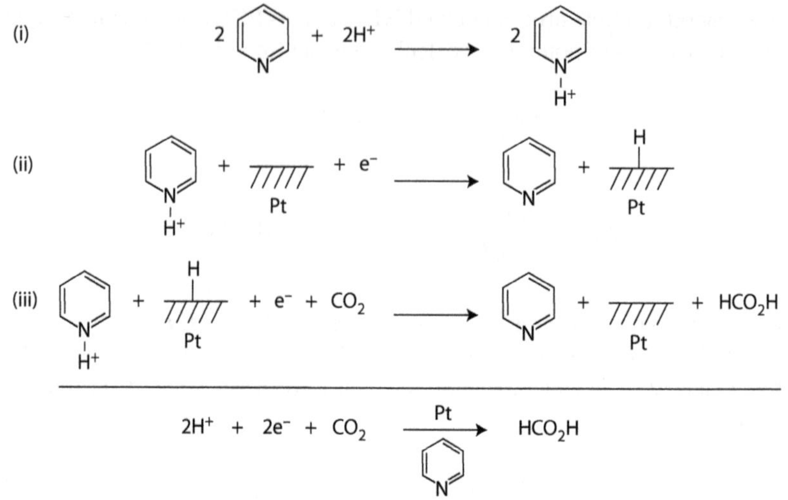

**FIGURE 8.7** Proposed mechanism of electrochemical reduction of $CO_2$ on Pt(111). (Reprinted from Ertem, M. Z., Konezny, S. J., Araujo, C. M. and Batista, V. S. *The Journal of Physical Chemistry Letters* 2013; 4: 745–748.)

the free energy analysis of step iii showed that a low free energy barrier of about 13 kcal/mol is needed to overcome, which also agrees well with the experimental values [42]. However, the authors did not present any calculations or comments on the $H_2$ formation from two Pt-H as a competitive path to the proposed PCHT mechanism. In summary, more detailed experiments are needed to determine the intermediates and reaction kinetics followed by theoretical calculations to have a complete understanding on a molecule level of this highly efficient $CO_2$ reduction system.

## 8.4 DESIGN OF NEW CATALYSTS

It is not always straightforward to find a new catalyst for the target reaction even after the determination of the reaction mechanism. Even for a very simple reaction, there are probably many elementary steps. Fortunately, there exist one or a few "descriptors" that govern the activity and selectivity of the target reaction which are composed of a series of sequent steps [47–49]. These descriptors, which are normally adsorption energies of intermediates, can be tuned for materials and easily applied for searching and designing new catalysts. A general rule is first introduced in the following.

The philosophy of descriptors comes from the famous "Sabatier principle," which states that a good catalyst binds the key intermediates neither too weakly nor too strongly [50]. Thus, there is a volcano curve indicating the highest activity as a function of the descriptors. Further, the number of descriptors, for example, adsorption energies of intermediates, can be reduced to only a few because of the scaling relations between different adsorption energies [51]. In a quantitative form of "Sabatier principle" applied in electrochemistry, it is stated that for "optimal reversible" electrocatalysis, the equilibrium potentials of all reaction steps must be equal to the overall equilibrium potential [50]. In other words, at the overall equilibrium potential, all the reaction steps become downhill in free energy as can be found from Figure 8.3d. This statement relates the activity of catalyst to equilibrium potentials of all reaction steps. From the theoretical point of view, the electrochemical reaction normally includes PT and ET, which are referred as PCET [52,53]. Based on whether PT and ET happen simultaneously or sequentially, PCET is divided into two types: concerted proton–electron transfer (CPET) and sequential proton–electron transfer (SPET) [52–54]. In molecular electrochemistry, SPET pathways are widely accepted [52], whereas in surface electrochemistry, CPET pathways are often assumed which is also the assumption of the CHE model. Here we show a two proton–electron transfer reaction in a concerted manner to see the relationship between descriptor and equilibrium potentials [50].

$$A + H^+ + e^- \leftrightarrow AH \tag{8.24}$$

$$AH + H^+ + e^- \leftrightarrow AH_2 \tag{8.25}$$

$$A + 2H^+ + 2e^- \leftrightarrow AH_2 \tag{8.26}$$

The equilibrium potentials of Equations 8.24 through 8.26 are defined as

$$E^0_{A,H^+/AH} = -\frac{G(AH) - G(A) - G(H^+)}{e} \tag{8.27}$$

$$E^0_{AH,H^+/AH_2} = -\frac{G(AH_2) - G(AH) - G(H^+)}{e} \tag{8.28}$$

$$E^0_{A,2H^+/AH_2} = -\frac{G(AH_2) - G(A) - 2G(H^+)}{2e} \tag{8.29}$$

If we redefine these two potentials relative to the RHE scale, they are written as

$$E^{0,RHE}_{A,H^+/AH} = -\frac{G(AH) - G(A)}{e} \tag{8.30}$$

$$E^{0,RHE}_{AH,H^+/AH_2} = -\frac{G(AH_2) - G(AH)}{e} \tag{8.31}$$

$$E^{0,RHE}_{A,2H^+/AH_2} = -\frac{G(AH_2) - G(A)}{2e} \tag{8.32}$$

And if we set the overall equilibrium potential relative to RHE scale (Equation 8.32) equals to zero, it leads to $G(AH_2) = G(A)$. Then Equation 8.31 can be rewritten as shown in Equation 8.33.

$$E^{0,RHE}_{AH,H^+/AH_2} = -\frac{G(AH_2)-G(AH)}{e} = -\frac{G(A)-G(AH)}{e} \tag{8.33}$$

Figure 8.8 shows the equilibrium potentials as a function of the descriptor: "$G(AH)$–$G(A)$" [50]. As shown in Figure 8.8, the principle mentioned above suggests that the optimal catalyst will be obtained at $G(AH)$–$G(A) = 0$, that is, the equilibrium potentials of each step (Equations 8.30 and 8.31) equal to the overall equilibrium potential (Equation 8.32). When $G(AH)$–$G(A) > 0$, reaction 8.24 has a overpotential and when $G(AH)$–$G(A) < 0$, reaction 8.25 also has an overpotential. Note that "Sabatier principle" does not take kinetics and activation barriers into account. Therefore, in some cases, the strategy for finding optimal catalyst using Sabatier principle may not be effective [32]. On the other hand, the criteria may not be exactly fulfilled in most cases because there are multiple elementary steps unlike the ideal case as mentioned above. In the following, we will introduce a calculation example of electrochemical reduction of $CO_2$ to explain the application of Sabatier principle in real case [55]. This calculation work is also based on the CHE model introduced above.

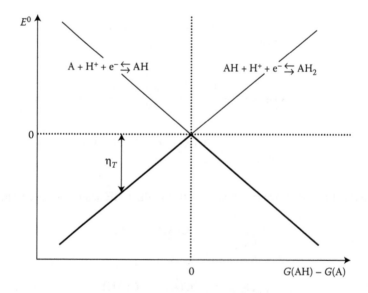

**FIGURE 8.8** Plot of the equilibrium potentials of Equations 8.30 and 8.31 of reactions (8.24) and (8.25) as a function of the "descriptor" $G(AH)$–$G(A)$. (Koper, M. T. M. Theory of multiple proton-electron transfer reactions and its implications for electrocatalysis. *Chemical Science* 2013; 4: 2710–2723. Reproduced by permission of The Royal Society of Chemistry.)

The concept of liming potential ($U_L$) should be first introduced. It states that at the potential ($U_L$) the specific elementary step of one reaction becomes exergonic. Therefore, at the $U_L$, $\Delta G = 0$, we can rewrite Equation 8.8 as follows:

$$\Delta G = \mu(AH^*) - \mu(A^*) - \left[\frac{1}{2}\mu(H_{2(g)}) - eU_L\right] = 0 \qquad (8.34)$$

So, $U_L$ is given by

$$U_L = -\frac{\mu(AH^*) - \mu(A^*) - 1/2\mu(H_{2(g)})}{e} \qquad (8.35)$$

as a function of the chemical potential of species. For example, that of CHO* can be calculated based on the following:

$$\mu(CHO^*) = E_B(CHO) + G_{corr}(CHO^*) \qquad (8.36)$$

where the $G_{corr}$ terms contain the zero-point, enthalpic (heat capacity), and entropic portions of the free energy, and $E_B$ is the binding energy. The binding energy of CHO or other hydrogenated species can all be related to the binding energy of CO because of the scaling relation [51].

$$E_B(CHO) = 0.88E_B(CO) + 2.03\,eV \qquad (8.37)$$

Based on these properties, we can obtain the profile of $U_L$ as a function of the binding energy of CO for each elementary step in the reduction of CO₂ as shown in Figure 8.9 [55]. The difference between $U_L$ and equilibrium potential is the overpotential for each elementary step and the most negative $U_L$ for specific binding energy of CO determines "theoretical overpotential" of the overall reaction. It can be found that the competitive path from adsorbed CO to COH* or CHO* dominate the theoretical overpotential in most cases. For metals which bind the CO strongly, such as Pt, the COH* route is more favored. Whereas for metals representing moderate binding energy with CO, such as Pd, Ni, and Cu, the CHO* route is more favored. The metal of Cu is the best electrocatalyst for the reduction of CO₂, because it shows the smallest theoretical overpotential. There is a vertical dotted line around −0.5 eV. It shows the equilibrium between adsorbed CO and gaseous CO. On the right side of this line, the CO which is produced in the system will be expected to desorb from the metal surface. Therefore, gaseous CO will be the main product without further reaction due to weak binding of CO such as on Au and on Ag. These data are in agreement with experimental findings [15]. Thus, a strategy of tuning the binding properties of CO on metals can be an effective guiding principle of the catalysts design for electrochemical reduction of CO₂.

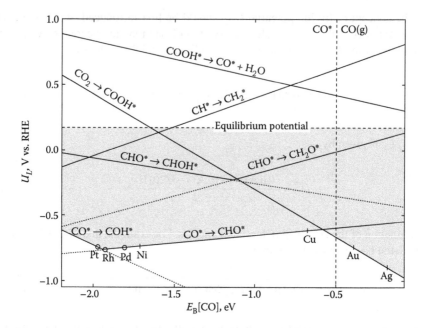

**FIGURE 8.9** Plot of limiting potentials ($U_L$) as a function of binding energy of CO on the elementary steps of electrochemical reduction of $CO_2$. (Reprinted with permission from Peterson, A. A. and Nørskov, J. K. Activity descriptors for $CO_2$ electroreduction to methane on transition-metal catalysts. *The Journal of Physical Chemistry Letters* 3: 251–258. Copyright 2012 American Chemical Society.)

## 8.5 CONCLUDING REMARKS AND OUTLOOK

In this chapter, some recent progresses in the theoretical electrochemical reduction of $CO_2$ are introduced. On the one hand, the thorough experimental studies for determining reaction pathways and intermediates are still needed, especially in an *in situ* way. On the other hand, there are some limitations for the proposed calculation methods prohibiting the exact mechanistic studies. The CHE model assumed the proton–electron transfer in a concerted manner that may be not always true in all the cases. In some calculations based on the homogeneous phase, the solvation model should be taken care deliberately. Finally, the double layer (Helmholtz plane and diffuse layer of electrolyte) on the electrode surface was not considered in most calculations, although it is an essential feature that determines the surface property.

## REFERENCES

1. Seshadri, G., Lin, C. and Bocarsly, A. B. A new homogeneous electrocatalyst for the reduction of carbon-dioxide to methanol at low overpotential. *Journal of Electroanalytical Chemistry* 1994; 372: 145–150.
2. Barton, E. E., Rampulla, D. M. and Bocarsly, A. B. Selective solar-driven reduction of $CO_2$ to methanol using a catalyzed p-GaP based photoelectrochemical cell. *Journal of the American Chemical Society* 2008; 130: 6342–6344.

3. Yan, Y., Zeitler, E. L., Gu, J., Hu, Y. and Bocarsly, A. B. Electrochemistry of aqueous pyridinium: Exploration of a key aspect of electrocatalytic reduction of CO$_2$ to methanol. *Journal of the American Chemical Society* 2013; 135: 14020–14023.

4. Yan, Y., Gu, J. and Bocarsly, A. B. Hydrogen bonded pyridine dimer: A possible intermediate in the electrocatalytic reduction of carbon dioxide to methanol. *Aerosol and Air Quality Research* 2014; 14: 515–521.

5. Keith, J. A. and Carter, E. A. Theoretical insights into pyridinium-based photoelectrocatalytic reduction of CO$_2$. *Journal of the American Chemical Society* 2012; 134: 7580–7583.

6. Keith, J. A. and Carter, E. A. Electrochemical reactivities of pyridinium in solution: Consequences for CO$_2$ reduction mechanisms. *Chemical Science* 2013; 4: 1490–1496.

7. Ertem, M. Z., Konezny, S. J., Araujo, C. M. and Batista, V. S. Functional role of pyridinium during aqueous electrochemical reduction of CO$_2$ on Pt(111). *The Journal of Physical Chemistry Letters* 2013; 4: 745–748.

8. Lim, C. H., Holder, A. M. and Musgrave, C. B. Mechanism of homogeneous reduction of CO$_2$ by pyridine: Proton relay in aqueous solvent and aromatic stabilization. *Journal of the American Chemical Society* 2013; 135: 142–154.

9. Rosen, B. A., Salehi-Khojin, A., Thorson, M. R., Zhu, W., Whipple, D. T., Kenis, P. J. A. and Masel, R. I. Ionic liquid-mediated selective conversion of CO$_2$ to CO at low overpotentials. *Science* 2011; 334: 643–644.

10. Blanchard, L. A., Gu, Z. and Brennecke, J. F. High-pressure phase behavior of ionic liquid/CO$_2$ systems. *The Journal of Physical Chemistry B* 2001; 105: 2437–2444.

11. Rosen, B. A., Haan, J. L., Mukherjee, P., Braunschweig, B., Zhu, W., Salehi-Khojin, A., Dlott, D. D. and Masel, R. I. *In situ* spectroscopic examination of a low overpotential pathway for carbon dioxide conversion to carbon monoxide. *Journal of Physical Chemistry C* 2012; 116: 15307–15312.

12. Rosen, B. A., Zhu, W., Kaul, G., Salehi-Khojin, A. and Masel, R. I. Water enhancement of CO$_2$ conversion on silver in 1-ethyl-3-methylimidazolium tetrafluoroborate. *Journal of the Electrochemical Society* 2013; 160: H138–H141.

13. Sun, L., Ramesha, G. K., Kamat, P. V. and Brennecke, J. F. Switching the reaction course of electrochemical CO$_2$ reduction with ionic liquids. *Langmuir* 2014; 30: 6302–6308.

14. Medina-Ramos, J., DiMeglio, J. L. and Rosenthal, J. Efficient reduction of CO$_2$ to CO with high current density using *in situ* or *ex situ* prepared Bi-based materials. *Journal of the American Chemical Society* 2014; 136: 8361–8367.

15. Hori, Y., Kikuchi, K. and Suzuki, S. Production of CO and CH$_4$ in electrochemical reduction of CO$_2$ at metal electrodes in aqueous hydrogencarbonate solution. *Chemistry Letters* 1985; 14: 1695–1698.

16. Li, C. W. and Kanan, M. W. CO$_2$ reduction at low overpotential on Cu electrodes resulting from the reduction of thick Cu$_2$O films. *Journal of the American Chemical Society* 2012; 134: 7231–7234.

17. Chen, Y. H., Li, C. W. and Kanan, M. W. Aqueous CO$_2$ reduction at very low overpotential on oxide-derived Au nanoparticles. *Journal of the American Chemical Society* 2012; 134: 19969–19972.

18. Lu, Q., Rosen, J., Zhou, Y., Hutchings, G. S., Kimmel, Y. C., Chen, J. G. G. and Jiao, F. A selective and efficient electrocatalyst for carbon dioxide reduction. *Nature Communications* 2014; 5, article no. 3242.

19. Kuhl, K. P., Cave, E. R., Abram, D. N. and Jaramillo, T. F. New insights into the electrochemical reduction of carbon dioxide on metallic copper surfaces. *Energy & Environmental Science* 2012; 5: 7050–7059.

20. Freund, H. J. and Roberts, M. W. Surface chemistry of carbon dioxide. *Surface Science Reports* 1996; 25: 225–273.

21. Hori, Y. and Suzuki, S. Electrolytic reduction of carbon dioxide at mercury electrode in aqueous solution. *Bulletin of the Chemical Society of Japan* 1982; 55: 660–665.

22. Bocarsly, A. B., Gibson, Q. D., Morris, A. J., L'Esperance, R. P., Detweiler, Z. M., Lakkaraju, P. S., Zeitler, E. L. and Shaw, T. W. Comparative study of imidazole and pyridine catalyzed reduction of carbon dioxide at illuminated iron pyrite electrodes. *ACS Catalysis* 2012; 2: 1684–1692.

23. Stalder, C. J., Chao, S. and Wrighton, M. S. Electrochemical reduction of aqueous bicarbonate to formate with high current efficiency near the thermodynamic potential at chemically derivatized electrodes. *Journal of the American Chemical Society* 1984; 106: 3673–3675.

24. Hori, Y., Electrochemical $CO_2$ reduction on metal electrodes. In *Modern Aspects of Electrochemistry*, Vayenas, C.; White, R.; Gamboa-Aldeco, M., Eds. Springer, New York: 2008; 42: 89–189.

25. Gattrell, M., Gupta, N. and Co, A. A review of the aqueous electrochemical reduction of $CO_2$ to hydrocarbons at copper. *Journal of Electroanalytical Chemistry* 2006; 594: 1–19.

26. Nørskov, J. K., Rossmeisl, J., Logadottir, A., Lindqvist, L., Kitchin, J. R., Bligaard, T. and Jónsson, H. Origin of the overpotential for oxygen reduction at a fuel-cell cathode. *The Journal of Physical Chemistry B* 2004; 108: 17886–17892.

27. Peterson, A. A., Abild-Pedersen, F., Studt, F., Rossmeisl, J. and Norskov, J. K. How copper catalyzes the electroreduction of carbon dioxide into hydrocarbon fuels. *Energy & Environmental Science* 2010; 3: 1311–1315.

28. Hori, Y., Murata, A. and Takahashi, R. Formation of hydrocarbons in the electrochemical reduction of carbon dioxide at a copper electrode in aqueous solution. *Journal of the Chemical Society, Faraday Transactions 1: Physical Chemistry in Condensed Phases* 1989; 85: 2309–2326.

29. Schouten, K. J. P., Kwon, Y., van der Ham, C. J. M., Qin, Z. and Koper, M. T. M. A new mechanism for the selectivity to C-1 and C-2 species in the electrochemical reduction of carbon dioxide on copper electrodes. *Chemical Science* 2011; 2: 1902–1909.

30. Nie, X. W., Esopi, M. R., Janik, M. J. and Asthagiri, A. Selectivity of $CO_2$ reduction on copper electrodes: The role of the kinetics of elementary steps. *Angewandte Chemie International Edition* 2013; 52: 2459–2462.

31. Schouten, K. J. P., Qin, Z., Gallent, E. P. and Koper, M. T. M. Two pathways for the formation of ethylene in CO reduction on single-crystal copper electrodes. *Journal of the American Chemical Society* 2012; 134: 9864–9867.

32. Koper, M. T. M. Analysis of electrocatalytic reaction schemes: Distinction between rate-determining and potential-determining steps. *Journal of Solid State Electrochemistry* 2013; 17: 339–344.

33. Calle-Vallejo, F. and Koper, M. T. M. Theoretical considerations on the electroreduction of CO to C2 species on Cu(100) electrodes. *Angewandte Chemie International Edition* 2013; 52: 7282–7285.

34. Montoya, J. H., Peterson, A. A. and Norskov, J. K. Insights into CC coupling in $CO_2$ electroreduction on copper electrodes. *ChemCatChem* 2013; 5: 737–742.

35. Haruta, M., Kobayashi, T., Sano, H. and Yamada, N. Novel gold catalysts for the oxidation of carbon monoxide at a temperature far below 0°C. *Chemistry Letters* 1987; 16: 405–408.

36. Haruta, M., Yamada, N., Kobayashi, T. and Iijima, S. Gold catalysts prepared by coprecipitation for low-temperature oxidation of hydrogen and of carbon monoxide. *Journal of Catalysis* 1989; 115: 301–309.

37. Takei, T., Akita, T., Nakamura, I., Fujitani, T., Okumura, M., Okazaki, K., Huang, J., Ishida, T. and Haruta, M., Chapter One - Heterogeneous catalysis by gold. In *Advances in Catalysis*, Bruce, C. G.; Friederike, C. J., Eds. Academic Press: 2012; 55: 1–126.

38. Zhu, W., Michalsky, R., Metin, Ö., Lv, H., Guo, S., Wright, C. J., Sun, X., Peterson, A. A. and Sun, S. Monodisperse au nanoparticles for selective electrocatalytic reduction of $CO_2$ to CO. *Journal of the American Chemical Society* 2013; 135: 16833–16836.

39. Zhu, W., Zhang, Y.-J., Zhang, H., Lv, H., Li, Q., Michalsky, R., Peterson, A. A. and Sun, S. Active and selective conversion of CO$_2$ to CO on ultrathin Au nanowires. *Journal of the American Chemical Society* 2014; 136: 16132–16135.

40. Kauffman, D. R., Alfonso, D., Matranga, C., Qian, H. and Jin, R. Experimental and computational investigation of Au-25 clusters and CO$_2$: A unique interaction and enhanced electrocatalytic activity. *Journal of the American Chemical Society* 2012; 134: 10237–10243.

41. Barton Cole, E., Lakkaraju, P. S., Rampulla, D. M., Morris, A. J., Abelev, E. and Bocarsly, A. B. Using a one-electron shuttle for the multielectron reduction of CO$_2$ to methanol: Kinetic, mechanistic, and structural insights. *Journal of the American Chemical Society* 2010; 132: 11539–11551.

42. Morris, A. J., McGibbon, R. T. and Bocarsly, A. B. Electrocatalytic carbon dioxide activation: The rate-determining step of pyridinium-catalyzed CO$_2$ reduction. *ChemSusChem* 2011; 4: 191–196.

43. Bianco, R., Hay, P. J. and Hynes, J. T. Theoretical study of O–O single bond formation in the oxidation of water by the ruthenium blue dimer. *The Journal of Physical Chemistry A* 2011; 115: 8003–8016.

44. Keith, J. A. and Carter, E. A. Quantum chemical benchmarking, validation, and prediction of acidity constants for substituted pyridinium ions and pyridinyl radicals. *Journal of Chemical Theory and Computation* 2012; 8: 3187–3206.

45. Hughes, T. F. and Friesner, R. A. Systematic investigation of the catalytic cycle of a single site ruthenium oxygen evolving complex using density functional theory. *The Journal of Physical Chemistry B* 2011; 115: 9280–9289.

46. Yasukouchi, K., Taniguchi, I., Yamaguchi, H. and Shiraishi, M. Cathodic reduction of pyridinium ion in acetonitrile. *Journal of Electroanalytical Chemistry and Interfacial Electrochemistry* 1979; 105: 403–408.

47. Nørskov, J. K., Bligaard, T., Logadottir, A., Bahn, S., Hansen, L. B., Bollinger, M., Bengaard, H., Hammer, B., Sljivancanin, Z., Mavrikakis, M., Xu, Y., Dahl, S. and Jacobsen, C. J. H. Universality in heterogeneous catalysis. *Journal of Catalysis* 2002; 209: 275–278.

48. Bligaard, T., Nørskov, J. K., Dahl, S., Matthiesen, J., Christensen, C. H. and Sehested, J. The Brønsted–Evans–Polanyi relation and the volcano curve in heterogeneous catalysis. *Journal of Catalysis* 2004; 224: 206–217.

49. Nørskov, J. K., Abild-Pedersen, F., Studt, F. and Bligaard, T. Density functional theory in surface chemistry and catalysis. *Proceedings of the National Academy of Sciences* 2011; 108: 937–943.

50. Koper, M. T. M. Theory of multiple proton-electron transfer reactions and its implications for electrocatalysis. *Chemical Science* 2013; 4: 2710–2723.

51. Abild-Pedersen, F., Greeley, J., Studt, F., Rossmeisl, J., Munter, T. R., Moses, P. G., Skúlason, E., Bligaard, T. and Nørskov, J. K. Scaling properties of adsorption energies for hydrogen-containing molecules on transition-metal surfaces. *Physical Review Letters* 2007; 99: 016105.

52. Huynh, M. H. V. and Meyer, T. J. Proton-coupled electron transfer. *Chemical Reviews* 2007; 107: 5004–5064.

53. Hammes-Schiffer, S. and Soudackov, A. V. Proton-coupled electron transfer in solution, proteins, and electrochemistry. *The Journal of Physical Chemistry B* 2008; 112: 14108–14123.

54. Koper, M. T. M. Theory of the transition from sequential to concerted electrochemical proton-electron transfer. *Physical Chemistry Chemical Physics* 2013; 15: 1399–1407.

55. Peterson, A. A. and Nørskov, J. K. Activity descriptors for CO$_2$ electroreduction to methane on transition-metal catalysts. *The Journal of Physical Chemistry Letters* 2012; 3: 251–258.

# 9 Photoelectrochemical Reduction of CO$_2$ Electroreduction

*Xiaomin Wang*

## CONTENTS

## 9.1 INTRODUCTION TO PHOTOELECTROCHEMICAL REDUCTION OF CO$_2$

### 9.1.1 GENERAL INTRODUCTION

Today, humans are facing three serious problems related to energy resources: shortage of energy, shortage of carbon resources, and the global warming problem. In the biosphere, photosynthesis has been employed for both conversion of CO$_2$ to organics and conversion of solar energy to chemical energy. Oil, coal, and natural gas, which were produced by photosynthesis in ancient times, are finally burnt as both major energy resources and chemical resources. For solving all three of these problems at once, one of the best solutions is the development of practical systems for converting $CO_2$ to useful chemicals using solar light. Research of photochemical and photoelectrochemical (PEC) reduction of $CO_2$ grew rapidly in the past few decades. This was a direct result by growing research from scientists who studied the increasing amount of $CO_2$ in the atmosphere and the global growing demand of fossil fuel.

The catalytic conversion of $CO_2$ to liquid fuels is a critical goal that would positively impact the global carbon balance by recycling $CO_2$ into usable fuels. As an extremely stable molecule generally produced by fossil fuel combustion and respiration, returning $CO_2$ to a useful state by activation/reduction is a scientifically challenging problem, requiring appropriate catalysts and energy input. This poses several fundamental challenges in chemical catalysis, electrochemistry, photochemistry, and semiconductor physics and engineering.

Photocatalysts for $CO_2$ reduction are one of the most important aspects of artificial photosynthesis. Consideration of the thermodynamics of $CO_2$ reduction gives an important strategy for constructing these photocatalysts [1,2]. Equations 9.1 through 9.6 indicate that the one-electron reduction of $CO_2$ (at pH 7 in an aqueous solution vs. normal hydrogen electrode [NHE]) is highly endothermic and that the product $(CO_2)$ is too reactive to handle with ease. On the other hand, the multielectron reduction of $CO_2$ can give stable and useful products with much lower energies, as shown in Reference [3]. However, one photon can usually induce the transfer of only one electron in photochemical reactions. Overcoming this obstacle is a key goal in the design of efficient photocatalysts for $CO_2$ reduction.

$$CO_2 + e^- \rightarrow CO_2^{\cdot-}, \quad E^0 = -1.90\ V \qquad (9.1)$$

$$CO_2 + 2H^+ + 2e^- \rightarrow CO + H_2O, \quad E^0 = -0.53\ V \qquad (9.2)$$

$$CO_2 + 2H^+ + 2e^- \rightarrow HCO_2H, \quad E^0 = -0.61\ V \qquad (9.3)$$

$$CO_2 + 4H^+ + 4e^- \rightarrow HCHO + H_2O, \quad E^0 = -0.48\ V \qquad (9.4)$$

$$CO_2 + 6H^+ + 6e^- \rightarrow CH_3OH + H_2O, \quad E^0 = -0.38\ V \qquad (9.5)$$

$$CO_2 + 8H^+ + 8e^- \rightarrow CH_4 + 2H_2O, \quad E^0 = -0.24\ V \qquad (9.6)$$

There are several ways to reduce $CO_2$ with the assistance of renewable solar energy, and these methods can be divided into three major categories: homogeneous photoreduction by a molecular catalyst, photoelectrochemical reduction by a semiconducting photocathode, and electrochemical reduction by an electrolyzer powered by commercial photovoltaic (PV) devices.

## 9.1.2 HOMOGENEOUS PHOTOCATALYTIC REDUCTION OF $CO_2$

A homogeneous $CO_2$ photoreduction system consists of a molecular catalyst, light absorption, sacrificial electron donor, and/or electron relay. When looking at these types of systems, the main figure of merit is the photochemical quantum yield, defined as

$$\text{Photochemical quantum yield } (\varnothing) = \left( \frac{\text{moles products}}{\text{absorbed photons}} \right)$$

$$\times \text{ (number of electrons needed for conversion)}$$

The generic mechanism of the photocatalytic reduction of $CO_2$ consists of a photosensitizer (P) capable of absorbing radiation in the ultraviolet (UV) or visible region and of the generation of an excited state (P\*). The excited state is reductively quenched by a sacrificial donor (D) generating a singly reduced photosensitizer (P–) and oxidized donor (D+). The choice of photosensitizer must be such that P– is able to transfer an electron efficiently to the catalyst species (cat) to generate the reduced catalyst species (cat⁻). In some cases, the photosensitizer and the catalyst are the same species. The cat⁻ is then able to bind $CO_2$ and proceed with the catalytic mechanism to release the intended products and regenerate cat. Common photosensitizers used in these systems include aromatics, $p$-terphenyl, phenazine, and polypyridine-coordinated transition metal complexes. The most common catalyst species include macrocycle complexes of Ni and Co, polypyridine Ru and Re catalysts, and suspended metal colloids.

Such metallomacrocycles strongly absorb visible light and do not require the addition of a photosensitizer. These systems do, however, suffer from low and low catalytic selectivity (CS) due to significant production of $H_2$. Tetraazamacrocyclics such as Ni(cyclam) have proved to reduce $CO_2$ efficiently and selectively to CO electrocatalytically while adsorbed on an Hg electrode. In purely photocatalytic systems, however, they tend to suffer from low CS and turnover number (TON). When Co is used as the metal center, the photocatalytic properties are improved. A well-studied catalyst, Re(bipy)(CO)$_3$X (where bipy = 2,2-bipyridine and X = Cl, Br), capable of selective production of CO without the use of a separate photosensitizer was developed by Lehn and coworkers [4,5]. With the addition of phosphate groups to the Re(bipy)(CO)$_3$X system (X = P(OEt)$_3$), one of the highest single-molecule quantum efficiencies was used for a homogeneous photocatalytic system. In a two-molecule system consisting of a 25:1 mixture of Re(bipy)(CO)$_3$(P(OMe)$_3$) as a photosensitizer and Re(bipy)(CO)$_3$MeCN as a catalyst, even higher quantum yields were achieved [6]. Although the Re catalyst systems mentioned above demonstrated remarkable, they lacked extended absorption in the visible region. With the use of solar photons, Ishitani and coworkers have addressed this issue by utilizing a bridging ligand to covalently attach a [Ru(bipy)$_3$]$^{2+}$-type photosensitizer, which absorbs strongly in the visible region, to a Re(bipy)(CO)$_3$X-type catalyst to create a supramolecular dyad complex.

Although photocatalytic reduction of $CO_2$ may become an important stepping stone to solar fuel production, much progress remains before it becomes practical as an industrial process. Currently, TONs remain in the hundreds and TOFs are typically in the tens per hour. More mechanistic work must be done in order to understand and increase the stability and rates of these systems. These quantitative measures of catalytic systems must also be scrutinized because they are dependent on catalyst concentration and can vary drastically depending on the concentrations and volumes chosen for the experiment. To compound the problem, different solvents, electron donors, photosensitizers, and light sources are employed by the various groups studying these photocatalysts.

Looking forward to improved systems with a more practical utility, some additional issues are considered. As many of the photocatalysts presently studied are metal complexes employing rare and expensive transition metals, it is especially important to raise the catalytic rate and long-term stability to make this process economically feasible. Another drawback to the reviewed systems is the use of a sacrificial donor to supply the electrons for the reduction process. Ideally, water would be the source of both the electrons and hydrogen atoms for $CO_2$ reduction catalysis in an artificial photosynthetic process. These issues are not trivial and will take considerable effort and creativity to solve.

### 9.1.3 HETEROGENEOUS PHOTOELECTROCHEMICAL REDUCTION OF $CO_2$

In a heterogeneous system, p-type semiconductor/liquid junctions are extensively studied as PV devices. The p-type semiconducting electrodes can act as photocathodes for photoassisted $CO_2$ reduction. Normally, there are four different types of photoassisted reduction of $CO_2$ using a semiconducting photocathode: (a) direct heterogeneous $CO_2$ reduction by a biased semiconductor photocathode, (b) heterogeneous $CO_2$ reduction by metal particles on a biased semiconductor photocathode, (c) homogeneous $CO_2$ reduction by a molecular catalyst through a semiconductor/molecular catalyst junction, and (d) heterogeneous $CO_2$ reduction by a molecular catalyst attached to the semiconductor photocathode surface [7–10].

Heterogeneous PEC reduction of $CO_2$ on semiconductor surfaces has been explored extensively in the last three decades. Both aqueous and nonaqueous solvents have been used for direct $CO_2$ PEC reduction on semiconductor surfaces, with the most commonly reported nonaqueous solvents being polypropylene carbonate, acetonitrile, dimethylformamide (DMF), dimethyl sulfoxide (DMSO), and methanol. The greatest difference between water and nonaqueous solvents is the solubility of $CO_2$. In nonaqueous solvents, the solubility is 7–8 times higher than in water. Methanol, for example, is known to be a physical absorber of $CO_2$ and is presently used in the rectisol processing industrial plants. Aqueous media further complicate $CO_2$ reduction, as different $CO_2$ hydration products are present in water. In water, $CO_2$ hydration occurs to form carbonic acid, which then undergoes stepwise dissociation to bicarbonate ($HCO^{3-}$) and carbonate ($CO_3^{2-}$). The predominant species is pH dependent: $CO_2$ is dominant at pH < 4.5; $HCO^{3-}$ is dominant at 7.5 < pH < 8.5; and $CO_3^{2-}$ is dominant at pH > 11.5. This, in turn, affects the thermodynamic potentials for generating certain products, as they are dependent on the form of $CO_2$ present in the solution.

Water, however, is commonly used as a proton source in aprotic solvents for $CO_2$ reduction. In one study, $CO_2$ solubility was shown not to change with the addition of up to 1% (500 mM) water to acetonitrile, but drastically decreased with higher water concentrations. DMF and DMSO performed better than polypropylene carbonate and acetonitrile when mixed with 1% water, as they were better able to suppress competing proton reduction processes.

The limitations of $CO_2$ solubility in water at standard pressures as well as its diffusion limitations set a maximum catalytic current density of 10 mA cm$^{-2}$ for electrochemical reduction of $CO_2$. $CO_2$ solubility, and thus maximum catalytic current, can be increased using high-pressure $CO_2$ environments. High-pressure $CO_2$ environments offer high catalytic current density and high selectivity over proton reduction for both metal and semiconductor electrodes. Gas-diffusion electrodes as well as other options have also been explored to increase $CO_2$ concentration, which is imperative to increase catalytic current densities.

Several metal electrodes have been used for catalytic reduction of $CO_2$ in both aqueous and nonaqueous media. The catalytic activity of a metal catalyst can be transferred to semiconductor photocathodes through discontinuous films of metals without sacrificing photovoltage. The objective of this approach is to incorporate the stability and catalytic activity of metal particles with semiconducting photocathodes. In a truly photoelectrocatalytic system, when a photoelectrode coupled to a catalyst (metal particles) is run under illumination, the Faradaic effciency (FE) versus applied potential has similar behavior to the catalyst alone with a positive shift in the onset voltage called the photovoltage shift. Apart from the photovoltage shift, catalytic activity and product distribution should not be affected by illumination.

On account of competing advantages of homogeneous catalysts (selectivity, tunability) and heterogeneous catalysts (robustness, easy separation of products from catalysts), there is considerable interest in "heterogenizing" homogeneous catalysts, by covalently linking them to surfaces. There have been several reports regarding surface modification of dark electrodes using the polymeric form of the molecular catalysts and/or enzymes for electrochemical reduction of $CO_2$ to various products. These modified dark electrodes have several advantages: control over the active site environment for better performance; prevention of aggregation or dimerization of the molecular catalyst, which leads to higher TONs; efficient charge transfer to the molecular catalyst; usability of water-insoluble molecular catalysts in aqueous media once anchored to electrodes; and stabilization of the catalyst and electrode [11,12]. The physical nature of these junctions is similar to a semiconductor/liquid junction with unbound molecular electrocatalysts [13]. Molecular catalyst surface-modified semiconductors can be subdivided into two categories: polymeric backbone attachment and direct anchoring to the semiconductor surface.

### 9.1.4 Prospect

Multijunction photoelectrolysis cells are presently used to overcome the high potential that is associated with water splitting. Four types of these cells are currently used: p/n junction photoelectrolysis cells, photoanode–PV cells, photocathode–PV cells, and PV photoelectrolysis cells.

Of note, the PV photoelectrolysis cell has a photovoltage that is independent of pH, which is important for pH-sensitive catalyst-mediated $CO_2$ reduction. This cell can also be easily modified with $CO_2$ reduction catalysts selective for specific products. In addition, this type of cell utilizes a majority of the solar spectra and has a high photovoltage, unlike wide band gap semiconductor photoelectrodes. It is possible for this technology to be used as a wireless, monolithic, two-compartment PV-type photoelectrolysis cell with single dual-face photoelectrode. The fabrication cost of the electrode may be high; however, the simple structure of the cell, the ease of product separation, and the robustness of the electrode could lead to a long operation time and an environmentally friendly process. There are two major challenges for these types of cells: identifying catalysts based on earth-abundant materials that have low overpotentials for $CO_2$ reduction and water oxidation, and finding a reliable and robust proton-exchange membrane. It is clear that the advantages of this type of system outweigh the cost. Therefore, it is imperative to explore and adopt water-splitting/hydrogen-generation technology for solar splitting of $CO_2$ for liquid fuel applications as well as to broaden the search for new robust, selective, and efficient catalytic systems. In the quest for new catalysts, combinatorial approaches will prove useful for both discovery and optimization of new catalysts.

## 9.2   CATALYSTS FOR PHOTOELECTROCATALYTIC SYSTEM

### 9.2.1   INTRODUCTION

The transformation of $CO_2$ into fuel by using solar light irradiation is an effective method because there is no addition of extra energy and no negative influence on the environment. The immediate requirement in this technology is to develop visible light-sensitive photocatalysts, which are prominent in $CO_2$ recycling. Different types of photocatalysts have been already introduced by many researchers in this technology. Some of the catalysts exhibited high conversion rates and selectivity under visible light irradiation, whereas other catalysts were not feasible for visible light response and presented low yield rates. The reaction mechanism of selective photocatalysts under light irradiation is illustrated in Figure 9.1. Researchers are still trying hard to advance the properties of catalysts in terms of solar fuel production. A concise classification of photocatalysts is also shown in Figure 9.2, based on the recent development in photocatalytic $CO_2$ transformation.

### 9.2.2   METAL OXIDE PHOTOCATALYSTS

Metal oxides as light-sensitive catalysts have been applied in several processes such as the breaking down of organic and inorganic materials to valuable products, water treatment, and self-cleaning processes. The leading approach of using metal oxide catalysts in $CO_2$ conversion to carbonaceous fuel was introduced by Inoue et al. [15] Variations in metal oxide photocatalysts have already been introduced by integrating foreign elements and other compounds with different supportive substrates [16–19] for lessening its particle size, controlling its crystal growth, and increasing its surface area and pore volume. Various metal oxide photocatalysts, which were frequently studied in many photocatalytic $CO_2$ reduction processes, are characterized in this section. A clear idea of metal oxide semiconductors used in various studies for photocatalytic $CO_2$ transformation is represented in Table 9.1.

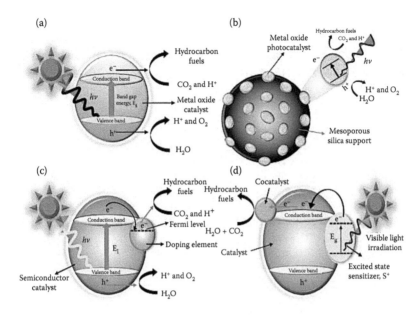

**FIGURE 9.1** Photocatalytic reaction mechanism of (a) an unmodified metal oxide semiconductor; (b) supported metal oxide semiconductor; (c) doped semiconductor; and (d) dyesensitized semiconductor under light irradiation. (Reprinted by permission from Macmillan Publishers Ltd. *RSC Adv.*, Das S, Daud WMA. A review on advances in photocatalysts towards $CO_2$ conversion, 4, 20856–20893, copyright 2014.)

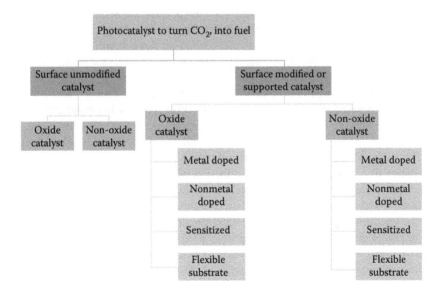

**FIGURE 9.2** Classification of photocatalysts on $CO_2$ transformation into fuel. (Reprinted by permission from Macmillan Publishers Ltd. *RSC Adv.*, Das S, Daud WMA. A review on advances in photocatalysts towards $CO_2$ conversion, 4, 20856–20893, copyright 2014.)

**TABLE 9.1**

**$CO_2$ Reduction Products and Corresponding Reduction Potential with Reference to NHE at pH 7**

| Product | Reaction | $E^0_{redox}$ (V vs. NHE) |
|---|---|---|
| Oxygen | $H_2O \rightarrow 0.5O_2 + 2H^+ + 2e^-$ | 0.82 |
| Methane | $CO_2 + 8H^+ + 8e^- \rightarrow CH_4 + 2H_2O$ | −0.24 |
| Ethane | $2CO_2 + 14H^+ + 14e^- \rightarrow C_2H_6 + 4H_2O$ | −0.27 |
| Carbon monoxide | $CO_2 + 2H^+ + 2e^- \rightarrow CO + H_2O$ | −0.51 |
| Methanol | $CO_2 + 6H^+ + 6e^- \rightarrow CH_3OH + H_2O$ | −0.39 |
| Ethanol | $2CO_2 + 12H^+ + 12e^- \rightarrow C_2H_5OH + 3H_2O$ | −0.33 |
| 1-Propanol | $3CO_2 + 18H^+ + 18e^- \rightarrow CH_3CH_2CH_2OH + 5H_2O$ | −0.31 |
| 2-Propanol | $3CO_2 + 18H^+ + 18e^- \rightarrow CH_3CH(OH)CH_3 + 5H_2O$ | −0.30 |
| Formaldehyde | $CO_2 + 4H^+ + 4e^- \rightarrow HCHO + H_2O$ | −0.55 |
| Acetaldehyde | $2CO_2 + 10H^+ + 10e^- \rightarrow CH_3CHO + 3H_2O$ | −0.36 |
| Propionaldehyde | $3CO_2 + 16H^+ + 16e^- \rightarrow CH_3CH_2CHO + 5H_2O$ | −0.32 |
| Acetone | $3CO_2 + 16H^+ + 16e^- \rightarrow CH_3COCH_3 + 5H_2O$ | −0.31 |
| Formic acid | $CO_2 + 2H^+ + 2e^- \rightarrow HCOOH$ | −0.58 |
| Acetic acid | $2CO_2 + 8H^+ + 8e^- \rightarrow CH_3COOH + 2H_2O$ | −0.31 |

*Source:* Reprinted with permission from Hong J, Zhang W, Ren J, Xu R. Photocatalytic reduction of $CO_2$: A brief review on product analysis and systematic methods. *Anal. Methods*, 5, 1086–1097. Copyright 2013 American Chemical Society.

Being an inert, corrosion-resistant, and inexpensive semiconductor, $TiO_2$ is a frequently used metal oxide in the photocatalytic transformation of $CO_2$ into hydrocarbon fuel. In nature, $TiO_2$ is established as three familiar minerals such as rutile, anatase, and brookite. Among them, rutile and anatase phases indicate better efficiency under light radiation. Mutually, anatase and rutile phases reveal definite band gap energy at around 3.2 and 3.0 eV, respectively. The rutile phase is capable of absorbing visible light because of its comparatively low band gap energy, whereas the anatase form only displays its response to UV irradiation. Thus, the rutile phase is not used due to its ineffective photoactivity; the best photoactivity can be achieved by combining anatase with a slight amount of rutile. The light absorption by a $TiO_2$ semiconductor is effective, whereas the semiconductor exhibits high surface area. The surface area of $TiO_2$ is indirectly proportional to the size of particle; conversely, the size of the particle is directly proportional to the band gap energy. The semiconductor shows high band gap energy as a result of the shifting of valence band energy to lower energies after a certain decrease in the particle size, whereas band energy is intensely shifted to higher energies. On the other hand, significantly narrow band gap energy leads to inadequate redox potential to achieve oxidizing and reducing reactions. Therefore, it is compulsory to fix a structure that exhibits a settlement between the particle size and band gap energy range. Tan et al. investigated a combination of 80% anatase and 20% rutile forms as a photoactive catalyst [20]. Commercial Degussa P25 is found to be slightly less active than modified TiO$_2$ (anatase and rutile) [21]. It is indicated that unmodified TiO$_2$ (P25) metal oxide semiconductors generally have high surface area as well as an affinity to organic impurities on their surfaces to facilitate the adsorption of organic compounds, which have the potential to perform as both an electron donor and a carbon source for the products [22]. Besides, this type of metal oxide catalyst expresses better methane production, compared with the Degussa P25 treated by calcination and washing where CO is the major product. In various studies, product dissimilarity is also observed due to the treatment procedure on anatase TiO$_2$ in several methods [23–25]. For instance, treatment by the acidic solutions (1 M nitric acid) is suitable for formic acid (FA) formation due to the protonation of reaction intermediates into the reaction medium. Treatment with 2-propanol acts as a hole scavenger in CO$_2$ photoreduction with anatase TiO$_2$, where the methane formation rate was considerably higher. Moreover, the synthesis of TiO$_2$ by the sol–gel method removes the presence of small brookite nanoparticles that normally arise in low-temperature formation reactions and considerably hinder the phase alteration to rutile at elevated temperatures.

Some prospective metal oxide photocatalysts other than TiO$_2$ were reported in many investigations. Inoue et al. studied three metal oxides, TiO$_2$, ZnO, and WO$_3$, to understand their photocatalytic characteristics under UV light radiation [15]. It was found that the yield of methyl alcohol increased as the conduction band becomes more negative with respect to the redox potential of H$_2$CO$_3$/CH$_3$OH, whereas methyl alcohol was not produced in the presence of WO$_3$ catalyst, which has a conduction band more positive than the redox potential of H$_2$CO$_3$/CH$_3$OH. However, WO$_3$ possesses low band gap energy (2.8 eV) with respect to visible light. W$_{18}$O$_{49}$ catalysts were synthesized in the form of nanowires with band gap energy of 2.7 eV for the photocatalytic conversion of CO$_2$ to methane under visible light irradiation. The diameter of W$_{18}$O$_{49}$ nanowires is approximately 0.9 nm and holds a large number of oxygen vacancies. These oxygen vacancies indicate an outstanding competency on photochemical carbon dioxide reduction over the visible light region.

Metal oxides such as Ga$_2$O$_3$, ZrO$_2$, and MgO were reported in various studies as having wide band gaps, compared with other photocatalytic metal oxides. In those studies, methane and hydrogen were used as reductants for photoreduction of CO$_2$, and CO was the major product. It was also observed that the surface bidentate formate species showed high photoactivity for photocatalytic reduction of CO$_2$ over ZrO$_2$ and MgO.

Recently, visible light-responsive and other oxide catalysts fabricated through various routes such as hydrothermal, solvothermal, and solid state reactions are widely used in the photochemical reduction of CO$_2$ because of high yield rates. It has been reported that hydrothermally prepared KNb$_3$O$_8$ and HNb$_3$O$_8$ nanobelts show high yields of methane, compared with the same catalysts derived from conventional solid state reactions and commercial TiO$_2$ (Degussa P25). In the case of hydrothermal synthesis, this nanobelt-like morphology and protonic acidity give higher photochemical activity for methane production through hydrogen bonding, which is facilitated by the separating and trapping of photogenerated carriers at the interlayer surfaces of HNb$_3$O$_8$ and KNb$_3$O$_8$. Ferroelectric materials can be used as remarkable substitutes for the standard semiconductor photocatalysts. Ferroelectric characteristics lead to driving the electrons and holes apart because they possess internal dipoles. Therefore, this phenomenon decreases the probabilities of the recombination of carriers and also inhibits the reaction of redox products, driving the equilibrium toward product formation. For this ferroelectric behavior, the polar compound $LiNbO_3$ was used by Stock and Dunn for photocatalytic $CO_2$ conversion to FA under an Hg lamp or direct natural sunlight irradiation [26].

### 9.2.3 MODIFIED/UNMODIFIED NON-OXIDE CATALYSTS

Non-oxide semiconductor photocatalysts are selected for $CO_2$ recycling because of their low band gap energy to facilitate the photocatalytic response to the visible light region, and high conversion efficiency is achieved for their unique photobehavior. It is investigated that semiconductor catalysts are one of the supreme sources for direct solar energy conversion. Many wide-ranging reviews already have been published on the advancement of oxide or non-oxide semiconductor photoactive materials. It was found that apart from the metal oxide semiconductor catalysts, metal sulfide semiconductor materials have effective photoactivity because of their outstanding ability to absorb the solar spectrum with high-energy yields [27]. Metal sulfides have comparatively high conduction band states more appropriate for enhanced solar responses than for metal oxide semiconductors, which are facilitated by the higher valence band states consisting of S 3p orbitals. It was reported that to absorb the entire UV and visible light region of solar irradiation, a $ZnS-AgInS_2-CuInS_2$ solid solution was capable due to an absorption edge of up to 800 nm. Cetyltrimethylammonium (CTA) with ZnS nanoparticles was used to inhibit the formation of bigger agglomerates. Different percentages of ZnS loading on porous $SiO_2$ matrices were also applied to investigate the improved photocatalytic yields. Cd–ZnS shows the highest activity than that of ZnS–CdS because Cd is more active in improving the quantum efficiency of the photochemical reduction of formate due to its ability to activate $CO_2$ more effectively in the photoreaction. Because of low band gap energy (3.0 eV) under the solar spectrum more competently, MnS semiconductor materials also have the distinct behavior of containing high reducing conduction band electrons which are satisfactorily active for encouraging the reduction of $CO_2$. That is why MnS semiconductors deliver higher quantum efficiency than other common catalysts such as $TiO_2$.

The use of catalysts for photochemical purposes without incorporating cocatalysts such as p-type GaP and p-type InP have also been studied due to narrow band gaps to facilitate the visible light response. Basically, in the PEC $CO_2$ conversion process, p-type metal phosphides (GaP) were mainly used in methanol production but needed remarkably high overpotentials. However, high faradaic efficiencies for the photoelectrochemical reduction of $CO_2$ to CO have also been investigated, whereas electrolytes are used in nonaqueous forms. Nevertheless, high overpotential is necessary to obtain high faradaic efficiency. It is found that p-type GaP was able to obtain faradaic efficiency of 100% while converting $CO_2$ to methanol at low potentials, which was more than 300 mV below the standard potential of 0.5 V versus the saturated calomel electrode (SCE). Furthermore, p-type GaP semiconductor exhibits low band gap energy (2.24 eV) and possesses high reducing conduction band electrons to ease the reduction of $CO_2$. In another study, for $CO_2$ photocatalytic conversions, it was detected that p-type InP provided high selectivity for FA in a PEC cell.

## 9.3 PRODUCT AND SYSTEMATIC METHODS OF PHOTOCATALYTIC REDUCTION OF $CO_2$

### 9.3.1 PRODUCTS OF PHOTOCATALYTIC REDUCTION OF $CO_2$

Because of leads to possible climate change and a serious impact on globe environment, there is an increasing need to mitigate $CO_2$ emissions using carbon-neutral energy sources. $CO_2$ conversion includes chemical transformation, photocatalytic reduction, electrochemical reduction, biological conversion, etc. Among these, conversion of $CO_2$ to value-added fuels or chemical products by direct use of sunlight is an attractive but a challenging process. This certainly exerts stress in product identification and quantification. Hence, many groups have practically adopted the routine analysis of one or two products in either gas or liquid phase for evaluation of photocatalytic efficiencies.

$CO_2$ photoreduction involves multielectron processes which can lead to a large variety of products ranging from CO, $CH_4$ to higher hydrocarbons in the gas phase, and various oxygenates in the liquid phase such as alcohols, aldehydes, and carboxylic acids. A few representative reactions leading to gas and liquid products are shown in Table 9.1. The identification and quantification of such products are essential to more accurate evaluations of the photocatalytic performance of the designing of better photocatalysts.

### 9.3.2 ANALYSIS OF GAS-PHASE PRODUCTS

Scientists focused on the gas-phase product analysis for photocatalytic reduction of $CO_2$. One set of methods is proposed for the analysis of a wide range of photoreaction products in gas and liquid phases with low detection limits. The effects of the most commonly used organic additives were also investigated. Though the analysis methods discussed and proposed here are in the context of photocatalytic reduction of $CO_2$, they are also applicable for other $CO_2$ reduction processes, such as electrochemical reduction, chemical conversion, biological transformation, and so on. Furthermore, at the current stage of research, analysis techniques with low detection limits are also important to identify the influence of carbon residues originating from photocatalyst synthesis which can possibly contribute to the reduction products.

$CH_4$ and CO are the major gaseous products of $CO_2$ photoreduction, whereas $H_2$ and/or $O_2$ may also be produced as byproducts from water splitting. In addition, higher alkanes as the photocatalytic reduction products of $CO_2$ was observed. Besides the products, $CO_2$ as the main species in the gas phase due to a low conversion needs to be quantified to determine the reaction progress and/or the extent of $CO_2$ dissolution in the liquid phase. Infrared spectroscopy (IR) or diffuse reflectance infrared Fourier transform spectroscopy (DRIFT) has been occasionally employed to verify the consumption of $CO_2$ and the generation of CO. Gas chromatography (GC) is by far the most commonly used method for quantification of these gas species. Besides GC, gas chromatography–mass spectroscopy (GC–MS) was also used for routine analysis of $CH_4$ and CO or for carbon source verification. Although a thermal conductivity detector (TCD) was the only detector used in GC as stated in some reports, it is suggested that a flame ionization detector (FID) should be used in order to achieve higher sensitivities for detection of low concentrations of CO and hydrocarbons. FID analysis is a destructive method. Therefore, the gas should first be analyzed by TCD for gases quantification followed by FID for the analysis of CO and alkanes. Many types of columns have been used, such as molecular sieves, carbon columns, and $Al_2O_3$. It has been noted that if a high concentration of $CO_2$ remains in the gas product stream, it can cause deactivation of the nickel catalyst in the methanizer and certain types of columns, including the most commonly used molecular sieve column, making frequent column regeneration necessary. Herein, a better design of GC configuration is proposed to separate and detect all the gases, including $CH_4$, $CH_3CH_3$, CO, $CO_2$, $O_2$, $N_2$, and $H_2$ with low detection limits as well as to avoid regular column regeneration and achieve longer lifetimes of the methanizer and detector.

Some research groups adopted the gas-phase reaction for $CO_2$ photoreduction. In a typical gas-phase reaction, the solid photocatalyst was initially well dispersed at the bottom of the reactor which contains $CO_2$ gas, followed by injecting a small amount of liquid water (e.g., 1 mL). Under a low-pressure condition, liquid water vaporizes and mixes with $CO_2$ gas. In these cases, the analysis methods proposed above can be used for accurate analysis of gas products. Gas-phase reduction was also carried out using $H_2$ or $CH_4$ instead of water to reduce $CO_2$. For these water-free gas–solid systems, a GC-based method is still widely used for the quantification of products. Nevertheless, it is worth noting that in situ DRIFT coupled with isotopically labeled $^{13}CO_2$ can be employed to monitor the intermediates or products adsorbed on the solid photocatalyst surface.

### 9.3.3 Analysis of Liquid-Phase Products

Although GC with TCD/FID is still the main technique used for the analysis of different types of liquid oxygenates (mainly alcohols), several other methods have been utilized, including GC–MS [29–31], high-performance liquid chromatography (HPLC), ion exchange chromatography (IEC), UV–visible (UV–Vis) spectroscopy (colorimetric assay) after reacting with chromotropic acid or Nash reagent, and nuclear magnetic resonance (NMR). The compounds that can be analyzed and their detection limits using various techniques, together with the associated limitations, are summarized in Table 9.2.

It is well known that alcohols (especially methanol) are frequently detected products from $CO_2$ photoreduction in the liquid phase. GC with an FID works well for the analysis of alcohols, and several kinds of columns have been used such as HP-5, DB-WAX, and polyethylene glycol (PEG) columns.

In many studies, organic compounds have been engaged in photoreduction of $CO_2$ for various reasons such as sacrificial reagents (e.g., methanol, triethanolamine),

## TABLE 9.2
## Comparison of Techniques Used for Liquid-Phase Analysis in $CO_2$ Photoreduction

| Technique | Compounds | Detection Limit | Limitations |
|---|---|---|---|
| GC | Alcohols | 3 μmol L$^{-1}$ | Much higher detection limits for |
|  | Aldehydes | 100 μmol L$^{-1}$ | aldehydes |
| HPLC | Carboxylic acids | 5 μmol L$^{-1}$ | Aldehydes need to be derivatized before analysis |
|  | Aldehydes | 0.07 μmol L$^{-1}$ |  |
| IEC | Carboxylic acids | 0.1 μmol L$^{-1}$ for HCOOH | For acids only |
|  |  | 10 μmol L$^{-1}$ for CH$_3$COOH |  |
| UV–Vis | HCHO after reaction with Nash's reagent or HCOOH | 0.17 μmol L$^{-1}$ 0.08 μmol L$^{-1}$ | Not applicable for other aldehydes and acids; HCOOH can only be analyzed when no other organics are present |
| $^1$H or $^{13}$C NMR | All oxygenates | Not available | High cost, difficult for quantification, although some work demonstrated quantification, mainly for product qualification, carbon source verification |
| GC/LC–MS | All oxygenates | Not available | Mainly for product qualification, carbon source verification |

*Source:* Reprinted with permission from Hong J, Zhang W, Ren J, Xu R. Photocatalytic reduction of $CO_2$: A brief review on product analysis and systematic methods. *Anal. Methods*, 5, 1086–1097. Copyright 2013 American Chemical Society.

solvents (e.g., acetonitrile, DMF), photocatalysts (e.g., Ru complexes) or photosensitizers (e.g., N3 dye, Rose bengal). The presence of such organic compounds in the reaction systems may affect the product analysis in the liquid phase.

Another group of possible reduction products, aldehyde analytical techniques needs to be developed for $CO_2$ reduction product screening. Besides Nash's colorimetric method, others also used GC with either TCD/FID with a Porapak T column or FID with a DB-WAX column 30 and BX-10 column to quantify aldehydes. Hence, only relatively concentrated aldehydes can be analyzed by the GC method, whereas the concentrations of aldehydes in the liquid phase of $CO_2$ photoreduction are usually much lower than such detection limits.

### 9.3.4 CARBON SOURCE VERIFICATION AND $O_2$ MONITORING

The contribution of carbon residues to the photoreduction products has been recently reported [32]. The *in situ* DRIFT study indicated that $^{12}$CO was the main product when $^{13}CO_2$ gas was used during surface photoreaction over the Cu(I)/TiO$_2$ catalyst.

The $^{12}C$ source was proven to be from the carbon residue on the photocatalyst, originating from the organics used during photocatalyst preparation. Based on this, the photocatalytic activities reported for $CO_2$ reduction in the literature may need to be verified if there was no tracking of carbon source or enough evidence for photocatalytic events. Thus, either carefully and systematically designed control experiments or carbon source tracking by NMR, GC–MS, or LC–MS with isotope $^{13}C$-labeled $CO_2$ as the reactant, is necessary. The carbon source in gas products such as methane and carbon monoxide can be verified by spectroscopic methods (e.g., DRIFT) or GC–MS, and in liquid oxygenates by $^{13}C$ NMR, GC–MS, or LC–MS. For example, Liu et al. studied the solvent effect on photocatalytic $CO_2$ reduction by using $^{13}CO_2$ with GC–MS and $^{13}C$ NMR to identify the source of carbon in CO and liquid products, respectively [33]. Yui et al. reported the use of $TiO_2$ (P25) for methane production. The carbon source of methane was confirmed from $CO_2$ based on a signal at $m/z = 17$ ($^{13}CH_4$) in GC–MS when $^{13}CO_2$ was used [22]. The photocatalytic activity of a titanium metal–organic framework in $CO_2$ reduction to produce FA was confirmed by using $^{13}C$ NMR. A signal at 165.30 ppm can be found, which was assigned to HCOO, whereas such a signal was not found when $^{12}CO_2$ was used. In another report, $^{13}C$-labeled HCOOH in the product was found to be easily detected by $^{13}C$ NMR, to confirm that the product was from the reduction of $CO_2$ rather than from other carbon sources on the used carbon nanoparticles. Using layered double hydroxides as a photocatalyst, the carbon source of CO and oxygen source of $O_2$ were also confirmed from $CO_2$ and $H_2O$, respectively, by using $^{13}CO_2$ and $H_2^{18}O$ with GC–MS method [34].

On the other hand, $O_2$ evolution is an important indicating factor to support photocatalytic $CO_2$ reduction, whereas water is used as the reducing agent for $CO_2$. Kudo and coworkers [35] reported that the production of $H_2$, $O_2$, and CO is stoichiometric over Ag-loaded $ALa_4Ti_4O_{15}$ (A = Ca, Sr, and Ba) photocatalysts. The ratio of the mole of reacted electrons to that of holes was found to be almost unity, indicating that the $CO_2$ reduction occurred photocatalytically with water as the reducing agent. To ascertain that $O_2$ is generated from the reaction rather than from air contamination, the volumetric (molar) ratio of $O_2$ to $N_2$ over reaction time should be monitored. In a recent study, the molar ratio of $O_2/N_2$ was found to gradually increase from 0.26 (the ratio in air) to 0.40 after light irradiation. In addition, the trend of this ratio was found to be similar to that of $CH_4$ production, indicating that $O_2$ was produced by photoreaction (see Figure 9.3).

In summary, major products in both gas and liquid phases from $CO_2$ photoreduction can be detected accurately with low detection limits by a combination of GC and HPLC methods. Figure 8.3 summarizes the analysis methods and the detection limits of the major chemical species in both gas and liquid phases in the absence of organic additives. The effects of several organic additives, including commonly used solvents, photosensitizers, and sacrificial reagents, in photoreaction were investigated. It has been found that alcohol analysis by GC methods is more sensitive to organic additives, whereas aldehyde and acid analyses by HPLC methods are not affected by most of the organics investigated. The importance of carbon source verification is highlighted and several techniques such as DRIFT, NMR, and GC–MS can be used.

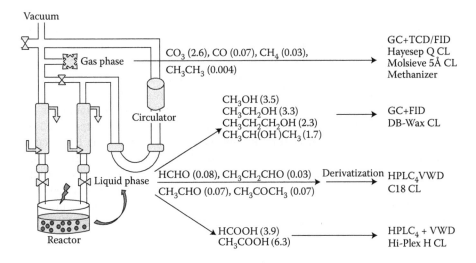

**FIGURE 9.3** Summary of analysis methods for gas and liquid-phase samples. Derivatization: aldehydes were derivatized with DNPH before performing HPLC analysis. The detection limits ($\mu$mol L$^{-1}$) are shown in brackets. (Reprinted by permission from Hong J, Zhang W, Ren J, Xu R. Photocatalytic reduction of $CO_2$: A brief review on product analysis and systematic methods. *Anal. Methods*, 5, 1086–1097. Copyright 2013 American Chemical Society.)

## 9.4 CONVERSION OF CARBON DIOXIDE INTO METHANOL

### 9.4.1 POTENTIAL ADVANTAGES OF CONVERTING $CO_2$ INTO METHANOL

At present, most of the commercial methanol is produced from synthetic gas (also called as syngas, which is a mixture of CO and $H_2$) on quite large-scale industrial plants in several millions tons per year capacity. Besides this, the processes such as selective oxidation of methane, catalytic gas-phase oxidation of methane, liquid-phase oxidation of methane, monohalogenation of methane, microbial and photochemical conversion of methane, etc. are also being employed to produce methanol. Nevertheless, production of methanol from $CO_2$ using solar energy to drive the reaction is highly attractive as it saves the natural fossil fuel resources.

Converting $CO_2$ into value-added chemicals, such as methanol, is both challenging and rewarding. The conversion of $CO_2$ into methanol using energy has been suggested to be one of the best ways of storing energy and solving both global warming and energy crisis problems to a great extent. The additional advantages in producing methanol from $CO_2$ include (i) high energy density by volume and by weight; (ii) no need of high pressures to store at room temperature like $H_2$ needs; (iii) relatively low toxic and safe to handle fuel, and shows limited risks in its distribution (nontechnical); (iv) no need to modify the internal combustion engines of the vehicles to use methanol; and (v) no impact on the environment during production and usage as methanol being a primary feedstock for many of the organic compounds, and a vital intermediate for several bulk chemicals used in day-to-day life products such as silicone, paint, and plastics. Furthermore, methanol is a green fuel and has almost half of the energy density in comparison to the mostly used fuel, gasoline (methanol: 15.6 MJ/L; gasoline: 34.2 MJ/L), also be employed directly in the fuel cells.

The methanol economy and uses of methanol conversion from $CO_2$ have been largely discussed, whereas an economy based on FA has been proposed by Ferenc [35]. However, the formation of renewable $H_2$ is the key step in the conversion of $CO_2$ into methanol, and there are already certain established technologies for producing $H_2$ from water (a carbon-neutral resource) using electricity. Based on several factors as discussed in several published review articles, the PV/electrolyzer approach shows unmatched potential for the $CO_2$ catalytic hydrogenation route. However, the biochemical, thermochemical, and PEC splitting of water into hydrogen would show better results if these processes are directly integrated with the $CO_2$ reduction reaction in comparison to the PV/electrolyzer-based process, where $H_2$ needs to be produced separately and then used for reducing $CO_2$ in a separate process.

### 9.4.2 PHOTOELECTROCHEMICAL REDUCTION OF $CO_2$ INTO METHANOL

Methanol can be prepared from $CO_2$ by catalytic hydrogenation and dehydration. These two reactions can also be performed together in a single step. All these reactions are equilibrium reactions, they occur almost at the same temperature and on the same catalyst. Table 9.3 lists the various catalytic systems employed for synthesis of methanol from $CO_2$ by following catalytic hydrogenation (i.e., thermochemical) routes [36,37]. In fact, the synthesis of methanol from syngas over copper–zinc oxide-based catalysts is a well-established process, and about 40 M tons of methanol per year is being produced every year at present by following this route. In this commercial process, about 3% $CO_2$ is supplied together with syngas to enhance the reaction rate. Because reverse water–gas shift (RWGS) is a reversible reaction, the same catalyst can be employed to carry out both reactions of water–gas shift (WGS) and RWGS. The best catalyst noted for these reactions has been a multicomponent Cu/ $ZnO/ZrO_2/Al_2O_3/SiO_2$ composition.

From the above, it can be understood that the activity of direct formation of methanol by the catalytic hydrogenation of $CO_2$ is not only influenced by active catalytic sites but also by the support material. The Cu supported on CuO/ZnO with 30/70 weight ratio provided a methanol yield of $3.63 \times 10^{-5}$ kg per square meter of the catalyst per hour at 250°C under a pressure of 75 atm., whereas pure Cu provided a yield of less than $10^{-8}$ kg per square meter of the catalyst per hour [38]. According to a theory, addition of wurtzite ZnO with an n-type semiconductor to Cu/CuO catalysts creates cation and anion lattice vacancies, which are responsible for improved adsorption and transformation of $CO_2$ as well as the enhancement of Cu dispersion on the catalyst support. The formate intermediate was found to adsorb at the interface between Cu and ZnO or Cu–O–Zn. By employing the mixtures of Cu/ $SiO_2 + ZnO/SiO_2$ as catalysts, it was found that ZnO creates Cu–Zn active sites for this reaction, and the morphology of Cu was found to undergo any change during the reaction. Further, ZnO is believed to stabilize many active sites by absorbing the impurities present in the syngas stream. A small amount of sulfur could be a poison

**TABLE 9.3**

**Catalysts Employed in the Catalytic Hydrogenation of $CO_2$ into Methanol**

| Catalyst | Catalyst Preparation Method | Reaction Temperature (°C) | $CO_2$ Conversion (%) | Methanol Selectivity (%) | Methanol Activity (mol kg$^{-1}$ cat. h) |
|---|---|---|---|---|---|
| Cu/Zn/Ga/SiO$_2$ | Co-impregnation | 270 | 5.6 | 99.5 | 10.9 |
| Cu/Ga/ZnO | Co-impregnation | 270 | 6.0 | 88.0 | 11.8 |
| Cu/ZrO$_2$ | Deposition–precipitation | 240 | 6.3 | 48.8 | 11.2 |
| Cu/Ga/ZrO$_2$ | Deposition–precipitation | 250 | 13.7 | 75.5 | 1.9 |
| Cu/B/ZrO$_2$ | Deposition–precipitation | 250 | 15.8 | 67.2 | 1.8 |
| Cu/Zn/Ga/ZrO$_2$ | Coprecipitation | 250 | n/a | 75.0 | 10.1 |
| Cu/Zn/ZrO$_2$ | Coprecipitation | 250 | 19.4 | 29.3 | n/a |
| Cu/Zn/ZrO$_2$ | Urea–nitrate combustion | 240 | 17.0 | 56.2 | n/a |
| Cu/Zn/ZrO$_2$ | Coprecipitation | 220 | 21.0 | 68.0 | 5.6 |
| Cu/Zn/ZrO$_2$ | Glycine–nitrate combustion | 220 | 12.0 | 71.1 | n/a |
| Cu/Zn/Al/ZrO$_2$ | Coprecipitation | 240 | 18.7 | 47.2 | n/a |
| Ag/Zn/ZrO$_2$ | Coprecipitation | 220 | 2.0 | 97.0 | 0.46 |
| Au/Zn/ZrO$_2$ | Coprecipitation | 220 | 1.5 | 100 | 0.40 |
| Pd/Zn/ZrO$_2$ | Incipient wetness | 250 | 6.3 | 99.6 | 1.1 |
| Ga$_2$O$_3$–Pd/SiO$_2$ | Incipient wetness | 250 | n/a | 70.0 | 7.9 |
| LaCr$_{0.5}$Cu$_{0.5}$O$_3$ | Sol–gel | 250 | 10.4 | 90.8 | n/a |

*Source:* Reprinted by permission from Macmillan Publishers Ltd. *Renew. Sust. Energ. Rev.*, Ganesh I. Conversion of carbon dioxide into methanol—a potential liquid fuel: Fundamental challenges and opportunities (a review), 31, 221–257, copyright 2014.

for Cu catalyst if ZnO is absent as a support, thus ZnO inhibits the sulfur poisoning of Cu active sites during this reaction.

The reaction activity, product yield, and catalyst life are not only found to be by the catalyst composition but also by the reaction conditions employed. As both catalytic hydrogenations of CO and $CO_2$ into methanol are exothermic reactions, methanol conversion was found to be increased upon increasing the reaction pressure and decreasing the reaction temperature according to the Le Chartelier's principle. As the equilibrium constant decreases with an increase in temperature, a low-temperature condition is preferred for methanol formation. However, increasing reaction temperature could also increase the reaction rates for both these hydrogenation reactions. Nevertheless, methanol formation has been found to be sensitive to optimal temperature ranges over different catalysts. Higher reaction temperatures could also rapidly reduce the activity and shorten the catalyst lifetime by promoting the sintering process and agglomeration of Cu on the catalyst surface. Catalysts also tend to undergo deactivation faster at high pressures again by the enhanced sintering process. A search for an ideal catalyst system that is very active under low pressures and low temperatures with long lifetime is still active.

In PEC cells, the semiconductor electrode immersed in the electrolyte is connected through an external circuit to a counter electrode. When this semiconductor electrode is illuminated by any light that is having energy higher than the band gap of this semiconductor, its electrons excite from the valance band to conduction band, and reach cathode counter electrode through an external wire. Furthermore, the electron–hole pairs thus formed are spatially separated by the semiconductor junction barrier, and are injected into the electrolyte at the respective electrodes to produce electrochemical oxidation and reduction reactions. As of now, not even a single semiconducting material has been identified which can be employed as a photoelectrode to split water into hydrogen and oxygen gases or reduce $CO_2$ into methanol using exclusively solar energy in an aqueous-based PEC cell with desired stability and efficiency. However, it needs greater energy input to make up losses due to band bending (necessary in order to separate charge at the semiconductor surface), resistance losses, and overvoltage potentials. When a semiconductor is placed in an electrolyte, partial differences between the two phases result in charging of the interface. This charging results in a perturbation of the energy levels of the semiconductor called "band-bending." Band bending is responsible for separation of electron–hole pairs in PEC reactions. Recombination and corrosion processes decrease the utilization of the electron–hole pairs generated on illumination. A major impediment to the exploitation of PEC cells in solar energy conversion and storage is the susceptibility of small band gap semiconductor materials to photoanodic and photocathodic degradation. The photoinstability is particularly severe for n-type semiconductors where the photogenerated holes, which reach the interface, can oxidize the semiconductor material itself. In fact, many of the semiconducting materials are predicted to exhibit thermodynamic instability toward anodic photodegradation. Whether a photoelectrode is stable or not depends on the competitive rates of the thermodynamically possible reactions such as the semiconductor decomposition reaction and the electrolyte reactions.

### 9.4.3 OXIDE SEMICONDUCTORS AS CATHODE MATERIALS

The photoreduction of $CO_2$ involves two main free radicals H and $CO_2^{3-}$, which are formed by taking electrons from photocathode (i.e., from semiconductor), when the electrons of this semiconductor are excited from the valence band to the conduction band by absorbing a photon having energy equal to or greater than that of the band gap of the semiconductor. $TiO_2$ being a very stable material against photocorrosion and possessing band edges amenable to water oxidation/electrolysis reaction, it has been considered as an ideal material to use in PEC reduction or photocatalytic reduction of $CO_2$. As the band gap energy of $TiO_2$ is high (3.2 eV), which is equivalent to UV light, it cannot capture a larger portion of sunlight; hence, the efficiency of the system involving pure $TiO_2$ photoelectrode does not exceed 4% as the percentage of UV light in the entire solar spectrum is only about 4%. Several researchers have tried to improve the light-absorbing capability of $TiO_2$ by doping it with several metals and nonmetals. For example, Cu was dispersed into an aqueous suspension of $TiO_2$ to enhance the photocatalytic activity and reduction of $TiO_2$ for reducing $CO_2$ into methanol [39–41]. Those distinct catalytic steps include (i) photoelectron–hole pair generation, charge separation, and trapping; (ii) oxidation and reduction reactions of the adsorbates; (iii) rearrangement and other surface reactions of formed intermediates; and (iv) the desorption of the products from the photocatalyst surface and the regeneration of the surface. The $TiO_2$ and $TiO_2$-supported Cu catalysts formed in a sol–gel route were also investigated for photocatalytic of $CO_2$.

Not only powder catalysts, but also the thin films of Cu-doped $TiO_2$, and Cu-containing zinc oxide surface-coated Pt electrodes were employed for photocatalytic $CO_2$ reduction reaction in an aqueous-based electrolyte [41–43]. The pulse technique has been used in a regular electrochemical reduction of $CO_2$ with a pair of Cu electrodes, which normally prevent the deactivation of hydrocarbon formation catalyst [44]. Furthermore, when the pulsed mode technique was employed to supply bias potential, the noted results were remarkable, and it increased the activity for methane and ethylene formation.

$CO_2$ was also reduced to methane, ethylene, and CO in $CO_2$-saturated aqueous electrolyte over the surface of illuminated p-Si semiconductor electrode surface modified with small metal particles of Cu, Ag, or Au [45]. These modified p-Si semiconductor electrodes produced products similar to the metal (Cu, Ag, or Au) electrodes, but at ca. 0.5 V (vs. NHE) more positive potentials than their corresponding metal electrodes, contrary to continuous metal-coated p-Si electrodes. These results clearly suggest that the metal-particle-coated p-Si electrodes not only possess high catalytic activity for electrode reactions, but also generate high photovoltages and thus work as ideal semiconductor electrodes. The formation of the dimeric and tetrameric products, namely oxalate, glyoxylate, glycolate, and tartrate, was also noted in the $CO_2$ reduction reaction when performed in an aqueous solution of tetramethylammonium chloride suspended with CdS or ZnS colloids. The performance of other semiconducting powders, including ZnO, SiC, $BaTiO_3$, and $SrTiO_3$, was also compared with those exhibited by CdS and ZnS colloids. The formation of HCOOH and HCHO was noted in the absence of tetramethylammonium ions. The relative quantum efficiencies of these semiconductors were found to be influenced by their band gap energies and conduction band potentials. The role and effectiveness of several hole acceptor (electron donor) compounds in this process were also studied, and it was found that the addition of one electron to a $CO_2$ molecule produces a $CO_2^{3-}$ radical anion. It can be concluded that the semiconductors, particularly CdS, SiC, and ZnS, with the higher negative conduction band potentials provide the best yields. Although the effectiveness of different semiconductors in reducing $CO_2$ is roughly in line with the order of increasing conduction band potentials, the variations, which are fairly small, may also be related to differences in particle size, which in quantum crystallites results in an increase in band gap and reduction in the rate of recombination of electron–hole pairs. The overall quantum yields exhibited by these six semiconductors followed this trend: ZnS > SiC > ZnO > CdS > $BaTiO_3$ > $SrTiO_3$.

From the above presented photocatalytic results, it can be concluded that although encouraging progress has been made toward the photocatalytic conversion of $CO_2$ using sunlight, further efforts are required for increasing sunlight-to-fuel photoconversion efficiencies.

### 9.4.4 NON-OXIDE P-TYPE SEMICONDUCTORS AS CATHODE MATERIALS

In 1978, for the first time, a photoelectrode made of a single crystal p-gallium phosphide (p-GaP) was employed in a PEC cell for converting $CO_2$ into FA, formaldehyde, and methanol. Unlike the reduction of $CO_2$ on metal cathodes, which stops essentially after two electron transfers because of high overpotential associated with FA reduction, the photoelectrolysis on p-GaP proceeds further to yield formaldehyde and methanol. After 90 h of irradiation, the concentrations of FA, formaldehyde, and methanol formed were estimated to be $5 \times 10^{-2}$, $2.8 \times 10^{-4}$, and $8.1 \times 10^{-4}$ M, respectively. In this process, the efficiency of the system was calculated using the formula suggested. Optical conversion efficiency is nothing but the efficiency of conversion of radiant energy into the chemical energy. The results clearly indicate that in contrast to the reduction of $CO_2$ on certain metal cathodes, the photoelectrolysis on p-GaP does not stop after FA formation, but proceeds further, yielding formaldehyde and methanol.

Subsequent to the above study, there were several studies aimed at reducing $CO_2$ into highly reduced products using different types of semiconducting materials in PEC cells [46]. As part of this, $CO_2$ was also reduced to methanol over n- and p-GaAs, n-Si, and p-InP semiconductors. Reduction at n-GaAs was found to be selective with nearly 100% Faradaic efficiency. However, these semiconductors are susceptible to corrosion. $CO_2$ was also reduced to HCOOH, HCHO, methanol, and $CH_4$ over various other semiconducting materials that include $WO_3$, $TiO_2$, ZnO, CdS, GaP, and SiC. HCHO and methanol were found to be the major products over SiC semiconductor catalyst after illumination for 7 h. The methanol yield was found to be increased as the conduction band becomes more negative with respect to the redox potential of $H_2CO_3$/methanol, whereas methanol was not produced at all in the presence of $WO_3$ catalyst that has a conduction band more positive than this redox potential. $CO_2$ was also found to be reduced electrocatalytically to HCOOH, HCHO, methanol, and $CH_4$ at unilluminated $TiO_2$ electrode or at the illuminated p-GaP electrode when both semiconductor electrodes were polarized at a potential of −1.5 V vs. SCE for 2 h, which indicates that electrons in the conduction bands of these semiconductors reduce $CO_2$ in aqueous solution. It has been suggested that at semiconductor electrodes, the charge transfer rates between photogenerated carriers in semiconductors and the solution species depend on the correlation of energy levels between the semiconductor and the redox agents in the solution. If the redox potential of solution species is more positive with respect to the conduction band level, then these species undergo improved reduction.

$CO_2$ also underwent reduction over a biological catalyst (a formate dehydrogenase enzyme) in a PEC cell over the surface of p-InP illuminated with a light source having wavelength range shorter than 900 nm (>1.35 eV). This enzyme catalyst performs two electron reduction of $CO_2$ to FA with the help of a mediator that couples the photogenerated electrons in the semiconductor with the enzyme catalyst. Although this process appears to be more analogous to natural photosynthesis, the former process is more efficient at light collection and more specific in the production of reduced carbon species. Among the various reaction conditions, the $CO_2$ pressure was found to be very critical for obtaining higher product yields. When $CO_2$ reduction reaction was performed over p-GaP and p-GaAs semiconductors in a PEC autoclave fitted with a quartz window, a cation exchange diaphragm, and a platinum counter electrode under the illumination with a 150 W Xe lamp, the formation of HCOOH, HCHO, and methanol was noted with a Faradaic efficiency of 80% on the surface of p-GaP at a cathodic bias of −1.00 V (vs. a standard silver electrode) in 0.5 M $Na_2CO_3$ solution under 8.5 atm. $CO_2$ pressure [47,48].

Although by following PEC routes, $CO_2$ could be converted into methanol using certain non-oxide semiconducting photoelectrodes (p-GaP) together with homogeneous organic molecular catalysts (pyridine), the recorded quantum efficiencies and the stability and durability of these routes have been rather low and insufficient for practicing on a commercial scale. This is due to the fact that these routes still need certain amount of external bias voltage from grid current, and the associated semiconductors show poor long-term stability against photocorrosion being non-oxide materials.

Among the stoichiometric, thermochemical, electrochemical, PEC, and photocatalytic routes developed so far for reducing $CO_2$ into value-added chemicals, only the electrochemical routes appear to be viable as these latter routes allow mild reaction conditions, most studied and understood systems only next to the thermochemical processes, and can be integrated with electricity that is derived from sunlight using any of the existing technologies such as PVs. However, the reaction efficiencies and product yields obtained in these electrochemical routes are also low at the moment. Nevertheless, these reaction efficiencies and product yields could be improved suitably (i) by suppressing the hydrogen evolution reaction (HER), which is the main reason for the noted low efficiency of the process, by using certain electrodes having high overpotentials toward HER; (ii) by reducing overpotentials associated with $CO_2$ reduction reaction into methanol using suitable molecular catalysts, such as room temperature ionic liquids (RTILs), pyridine, and a mixture of nitroso-R salt+Co or Cu sulfate+methanol, which can be employed in conjunction with electrodes; and (iii) by employing electricity that is produced from sunlight using cheaper technologies such as polymer-based PVs. Furthermore, the HER could also be suppressed by using certain organic molecular catalysts, such as triethanolamine, in the anode compartment of the electrochemical reaction.

## 9.5   CONCLUSIONS AND OUTLOOK

Research in the field of photochemical and PEC reduction of $CO_2$ has grown rapidly now. It is a response by physical scientists and engineers to the increasing amount of $CO_2$ in the atmosphere and the steady growth in global fuel demand. $CO_2$ is an extremely stable molecule generally produced by fossil fuel combustion and respiration. Returning $CO_2$ to a useful state by activation/reduction is a challenging task in chemical catalysis, electrochemistry, photochemistry, and semiconductor physics and engineering.

PEC which uses both homogeneous and heterogeneous system for $CO_2$ reduction has been reviewed. In homogeneous systems, some photocatalysts using transition metal complexes show outstanding performance, such as high absorbance in the visible region, high quantum yields, and high product selectivities. $CO_2$ reduction using semiconductor photocatalysts, such as metal oxide semiconductors $TiO_2$, has been increasing rapidly. Organic contaminants have paid attention to both carbon sources and reductants of the reaction products for photocatalytic reduction of $CO_2$.

Development of an efficient process for converting $CO_2$ into methanol or to any other value-added chemical using exclusively solar energy is of great importance, as this process can indeed deal with (i) the $CO_2$-associated global warming problem, (ii) depletion of fossil fuels, and (iii) the problems associated with storing of energy (electricity as well as solar energy) in the form of high energy density liquid fuels for future applications.

## REFERENCES

1. Kumar B, Llorente M, Froehlich J et al. Photochemical and photoelectrochemical reduction of $CO_2$. *Ann. Rev. Phys. Chem.*, 2012, 63, 541–569.
2. Kedzierzawski P, Augustynski J. Poisoning and activation of the gold cathode during electroreduction of $CO_2$. *J. Electrochem. Soc.*, 1994, 141, 58–60.
3. Saveant J-M. Molecular catalysis of electrochemical reactions: Mechanistic aspects. *Chem. Rev.*, 2008, 108, 2348–2378.
4. Hawecker J, Lehn J-M, Ziessel R. Efficient photochemical reduction of $CO_2$ to CO by visible light irradiation of systems containing Re(bipy)(CO)3X or Ru(bipy)32+ – $CO_2$+ combinations as homogeneous catalysts. *J. Chem. Soc. Chem. Commun.*, 1983, 9, 536–538.
5. Hawecker J, Lehn J-M, Ziessel R. Photochemical and electrochemical reduction of carbon dioxide to carbon monoxide mediated by (2,2'-bipyridine)tricarbonylchlororhenium(I) and related complexes as homogeneous catalysts. *Helv. Chim. Acta*, 1986, 69, 1990–2012.
6. Takeda H, Koike K, Inoue H, Ishitani O. Development of an efficient photocatalytic system for $CO_2$ reduction using rhenium(I) complexes based on mechanistic studies. *J. Am. Chem. Soc.*, 2008, 130, 2023–2031.
7. Bockris JOM, Wass JC. On the photoelectrocatalytic reduction of carbon dioxide. *Mater. Chem. Phys.*, 1989, 22, 249–280.
8. Junfu L, Baozhu C. Photoelectrochemical reduction of carbon dioxide on a p+/p−Si photocathode in aqueous electrolyte. *J. Electroanal. Chem.*, 1992, 324, 191–200.
9. Beley M, Collin J-P, Sauvage J-P, Petit J-P, Chartier P. Photoassisted electro-reduction of $CO_2$ on p-GaAs in the presence of Ni cyclam2+. *J. Electroanal. Chem.*, 1986, 206, 333–339.

10. Aurian-Blajeni B, Taniguchi I, Bockris JOM. Photoelectrochemical reduction of carbon dioxideusing polyaniline-coated silicon. *J. Electroanal. Chem.*, 1983, 149, 291–293.

11. Dominey RN, Lewis NS, Bruce JA, Bookbinder DC, Wrighton MS. Improvement of photoelectrochemical hydrogen generation by surface modification of p-type silicon semiconductor photocathodes. *J. Am. Chem. Soc.*, 1982, 104, 467–482.

12. Lewis NS. Chemical control of charge transfer and recombination at semiconductor photoelectrode surfaces. *Inorg. Chem.*, 2005, 44, 6900–6911.

13. Kumar A, Wilisch WCA, Lewis NS. The electrical properties of semiconductor/metal, semiconductor/liquid, and semiconductor/conducting polymer contacts. *Crit. Rev. Solid State Mater. Sci.*, 1993, 18, 327–353.

14. Das S, Daud WMA. A review on advances in photocatalysts towards CO$_2$ conversion. *RSC Adv.*, 2014, 4, 20856–20893.

15. Inoue T, Fujishima A, Konishi S, Honda K. *Nature*, 1979, 277, 637–638.

16. Das S, Daud WMAW. Photocatalytic CO$_2$ transformation into fuel: A review on advances in photocatalyst and photoreactor. *Renew. Sust. Energy. Rev.*, 2014, 39(14), 765–805.

17. Magdesieva T, Yamamoto T, Tryk D, Fujishima A. Electrochemical reduction of CO$_2$ with transition metal phthalocyanine and porphyrin complexes supported on activated carbon fibers. *J. Electrochem. Soc.*, 2002, 149(6), D89–D95.

18. Tahir M, Amin NS. *Appl. Catal., B*, 2013, 142–143, 512–522.

19. Van Grieken R, Aguado J, López-Mu~noz M, Marug´an J. *J. Photochem. Photobiol., A*, 2002, 148, 315–322.

20. Tan SS, Zou L, Hu E. *Catal. Today*, 2006, 115, 269–273.

21. Fotou GP, Pratsinis SE. *Chem. Eng. Commun.*, 1996, 151, 251–269.

22. Yui T, Kan A, Saitoh C, Koike K, Ibusuki T, Ishitani O. *ACS Appl. Mater. Interfaces*, 2011, 3, 2594–2600.

23. Kaneco S, Kurimoto H, Shimizu Y, Ohta K, Mizuno T, *Energy*, 1999, 24, 21–30.

24. Kaneco S, Kurimoto H, Ohta K, Mizuno T, Saji A. *J. Photochem. Photobiol., A*, 1997, 109, 59–63.

25. Dey G, Belapurkar A, Kishore K. *J. Photochem. Photobiol., A*, 2004, 163, 503–508.

26. Stock M, Dunn S. *Ferroelectrics*, 2011, 419, 9–13.

27. Zhang K, Guo L. *Catal. Sci. Technol.*, 2013, 3, 1672–1690.

28. Hong J, Zhang W, Ren J, Xu R. Photocatalytic reduction of CO$_2$: A brief review on product analysis and systematic methods. *Anal. Methods*, 2013, 5, 1086–1097.

29. Yuan L, Xu YJ. Photocatalytic conversion of CO$_2$ into value-added and renewable fuels. *Appl. Surf. Sci.*, 2015, 342, 154–167.

30. Liu YY, Huang BB, Dai Y, Zhang XY, Qin XY, Jiang MH, Whangbo MH. Selective ethanol formation from photocatalytic reduction of carbon dioxide in water with BiVO$_4$ photocatalyst. *Catal. Commun.*, 2009, 11, 210–213.

31. Qin S, Xin F, Liu Y, Yin X, Ma W. Photocatalytic reduction of CO$_2$ in methanol to methyl formate over CuO–TiO$_2$ composite catalysts. *J. Colloid Interface Sci.*, 2011, 356, 257–261.

32. Yang CC, Yu YH, van der Linden B, Wu JCS, Mul G. Artificial photosynthesis over crystalline TiO$_2$-based catalysts: Fact or fiction? *J. Am. Chem. Soc.*, 2010, 132, 8398–8406.

33. Liu BJ, Torimoto T, Matsumoto H, Yoneyama H. Effect of solvents on photocatalytic reduction of carbon dioxide using TiO$_2$ nanocrystal photocatalyst embedded in SiO$_2$ matrices. *J. Photochem. Photobiol., A*, 1997, 108, 187–192.

34. Teramura K, Iguchi S, Mizuno Y, Shishido T, Tanaka T. Photocatalytic conversion of CO$_2$ in water over layered double hydroxides. *Angew. Chem. Int. Ed.*, 2012, 51, 8008–8011.

35. Lizuka K, Wato T, Miseki Y, Saito K, Kudo A. Photocatalytic reduction of carbon dioxide over Ag cocatalyst-loaded ALa$_4$Ti$_4$O$_{15}$ (A = Ca, Sr, and Ba) using water as a reducing reagent. *J. Am. Chem. Soc.*, 2011, 133, 20863–20868.

36. Joo F, Laurenczy G, Karady P, Elek J, Nadasdi L, Roulet R. Homogeneous hydrogenation of aqueous hydrogen carbonate to formate under mild conditions with water soluble rhodium(I)-and ruthenium(II)-phosphine catalysts. *Appl. Organomet. Chem.*, 2000, 14, 857–859.

37. Ganesh I. Conversion of carbon dioxide into methanol—a potential liquid fuel: Fundamental challenges and opportunities (a review). *Renew. Sust. Energ. Rev.*, 2014, 31, 221–257.

38. Omae I. Aspects of carbon dioxide utilization. *Catal. Today*, 2006, 115(1–4), 33–52.

39. Herman RG, Klier K, Simmons GW, Finn BP, Bulko JB, Kobylinski TP. Catalytic synthesis of methanol from $CO-H_2$. I. Phase composition, electronic properties, and activities of the $Cu/ZnO/M_2O_3$ catalysts. *J. Catal.*, 1979, 56(3), 407–429.

40. Ganesh I, Sekhar PSC, Padmanabham G, Sundararajan G. Influence of Lidoping on structural characteristics and photocatalytic activity of ZnO nanopowder formed in a novel solution pyro-hydrolysis route. *Appl. Surf. Sci.*, 2012, 259, 524–537.

41. Hirano K, Inoue K, Yatsu T. Photocatalyzed reduction of $CO_2$ in aqueous $TiO_2$ suspension mixed with copper powder. *J. Photochem. Photobiol. A: Chem.*, 1992, 64(2), 255–258.

42. Hemminger JC, Carr R, Somorjai GA. Photoassisted reaction of gaseous water and carbon dioxide adsorbed on $SrTiO_2$ (111) crystal face to form methane. *Chem. Phys. Lett.*, 1978, 57(1), 100–104.

43. Yang YX, Evans J, Rodriguez JA, White MG, Liu P. Fundamental studies of methanol synthesis from $CO_2$ hydrogenation on Cu(111), Cu clusters, and Cu/ZnO(000(1)overbar). *Phys. Chem. Chem. Phys.*, 2010, 12(33), 9909–9917.

44. Ichikawa S, Doi R. Hydrogen production from $H_2O$ and conversion of $CO_2$ to useful chemicals by room temperature photoelectrocatalysis. *Catal. Today*, 1996, 27, 271–277.

45. Sakka S, Kamiya K, Makita K, Yamamoto Y. Formation of sheets and coating films from alkoxide solutions. *J. Non-Crystal Solids*, 1984, 63(1 and 2), 223–235.

46. Hinogami R, Nakamura Y, Yae S, Nakato Y. An approach to ideal semiconductor electrodes for efficient photoelectrochemical reduction of $CO_2$ by modification with small metal particles. *J. Phys. Chem. B*, 1998, 102, 974–980.

47. Gerischer H. Electrochemical photo and solar cells principles and some experiments. *J. Electroanal. Chem. Interfacial Electrochem.*, 1975, 58(1), 263–274.

48. Aurian-Blajeni B, Halmann M, Manassen J. Electrochemical measurements on the photoelectrochemical reduction of aqueous carbon dioxide on p-gallium phosphide and p-gallium arsenide semiconductor electrodes. *Sol. Energy Mater.*, 1983, 8, 425–440.

# 10 Challenges and Perspectives of CO$_2$ Electroreduction

*Mengyang Fan and Jinli Qiao*

## CONTENTS

## 10.1  INTRODUCTION

The electrocatalytic reduction of CO$_2$ has a long history dating from the nineteenth century. However, this topic becomes more attractive from both academia and industry, because the excessive burning of fossil fuels leads to the extra emission of CO$_2$, which has exceeded nature's CO$_2$ recycle capability. To deal with the extra CO$_2$ recycle, the storage of renewable energy through the electrochemical reduction of CO$_2$ is an attractive strategy to transform the current linear utilization of carbon fuels: the process could be operated under normal condition; the electricity source used to reduce the CO$_2$ could be renewable green energy; the reaction system can be built up to a continuous one and scaled up to realize the industrial application.

However, to reach the final goal of CO$_2$ electroreduction, challenges and opportunities exist at the same time. For the CO$_2$ reduction reaction itself, optimal catalysts with high catalyzed activity, selectivity, and stability should be developed, which can be used to reduce energy consumption during the $CO_2$ electroreduction process. For the reaction system, the electrode and electrolyte parts should be moderate to adapt the application, and the system should be designed to a more compact structure, which can be easily scaled up. Another significant part in $CO_2$ electroreduction process is the energy used for the $CO_2$ conversion. This energy source should be green electricity and will not generate additional $CO_2$ in the process. However, such green energy is not feasible to obtain because of the low power density and the limited power generation conditions. Therefore, all these main factors cause major challenges in $CO_2$ electroreduction and need to be taken into consideration and treated as the key points in future research.

## 10.2   INSUFFICIENT CATALYST ACTIVITY, SELECTIVITY, AND STABILITY OF $CO_2$ ELECTROREDUCTION

When judging the properties of catalysts, the catalyzed activity, production selectivity, and electrode stability are the main parameters to be investigated. Developing some novel kinds of effective catalysts for $CO_2$ electroreduction is the primary task. In current states, some of the developed catalysts may have one prominent feature of the three main aspects; however, synthesizing catalysts that function well in comprehensive performance is a big challenge.

### 10.2.1   High Overpotential and Low Current Efficiency

Overpotential and current efficiency are main aspects to judge the catalyst activity. Catalysts that perform the low overpotential and high current efficiency could be considered as optimal ones within a certain range; on the contrary, those with high overpotential and low current efficiency in $CO_2$ electroreduction seem to be insufficiently active. Hara Kohjiro and his coworkers have compared different metal catalysts in different groups in their research, which can be seen in Table 10.1, and have classified these metals into four main groups on the basis of their electrocatalytic activities [1]. Group (1) includes Ti, Nb, Ta, Mo, Mn, and Al, which has hydrogen as the prominent product formed by water reduction. In group (2), which includes Zr, Cr, W, Fe, Co, Rh, Ir, Ni, Pd, Pt, C, and Si, there are formic acid and CO products reduced from $CO_2$ reduction when the pressure of $CO_2$ increased high up to 30 atm. However, when the $CO_2$ pressure was 1 atm, hydrogen was still the dominant product. In addition, the overpotentials of $CO_2$ reduction on these metals are higher than $-1.45$ V vs. Ag/AgCl. Meanwhile, the faradaic efficiencies of $CO_2$ reduction on these metals in group (2) are not that satisfactory. Interestingly, when the current density was 163 mA cm$^{-1}$, the faradaic efficiency of reduced formic acid on Pd was 35.4%; however, when the current density increased to 500 mA cm$^{-1}$, the faradaic efficiency of formic acid production increased to 44%. It can be seen that the production efficiency could not be improved under lower current density, which results in low current efficiency. This is one of the least known aspects of catalyst activity which needs to be investigated in the future. Metal Ag, Au, Zn, In, Sn, Pb, and Bi are classified in group (3) [1]. In this group, CO and formic acid are the major products. The faradaic

**TABLE 10.1**

**Electrochemical Reduction of CO$_2$ under a Pressure of 30 atm on Various Electrodes at 163 mA cm$^{-2}$**

| Group | Electrode | E$^a$ (V) | CH$_4$ | C$_2$H$_6$ | C$_2$H$_4$ | CO | HCOOH | H$_2$ | CO$_2$ red$^b$ | Total | PCD(CO$_2$red.)$^c$ (mA cm$^{-2}$) |
|---|---|---|---|---|---|---|---|---|---|---|---|
| | | | | | | Faradaic Efficiency (%) | | | | | |
| 4 | Ti | −1.57 | 0.18 | 0.01 | 0.08 | Trace | 4.6 | 80.8 | 4.9 | 85.7 | 8 |
| | Zr | −1.73 | 0.13 | 0.01 | 0.01 | 32.5 | 7.6 | 44.2 | 40.3 | 84.5 | 65.7 |
| 5 | Nb | −1.45 | 0.56 | 0.05 | 0.01 | n$^d$ | 3.5 | 81.4 | 4.1 | 85.5 | 6.7 |
| | Ta | −1.51 | 0.55 | 0.05 | Trace | Trace | 7.6 | 74.4 | 8.2 | 82.6 | 13.4 |
| 6 | Cr | −1.49 | 0.53 | 0.05 | 0.07 | 11.8 | 8.2 | 68.6 | 20.7 | 89.3 | 33.7 |
| | Mo | −1.34 | 0.4 | 0.05 | 0.03 | n | 6.5 | 83.3 | 7 | 90.3 | 11.4 |
| | W | −1.61 | 0.38 | 0.04 | 0.01 | Trace | 31.9 | 53.1 | 32.3 | 85.4 | 52.6 |
| 7 | Mn | −1.69 | 0.68 | 0.1 | 0.06 | 2.8 | 2.8 | 78.8 | 6.5 | 85.3 | 10.6 |
| 8 | Fe | −1.63 | 2.03 | 0.4 | 0.16 | 4.2 | 28.6 | 51.6 | 35.4 | 87.0 | 57.7 |
| 9 | Co | −1.54 | 3.09 | 0.17 | 0.38 | 15.8 | 21.9 | 46.9 | 41.5 | 88.4 | 67.6 |
| | Rh | −1.41 | 0.26 | 0.03 | 0.01 | 61 | 19.5 | 13.1 | 80.8 | 93.9 | 131.7 |
| | Ir | −1.55 | 0.62 | 0.05 | 0.05 | 17.5 | 22.3 | 48.3 | 40.5 | 88.8 | 66 |
| | Ni | −1.59 | 0.72 | 0.08 | 0.11 | 33.5 | 31.3 | 26 | 65.7 | 91.7 | 107.1 |
| 10 | Pd | −1.56 | 0.13 | 0.01 | Trace | 46.1 | 35.6 | 12.8 | 81.8 | 94.6 | 133.3 |
| | Pd$^e$ | −1.76 | 0.21 | 0.01 | 0.02 | 35.2 | 44 | 13.8 | 79.4 | 93.2 | 397 |
| | Pt | −1.48 | 0.22 | 0.02 | Trace | 6.1 | 50.4 | 33.6 | 56.7 | 90.3 | 92.4 |
| | Cu | −1.64 | 9.95 | 0.06 | 3.74 | 20.1 | 53.7 | 2.5 | 87.6 | 90.1 | 142.8 |
| 11 | Ag | −1.48 | 0.2 | 0.01 | Trace | 75.6 | 16.8 | 3.9 | 92.6 | 96.5 | 150.9 |
| | Au | −1.3 | 0.21 | 0.02 | 0.11 | 64.7 | 11.8 | 15.4 | 76.8 | 92.2 | 125.2 |

(*Continued*)

# TABLE 10.1 (Continued)
## Electrochemical Reduction of $CO_2$ under a Pressure of 30 atm on Various Electrodes at 163 mA cm$^{-2}$

| Group | Electrode | $E^a$ (V) | Faradaic Efficiency (%) | | | | | | | | PCD($CO_2$,red.)[c] (mA cm$^{-2}$) |
| --- | --- | --- | --- | --- | --- | --- | --- | --- | --- | --- | --- |
| | | | $CH_4$ | $C_2H_6$ | $C_2H_4$ | CO | HCOOH | $H_2$ | $CO_2$ red[b] | Total | |
| 12 | Zn | -1.7 | 0.31 | 0.03 | Trace | 48.7 | 40.5 | 2.8 | 89.5 | 92.3 | 145.9 |
| | Al | -1.97 | 0.66 | 0.01 | n | n | 1.3 | 86.5 | 2 | 88.5 | 3.3 |
| 13 | In[f] | – | 0.28 | Trace | 0.04 | 3.8 | 90.1 | 5.6 | 90.5 | 99.1 | 147.5 |
| | C[g] | -1.68 | 0.45 | 0.03 | 0.04 | 44 | 30.2 | 15.6 | 74.7 | 90.3 | 37.4 |
| | C | -2.14 | 0.66 | 0.02 | 0.05 | 3.6 | 6.8 | 75.5 | 11.2 | 86.7 | 18.3 |
| 14 | n-Si | -2.04 | 0.87 | 0.01 | 0.02 | 2 | 46.3 | 40.6 | 49.2 | 89.8 | 80.2 |
| | Sn | -1.39 | 0.06 | Trace | Trace | 8 | 92.3 | 1.3 | 100.4 | 101.7 | 163 |
| | Pb | -1.57 | 0.2 | 0.01 | Trace | Trace | 95.5 | 1.2 | 95.7 | 96.9 | 156 |
| 15 | Bi[h] | -1.42 | 0.17 | 0.01 | Trace | 3.3 | 82.7 | 6.3 | 86.2 | 92.5 | 140.5 |

*Source:* Reprinted from *Journal of Electroanalytical Chemistry*, 391(1), Hara, K., Kudo, A., and Sakata, T., Electrochemical reduction of carbon dioxide under high pressure on various electrodes in an aqueous electrolyte, 141–147, Copyright 1995, with permission from Elsevier.

*Note:* Reaction temperature, 25°C; electrolyte, 0.1 mol dm$^{-3}$ KHCO$_3$; charge passed, 300°C.

[a] Corrected with an IR compensation instrument (vs. Ag/AgCl).
[b] Total faradaic efficiency for $CO_2$ reduction.
[c] Partial current density for $CO_2$ reduction.
[d] Not detected.
[e] Current density, 500 mA cm$^{-2}$.
[f] Current density, 200 mA cm$^{-2}$.
[g] Current density, 50 mA cm$^{-2}$.
[h] Current density, 150 mA cm$^{-2}$.

efficiency of formic acid was enhanced by the increase of $CO_2$ pressure. However, by observing the production of $CO_2$ reduction, only Au and Sn have the overpotential more positive than −1.4 V vs. Ag/AgCl, which is not that sufficient for $CO_2$ reduction when this technology needs to be enlarged in the future. Cu was classified in the fourth group (group [4]) [1]. The overpotential of $CO_2$ production is about −1.64 V vs. Ag/AgCl, not higher than Sn and Au, and the selectivity of the reduction products on Cu electrodes depends strongly on current density and $CO_2$ pressure. From what has been discussed above, it could not be ignored that insufficient catalyst activity, including the high overpotential and low current efficiency, especially under normal $CO_2$ pressure, is one of the biggest challenges in the investigation of $CO_2$ electroreduction.

## 10.2.2 INSUFFICIENT SELECTIVITY OF PRODUCTIONS

Selectivity is another key aspect to judge the properties of catalysts. The final goal of $CO_2$ electroreduction investigation is to find some kinds of catalysts which can catalyze the $CO_2$ reduction reaction to produce more target products under low energy depletion. Therefore, in most research of $CO_2$ reduction, the production selectivity is one of the significant points to be investigated when evaluating the properties of catalysts.

In general, formic acid (HCOOH) and methane are the most desired target products [2,3]. Further, according to the $CO_2$ reduction mechanism, HCOOH is a product which is much easier to obtain [4]. In Table 10.1, Sn, Cu, and Pb are the kinds of metals which could catalyze $CO_2$ reduction to get a relatively high HCOOH faradaic efficiency. However, in most previous work, the HCOOH selectivity of $CO_2$ electroreduction is not that satisfied, especially under normal temperature and pressure conditions. Table 10.2 shows the faradaic efficiency of formate on different Cu and Sn catalysts or metal electrodes. It could be seen that most of the formate faradaic efficiencies on these Cu- and Sn-based catalysts are lower than 50%. A higher faradaic efficiency can be obtained merely when the electrolysis potential is high enough or the $CO_2$ pressure high up to some extent [1,10,19]. In Kaneco's paper, the Cu foil could catalyze $CO_2$ reduction in methanol electrolyte and the HCOOH efficiency reaches about 85% when the $CO_2$ pressure is 10 atm [10]. In addition, Hara reported that the faradaic efficiency of HCOOH in 0.1 M $KHCO_3$ saturated with 30 atm $CO_2$ could reach high up to 92.35% [1]. However, this high-pressure condition for $CO_2$ catalyzed electroreduction decreases the durability of catalysts and increases the application costs. In this case, developing efficient catalysts with high selectivity for target products under normal conditions is one of the key aspects to reduce costs when the $CO_2$ electroreduction technique is applied to the reality.

## 10.2.3 POOR STABILITY OF THE ELECTRODE

Catalyst deactivation caused by the poisonous formation and deposition on the electrode surface is the foremost reason that results in the poor stability of catalysts in the $CO_2$ electroreduction [21–26]. The main causes of the limited stability on Cu electrode were summarized by Hori as (1) heavy metal impurities contained in electrolyte solution; (2) very small amount of organic substances contained in water; and (3)

**TABLE 10.2**

**Faradaic Efficiency of Formate on Different Cu and Sn Catalysts or Electrode**

| Catalysts/ Electrode | Electrolyte | Electrolysis Potential | Faradaic Efficiency (Formate) | References |
|---|---|---|---|---|
| Cu foil | 1.5 M KCl | −1.36 V vs. normalized hydrogen electrode (NHE) | 31.6% | [5] |
| Cu foil | 0.1 M KHCO$_3$ | −1.41 V vs. NHE | 9.7% | [6] |
| Cu foil | LiClO$_4$ (40 atm) | −4.1 V vs. Ag quasi-reference electrode (QRE) | 46.7 % | [7] |
| Cu foil | 0.08 M LiOH in methanol | −2.0 V vs. Ag/AgCl | ~35% | [8] |
| Cu foil | 0.08 M CsOH in methanol | −4.5 V vs. Ag/AgCl | ~10% | [9] |
| Cu foil | 0.5 M CsOH in methanol (10 atm) | −3.5 V vs. Ag QRE | ~85% | [10] |
| Sn/Cu mesh | 0.45 M KHCO$_3$ | 3.9–5.9 V | 36% | [11] |
| Cu$_2$O/carbon cloth | 0.5 M NaOH | −1.70 V vs. SCE | – | [12] |
| Cu$_2$O/Zn | 0.3 M KOH in methanol | −3.0 V vs. Ag QRE | ~35% | [13] |
| CuO/Zn | 0.3 M KOH in methanol | −3.0 V vs. Ag QRE | ~35% | [13] |
| Cu$_2$O/Cu | 0.5 M NaHCO$_3$ | −0.75 V vs. relative standard hydrogen electrode (RHE) | ~33% | [14] |
| Sn | 0.5 M KHCO$_3$ | −1.40 V vs. SCE | ~65% | [15] |
| Sn granules | 0.5 M KHCO$_3$ | −2.0 V vs. SCE | 47% | [16] |
| Sn/GDE | 0.35 M Na$_2$SO$_4$ | −1.80 V vs. SCE | 57% | [17] |
| Sn powder/GDE | 0.5 M NaHCO$_3$ | −1.6 V vs. NHE | ~70% | [18] |
| Sn | 0.5 M KHCO$_3$ (high pressure) | −1.8 V vs. Ag/AgCl | 99.2% | [19] |
| Sn | 0.1 M KHCO$_3$ (30 atm) | – | 92.35 | [1] |
| Sn/SnO$_x$ | 0.5 M NaHCO$_3$ | −0.70 V vs. RHE | ~40% | [20] |

some intermediate products formed during the CO$_2$ reduction [26]. Research shows that the Sn particles exhibited more stable operating times than the tin-coated Cu mesh cathode (formate efficiency dropped from 50% to 20% in minutes) [11]. In addition, the catalysis condition (such as catalysis potential or current density) is another factor that may affect the durability of catalysts. For instance, the pulse electrolysis method could alleviate the deactivation of the electrode [27]. The pretreatment on Cu electrode by applying a potential to change the electrode surface can significantly prolong the durability of the electrode. However, the detailed mechanism of CO$_2$ reduction on such changed electrode surface is still unclear and needs to be discussed.

When discussing about Sn catalysts, four main factors of the deactivation on pure Sn electrode are summarized as (1) cathodic degradation of the catalyst surface [28], (2) deposition of noncatalytic species from reaction intermediates in the reduction of the pollutant species [29], (3) organometallic complex formation on the electrode surface [30], and (4) deposition of noncatalytic metallic species from contaminants in the electrolyte [31]. Agarwal conducted experiments on Sn and proprietary Sn alloy electrode using a flow-through reactor at a gas/solid/liquid interface. The Sn electrode performed better durability than Cu electrode. However, the reason for the color change appearing on the surface during the electrode degradation was still unclear [32].

Therefore, considering the enhancement of catalyst stability, the catalyst category, surface morphology, and the electrocatalysis condition are the factors that should be taken into account.

## 10.3   HIGH COST DURING THE $CO_2$ ELECTROREDUCTION PROCESS

It is well known that the conversion of $CO_2$ requires energy because of the molecule stability of $CO_2$. During the $CO_2$ electroreduction process, the electricity cost is a main input and the electric power is a main consumption, so the energy cost should be minimized in the development of $CO_2$ reduction. In addition, during the electroreduction process, green energy is required to prevent extra $CO_2$ emission, which will also increase the operation expense. Moreover, chemical costs such as the electrolyte and the waste disposal is another major spending that cannot be ignored.

### 10.3.1   ELECTRICAL ENERGY COSTS

One big challenge facing $CO_2$ electroreduciton could be obviously seen as the low cathode reaction kinetic which is not favorable to production selectivity and current efficiency [33–36]. The geometric current density should be high enough to support an industrial electroreduction of $CO_2$ for two main causes. For one aspect, there is a current loss according to Faraday's law. For another aspect, in Oloman's research, the cost of capital reactors for $CO_2$ electroreduction is about US\$ $30 \times 10^3$ m$^{-2}$ of the electrode area [33]. Therefore, the geometric current density should be at least 1 kA m$^{-2}$ with 50% current efficiency to meet the need of the capital costs. The total installed capital cost of Oloman's system is about 170 million US\$, the product value could reach \$ 1000 t$^{-1}$ HCOOH, and the output of HCOOH is about 350 t day$^{-1}$. When using renewable electricity, the energy cost is about \$ 0.01–0.10 kW h$^{-1}$, and there is a little return on investment. However, the electricity price for the renewable sources should be higher in practice. Therefore, the electricity cost is a huge input, and it is a big challenge that $CO_2$ electroreduction technology is facing.

### 10.3.2   CHEMICAL CONSUMPTIONS

Chemical consumption is another huge investment of $CO_2$ reduction apart from the electrical energy cost. Sridhar developed a systematic modeling of the entire value chain of $CO_2$ reduction and utilization [32]. In his investigation, Sridhar added supporting systems analysis for $CO_2$ conversion, including $CO_2$ capture and separation, of which the price is higher than previous analysis to some extent. As a result, the chemical consumptions are main parts that cannot be ignored. In Sridhar's work, his group compared two process scenarios: (1) using a consumable chemical to reduce the total energy costs and (2) using a portion of waste water instead of the consumable

**TABLE 10.3**

**General Inputs for Scenarios 1 and 2, Not Specific to the Product or Electrochemical Reaction**

| Parameter[a] | Minimum | Most Likely | Maximum |
|---|---|---|---|
| Emissions source output (tpd) | 0.8 | 1 | 1.2 |
| Electricity price ($ [kW h]$^{-1}$) | 0.06 | 0.07 | 0.08 |
| CAPEX MEA plant ($10^6$\$ [tpd]$^{-1}$) | 0.6 | 0.7 | 1.4 |
| CAPEX diversion and delivery equipment ($10^6$\$ [tpd]$^{-1}$) | 0.10 | 0.15 | 0.20 |
| CAPEX product purification and storage ($10^6$\$ [tpd]$^{-1}$) | 0.75 | 1.00 | 1.20 |
| $CO_2$ price ($ t$^{-1}$) | 0 | 0 | 0 |
| NaOH price ($ t$^{-1}$) | 150 | 165 | 180 |
| NaCl price ($ t$^{-1}$) | 75 | 101 | 130 |
| NaCl price ($ t$^{-1}$) | 72 | 95 | 140 |
| $O_2$ price ($ t$^{-1}$) | 50 | 150 | 300 |
| HCl price ($ t$^{-1}$) | 72 | 150 | 300 |
| $H_2SO_4$ price ($ t$^{-1}$) | 72 | 150 | 300 |
| $H_2$ price ($ t$^{-1}$) | 2300 | 2700 | 5000 |
| $H_2O$ price ($ t$^{-1}$) | 1 | 2 | 3 |
| Nominal discount factor | 0.03 | 0.04 | 0.05 |
| Financing interest rate | 0.04 | 0.05 | 0.06 |
| Electrode lifetime (year) | 0.4 | 0.5 | 0.8 |
| Plant lifetime (year) | 25 | | |

*Source:* Agarwal, A. S., Zhai, Y., Hill, D., and Sridhar, N.: The electrochemical reduction of carbon dioxide to formate/formic acid: Engineering and economic feasibility. *ChemSusChem.* 2011. 4(9). 1301–1310. Copyright Wiley-VCH Verlag GmbH & Co. KGaA. Reproduced with permission.

[a]  tpd: ton per day, $ (kWh)$^{-1}$: US$ per kWh of electricity, $ t$^{-1}$: US$ per ton.

chemicals, which increases the energy consumption. The detailed inputs of these two scenarios are listed in Table 10.3 [32]. Sridhar concluded that both the scenarios are profitable after 25 years of operation, which is comparatively long.

Thus, we need to find a way to reduce the inputs of $CO_2$ electroreduction to shrink the process time to gain economic profits to realize the practical utilization of $CO_2$ electroreduction in three main parts: (1) to reduce the electrical energy consumption, (2) to improve the stability of catalysts, and (3) to study the pathways of $CO_2$ reduction [32].

## 10.4  HIGH ENERGY CONSUMPTION AND LIMITED AVAILABILITY OF GREEN ELECTRICITY SOURCES

The energy consumption during $CO_2$ electroreduction process is high because of the insufficient catalyzed activity of catalysts or electrodes. Meanwhile, the energy used in $CO_2$ conversion should be nonfossil energy or $CO_2$-free electricity. However, the green energy resources are not that popular as the fossil fuels, so that the utilization of renewable electricity is limited and the price is not economical.

## 10.4.1  HIGH ENERGY CONSUMPTION

As discussed above, the energy consumption of CO$_2$ electroreduction is high mainly resulting from the low faradaic efficiency of catalyzed reduction process. Therefore, the electricity cost is a major factor to reduce the production cost of formic acid during the electrolysis of CO$_2$. Table 10.4 lists the cost of different energy sources [37]. It could be seen that the construction costs of the fossil fuels are comparably cheap, and the electricity generation cost of coal is not high. But the utilization of fossil fuels may lead to a new round of extra CO$_2$ emission, which is in contradiction with our intention. Nuclear energy seems to be more competitive considering both the construction and production costs. Other sustainable energy such as wind and solar energy may cost higher during the generation process. The production cost of wind energy is expected to decline in the future. The market price of HCOOH is about $ 1200 t$^{-1}$. Therefore, if the wind energy price decreases to $ 0.04 kW h$^{-1}$, the maximum energy consumption in CO$_2$ process should be high as about 30,000 kW h t$^{-1}$ when we ignore the other capital investments [37]. However, the electricity consumption is higher than the ideal value and we cannot neglect the installed fees and chemical costs. So even if the wind energy price goes down, the HCOOH production cost is still much higher than the commercial price of formic acid sold in market. This means that the electroreduction of CO$_2$ technology is still immature and cannot be realized for commercial utilization at the present stage.

## 10.4.2  LIMITED GREEN ELECTRICITY SOURCES

Green electricity technologies include solar thermal, wind energy, biomass, geothermal, tidal, and nuclear energy [38,39]. For the most competitive wind energy, electricity is the dispersed source and needs to be transmitted, stored, and regulated before utilization. For biomass and solar energy, they need more land for the system

---

**TABLE 10.4**
**Construction Costs and Electricity Generation Costs of Different Energy Sources**

|  | Construction Cost (US$/kWe) | Electricity Generation Costs (US¢/kWh) |
|---|---|---|
| Coal | 1000–1500 | 2.5–5 |
| Natural gas | 400–800 | 3.7–6 |
| Nuclear | 1000–2000 | 2.1–3.1 |
| Wind | 1000–2000 | 3.5–9.5 (mostly under 6.0) |
| Solar | 3000–10,000 | 15–40 |

*Source:* Reprinted with permission from Olah, G. A., Prakash, G. S., and Goeppert, A. Anthropogenic chemical carbon cycle for a sustainable future. *Journal of the American Chemical Society* 2011; 133(33): 12881–12898. Copyright 2011 American Chemical Society.

and plantations so that the power densities of these two green energy sources are low [38,40]. Besides, the existing grids could not manage the unstable loads of such renewable energy, the electricity grids should build advanced system with loads of conditioner to adapt to the new energy resources, which is another major input for the green electricity. Nuclear energy, which is most commonly used, is produced by $^{235}U$. There are about 500 plants that produce nuclear energy in the world [38,41]. However, waste disposal and nuclear leakage are dangerous, and they are big hazards in nuclear energy operation.

Apart from green energy sources, many other factors will also affect the development of green electricity, such as the public policies and regulations, the technology levels, and the consumer of green energy product. For the government policies and regulations, if the government supports the renewable energy and makes some policies to encourage the green energy utilization and provides some bonus for those who uses the green electricity, the awareness of renewable power would be increased among people. The improvement of technology level is important for the decline of green energy costs because the promotion of technology could have a positive effect on the storage and transmission of green electricity and makes it less expensive and more convenient.

## 10.5   PROPOSED RESEARCH DIRECTIONS

To develop the $CO_2$ electroreduction technology, the optimization of catalysts and reaction system is the main direction. We are aiming to investigate catalysts with high activity, selectivity, and stability of $CO_2$ electroreduction, and at the same time, we need to optimize the working electrode and the electrolyte.

### 10.5.1   Optimizing the Activity, Selectivity, and Stability of Catalysts

The catalyst property is the key factor in $CO_2$ electroreduction technology. Future research should be focused on the three main aspects of catalyst property: the catalyzed activity, production selectivity, and catalyst stability. To realize the commercial application of $CO_2$ electroreduction, the catalysts at the present stage do not meet the need of the utilization, and inventions are desirable. Keeping this in mind, developing new catalysts with high activity, selectivity, and stability is a significant target. Two principles should be mentioned: (1) the catalysts should have high production faradaic efficiency and low hydrogen evolution reaction, and (2) the catalysts should have high active surface area which may provide more active sites. With respect to these principles, the composite catalysts synthesized by materials with different advantages and the nanoscale catalysts with special three-dimensional structures could be developed and investigated in the future.

Moreover, $CO_2$ electroreduction is a complicated process, and the theory of the reaction is still not clear. For a target research about $CO_2$ reduction reaction, better understanding of this reaction process could provide us more detailed information about the decided step of reaction and lead us to do more targeting and effective research to improve catalysts.

## 10.5.2   SCALING UP THE REACTORS AND REACTION SYSTEMS

To realize the commercial application of $CO_2$ electroreduction, a continuous system on a large scale should be designed and optimized. The system partially includes the electrode design and the electrolyte choice. For electrode, the gas diffusion electrodes (GDEs) used in full cells are relatively good choice which could improve the mass transfer process. Some researchers used GDE for $CO_2$ reduction which is promising but also with several defects, including lacking electrode stability and the deficiency of the potential for electrochemical reactions in the pores of GDE [33,42,43]. In the future, different supporting materials such as graphene, carbon nanotubes, and other materials which could improve the conductivity and electrode surface area need to be investigated. For the electrolyte, the most commonly used type in the current state is $KHCO_3$ solution; however, some waste waters from factories could be used as the electrolyte of $CO_2$ electroreduction. If this comes into existence, the chemical costs of $CO_2$ electroreduction could be greatly reduced and the waste disposal of some plants could be solved at the same time.

In short, research on the electrochemical reduction of $CO_2$ powered by renewable electricity is justified by robust economic and environmental reasons. It shows potential for implementation on a large scale, with the predicted feasibility of the products decreasing according to the series $CO \approx HCOOH > CH_3OH \gg C_2H_4 > CH$ [44]. Although these reactions require higher energy exchanges provided by strong reducing reagents ($H_2$, metals) and/or electricity, auxiliary processes such as $CO_2$ capture or distillation may be equally or more limiting than the energy consumption by electrolyzers currently [32]. From the environment point of view, using emission-free energy sources is the only sensible approach to $CO_2$ recycling. However, the main limiting phenomena affecting $CO_2$ electrolyzers in terms of durability, current efficiency, energy efficiency, faradaic efficiency, and overvoltage must be achieved, for example, to reach a similar energy storage capability as the electrochemical production of $H_2$ from water splitting which has been studied for decades and commercially available up to the megawatt range [45]. It is interesting to note that the conversion of $CO_2$ to CO and HCOOH has led the race toward practical use, although both still exhibit major performance gaps. Present energy efficiencies are low, mainly due to high overvoltages, and durability is not yet a developed research area. Besides the development of more efficient electrocatalysts, research advances on the fronts of electrode and electrolyzer design, coupled with the optimization of methods for the preparation of electrodes, are expected to push forward the electrochemical reduction of $CO_2$ on its way to viability.

## REFERENCES

1. Hara, K., Kudo, A., and Sakata, T. Electrochemical reduction of carbon dioxide under high pressure on various electrodes in an aqueous electrolyte. *Journal of Electroanalytical Chemistry* 1995; 391(1): 141–147.
2. Qiao, J., Liu, Y., Hong, F., and Zhang, J. A review of catalysts for the electroreduction of carbon dioxide to produce low-carbon fuels. *Chemical Society Reviews* 2014; 43(2): 631–675.

3. Hori, Y., Wakebe, H., Tsukamoto, T., and Koga, O. Electrocatalytic process of CO selectivity in electrochemical reduction of $CO_2$ at metal electrodes in aqueous media. *Electrochimica Acta* 1994; 39(11): 1833–1839.

4. Bard, A. J., Parsons, R., and Jordan, J. *Standard Potentials in Aqueous Solution.* 1985; New York: Marcel Dekker, Inc.

5. Hori, Y., Murata, A., Takahashi, R., and Suzuki, S. Enhanced formation of ethylene and alcohols at ambient temperature and pressure in electrochemical reduction of carbon dioxide at a copper electrode. *Journal of the Chemical Society, Chemical Communications* 1988; 1(1): 17–19.

6. Hori, Y., Murata, A., and Takahashi, R. Formation of hydrocarbons in the electrochemical reduction of carbon dioxide at a copper electrode in aqueous solution. *Journal of the Chemical Society, Faraday Transactions 1: Physical Chemistry in Condensed Phases* 1989; 85(8): 2309–2326.

7. Saeki, T., Hashimoto, K., Kimura, N., Omata, K., and Fujishima, A. Electrochemical reduction of $CO_2$ with high current density in a $CO_2$ + methanol medium II. CO formation promoted by tetrabutylammonium cation. *Journal of Electroanalytical Chemistry* 1995; 390(1): 77–82.

8. Kaneco, S., Iiba, K., Suzuki, S. K., Ohta, K., and Mizuno, T. Electrochemical reduction of carbon dioxide to hydrocarbons with high faradaic efficiency in LiOH/methanol. *The Journal of Physical Chemistry B* 1999; 103(35): 7456–7460.

9. Kaneco, S., Iiba, K., Hiei, N. H., Ohta, K., Mizuno, T., and Suzuki, T. Electrochemical reduction of carbon dioxide to ethylene with high Faradaic efficiency at a Cu electrode in CsOH/methanol. *Electrochimica Acta* 1999; 44(26): 4701–4706.

10. Kaneco, S., Iiba, K., Katsumata, H., Suzuki, T., and Ohta, K. Electrochemical reduction of high pressure carbon dioxide at a Cu electrode in cold methanol with CsOH supporting salt. *Chemical Engineering Journal* 2007; 128(1): 47–50.

11. Li, H. and Oloman, C. The electro-reduction of carbon dioxide in a continuous reactor. *Journal of Applied Electrochemistry* 2005; 35(10): 955–965.

12. Chang, T. Y., Liang, R. M., Wu, P. W., Chen, J. Y., and Hsieh, Y. C. Electrochemical reduction of $CO_2$ by $Cu_2O$-catalyzed carbon clothes. *Materials Letters* 2009; 63(12): 1001–1003.

13. Ohya, S., Kaneco, S., Katsumata, H., Suzuki, T., and Ohta, K. Electrochemical reduction of $CO_2$ in methanol with aid of CuO and $Cu_2O$. *Catalysis Today* 2009; 148(3): 329–334.

14. Li, C. W. and Kanan, M. W. $CO_2$ reduction at low overpotential on Cu electrodes resulting from the reduction of thick $Cu_2O$ films. *Journal of the American Chemical Society* 2012; 134(17): 7231–7234.

15. Hori, Y., Kikuchi, K., and Suzuki, S. Production of CO and $CH_4$ in electrochemical reduction of $CO_2$ at metal electrodes in aqueous hydrogencarbonate solution. *Chemistry Letters* 1985; 14(11): 1695–1698.

16. Köleli, F., Atilan, T., Palamut, N., Gizir, A. M., Aydin, R., and Hamann, C. H. Electrochemical reduction of $CO_2$ at Pb-and Sn-electrodes in a fixed-bed reactor in aqueous $K_2CO_3$ and $KHCO_3$ media. *Journal of Applied Electrochemistry* 2003; 33(5): 447–450.

17. Mahmood, M. N., Masheder, D., and Harty, C. J. Use of gas-diffusion electrodes for high-rate electrochemical reduction of carbon dioxide. I. Reduction at lead, indium-and tin-impregnated electrodes. *Journal of Applied Electrochemistry* 1987; 17(6): 1159–1170.

18. Prakash, G. S., Viva, F. A., and Olah, G. A. Electrochemical reduction of $CO_2$ over Sn-Nafion® coated electrode for a fuel-cell-like device. *Journal of Power Sources* 2013; 223: 68–73.

19. Mizuno, T., Ohta, K., Sasaki, A., Akai, T., Hirano, M., and Kawabe, A. Effect of temperature on electrochemical reduction of high-pressure $CO_2$ with In, Sn, and Pb electrodes. *Energy Sources* 1995; 17(5): 503–508.

20. Chen, Y. and Kanan, M. W. Tin oxide dependence of the $CO_2$ reduction efficiency on tin electrodes and enhanced activity for tin/tin oxide thin-film catalysts. *Journal of the American Chemical Society* 2012; 134(4): 1986–1989.

21. Kaneco, S., Iiba, K., Ohta, K., and Mizuno, T. Electrochemical reduction of carbon dioxide on copper in methanol with various potassium supporting electrolytes at low temperature. *Journal of Solid State Electrochemistry* 1999; 3(7–8): 424–428.

22. Kyriacou, G. and Anagnostopoulos, A. Electroreduction of $CO_2$ on differently prepared copper electrodes: The influence of electrode treatment on the current efficiencies. *Journal of Electroanalytical Chemistry* 1992; 322(1): 233–246.

23. Liu, C., Cundari, T. R., and Wilson, A. K. $CO_2$ reduction on transition metal (Fe, Co, Ni, and Cu) surfaces: In comparison with homogeneous catalysis. *The Journal of Physical Chemistry C* 2012; 116(9): 5681–5688.

24. Hossain, A. M., Nagaoka, T., and Ogura, K. Electrocatalytic reduction of carbon dioxide by substituted pyridine and pyrazole complexes of palladium. *Electrochimica Acta* 1996; 41(17): 2773–2780.

25. Friebe, P., Bogdanoff, P., Alonso-Vante, N., and Tributsch, H. A real-time mass spectroscopy study of the (electro) chemical factors affecting $CO_2$ reduction at copper. *Journal of Catalysis* 1997; 168(2): 374–385.

26. Hori, Y., Konishi, H., Futamura, T., Murata, A., Koga, O., Sakurai, H., and Oguma, K. "Deactivation of copper electrode" in electrochemical reduction of $CO_2$. *Electrochimica Acta* 2005; 50(27): 5354–5369.

27. Yano, J., Morita, T., Shimano, K., Nagami, Y., and Yamasaki, S. Selective ethylene formation by pulse-mode electrochemical reduction of carbon dioxide using copper and copper-oxide electrodes. *Journal of Solid State Electrochemistry* 2007; 11(4): 554–557.

28. Chen, Y., Li, C. W., and Kanan, M. W. Aqueous $CO_2$ reduction at very low overpotential on oxide-derived Au nanoparticles. *Journal of the American Chemical Society* 2012; 134(49): 19969–19972.

29. Chiacchiarelli, L. M., Zhai, Y., Frankel, G. S., Agarwal, A. S., and Sridhar, N. Cathodic degradation mechanisms of pure Sn electrocatalyst in a nitrogen atmosphere. *Journal of Applied Electrochemistry* 2012; 42(1): 21–29.

30. Kapusta, S. and Hackerman, N. The electroreduction of carbon dioxide and formic acid on tin and indium electrodes. *Journal of the Electrochemical Society* 1983; 130(3): 607–613.

31. Wu, J., Risalvato, F. G., Ke, F. S., Pellechia, P. J., and Zhou, X. D. Electrochemical reduction of carbon dioxide I. Effects of the electrolyte on the selectivity and activity with Sn electrode. *Journal of the Electrochemical Society* 2012; 159(7): F353–F359.

32. Agarwal, A. S., Zhai, Y., Hill, D., and Sridhar, N. The electrochemical reduction of carbon dioxide to formate/formic acid: Engineering and economic feasibility. *ChemSusChem* 2011; 4(9): 1301–1310.

33. Oloman, C. and Li, H. Electrochemical processing of carbon dioxide. *ChemSusChem* 2008; 1(5): 385–391.

34. Chaplin, R. P. S. and Wragg, A. A. Effects of process conditions and electrode material on reaction pathways for carbon dioxide electroreduction with particular reference to formate formation. *Journal of Applied Electrochemistry* 2003; 33(12): 1107–1123.

35. Vassiliev, Y. B., Bagotsky, V. S., Osetrova, N. V., Khazova, O. A., and Mayorova, N. A. Electroreduction of carbon dioxide: Part I. The mechanism and kinetics of electroreduction of $CO_2$ in aqueous solutions on metals with high and moderate hydrogen overvoltages. *Journal of Electroanalytical Chemistry and Interfacial Electrochemistry* 1985; 189(2): 271–294.

36. Jitaru, M., Lowy, D. A., Toma, M., Toma, B. C., and Oniciu, L. Electrochemical reduction of carbon dioxide on flat metallic cathodes. *Journal of Applied Electrochemistry* 1997; 27(8): 875–889.

37. Olah, G. A., Prakash, G. S., and Goeppert, A. Anthropogenic chemical carbon cycle for a sustainable future. *Journal of the American Chemical Society* 2011; 133(33): 12881–12898.

38. Hoffert, M. I., Caldeira, K., Benford, G., Criswell, D. R., Green, C., Herzog, H., Jain, A. K. et al. Advanced technology paths to global climate stability: Energy for a greenhouse planet. *Science* 2002; 298(5595): 981–987.

39. Johansson, T., Kelly, H., Reddy, A., Williams, R. *Renewable Energy: Sources for Fuels and Electricity.* 1993; Washington, DC: Island Press.

40. Kheshgi, H. S., Prince, R. C., and Marland, G. The potential of biomass fuels in the context of global climate change: Focus on transportation fuels 1. *Annual Review of Energy and the Environment* 2000; 25(1): 199–244.

41. Kuliasha, M. A., Zucker, A., and Ballew, K. J. *Technologies for a Greenhouse-Constrained Society.* 1992; Boca Raton, FL: CRC Press.

42. Fan, M., Bai, Z., Zhang, Q., Ma, C., Zhou, X. D., and Qiao, J. Aqueous $CO_2$ reduction on morphology controlled $Cu_xO$ nanocatalysts at low overpotential. *RSC Advances* 2014; 4(84): 44583–44591.

43. Qiao, J., Fan, M., Fu, Y., Bai, Z., Ma, C., Liu, Y., and Zhou, X. D. Highly-active copper oxide/copper electrocatalysts induced from hierarchical copper oxide nanospheres for carbon dioxide reduction reaction. *Electrochimica Acta* 2015; 153: 559–565.

44. Sridhar, N., Agarwal, A. S., Guan, S., and Rode, E. Degradation of electrocatalysts used in the reduction of $CO_2$: A review. Meeting Abstracts. *The Electrochemical Society* 2012; 1(21): 2162–2162.

45. Martín, A. J., Larrazába, G. O., and Pérez-Ramírez, J. Towards sustainable fuels and chemicals through the electrochemical reduction of $CO_2$: Lessons from water electrolysis. *Green Chemistry* 2015; 17: 5114–5130.

# Index